Material Characterization Using Ion Beams

NATO ADVANCED STUDY INSTITUTES SERIES

A series of edited volumes comprising multifaceted studies of contemporary scientific issues by some of the best scientific minds in the world, assembled in cooperation with NATO Scientific Affairs Division.

Series B: Physics

RECENT VOLUMES IN THIS SERIES

The series is published by an international board of publishers in conjunction with NATO Scientific Affairs Division

A	Life Sciences	Plenum Publishing Corporation
B	Physics	New York and London
C	Mathematical and Physical Sciences	D. Reidel Publishing Company Dordrecht and Boston
D	Behavioral and Social Sciences	Sijthoff International Publishing Company Leiden
E	Applied Sciences	Noordhoff International Publishing Leiden

Material Characterization Using Ion Beams

Edited by

J. P. Thomas

Nuclear Physics Institute
Université Claude Bernard (Lyon I)
Villeurbanne, France

and

A. Cachard

Department of Physics of Materials
Université Claude Bernard (Lyon I)
Villeurbanne, France

PLENUM PRESS • LONDON AND NEW YORK
Published in cooperation with NATO Scientific Affairs Division

Library of Congress Cataloging in Publication Data

Nato Advanced Study Institute on Material Characterization Using Ion Beams, Aleria, France, 1976.
Material characterization using ion beams.

(NATO advanced study institutes series: Series B, Physics; v. 28)
"Lectures presented at the NATO Advanced Study Institute on Material Characterization Using Ion Beams held in Aleria, Corsica, August 29-September 12, 1976."
Includes index.
1. Ion bombardment–Addresses, essays, lectures. 2. Materials–Analysis–Addresses, essays, lectures. 3. Solid state physics–Addresses, essays, lectures. I. Thomas, J. P. II. Cachard, A. III. Title. IV. Series.

QC702.7.B65N37 1976	530.4	77-13269

ISBN 978-1-4684-0858-4   ISBN 978-1-4684-0856-0 (eBook)
DOI 10.1007/978-1-4684-0856-0

Lectures presented at the NATO Advanced Study Institute
on Material Characterization Using Ion Beams held in
Aleria, Corsica, August 29–September 12, 1976

© 1978 Plenum Press, New York
Softcover reprint of the hardcover 1st edition 1978

A Division of Plenum Publishing Corporation
227 West 17th Street, New York, N.Y. 10011

A.S.I.M.S. ACTIVITIES

The Advanced Study Institute on Materials Science was created in 1973 in the Département de Physique des Matériaux of the University of Lyons.

The aim of this Institute was to organize specialized courses on material science at a doctoral level in Lyons and International Summer Schools on the same subject at a post-doctoral level in Corsica.

Here is the list of the Summer Schools which have been done with NATO as the main sponsor:

Radiation Damage Processes in Materials
1 - 15 September, 1973
Director: C.H.S. Dupuy
Proceedings edited by C.H.S. Dupuy
(Noodhoff International Publishing Company)

Physics of Nonmetallic Thin Films
29 August - 12 September, 1974
Director: C.H.S. Dupuy
Proceedings edited by C.H.S. Dupuy and A. Cachard
(Plenum Publishing Company)

Electrode Processes in Solid State Ionics
Theory and Application to Energy Conversion and Storage
28 August - 9 September, 1975
Director: M. Kleitz
Proceedings edited by M. Kleitz and J. Dupuy
(D. Reidel Publishing Company)

PREFACE

The extensive use of low-energy accelerators in non-nuclear physics has now reached the stage where these activities are recognized as a natural field of investigation.

Many other areas in physics and chemistry have undergone similarly spectacular development: beam foil spectroscopy in atomic physics, studies in atomic collisions, materials implantation, defects creation, nuclear microanalysis, and so on.

Now, this most recent activity by itself and in its evident connection with the others has brought a new impetus to both the fundamental and the applied aspects of materials science. A summer school on "Material Characterization Using Ion Beams" has resulted from these developments and the realization that the use of ion beams is not restricted to accelerators but covers a wide energy range in the developing technology.

The idea of the ion beam as a common denominator of many activities dealing with surface and near-surface characterization was enthusiastically received by many scientists and a school on this subject received the positive endorsement of NATO.

The Advanced Study Institute on Materials Science has assumed for us the status of an "institution" leading to better contact among the many laboratories engaged in this field. The fourth Institute in this series was held in Aleria, Corsica, between August 22 and September 12, 1976.

It is a pleasure to acknowledge the efforts of all the people who participated in making this summer school a success. First of all, we must cite not only the lecturers but also all those who acted as efficient discussion leaders during the panels: they managed to maintain the constant interest of the attendees.

We are very grateful to the local authorities in Aleria, the mayor, Mr. Carlotti, and Mr. V. Carlotti, general councilor, who made it possible for the school to take place under the best of conditions.

The "Mission Régionale" and especially Mr. J. de Rocca Serra was of great help to the organization, together with Miss M. L. Carlotti, who took care of social events, among them a trip to Bavella and a Corsican evening.

The penitentiary administration, and especially Mr. Talbert and Mr. Bonaldi, did their best to make our stay in the "Village de Vacances" as fruitful as possible. Many thanks are due to them and to all the staff of the V.I.V.E. of Casabianda.

As noted above, NATO was the main sponsor of this meeting, and once again we are very grateful to it. Our respective departments, the "Departement de Physique des Matériaux" and "l'Institut de Physique Nucléaire" of the University of Lyons were also of great help in the organization.

The Organizing Committee

J. P. Thomas
A. Cachard
C.H.S. Dupuy
M. Fallavier
G. Chassagne
J. Remilleux

CONTENTS

INTRODUCTION

J.P. THOMAS

Institut de Physique Nucléaire and IN2P3
Villeurbanne, France

The idea to hold a summer school devoted to the use of ion beams in material science came for the people of our two departments, from the acknowledgement of a favorable local and international context. The local context concerned obviously the continuity of the spirit of the ASIMS meetings, previously invoked in the preface, and, more than that, the mutual interest in this field of research of the Institut de Physique Nucléaire and the Département de Physique des Matériaux. The international context, certainly more significant, showed evidence of a strong need to have a full review of well established techniques dealing with ion beams. As a matter of fact, following an increasing period of meetings on the subject, a more didactic presentation and discussion of these techniques seemed very feasible at this time in the form of a summer school.

Taking account of my feeling on the subject before the conference and my observations during this event, I still realise that the task of such an introduction is not an easy one. Nevertheless I will keep the way I followed at the opening session, trying to make several comments about the title, about the topics, and probably the most difficult, about what should remain from such a gathering of specialists.

About the title first : necessarily short, it so appears as somewhat ambitious especially if we intend to discuss the use of ion beams in a large energy range. Instead of setting an arbitra-

ry upper limit to it, we will not consider in these lectures the
activation analysis techniques in the sense of radioactive isotope
formation (generally dealing with high energy). Nevertheless it
is rather difficult to introduce other techniques without referring
somehow to the one which is certainly the starting point of the
extensive use of ion beams for material characterization. As
characteristics of activation analysis we will particulary mention
that the determination of impurities at the substrate level, with
very high sensitivity, applies mainly for the bulk of the material.
If this feature is still of great value in characterization, recent
years have shown that interest has shifted from bulk problems
to surfaces and thin films, and more than average composition
three dimensional profiles are more and more required. These
new needs must be considered as enhancing, in the definition of
the characterization (1), the determination of the structure aside
those of the composition. Similarly it is worthwhile to note the
acknowledgement of the whole concept of defect, more and more
related to a discontinuity of the material, and, in which the physical aspect is as important as the chemical one.

Still referring to activation analysis, the other methods,
often qualified of "prompt" at low energy (MeV range and less)
point out the first evolution concerning the "applied" use of ion
beams. This evolution can be related to those of nuclear and
atomic physics. In the good old days of nuclear physics, the
basic instrument was the cyclotron, in the 10 to 100 MeV range,
and very limited shifts of protons, deuterons and α-particles
were available for activation analysis. By comparison, the electrostatic accelerator V.D.G type is operating in the MeV range,
with an increasing variety of projectiles, and nuclear physicists
are no more the majority of users. The conjunction of the availability of such machines with these new needs we discussed previously explains the development of material research teams
using these accelerators. As pointed our by the book of J.F.
ZIEGLER "New uses of ion accelerators"(2) or more recently by
the french prospect report on non-nuclear developed through the
use of accelerators(3), this field of research has shown is matu-
rity. Nuclear physics going up to higher energy, it is easy to ca-
ricature showing the opposite evolution of the methods of material
characterization. In fact it is true that there is a tremendous development of low energy machines (down to several keV) but the
explanation must be found in technological improvement, allowing with this decrease in energy surface studies with easier to
handle devices.

Thus, opposite directions seem the ways nuclear physics and
our field of interest are taking. Nevertheless if interactions are

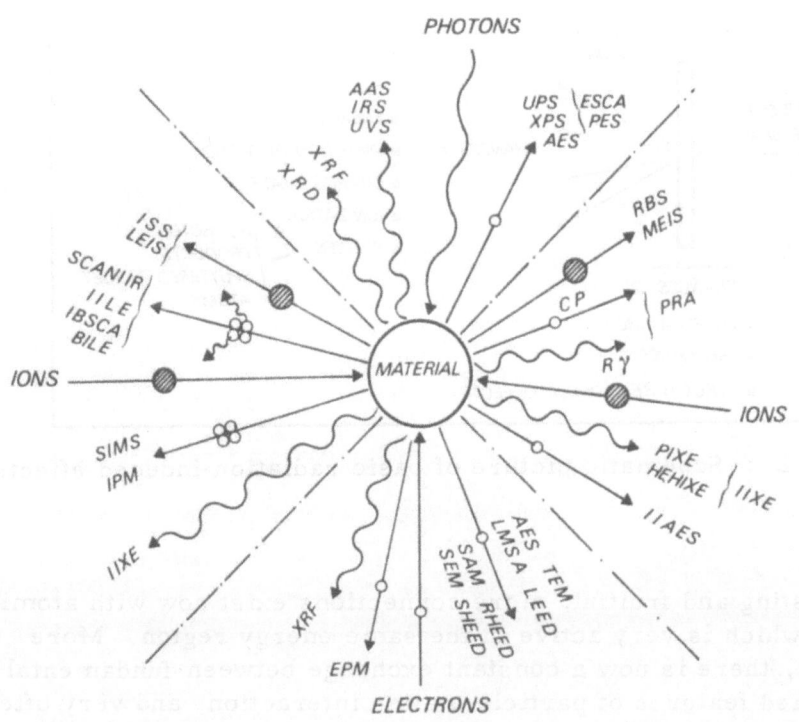

Fig. 1

AES: Auger electron spectroscopy; BILE: Beam induced light emission;
EPM: Electron probe microscopy; ESCA: Electron spectroscopy for chem-
ical analysis; HEHIXE: High energy heavy ions X-rays emission; IBSCA:
Ion beam spectrochemical analysis; IIAES: Ion induced Auger electron
spectroscopy; IILE: Ion induced light emission; IIXE: Ion induced
X-ray emission; IPM: Ion probe microscopy; IRS: Infra-red spectro-
scopy; ISS: Ion scattering spectroscopy; LEED: Low energy electron
diffraction; LETS: Low energy ion scattering; LMSA: Le Gressus-
Massignon-Sopizet-Auger; MEIS: Medium energy ion scattering; PES:
Photo electron spectroscopy; PIXE: Particle induced X-ray emission;
PRA: Prompt radiation analysis; RBS: Rutherford backscattering; RHEED:
Reflection high energy electron diffraction; SCANIR: Surface composition
by analysis of neutral and ion impact radiation; SEM: Scanning electron
microscopy; SHEED: Secondary high energy electrons diffraction; SAM:
Scanning Auger microscopy.

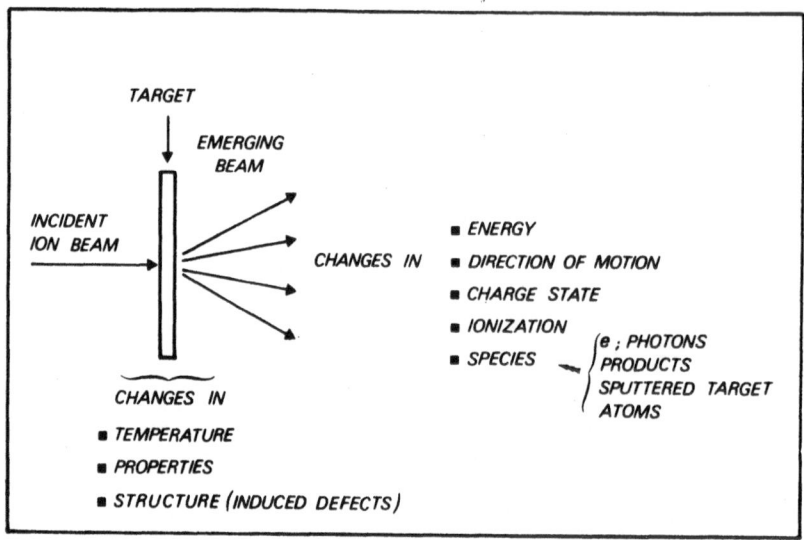

Fig. 2 : Schematic picture of basic radiation-induced effects.

still existing and fruitful, more connections exist now with atomic
physics which is very active in the same energy region. More
than that, there is now a constant exchange between fundamental
and applied features of particle matter interaction and very often
the same people do the two kinds of experiments with the same
equipment. This becomes apparent from the communications
presented at the meetings on atomic collisions (4) and ion beam
surface layer analysis (5).

About the topics now : although described as ambitious our
title suffers, on the other hand, to be somewhat restrictive especially after having emphasized the interest of surface analysis.
In this matter, it must be recognized that more than ion beams
the general term of radiation would be adequate to give a good
picture of the present universe of characterization. As a picto-
rial representation of the techniques so involved we can adopt the
allegorical form of the bright sun appearing on Figure 1. Although
the means of characterization are not so numerous as suggested
by the thirty-one initials found in the current literature, a coherent meeting cannot afford to treat all of them. Nevertheless
extending the discussion to the most significant methods not related to ion beams is obviously of great interest. If we want now
to practice this esoteric vocabulary for characterization, a sim-

pler picture based on basic physical processes may help. In this
way, radiation-induced effects can be described in terms of changes
as represented in Figure 2. The most important of these
changes concerns of course the emerging beam (not taken in the
restrictive sense of transmitted beam) because this is the matter
of the methods of characterization. Nevertheless the changes
induced in the target itself have always to be considered, either,
as another means of characterization, as a possible perturbation
for the analysis, or at last as a way to obtain a new material to
be characterized.

It is hardly conceivable to not consider, among these changes,
the slowing down of the beam as one of fundamental importance
for the matter of this book and especially for in-depth analysis.
Thus, elements of the stopping theory will naturally be given prior
to go into details of the techniques and methods. Since exhaustive
reviews have been done about the wide subject of collisions (6),
the lectures of Dr.CHU will be more concerned with the practical
aspect of the energy loss and related phenomena. In this way, a
presentation of usable theory combined with useful references
will have the ambition to offer relevant data for analysis.

Ion beam production and manipulation will also appear in the
first chapter, since, despite the variety of set-ups, the problem of
the quality of ion beams is common to all the techniques. In Dr.
FREEMAN's lecture, the discussion will go from general accelerator concepts to a detailed description of the phenomena occurring
in ion beam production and transport leading to a scope for substantial improvements.

For the central subject, methods and techniques of characterization, we decided to have two chapters - surface and in-depth
analysis - simply following an increasing of the incident beam energy. Of these two key words, one must be aware that the first one,
though commonly used, is somewhat ambiguous and it may deserve
some comment. As a matter of fact, the questions related to the
concept of surface are numerous, but among them the physical
question of how deep is the surface phase studied is primordial.
Defining the surface as the outermost layer bounding the solid suggests a minimum probing depth as the criterion but is it not then
a too restrictive definition ? When doing in-depth analysis, the
methods being destructive or nondestructive, does the depth resolution lead to the same information in each case ? In order to
answer these fundamental questions and particularly the first
one, it must be recognized that some techniques not directly related to ion beams can hardly be ignored due to their impact on the

discipline. Auger spectroscopy and ESCA, the most representative, will appear in the lecture as "complementary techniques"
which is of course a fairly egocentric point of view ! Dr. TOUSSET
will present these techniques of electron spectroscopy to end the
surface studies chapter following Dr. BRONGERSMA who will
make the first reference to the scattering process and Dr. VAN
DER WEG who will introduce the use of light emission.

As a first contribution to the in-depth analysis chapter, Dr.
BLAISE's lecture - ion microanalysis - will make the transition
surface - near surface - since the sputtering phenomenon is part
of the ion emission mechanism. This phenomenon will be discussed with its complement, the ionization, through this lecture where
fundamental aspects of the technique will be treated in great detail.

The principles of ion-induced X-ray emission as well as the
general analytical applications will appear in Dr. FOLKMANN's
lecture while Dr. CAIRNS will present more generally the relative merits of these particles but also of the electrons in the characterization. The remaining interest of electrons will more precisely appear in the description of a new sophisticated complementary technique : controlled atmosphere electron microscopy.

The backscattering technique as well as the use of nuclear
reactions for analysis do not require special comments at this
place since they are probably the best known techniques of characterization developed using ion accelerators. Dr. MAYER will
present Rutherford backscattering emphasizing its easy use for applications and restricting it to alpha-particles in the MeV range.
Dr. VERBEEK will describe the extension of the technique to lower
energies with the subsequent advantages and disadvantages. For
nuclear reactions, Dr. RIGO's lecture will particularly emphasize the specificity of light element determination (including
isotopic analysis).

Thus, X-ray emission, backscattering and nuclear reactions, as
complementary techniques, constitute the trilogy of the powerful
means of characterization accelerators can offer.

Concerning the determination of the elemental composition of
the surface and near surface, as well as the identification of the
species attached to these surfaces, nearly all the significant techniques are represented. For crystalline materials one may obtain unique structural information when channeling effects of ion
beams occur. A general and comprehensive description of a phenomenon which has turned from a specifically fundamental interest
to a widely applied one will be given by Dr. DAVIES. Two other
lectures will be devoted to specific applications : Dr. MAZZOLDI
will be in charge of the description of the impurities location in

crystal lattices while the analysis of defects will be given in Dr. RIMINI's lecture. (The channeling technique has strongly contributed to the impact of ion implantation (7) in material science). We must also particulary emphasize the recent sophisticated use of channeling in surface studies reported by Dr. DAVIES in the last lecture. With the improvement of vacuum conditions, the possibility of single and double alignment of the technique has led to new data on surface atoms which deserve to be compared to other surface techniques.

Since the lectures, briefly presented here, will remain the only witnesses of a two weeks summer school, the introduction may suffer last additional comments arising from personal appraisal of the live meeting. As a main remark, it is quite clear that the development of the techniques and methods of characterization is strongly oriented by current problems of materials of economic interest. As the most representative of these materials, silicon, must still be reported, most of the development of ion implantation and ion analysis having been done since the beginning of the sixties for its applications. Compound semiconductors, amorphous glasses and more generally nonmetallic thin films (8), as existing or potential components of electronics technology, follow the same way. Research in metals and alloys (9) is important in nuclear reactor technology due to crucial concerns in energy production : most of the characterization is then devoted to induced-defect analysis and light-element profiling on suitable materials for irradiation.

As a consequence of this increase of the means of characterization there is a revival of some fundamentals in particle-matter interaction. An unusually large part of atomic collision people have allowed spirited discussions on the basis of these techniques. These discussions like those dealing with the previous aspects or with the respective merits of the techniques were expressed through numerous panels : the concluding lecture will try to summarize the trends which appeared since no publications have been possible for these informal presentations.

Exciting and controversial results are now appearing at a tremendous rate from the different techniques and the necessary background on each of them is more and more difficult to acquire. If this book can contribute to an easier communication between specialists of sophisticated (and expensive) techniques, we think that it is worth our efforts. Despite or perhaps because of the expensive character of these techniques and also considering their complementarity there is a strong indication of a concentration

for an in-situ multicharacterization, in the future. According
to the respective quality of the systems described here, a personal feeling is that there will still be little room to connect an
accelerator to these new devices.

REFERENCES

(1)   Characterization of materials, Publication MAB-229-M
         National Academy of Sciences, Washington, D. C.
         (March 1967)

(2)   "New uses of ion accelerators", Ed. by J. F. ZIEGLER
         Plenum Press, N. Y., (1975)

(3)   Rapport de prospective : "Physique autour des accéléra-
         teurs dans des domaines autres que nucléaires. Applica-
         tions". DGRST Report 1977, edited by "La Documen-
         tation Française", 29-31, quai Voltaire, 75340 Paris
         Cedex 07

(4)   "Atomic Collisions in Solids", ed. by F. W. SARIS and
         W. F. VAN DER WEG, North-Holland Publ. Co.,
         Amsterdam, (1976)

(5)   "Ion Beam Surface Layer Analysis", ed. by O. MEYER
         G. LINKER, F. KAPPELER, Plenum Press, N. Y.,
         (1976)

(6)   For example P. SIGMUND in "Radiation Damage Pro-
         cesses in Material", ed. by C. H. S. DUPUY, Noordhoff
         Intern. Publ., Leyden, The Netherlands, (1975)

(7)   "Ion Implantation, Sputtering and their Applications",
         P. D. TOWNSEND, J. C. KELLY and N. E. W. HARTLEY
         Academic Press, London, (1976)

(8)   "Physics of Non-Metallic thin Films", ed. by C. H. S.
         DUPUY and A. CACHARD, Plenum Press, N. Y., (1976)

(9)   "Applications of Ion Beams to Metal",
         ed. by S. T. PICRAUX, E. P. EERNISSE, F. L. VOOK,
         Plenum Press, N. Y., (1974)

Part I

ION BEAMS: PRODUCTION AND

INTERACTION WITH MATTER

ENERGY LOSS OF CHARGED PARTICLES

W. K. Chu

IBM System Products Division, East Fishkill

Hopewell Junction, New York   12533

1.   INTRODUCTION

When a fast charged particle penetrates into a material, it
loses its energy, mainly in exciting and ionizing the target electrons and in transferring momentum to target nuclei during collisions.   Energy loss and scattering have been an important source
of information on the constitution of atoms.   Theoretical and experimental work on this subject has been going on constantly since
the beginning of this century.   In the early stages, information on
energy loss was needed mainly for applications to nuclear physics--
in the study of nuclear reactions, the depth to which fission fragments penetrate, and radiation health physics.   Recently the application of ion beams to materials, for example in ion implantation,
ion backscattering analysis, and ion sputtering, has provided additional motivation to the study of energy loss of charged particles
and related subjects.   Since material characterization by use of
ion beams is the concern of this NATO Advanced Study Institute, it
is necessary to include the subject of energy loss to support the
other lectures.

In 1973, during the Advanced Study Institute on Radiation
Damage Processes in Materials, Peter Sigmund presented a very
complete review on energy loss of charged particles.   In his
lecture notes, he covered both the classical and the quantum theories
of stopping.   Both mathematical treatments and physical assumptions
are presented in great detail, and a very complete list of references is given. In 1974, Sigmund presented a brief survey of recent
theoretical and experimental work on energy loss and range of charged
particles in matter.   Both of the above two publications give a
comprehensive and illuminating discussion and a very complete

documentation on the subject of energy loss of charged particles
in matter.

The existence of the above two publications makes my job of
lecturing on the same subject both easy and difficult. It is easy
because I can rely heavily on Sigmund's reviews, referring most of
the theoretical discussion to his notes. It is difficult because
to repeat the same lecture and publish similar lecture notes would
be meaningless; besides, I do not think I can surpass Sigmund's
fine reviews on this subject. Therefore I would like to address the
subject of energy loss differently, by presenting a basic overview
of the questions, What is energy loss? How does it depend on various
parameters? Why does energy loss behave as it does? The main purpose of this lecture is to provide general concepts to those who use
ion beams in material analysis. Therefore, after looking at the
general behavior of energy loss, we will use a combination of theories
and measurements to form general guidelines for interpolating and
extrapolating information where no measurement is available. We
will also look at some experimental methods for obtaining information on energy loss.

There are several differences between these lecture notes and
Sigmund's. These notes emphasize the elemental understanding of
general trends and of energy loss information applicable directly
to experiments. Sigmund's notes emphasize more advanced understanding of theories, detailed formulations, and mathematical tools
to solve problems. It is always helpful to study both sets of notes
together.

2. DEFINITIONS

First let me start with some tutorial definitions and descriptions. The projectile is described by its atomic mass M_1, its
atomic number Z_1, and its initial energy E. When the projectile
penetrates a target material, it loses its energy to the target
material by several processes. Let us assume, for the time being,
that the target material is made of a single element of atomic
number Z_2 with mass number M_2, and that it has mass density ρ and
atomic density N. The amount of energy loss from the projectile to
the target is given schematically in Fig. 1. For a very thin
layer, the amount of energy loss is proportional to the layer
thickness, and the energy loss of the projectile with energy E to
the target is defined as

$$\frac{dE}{dx}(E) = \lim_{\Delta x \to 0} \frac{\Delta E}{\Delta x} \tag{1}$$

In a transmission experiment to measure the energy loss, or in any

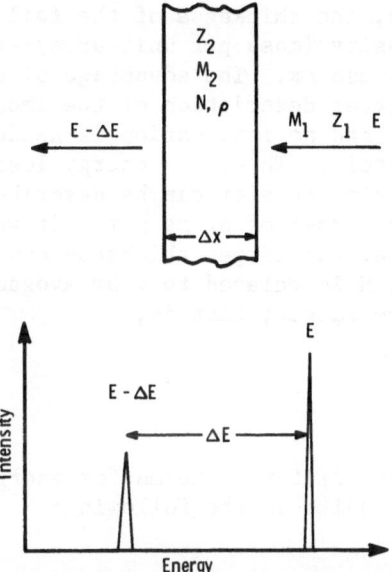

Fig. 1. Schematic diagram of an energy loss problem
in a transmission experiment. The projectile
has mass M_1 and atomic number Z_2 for energy
E losses ΔE to the target layer Δx. The
target has atomic number Z_2, mass number M_2,
mass density ρ, and atomic density N.

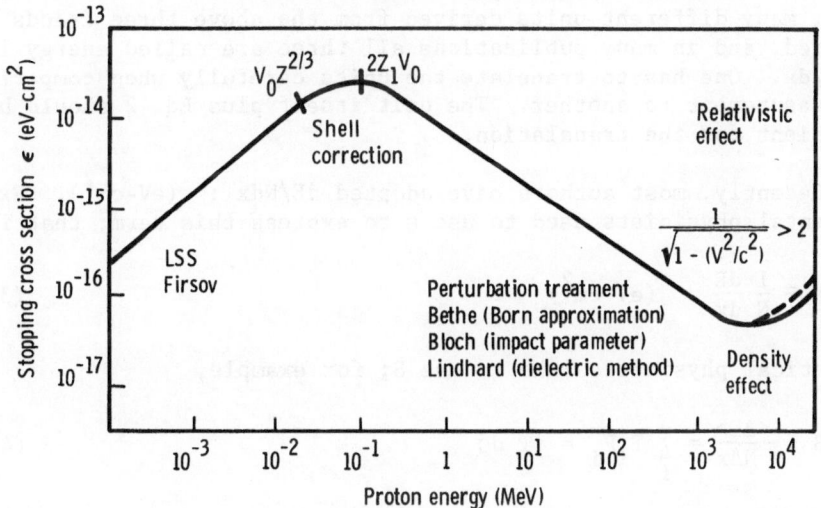

Fig. 2. Stopping cross section of protons in silicon. The
general shape is described by various theories for
various energy regions.

thin foil experiments, the thickness of the foil Δx is often defined by an areal density (mass per unit area)--that is, $(\rho \Delta x)$
rather than the thickness Δx. The advantage of using $(\rho \Delta x)$ is
that $(\rho \Delta x)$ gives a better description of the amount of material in
the thin target foil, and no information is needed on the packing
density and Δx separately. Therefore energy loss is often expressed
as $(dE/\rho dx)$. The packing density can be described either by mass
per unit volume or by number of atoms per unit volume. As an
alternative, therefore, the target thickness can be described by
using $N\Delta x$. Note that N is related to ρ by Avogadro's number N_0 and
the mass number of the target; that is,

$$N \equiv \frac{N_0 \rho}{M_2} \tag{2}$$

Now we have given three different terms for energy loss. We list
their frequently used units in the following:

$$\frac{dE}{dx} : \quad eV/\overset{o}{A}, \ MeV/cm, \ \ldots$$

$$\frac{dE}{\rho dx} : \quad eV/(\mu g/cm^2), \ keV/(\mu g/cm^2), \ MeV/(mg/cm^2), \ \ldots$$

$$\frac{dE}{Ndx} : \quad eV/(atoms/cm^2), \ eV\text{-}cm^2, \ \ldots$$

In the literature, especially some of the earlier experimental
works, many different units derived from the above three trends have
appeared, and in many publications all three are called energy loss
or dE/dx. One has to translate the units carefully when comparing
one measurement to another. The unit itself plus Eq. 2 should be
sufficient for the translation.

Recently, most authors have adopted dE/Ndx : $(eV\text{-}cm^2)$. Experimental physicists used to use ε to express this term; that is,

$$\varepsilon \equiv \frac{1}{N} \frac{dE}{dx} \quad (eV\text{-}cm^2) \tag{3}$$

Theoretical physicists like to use S; for example,

$$S \equiv \frac{\langle \Delta E \rangle}{N\Delta x} = \sum_i T_i P_i = \int T \ d\sigma \tag{4}$$

In Eq. (4), $\langle \Delta E \rangle$ is the average energy lost to a target of thickness
$N\Delta x$; T_i is the kinetic energy transferred to the i-th electron with
probability P_i; and the integral extends over all possible energy
losses in individual collisions. The last term of Eq. 4 describes

the cross section $d\sigma$ for the energy loss (stopping process), which therefore is often called the stopping cross section. From the dimension point of view, Eqs. 3 and 4 are identical and mean the same thing.

The advantage of using the stopping cross section (ε or S) rather than dE/dx is obvious. Especially in the systematic study, ε gives description of energy loss on an atom-to-atom basis, which permits convenient extrapolation, whereas dE/dx changes from material to material even for the same material with different density.

3.   WHAT CAUSES ENERGY LOSS?

When a charged particle encounters the target material, many things happen to the target and to the projectiles. On a large scale, the target material undergoes a physical or chemical change: radiation damage and/or defect formation, sputtering on the target surface, and a general increase in the target temperature. On a small scale, the target atom could be ionized or its electrons could be in a excited state because of collision and subsequently emitting x-ray, or Auger electrons. If the projectile is energetic enough, the target nucleus could undergo transmutation and then emit nuclear reaction products.

Nearly all these changes in the target are at the expense of the projectile energy, directly or indirectly. The projectile it-self could also undergo one or more changes because of its inter-action with the target. One thing is certain, that the projectile loses its energy to the target material. Energy loss of the pro-jectile to the target is only a very small aspect of the problem, which involves the interaction of the charged particle with a target atom or target material.

A charged particle interacts with the target material mainly by the following processes:

1.   Excitation and ionization of the target electrons
2.   Nuclear motion of the target atoms
3.   Generation of photons (bremsstrahlung, Cerenkow radiation)
4.   Nuclear reactions

The first two processes are important here because they have a general behavior and for a broad energy region they are the major contributors to energy loss. The third process is also a general property in the slowing down of charged particles, but this term is important only for extremely fast particles (relativistic velocities). The last term is very specific, depending on the combination of

target, projectile, and its energy, and it has to be treated indi-
vidually. Of the above three general energy loss processes, 1, 2,
and 3, the first is also called the electronic energy loss, and is
important for nearly all energies. The second process, called nuclear
energy loss, is important only for low-velocity projectiles. The
contribution of the third process is appreciable only at relativistic
velocities and can be ignored for all practical purposes.

4. GENERAL BEHAVIOR OF ENERGY LOSS

We have just discussed various contributions to energy loss.
In this section we will present a crude picture of energy dependence
of energy loss. Figure 2 gives a schematic diagram of the stopping
cross section of protons in silicon. The selection of projectile
and target is arbitrary. Other selections will produce the similar
general shape of the curve, with some modification in shape and in
scale. There are various names and terms labeled in Fig. 2, which
I have not yet defined or described here. They are presented in
Fig. 2 to relate various theories to the region of applicability.
These theories will not be described in detail in this lecture.
However, they will be discussed, less rigorously, to show physical
trends.

Figure 2 covers a very broad region of energies. It starts
from a fraction of a keV to many GeV. This curve has a peak near
100 keV and a dip around 50 GeV. The increase of energy loss at
relativistic velocity region is due to photon emission, bremsstrahlung,
and Cerenkow radiation. In this study we are more interested in the
nonrelativistic region. For ion beam analysis, the energy region
could be very narrow in the log diagram of Fig. 2. In future
lectures we will limit our interest to a few parts of this curve.

In Fig. 2, nuclear energy loss is small when compared to electronic energy loss, even in the very low-energy region. For example,
according to Lindhard, Scharff, and Schiøtt (LSS theory, 1961, 1963)
nuclear stopping accounts for 2% of the total stopping for
protons at 10 keV in silicon and 16% for protons at 1 keV. With
respect to energy loss, nuclear energy loss can be ignored, especially for light ions at medium and high velocities. However,
nuclear energy loss produces different effects in the target; in
studies of radiation damage, therefore, one should focus attention
on the nuclear stopping cross section.

In the low-energy region, the stopping cross section is proportional to the velocity, until it reaches a maximum and starts to
decrease again. Various regions will be discussed in separate
sections later.

For the time being, we would like to study what happens for other projectiles. In Fig. 3, we start by repeating the proton energy loss in silicon, simply changing the coordinates by taking ε/Z_1^2 for the normalized stopping cross section and by taking E/M_1 for the velocity parameter. For protons as projectiles, $Z_1 = M_1 = 1$; the curve for protons in Si in Fig. 3 is identical with that of Fig. 2. One might object to the E/M_1 at relativistic velocities, but that is a minor detail. The advantage of presenting the same Fig. 2 in a different normalized parameter allows us to see how the trend is affected by changing projectiles. The values of stopping cross sections in Figs. 2 and 3 are obtained from Northcliffe and Schilling (1970).

Several features can be observed from Fig. 3:

1. At high velocities, ε/Z_1^2 tends to converge to a single curve.
2. For higher Z_1, the peak position of ε/Z_1^2 shifts to higher velocities.
3. At lower velocities, curves in the ε/Z_1^2 are nearly parallel.
4. At very low velocities, the nuclear stopping cross section starts to influence the curves. The larger the mass number of the projectile, the larger the influence.

The above observation is purely empirical and phenomenological. However, various theories at various velocity regions give detailed descriptions of the behavior of stopping cross section as a function of the projectile and its velocity, and as a function of target atoms.

Ion beam analysis is most commonly applied to light ions at high velocities. We will start there, and then discuss the low velocity region, where most of the ion implantation work is done and nuclear stopping enters in.

5. HIGH-VELOCITY REGION (BETHE-BLOCH FORMULA)

When light projectiles such as protons, helium ions, and light nuclei move at high velocity, such that their orbital electrons are totally stripped, the Coulomb interaction between a given target medium and the projectiles scales with Z_1^2. The energy loss becomes a property of the stopping medium. A very accurate formula in this velocity region can be derived by various approaches. This formula, generally called the Bethe or Bethe-Bloch formula, has the form

$$\varepsilon = S_e = \frac{4\pi Z_1^2 Z_2 e^4}{mv^2}\left[\log\frac{2mv^2}{I} + \ldots\right] \tag{5}$$

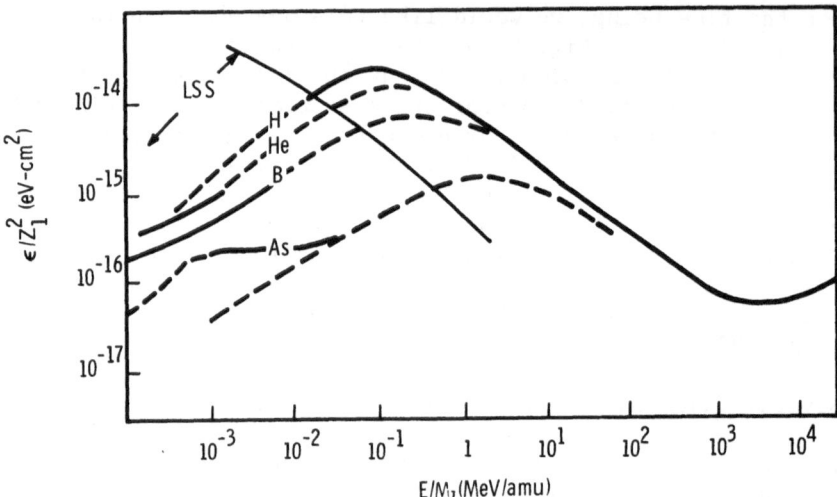

Fig. 3. Family of curves of stopping cross section for heavy ions
in silicon. Based on Northcliffe and Schilling's tabulation
(1970).

where m is the mass of electrons and I is the mean excitation and
ionization energy, a property of the target atom Z_2. To a
zeroth-order approximation, Bloch establishes a simple relation
based on the Thomas-Fermi model of atoms, namely that
$I \simeq 10$ eV $\times Z_2$. The quantity I will be discussed in a later
section.

5.1. Classical Derivation

Equation 5 can be understood from a very crude treatment. On
the basis of classical mechanics, the projectile of point charge
Z_1, moving with velocity v, interacts with individual target nuclei
by Coulomb's Law. The scattering is governed by the equation of
Kepler motion,

$$\tan \frac{\theta}{2} = \frac{b}{2p} \tag{6}$$

where θ is the scattering angle in the center-of-mass system, p is
the impact parameter, and b is the collision diameter, defined by
the Coulomb interaction as

$$\frac{Z_1 Z_2 e^2}{b} = \frac{M}{2} v^2 \tag{7}$$

M is the reduced mass in binary collision between projectile and target nucleus.

$$M = \frac{M_1 M_2}{M_1 + M_2} \tag{8}$$

The energy transfer from the projectile to a target nucleus is given by

$$T = T_{max} \sin^2 \frac{\theta}{2} \tag{9}$$

where the maximum energy transfer from the projectile with energy E to a target nucleus is

$$T_{max} = \frac{4M_1 M_2}{(M_1 + M_2)^2} E \tag{10}$$

Eliminating θ from Eqs. 6 and 9 gives

$$T = \frac{T_{max}}{1 + (2p/b)^2} \tag{11}$$

and the differential Rutherford scattering cross section gives

$$d\sigma = \pi \left(\frac{b}{2}\right)^2 \frac{\cos \theta/2}{\sin^3 \theta/2} d\theta = \pi \left(\frac{b}{2}\right)^2 T_{max} \frac{dT}{T^2} \tag{12}$$

Substituting Eqs. 7 and 12 into the definition of energy loss (Eq. 4) yields:

$$S_n = \frac{2\pi z_1^2 z_2^2 e^4}{M_2 v^2} \log \frac{T_{max}}{T_{min}} \tag{13}$$

Equation 13 gives the energy loss of the projectile for a single nucleus M_2 with charge Z_2, where T_{min} is some finite minimum energy transfer assigned to prevent divergence.

When the process is repeated for the energy loss to a single electron, the same derivation follows except that $M_2 \rightarrow m$ and $Z_2 \rightarrow 1$ for the derivation, and the electronic stopping per target electron becomes

$$S_e \text{ (per electron)} = \frac{2\pi z_1^2 e^4}{mv^2} \log \frac{T_{max}}{T_{min}} \qquad (14)$$

The total electronic energy loss per atom will be Z_2 times the above equation, simply because there are Z_2 electrons in an atom. Therefore,

$$S_e \text{ (per atom)} = \frac{2\pi z_1^2 z_2 e^4}{mv^2} \log \frac{T_{max}}{T_{min}} \qquad (15)$$

if we ignore the small difference inside the log term in Eqs. 13 and 15. The nuclear energy loss is much less than the electronic energy loss, because it gives

$$\frac{S_n}{S_e} \simeq \frac{Z_2 m}{M_2} \simeq \frac{1}{2} \frac{\text{mass of electron}}{\text{mass of proton}} \simeq \frac{1}{4000} \qquad (16)$$

Equation 16 indicates that the electron stopping cross section (Eq. 15) dominates the stopping process. Equation 16 has a form very similar to that of Eq. 5, which can be approached by several methods.

The classic treatment given in Eqs. 6-15 only provides a feeling for the problem. Various treatments, especially to the logarithmic term, have different levels of sophistication. A very systematic listing of elementary and quantum mechanical treatments to the stopping theory is given by Sigmund (1973), along with reference listings and some derivations.

Different corrections to Bethe's formula will be needed to complete Eq. 5. We have

$$\varepsilon = \frac{4\pi z_1^2 z_2 e^4}{mv^2} \left[\log \frac{2mv^2}{I} - \log \left(1 - \frac{v^2}{c^2} \right) - \frac{v^2}{c^2} - \frac{C}{Z_2} - \frac{\delta}{2} \right] + \dots \qquad (17)$$

The term C/Z_2 accounts for the shell correction, which is more important at the lower-energy end of the Bethe region, where the nonparticipation of the inner-shell electrons in the stopping process influences the curve. The term $\delta/2$, called the density effect, which is important only at the high-energy end, reduces the energy loss from the dashed line to the solid line at the high-energy end of Fig. 2. This subject has been reviewed by Crispin and Fowler (1970). Both C and δ are functions of the target atoms as well as the projectile velocity.

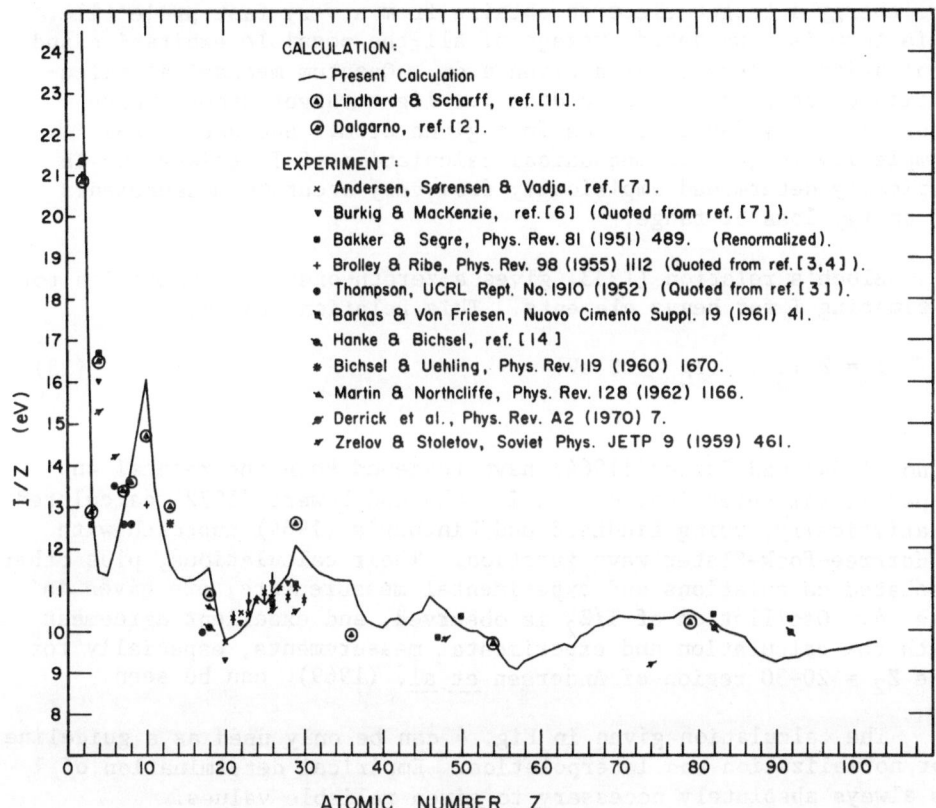

Fig. 4. Calculation of mean excitation energy (Chu and Powers 1970) by Lindhard and Scharff's theory with a Hartree-Fock-Slater charge distribution. The calculation I/Z vs. atomic number Z reveals structure, as was observed in many experimental measurements cited in the figure. Used by permission.

The relativistic correction is contained in two remaining correction terms of Eq. 17. The overall shape of Eq. 17 is that given in the intermediate and high-energy regions of Fig. 2.

The terms not specified in Eq. 17 are terms containing Z_1 to a higher power (>2), and will be discussed briefly in Section 5.6.

5.2. Mean Excitation and Ionization Energy

In this section we will focus our attention on the mean excitation and ionization energy, I, of Eqs. 5 and 17. I, the stopping property of a given target Z_2, is some kind of average of the energy a target electron obtains from a very fast projectile. This term is a weighted average of all the possible excitation and ionization processes for a given atom. Quantum mechanical calculation of this value can be done for light target atoms, where there are very few electrons in a given atom. Because of the complexity in quantum mechanical calculation of I, this value is typically determined empirically from very accurate measurements of energy loss or range.

Bloch's relation (1933) gives a zeroth-order approximation for estimating I for heavy elements. This relation states:

$$I = Z_2 I_0 \qquad (I_0 \simeq 10 \text{ eV}) \tag{18}$$

Fano (1964) and Turner (1964) have reviewed both theoretical and experimental determinations of I. Chu and Powers (1972) calculated I statistically, using Lindhard and Winther's (1964) approach with a Hartree-Fock-Slater wave function. Their calculations, plus other isolated calculations and experimental measurements, are given in Fig. 4. Oscillation of I/Z_2 is observed, and excellent agreement with the calculation and experimental measurements, especially for the $Z_2 = 20$–30 region of Andersen et al. (1969), can be seen.

The calculation given in Fig. 4 can be only used as a guideline for normalization and interpolation. Empirical determination of I is always absolutely necessary to given reliable values.

5.3. Shell Correction

Equation 17 contains a term C/Z_2, which is the correction for the nonparticipation of inner-shell electrons to the stopping power. When a projectile is extremely fast, all target electrons contribute to the stopping power in a more predictable manner. When the projectile is not so fast, the inner-shell electrons contribute less to the stopping power, and the shell correction term enters in, to account for the fact that expressing the stopping problem with a single variable I is insufficient. Since the correction is due to inner shells, it can be expressed as

$$\frac{C}{Z_2} = \frac{C_K + C_L + \dots}{Z_2} \tag{19}$$

where C_K, C_L mean corrections to the K and L shells. Methods for treating shell corrections have been described by Walske (1952, 1957). Empirically speaking, a careful measurement of energy loss should yield the values of C/Z_2 and I. If one ignores the density effect, a measurement of ε vs. E enables one to extract a single energy-dependent parameter X following Bichsel's treatment (1964), where

$$X = \log I + \frac{C}{Z_2} \qquad (20)$$

The value of mean excitation and ionization energy I is independent of energy; the shell correction C/Z_2 is a function of energy, and its value is supposed to be zero at very high energies. In principle, the above two boundaries make it possible to separate the two terms. In practice, any small error in experimental measurements will be amplified in the extraction of C/Z_2 and I. Once C/Z_2 and I are determined, values of ε vs. E can be obtained for the very broad energy region and for several light projectiles.

So much for the energy dependence and target dependence of ε in Bethe's region. In the next few sections we will discuss the projectile dependence of energy loss.

5.4. Velocity Criteria

Bohr (1948) has given velocity criteria for applying classical concepts such as the collision diameter and the impact parameter to Rutherford scattering and to the classical derivation of the Bethe formula. Classical treatment is justifiable if the collision diameter b defined in Eq. 7 is much larger than de Broglie's wave length. From Eq. 7 we have

$$b = \frac{2Z_1 Z_2 e^2}{Mv^2} \qquad (21)$$

and de Broglie's wave length λ is defined as

$$\lambda = \frac{\hbar}{Mv} \qquad (22)$$

For $b \gg \lambda$,

$$\frac{2Z_1 Z_2 e^2}{\hbar v} \gg 1 \qquad (23)$$

Equation 23 is the necessary and sufficient condition for the application of the classical treatment of scattering.

For proton-electron scattering, Eq. 23 reduces to

$$\frac{2e^2}{\hbar v} \gg 1 \tag{24}$$

with the Bohr velocity defined as

$$v_o = \frac{e^2}{\hbar} = \frac{c}{137} \tag{25}$$

Equation 24 gives

$$v \ll 2v_o \tag{26}$$

The proton velocity that is equivalent to v_o amounts to 25 keV; for $v = 2v_o$, therefore, the proton energy is about 100 keV.

Complementary to Eq. 23 is a region in which quantum mechanical perturbation methods may be used if

$$\frac{2Z_1 Z_2 e^2}{hv} \ll 1 \tag{27}$$

For heavy ions interacting with target electrons ($Z_2 = 1$), this equation becomes

$$v \gg 2Z_1 v_o \tag{28}$$

A further condition to Eq. 28 is that the ion starts to carry an appreciable number of electrons when its velocity becomes

$$v < v_o Z_1^{2/3} \tag{29}$$

in collisions that are not too close.

Equation 29 gives another characteristic velocity:

$$v_1 = v_o Z_1^{2/3} \tag{30}$$

When $v < v_1$ (Eq. 29), the equivalent charge Z_1^* of the ion, based on Bohr (1948), is

$$Z_1^* = Z_1^{1/3} \left| \frac{v}{v_o} \right| \tag{31}$$

which is very small at low velocities; we are dealing with nearly neutral projectiles colliding with target atoms. The electronic

stopping increases with v; this is because in the Bethe stopping
formula Z_1 is replaced by Z_1^*, and therefore Z_1^{2*}/v^2 is independent
of velocity, and the remaining part of Bethe's equation is a slowly
increasing function of velocity. If v is large compared with
v_1, then $Z_1^* \rightarrow Z_1$ and Bethe's formula gives the general decreasing
to ε when v increases. This type of argument produces the general
shape of ε versus E or v (Fig. 2).

5.5. Charge State of Heavy Ions, Z_1^*, and Scaling of Energy Loss

     In earlier sections we learned that the Z_1^2 dependence in
Bethe's formula is correct only when the perturbation treatment is
valid and the projectile is totally ionized on the condition expressed by Eq. 28. To fulfill Eq. 28, projectiles must have the
following energies:

$$E > 100 \text{ keV for protons}$$
$$E > 1.6 \text{ MeV for } {}^4\text{He ions} \tag{32}$$
$$E > (M_1 Z_1^2) \times 0.1 \text{ MeV for heavy ions}$$

At lower energies, projectiles are not fully ionized. The charge
states of moving heavy ions have been studied and reviewed by
Betz (1972) and by Datz (1975).

     Equation 31 gives an estimation of the charge state for low
velocities. Betz (1972) has given a semiempirical expression

$$Z_1^* = Z_1 \left[1 - A_0 \exp \left(\frac{-v}{v_0 Z_1^\gamma} \right) \right] \tag{33}$$

where A_0 and γ are adjustable parameters having values close to
$A_0 \simeq 1$ and $\gamma \simeq 2/3$.

     Ward et al. (1976) have recently proposed a Z_1^* based on measure-
ments of the energy loss of five different ions in six different
metal targets. Their formula shows, for $v > 2v_0$,

$$Z_1^* = Z_1 \left[1 - A(Z_1) \exp \left(\frac{-0.879v}{v_0 Z_1^{0.65}} \right) \right] \tag{34}$$

$$A(Z_1) = 1.035 - 0.4 \exp (-0.16 Z_1) \tag{35}$$

When Z_1^* is known, Bethe's formula provides a method of extrapolation
of energy loss of a projectile A to another projectile B at the
same velocity v in the same target medium Z_2. That is,

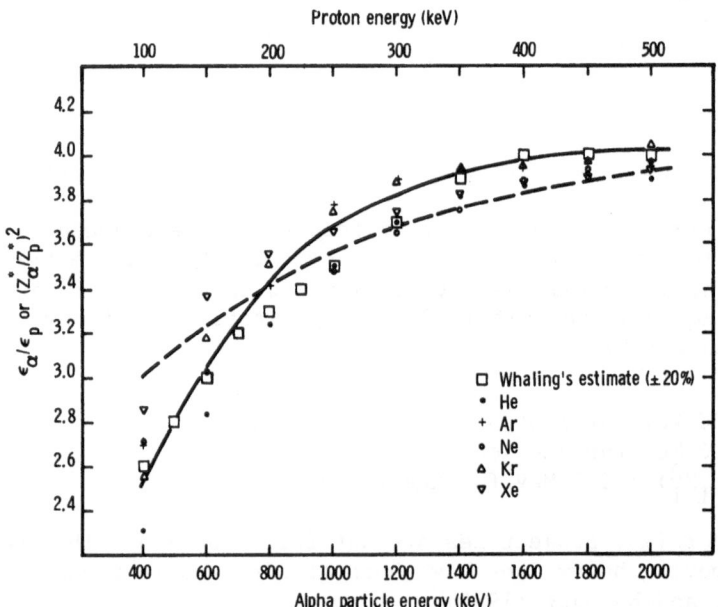

Fig. 5. Stopping cross section ratio or effective charge
ratio of α-particles to protons evaluated at the
same velocity. See Section 5.5 and Table 1.

$$\frac{1}{Z_A^{*2}}\left(\frac{dE}{dx}\right)_{Z_A,v} = \frac{1}{Z_B^{*2}}\left(\frac{dE}{dx}\right)_{Z_B,v} \tag{36}$$

For example, when one data base is for protons, the above relation
becomes

$$\frac{1}{Z_1^{*2}}\left(\frac{dE}{dx}\right)_{Z_1,v} = \left(\frac{dE}{dx}\right)_{proton,v} \tag{37}$$

The ratio of α particle energy loss to proton energy loss provides
Z_1^* for He ions. More rigorously, it provides $(Z_\alpha^*/Z_p^*)^2$.

The effective charge of protons (Z_p^*) is usually accepted as being given by the parameterization by Booth and Grant (1965):

$$(Z_p^*)^2 = [1 - \exp(-150E_p)] \exp[-0.835 \exp(-14.5E_p)] \quad (38)$$

where E_p is the energy of protons, in MeV. The effective charge for helium ion is empirically given by Ward et al. (1976) as

$$(Z_\alpha^*)^2 = 4 - 3.04 \exp(-1.77E_\alpha) \quad (39)$$

where E_α is the energy of ^{4}He ions, in MeV. The above equation is valid only for $E_\alpha > 0.4$ MeV. Large uncertainties are expected at the lower bound of Eqs. 38 and 39.

Calculations of Eqs. 38 and 39 and their ratios are given in Table 1. The ratios should be equivalent to the energy loss ratios of helium ions to protons at the same velocities.

Values of energy loss ratio have been studied and listed many times, for example by Whaling (1964), Bourland, Chu, and Powers (1971), Bourland and Powers (1971), and Ward et al. (1971).

Figure 5 presents an example of the $\varepsilon_\alpha/\varepsilon_p$ ratio versus E. This ε ratio is equivalent to the effective charge ratio. Whaling's estimates of this ratio are denoted by the squares, those of Bourland and Powers by the solid curve. The measurements of energy loss in noble gas are taken from Chu and Powers (1971) and from earlier measurements cited in their reference lists. The ratio of Eqs. 39 and 38 is given as a dashed line in Fig. 5. With proper averaging from this figure, it seems that one can scale energy loss from protons to helium ions or vice versa with a probable error of ±5%.

5.6. Higher-Order Corrections of Z_1

The Bethe formula (Eq. 17) is based on quantal perturbation theory to the first order, which scales with Z_1^2. Corrections to Z_1^3, Z_1^4, ..., would be required to account for higher order. The Barkas effect is the difference in stopping of positive and negative particles (Z_1^3 correction has opposite sign). Lindhard (1976), in a brief review of this subject, approaches the problem by classical Rutherford scattering calculation. Recent measurements by Andersen et al. (1976) for H, He, and Li ions in several different targets seem to confirm theoretical predictions of the correction. The amount of correction is very small, about 2%; therefore very accurate measurements are required to evaluate this correction.

Table 1. Effective Charge of Protons and Helium Ions
          at the Same Velocities (Eqs. 38 and 39)

E_p (MeV)	$(Z_p^*)^2$	E_α (MeV)	$(Z_\alpha^*)^2$	$(Z_\alpha^*/Z_p^*)^2$
0.10	0.822	0.4	2.502	3.044
0.15	0.910	0.6	2.949	3.241
0.20	0.955	0.8	3.262	3.416
0.25	0.978	1.0	3.482	3.560
0.30	0.989	1.2	3.637	3.677
0.35	0.995	1.4	3.745	3.764
0.40	0.998	1.6	3.821	3.829
0.45	0.999	1.8	3.874	3.878
0.50	0.999	2.0	3.912	3.916
0.55	1.000	2.2	3.938	3.938
0.60	1.000	2.4	3.957	3.957
0.65	1.000	2.6	3.970	3.970
0.70	1.000	2.8	3.979	3.979
0.75	1.000	3.0	3.985	3.985
0.80	1.000	3.2	3.989	3.989
0.85	1.000	3.4	3.993	3.993
0.90	1.000	3.6	3.995	3.995
0.95	1.000	3.8	3.996	3.996
1.00	1.000	4.0	3.997	3.997

The last column is also plotted in Fig. 5.

6.   LOW-VELOCITY REGION

At low velocities, the Bethe formula does not apply to electronic stopping because the electrons in the inner shell contribute
less to the stopping power. This reduction gives very large
correction. Also, at very low velocities the neutralization
probability becomes so large that the collision between the projectiles and the surrounding electrons is almost elastic in a
reference frame moving with the ion. The energy loss becomes
proportional to the velocity of the projectile. Lindhard, Scharff,
and Schiøtt (1963, abbreviated as LSS) and Firsov (1957 a, b; 1959)
gave a theoretical description for this energy region. The LSS
expression is based on elastic scattering of free target electrons
in the static field of a screened point charge. Firsov's expression
is based on a simple geometric model of momentum exchange between
the projectile and the target atom during the interpenetration of
electron clouds. Both theories adequately describe the general
behavior of the stopping power with regard to the energy dependence
and the magnitude of the stopping power.

6.1. Firsov Theory

The beauty of Firsov's approach (1957a,b) is its simplicity.
The geometric model of the interaction of two atoms can easily be
modified to a more realistic atomic structure. Firsov's approach
is through the momentum transfer of electrons from one atom to
another. An imaginary surface S is constructed halfway between the
two atoms, or at the position of the minimum potential. As the
projectile interpenetrates the target atom, electrons of one atom,
upon reaching the surface S, are assumed to transfer a momentum mv
to the other atom. The total momentum transfer per unit time is
given by assuming an electron flux $1/4$ nv_e, where n is the localized
electron density and v_e is the velocity of the electron in the atom.
The total energy loss in the collision is related to the impact
parameter by

$$T(b) = \frac{1}{4} mv \int_{-\infty}^{+\infty} dx \int_S ds\, nv_e \qquad (40)$$

and the stopping cross section becomes

$$S_e = \frac{1}{N} \frac{dE}{dx} = \int_{b_o}^{\infty} 2\pi b\, db\, T(b) \qquad (41)$$

From here on, it is up to the user to determine what kind of n, v_e
to be used and how to integrate over the plane S, distance x
and impact parameter b.

6.2.   LSS Theory

The beauty of the LSS approach is that by making extensive use of the Thomas-Fermi model, similarities among different stopping systems can be obtained, and therefore energy loss can be scaled from one system to the other with adequate accuracy. The reduction of energy, distance, and energy loss into a set of universal units, plus the proper treatment of nuclear energy loss and range study, has made the LSS theory one of the most influential theories on low-energy ion implantation. Because of the complexity of the problem and some necessary but crude approximations in the treatment, the theoretical calculations can be expected to differ from the experimental measurements.

As for electronic energy loss, the major contributor to the difference between theoretical trend and experimental measurement is the charge distribution of a target atom. The smooth distribution described by the Thomas-Fermi model has great significance to the general trend of the interaction of charged particles, and allows a smooth scaling from one system to another. In the LSS method, the Thomas-Fermi statistical treatment makes it possible to generalize the problems and reduce the parameters so as to form an overall picture of the stopping problem, whereas a Hartree-Fock-Slater charge distribution is very specific and causes stopping to fluctuate about the norm (LSS). Realizing the difference between the Thomas-Fermi and Hartree-Fock-Slater charge distribution enables one to adjust the energy loss theroy as necessary.

The electronic energy loss as low velocities can be written from the LSS theory,

$$S_e = \frac{1}{N}\frac{dE}{dx} = \eta \; \frac{8\pi e^2 a_o}{N} \; \frac{Z_1 Z_2}{Z} \; \frac{v}{v_o} \tag{42}$$

where $Z = \left(Z_1^{2/3} + Z_2^{2/3}\right)^{3/2}$ and $\eta \approx Z_1^{1/6}$. This relation, according to LSS, is applicable for $v < v_o Z_1^{2/3}$.

In this section I will give the LSS calculation graphically to show the magnitude of the nuclear-to-electronic stopping and list the conversion factors for the translation from the physical system to the dimensionless universal system.

The energy and depth, in LSS dimensionless units, are

$$\varepsilon = E \; \frac{aM_2}{Z_1 Z_2 e^2 (M_1 + M_2)} \tag{43}$$

$$\rho = RNM_2 4\pi a^2 \frac{M_1}{(M_1 + M_2)^2} \tag{44}$$

where

$$a = 0.8853a_o Z^{-1/3}$$

and

$$Z = \left(Z_1^{2/3} + Z_2^{2/3}\right)^{3/2}$$

Using dimensionless units, one can express the electronic stopping as

$$\left(\frac{d\varepsilon}{d\rho}\right)_e = k\varepsilon^{1/2} \tag{45}$$

where k is a constant that depends on Z_1, Z_2, M_1, and M_2:

$$k \simeq \frac{0.0793Z_1^{2/3}Z_2^{1/2}(M_1 + M_2)^{3/2}}{Z^{1/2}M_1^{3/2}M_2^{1/2}} \tag{46}$$

The range of k is from 0.1 to 0.2, unless the projectile is much lighter than the target atom. Values of E/ε, R/ρ, and k for several systems are given in Table 2.

Nuclear stopping is a relatively small effect in the Rutherford collision region, and its contribution to the total stopping cross section is significant only at very low velocities. However, its significance in the theory of radiation effects, such as radiation damage, sputtering, and the relation between projected range and range, makes the study of nuclear collision important. The scattering cross section for heavy projectiles in the Thomas-Fermi region and the excessive-screening region has been reviewed by Sigmund (1972 a, b).

From the Thomas-Fermi atom model, the differential cross section can be written

$$d\sigma = \pi a^2 \frac{dt^{1/2}}{t} f(t^{1/2}) \tag{47}$$

where t is a reduced variable that contains both dimensionless energy ε and the scattering angle θ; that is,

$$t^{1/2} = \varepsilon \sin\left(\frac{\theta}{2}\right) \tag{48}$$

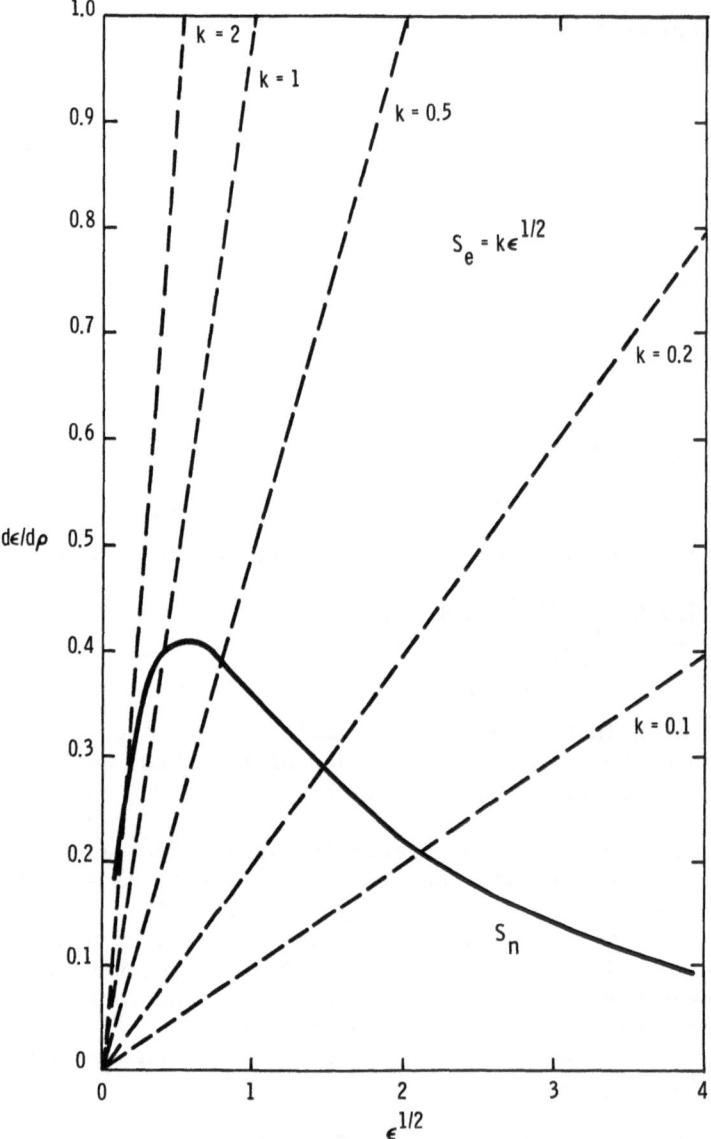

Fig. 6. Nuclear stopping (solid curve) and electronic stopping (dashed curves) as calculated by the LSS theory for various
values of k. Light ions in a solid have
a large k. Heavy ions in a solid have k
around 0.1.

This type of cross section leads to a nuclear stopping cross section in the form

$$\left(\frac{d\varepsilon}{d\rho}\right)_n = f(\varepsilon) \tag{49}$$

which is given in Fig. 6. In this figure electronic stopping cross sections with various k values are also given. From Table 2 and Fig. 6, one can easily estimate at what energy region the nuclear stopping is comparable to electronic stopping. For example, for protons bombarding Si, k = 2.08, and electronic stopping dominates the total energy loss even down to the energy region below 1 keV.

Table 2. LSS Conversion Factors

Ion	Target	E/ε (keV)	R/ρ (Å)	k
H	Si	1.163	1480	2.08
He	Si	2.674	492	0.45
B	Si	8.850	313	0.24
As	Si	209	590	0.12
Sb	Si	515	928	0.11
H	Ge	3.32	6730	5.47
As	Ge	298	668	0.16
Sb	Ge	656	835	0.14
H	Au	10.75	23540	15.1
As	Au	592	883	0.46
Sb	Au	1136	837	0.22

Kalbitzer et al. have given a simple universal formula to fit the published data on nuclear stopping power:

$$\frac{d\varepsilon}{dp_n} = \frac{1.7\varepsilon^{1/2} \ln (\varepsilon + e)}{1 + 6.8\varepsilon + 3.4\varepsilon^{3/2}}$$

This formula applies to the range of $10^{-4} \leq \varepsilon \leq 10^2$; it gives a value much lower than that obtained by use of the LSS theory.

7. MEDIUM VELOCITY REGION

This is the velocity region in which maximum energy loss occurs. The Bethe formula does not work well in this region, because of the charge neutralization of the projectile and the reduction in the participation of inner-shell electrons in the stopping power. Accurate knowledge of shell correction for the target atoms may push the applicability of the Bethe formula into the medium-velocity region, but when the $2mv^2 \simeq I$, the value of Eq. 17 depends entirely on the value of $(-C/Z)$. Neither the LSS theory nor the Firsov theory works in this region. Accurate knowledge of energy loss in this region can be obtained only by measurement or by some semiempirical methods involving inter-polation and extrapolation:

(1) The scaling of energy loss from the charge state of the projectile (discussed in §5.5) makes it possible to scale from one projectile to another for a given target and a given velocity.

(2) The three-parameter-curves fit of ε vs. E, described by Brice (1972), makes it possible to interpolate energy loss values from one energy region to another for a given combination of projectile and target.

(3) For a given projectile at a given velocity, the scaling of energy loss from one target to another can be made from a calculation of energy loss for a given charge distribution (Ziegler and Chu 1974).

These three guidelines are generally applied in the medium-velocity region; however, they can be applied in the other regions as well.

The accuracy of the interpolation and extrapolation depends on the amount and the accuracy of the data base one uses for interpolation. Each case will be different. In general, an

accuracy of ±10% can be reached by proper interpolation.
Ziegler (next chapter) will give detailed information on the
practice of interpolation and extrapolation.

Accurate values of energy loss always come from careful
measurements. In the next section I will briefly discuss experimental methods of measuring energy loss.

8. MEASUREMENTS OF ENERGY LOSS

There are several different methods of measuring energy loss.
One is to prepare thin foil or thin film, measure the film thickness, and measure the total energy loss in the film, to obtain the
stopping cross section. Another is to make an indirect measurement
of a physical quantity that is related to the stopping cross section
in a predictable way--range, backscattering yield, or Doppler shift
of γ emission of compound nuclei in a target medium. All of these
are related to energy loss, and therefore an accurate measurement
of any one of them will yield information on energy loss.

The measurement of energy loss is to be treated in a chapter,
"Energy Loss of High Velocity Ions in Matter," in a volume entitled
Methods of Experimental Physics (Richards, to be published). The
examples presented in the lecture here recorded will be published
in that volume. In the present publication the various methods
will be discussed briefly.

8.1. Transmission Measurement

The principle of this measurement (Fig. 1) is to prepare a
uniform, self-supported thin foil of the target material and
carefully measure the energy with and without the foil, to determine the amount of energy loss in the foil.

A thin foil is usually prepared by vacuum-evaporating the
target material onto a plastic substrate, which is subsequently
dissolved; an example of this method is given by Valenzuela and
Eckardt (1971). Foils of some elements can be obtained commer-
cially. The thickness of the foil is usually determined by
measuring the mass of the foil on a microbalance or a quartz
oscillator during the evaporation, and measuring the area of the
foil. The mass per unit area is equivalent to the density times
the thickness of the foil. It can also be obtained by calibration
of the observed energy loss against a given projectile whose
stopping cross section in this element is known. Target mass
per unit area can be expressed in $\rho\Delta x$ or $N\Delta x$, to an accuracy

ranging from ±0.1% for thick foils to ±2% for thin. Thick foils, on
the order of 10 mg/cm^2, are suitable for protons with high energy
(Bethe region), but are too thick for other projectiles, especially
at medium or low energies. Foils on the order of fractions of
a mg/cm^2 are difficult to prepare. Therefore the measurement of
transmission by use of thin foils is useful only for protons in the
low-energy region, where energy loss is not maximum.

The energy of the projectile can be measured by any of various
instruments: solid state detectors, electrostatic and magnetic
analyzers, and so on. The various methods give measurements with
various probable errors. Typically, a transmission experiment gives
an accuracy of 2 to 5%. However, a few methods yield probable errors
of less than 1%. Andersen et al. (1966) developed a highly accurate
calorimetric method, involving the measurement of the amount of
heat that the projectile gives to the foil.

The transmission methods can be applied not only to solids,
but also to gaseous targets. Typically a gaseous target is contained
in a differentially pumped gas cell or in a sealed gas cell with
thin windows at the ends. The thickness of the target, $N\Delta x$, is
related to the physical length of the gas cell, Δx, and the number
of gas atoms per unit volume, N, which is directly related to the
pressure of the gas in the cell. Such a measurement is described
by Bourland et al. (1971).

Transmission measurements can also be made on a thin film
supported by a thick substrate. Usually in such a case a beam can
be transmitted through the film but not the substrate. The difficulty
of making a one-way measurement can be overcome by using a response
that is strongly dependent on the projectile energy. Johansen et al.
(1971) used the energy shift of the x-ray yield induced by bombarding
a gold substrate with protons to measure the energy loss of protons
on a thin elemental layer evaporated on gold. Leminen and Anttila
(1971) have used the ^{27}Al (p, γ) reaction at various resonance
energies to measure the energy loss and energy straggling of 0.6- to
2-MeV protons in Fe, Co, and Sb. A very thin Al layer, 4 μg/cm^2,
was evaporated onto a Ta backing. The energy loss of the protons
was determined by measuring the γ yield first through a layer of Al
and then through the Al layer covered with an absorbing layer. The
shift in the centroids of the γ yield curve gives the mean energy
loss of the protons in the absorbing layer, and the broadening of
the curve gives the energy straggling.

This method is easy to use and gives excellent results, but it
is very specific. It works only for projectiles in an energy region
in which some kind of sharp resonance exists. It is good for protons,
for which (p, γ) can be used over a broad energy region for several
elements.

The transmission measurement of energy loss can also be extended
to measure the energy loss of a thin layer on top of a solid state
detector. This idea is very similar to the idea of placing a self-
supported foil right in front of a solid state detector. Since a
layer on the detector can be made very thin (a few hundred angstroms),
this method can be used for heavy ions at very low energies. Such
ions cannot penetrate a self-supported thin foil, but they can
penetrate a very thin film or dead layer on a detector. The pulse
height of a silicon detector then corresponds not to the total energy
of a particle entering the detector, but rather to the electronic
energies. The loss of nuclear energy cannot be measured directly.
Graham and Kalbitzer (1976) have used the above facts in developing a
simple and novel method that gives direct measurement of electronic
energy loss at very low energies (<60 keV) where nuclear energy loss
is also deduced.

8.2. Backscattering Measurement

Backscattering is one of the most often used methods of determining energy loss in solid targets. For a thin film on a thick
substrate, the energy of projectiles backscattered from the thin film
surface, or from the substrate surface if no thin film is present,
will differ from that of projectiles scattered from the same element
at the interface. The difference is attributed to the energy loss
in the incident and the outgoing paths.

This method was developed by Warters (1953) to find the stopping
cross section of protons in Li. Several others have used it; see, for
example, Chu and Powers (1969), and Lin, Olsen, and Powers (1973).
The method is familiar to most of those who use backscattering as an
analytical tool. Backscattering analysis and its applications will
be discussed in a separate chapter of this book (Mayer et al.). In
one of the examples in that chapter, the thickness of a thin elemental
film is determined by backscattering, for known dE/dx. Our problem is
just the inverse. We wish to determine the energy loss of a given
projectile in a thin film of known thickness.

8.3. Measurement By Backscattering Yield

This method has the advantage of avoiding the determination of
film thickness $N\Delta x$, which is necessary for all the methods described
so far. The method was originated by Wenzel and Whaling (1952), who
were measuring the proton stopping cross section of ice. Similar
methods have been used, for example by Bethge and Sandner (1965) on
heavy ions, and by Scherzer et al. (1976) on helium ions. The method
is based on the fact that the backscattering yield is directly related
to the stopping cross section. If the scattering cross section is
known, as is the case for Rutherford scattering, the stopping cross
section can be calculated from a measurement of the thick target yield.

8.4. Measurement By Doppler Shift Attenuation

This experimental method for measuring energy loss is to some
extent an inversion of the well known Doppler shift attenuation (DSA)
method for determining lifetime. For the lifetime measurement, dE/dx
must be known. By inverting the problem--that is, by knowing the
lifetime of a γ-emitting reaction and measuring the Doppler shift
attenuation of the γ ray--one can obtain dE/dx. This method has
been described in several publications, such as Neuwirth et al. (1969).

The γ-ray lifetime of an excited nuclear state can be measured
by allowing a moving excited nucleus to slow down in some absorbing
medium. The Doppler shift of the γ ray emitted is attenuated in a
manner that depends on the lifetime of the γ-emitting nuclei in the
medium. Therefore the Doppler spectrum of an isotropically emitted
γ ray is a direct measurement of the stopping cross section of the
γ-emitting nucleus in its surrounding medium.

The method is very specific: it works only for a nucleus that
is emitting γ rays within a limited recoil energy region. Therefore
it does not apply to all dE/dx problems. However, because of its high
relative accuracy, and because the study is for a specific ion at
a well defined energy in various media, it is best applied when
target medium is the parameter. It is very powerful in applications
of Bragg's rule--for example, the work of Neuwirth et al. (1975)--and
also in the study of Z_2 structure in dE/dx, as in the work of Pietsch
et al. (1976).

9. CONCLUSIONS

1.   Knowledge of energy loss is needed in many fields of research.
2.   Energy loss can be determined with high accuracy only by care-
      ful measurements.
3.   Energy loss can usually be determined with moderate accuracy
      (± 10-20%) by interpolating from the available data in accordance
      with current theory.
4.   There are three parameters for interpolation and extrapolation:
      $\varepsilon = \varepsilon(Z_1, Z_2, E)$. Work on one parameter at a time in extracting
      information.
5.   By interpolating or extrapolating on the basis of energy E or
      velocity, energy loss for high and low velocities can be deter-
      mined with great accuracy, $\lesssim 1$-2%. For medium velocity, at
      which energy loss is highest, there is no adequate theory to
      describe this region.
6.   The scaling of energy loss vs. Z_2 is highly specific because
      of the oscillatory structure of $\bar{\varepsilon}(Z_2)$ vs. Z_2. At high velocity,
      the Z_2 structure of $\varepsilon(Z_2)$ is absorbed in the Z_2 oscillation of
      the mean excitation and ionization energy I.

7. At low velocity, for which the Z_2 oscillation of εZ_2 is very
    pronounced, various theoretical treatments exist but are not
    easy to apply.
8. To scale energy loss vs. Z_1, the energy loss of two different
    projectiles at the same velocity in a given target must be
    compared. Scaling based on the effective charge is typically
    accurate to ±10%. Some evidence on Z_1 oscillation exists,
    but so little systematic study has been done that no recom-
    mendation can be given at present.
9. To extract $\varepsilon(Z_1, A_2, E)$, it is always advisable to extrapolate
    or interpolate from more than one parameter to see if the results
    are consistent.
10. In practice, the method of extracting information always depends
    on the available data base and on how well the theory applies in
    the region of interest.

This introductory lecture has been intended to provide a general
feeling for the physical parameters and their influence on energy loss.
In the next chapter, Ziegler will demonstrate methods for establishing
a data base on energy loss by curve fitting and by interpolation and
extrapolation from available measurements in accordance with current
theory.

REFERENCES

Andersen, H. H., J. F. Bak, H. Knudsen, P. Møller Petersen,
    and B. R. Nielsen, to be published.

Andersen, H. H., A. F. Garfinkel, C. C. Hanke, and H. Sørensen (1966),
    Mat. Fys. Medd. Dan. Vid. Selsk. 35, no. 4.

Andersen. H. H., H. Sørensen, and P. Vajda (1969), Phys. Rev. 180,
    373.

Bethge, K., and P. Sandner (1965), Phys. Lett. 19, 241.

Betz, H. D. (1972), Rev. Mod. Phys. 44, 465.

Bloch, F. (1933a), Ann. Phys. (5), 16, 287.

Bloch, F. (1933b), Z. Physik 81, 363.

Bohr, Aa. (1948), Mat. Fys. Medd. Dan. Vid. Selsk. 24, no. 19.

Booth, D. W., and I. S. Grant (1965), Nucl. Phys. 65, 481.

Bourland, P. D., W. K. Chu, and D. Powers (1971), Phys. Rev. B3,
    3625.

Bourland, P. D., and D. Powers (1971), Phys. Rev. B3, 3625.

Brice, D. K. (1972), Phys. Rev. A6, 1791.

Chu, W. K., and D. Powers (1969), Phys. Rev. 187, 478.

Chu, W. K., and D. Powers (1971), Phys. Rev. B4, 10.

Crispin, A., and G. N. Fowler (1970), Rev. Mod. Phys. 42, 290.

Datz, S. (1975), "On the States of Ions Penetrating Solids," in
    Radiation Damage Processes in Materials (proceedings of the
    NATO Study Institute on Radiation Damage Processes in Materials,
    Aleria, Corsica, France, August 27 to September 9, 1973),
    Noordhoff, Leyden, 1975, pp. 119-168.

Fano, U. (1964), in Studies in Penetration of Charged Particles
    in Matter, National Academy of Sciences--National Research
    Council, Publ. 1133, Washington, D. C.

Firsov, O. B. (1957a) Zh. ETF 32, 1464; Engl. transl. (1957), Sov.
    Phys. JETP 5, 1192.

Firsov, O. B. (1957b), Zh. ETF 33, 696; Engl. transl. (1958), Sov.
    Phys. JETP 6, 534.

Firsov, O. B. (1959), Zh. ETF 36, 1517; Engl. transl. (1959), Sov.
    Phys. JETP 9, 1076.

Grahmann, H., and S. Kalbitzer (1976), Nucl. Instrum. Methods
    132, 119.

Johansen, A., S. Steenstrup, and T. Wohlenberg (1971), Rad. Effects
    8, 31.

Kalbitzer, S., H. Oetzmann, H. Grahmann, and A. Feuerstein, to be
    published in Z. Physik A.

Leminen, E., and A. Anttila (1971), Ann. Acad. Sci. Fennicae,
    Series A, VI, Physica, 370.

Lin, W. K., H. G. Olson, and D. Powers (1973), Phys. Rev. B8,
    1881.

Lindhard, J. (1976), Nucl. Instrum. Methods 132, 1.

Lindhard, J., and M. Scharff (1961), Phys. Rev. 124, 128.

Lindhard, J., M. Scharff, and H. E. Schiøtt (1963), Mat. Fys. Medd.
    Dan. Vid. Selsk. 33, no. 14.

Lindhard, J., and Aa. Winther (1964), Mat. Fys. Medd. Dan. Vid.
    Selsk. 33, no. 14.

Neuwirth, W., U. Hauser, and E. Kuehn (1969), Z. Physik 220, 241.

Neuwirth, W., W. Pietsch, K. Richter, and U. Hauser (1975), Z.
    Physik A275, 215.

Northcliffe, L. C., and R. F. Schilling (1970), Nuclear Data Tables
    7A, 233.

Pietsch, W., U. Hauser, and W. Neuwirth (1976), Nucl. Instrum.
    Methods 132, 79.

Richards, P., ed., Methods of Experimental Physics, Academic Press,
    New York, in publication.

Scherzer, B. M. U., P. Børgesen, M.-A. Nicolet, and J. W. Mayer
    (1976), Proceedings of the International Conference on Ion
    Surface Layer Analysis, Karlsruhe, Germany, 1975, Plenum Press,
    New York.

Sigmund, P. (1972a), Rev. Roum. Phys. 17, 823.

Sigmund, P. (1972b), in Physics of Ionized Gases (M. Kurepa, ed.),
    Institute of Physics, Belgrade, p. 137.

Sigmund, Peter (1974), "Energy Loss of Charged Particles in Solids,"
    in Radiation Damage Processes in Materials (proceedings of
    Radiation Damage Processes in Materials, Aleria, Corsica, France,
    August 27 to September 9, 1973), Noordhoff, Leyden, 1975,
    pp. 3-118.

Sigmund, Peter (1974), 5th International Congress on Radiation
    Research, Seattle, Washington, USA, July 1974.

Turner, J. E. (1964), in Studies in Penetration of Charged Particles
    in Matter, National Academy of Sciences--National Research
    Council, Publ. 1133, Washington, D. C.

Valenzuela, A., and J. C. Eckardt (1971), Rev. Sci. Instrum. 42, 127.

Walske, M. C. (1952), Phys. Rev. 88, 1283.

Walske, M. C. (1957), Phys. Rev. 101, 940.

Ward, D., J. S. Forster, H. R. Andrews, I. V. Mitchell, G. C. Ball,
    W. G. Davies, and G. J. Costa (1976), unpublished, AECL-5313,
    Chalk River, Canada.

Warters, W. D. (1953), Ph. D. thesis, California Institute of
    Technology, Pasadena.

Wenzel, W. A., and W. Whaling (1952), Phys. Rev. 87, 499.

Whaling, W. (1958), Ency. Phys. (S. Fluegge, ed.), 34, 193;
    Springer, Berlin, Goettingen, Heidelberg.

Ziegler, J. F., and W. K. Chu (1974), Atomic Data and Nuclear
    Data Tables 13, 463.

SOME GENERAL CONSIDERATIONS OF ION BEAM PRODUCTION

AND MANIPULATION

J. H. Freeman

A.E.R.E., Harwell, Oxon. England

SUMMARY

Some of the general features which govern the production and
manipulation of ion beams are briefly reviewed. Consideration is
given to design factors which are basic to most types of accelerator
and to the ion beam qualities which define their performance. The
mechanism of ionisation and ion beam formation is outlined and in
conclusion some specific examples of ion beam behaviour of particular
interest are described.

1. INTRODUCTION

   In this contribution to the study of material characterisation
using ion beams we are concerned with the production and manipulation
of the required flux of charged particles. Although this subject
appears central to the entire proceedings it is in fact worth
questioning on two counts, the justification for a detailed consideration.

   Firstly, it is worth asking why we need to include an account
of the underlying scientific principles, or the experimental procedures, which ensure that the required beam of ions reaches the
surface of the sample to be characterised. For example, in a
similar consideration of the use of electrons in microscopy it
would not normally be considered appropriate to review the design,
or behaviour, of the heated cathode which lies at the heart of the
electron gun. The simple answer to this question lies in the im-
perfect performance which is a characteristic of so many experiments
involving the use of ion beams. Although it must be emphasised that
commercial equipment of a much improved nature is increasingly

available the problem of producing and manipulating ion beams is
inherently much more complex than that for electrons. It remains
evident that the quality of the ion beam, as measured by a number
of factors (intensity, purity, resolution, stability etc., etc.)
frequently provides the ultimate limitation on the experimental
capability of accelerators used for surface analysis, or for other
charged particle studies. It is also apparent that such limitations
as do exist are rarely fundamental. In a number of cases there is
scope for substantial improvements. It is therefore in this sense
that a consideration of ion beam production and manipulation could
be worthwhile.

The second question of relevance relates, however, to the broad
scope of these proceedings. Although all of the analytical applications to be described involve the use of charged particles in one
way or another, even a cursory consideration of the range of topics
indicates an extended gamut of ion-beam requirements with little
evidence of a connecting theme or a common set of problems. Thus
the experimental facilities range from relatively simple, low-energy
laboratory installations to very high-energy, complex nuclear physics
accelerators. Since these disparate experimental objectives with
their wide variety of machines do not manifest a common set of
problems, or unresolved requirements, the only practicable basis for
a review of ion beam production and manipulation is to restrict the
consideration to a description of some general principles. We
therefore consider below an elementary outline of accelerator performance in broad terms and then consider in rather more detail
some features of particular interest. The review is however limited
insofar as we exclude the design of the target stage and the specific
problems which may be associated with the measurement of the flux of
primary, or secondary, charged particles in the vicinity of the
surface of the bombarded specimen.

2.   GENERAL ACCELERATOR CONCEPTS

In all except the most elementary ion accelerators we are concerned with four basic elements:- the ion source; the beam analysis;
the beam manipulation and the target assembly. It is convenient to
consider these separately but it cannot be too strongly emphasised
that the ultimate performance of any experimental ion-beam facility
is largely a measure of the successful matching of the various components. For example, it is the failure to observe this rather
obvious maxim which accounts for some of the very marked discrepancies
between the claims made for ion sources on the basis of simple laboratory measurements and their subsequently observed performance on real
accelerators. It is possible to arrange the different elements of
the accelerator in a variety of ways and the principal configurations
of machines incorporating beam analysis are shown in Fig. 1. Each
layout has some advantages and some drawbacks, but by far the com-
monest for surface studies is the first shown in Fig. 1. This is

CONVENTIONAL ACCELERATOR

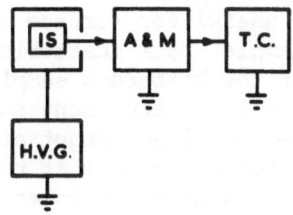

Convenient layout , access restricted to ion source only. Large analyser required for high energy beams, machine performance commonly deteriorates at low energies. Energy programming of ion beam requires analyser adjustment .

ACCELERATOR WITH BEAM ANALYSIS AT HIGH VOLTAGE

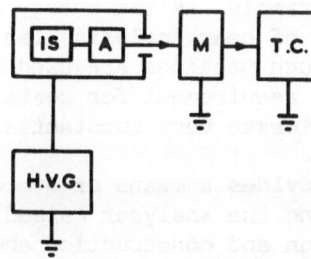

Analysis at constant low energy.
Small analyser adequate.
No deterioration of performance over wide energy range.
Beam energy programming by H.T. adjustment only.
Analyser or beam analysis flight tube must be at high voltage.

ACCELERATOR WITH TARGET CHAMBER AT HIGH VOLTAGE

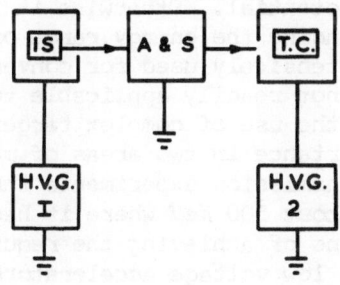

Restricted access to target but convenient method of increasing accelerator energy range.
Target does not require high stability H.T. supply.
Beam energy programming by target H.T. adjustment only.
Wide energy range with small analyser.
High resolution and intense beam currents readily obtained.

Fig. 1: Accelerator configurations (I.S. - Ion source; A. - Analysis; M. - Manipulation; T.C. - Target chamber; H.V.G. - High voltage generator)

the simple and conventional form of accelerator with the ion source
at high voltage and the analyser and target chamber at earth potential.   The advantages of such a system are that the source and its
power supplies are electrically and physically isolated from the
rest of the system and that there is thus unrestricted access to the
target chamber.   This is particularly useful for surface analysis
experiments which may involve, for example, back scattering, beam
transmission or microscopic studies of the bombarded sample.   It
greatly simplifies the use of complex related equipment such as
motorised goniometers, or temperature-controlled target stages.   Yet
another advantage of this accelerator layout is that it readily lends
itself to the use of multiple beam lines with a consequent increase
in flexibility or experimental efficiency.   This can take the normal
form of a multiport analyser, or alternatively the arrangement used
on the Cockcroft Walton accelerator at Harwell can be employed
(Fig. 2).

        In spite of its advantages the simple conventional layout also
has some inherent drawbacks.   For example, as the energy and/or the
ion mass is increased, the problems of beam analysis also increase.
A more subtle factor is that when such machines are used below their
normal operating voltage - a common requirement for certain surface
studies - the performance may deteriorate very substantially.

        The second layout of Fig. 1 provides a means of overcoming
these problems at the price of having the analyser raised to high
voltage.   This complicates the design and construction and is only
applicable to certain types of ion accelerator but because of its
advantages, it is a solution which is becoming increasingly popular
in the related field of ion implantation.   It maintains the flexibility of the earthed target chamber.

        This last advantage is lost on the third layout of Fig. 1 since
it has the target chamber at high potential.   Otherwise it is a very
convenient and low cost way of extending the energy range of an
accelerator and it has also been extensively used for conventional
ion doping studies.   It clearly is not readily applicable to surface
characterisation studies involving the use of complex target assem-
blies but it is potentially of importance in two areas of material
characterisation.   Firstly, for transmission experiments through
very thin foils at energies below about 500 keV where it has the
advantage of providing a simple means of achieving the required
energy on a small accelerator (such low voltage accelerators may
have high fluxes of doubly and triply-charged ions which further
increase the available energy).   When the excited, transmitted ion
is to be studied, there is an additional benefit that it is decel-
erated after transmission.

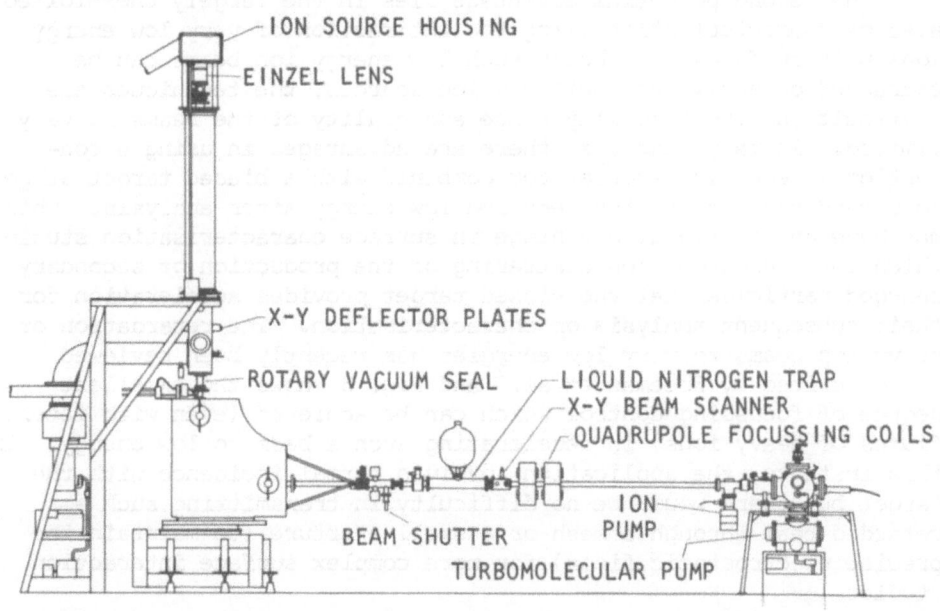

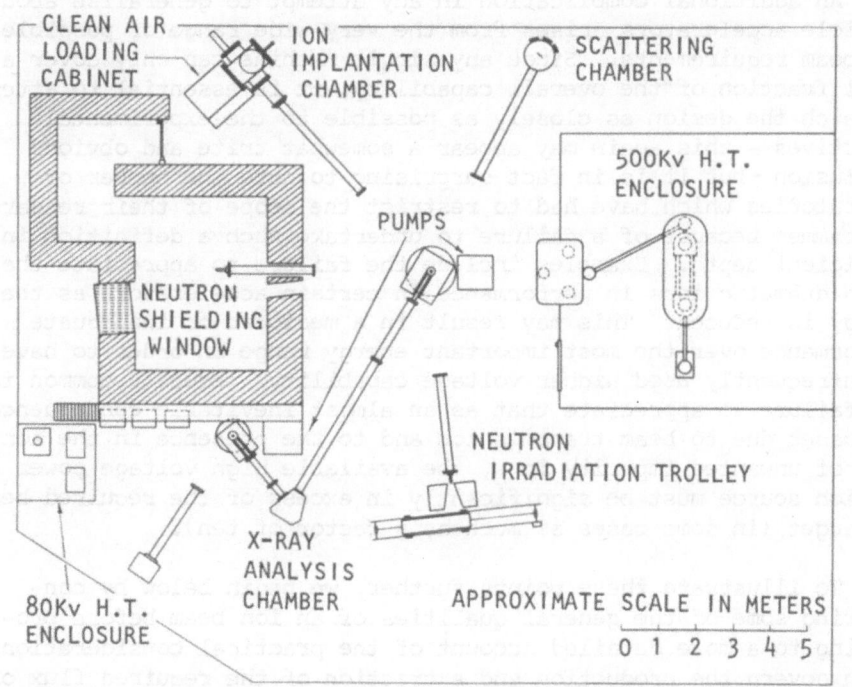

Fig. 2: Harwell 500 kV Cockcroft Walton accelerator illustrating
        multiple beam line arrangement

The second potential advantage lies in the largely unexploited area of characterisation using the interaction of very low energy ions with surfaces. Although such low energy ion beams can be extracted directly from suitable ion sources, the techniques are difficult and the intensity range and quality of the beams is very limited. In many respects, there are advantages in using a conventional heavy ion accelerator combined with a biased target stage to retard the beam to the required low energy after analysis. This may have an additional advantage, in surface characterisation studies which involve either ion scattering or the production of secondary charged particles, that the biased target provides acceleration for their subsequent analysis or characterisation. The retardation of heavy ion beams to very low energies has recently been reviewed and described by Freeman et al.[1] and Fig. 3 shows the excellent degree of focussing control which can be achieved (even with intense fluxes of heavy ions) in decelerating such a beam to low energy. In this instance, the application involves normal incidence with the target but there would be no difficulty in transmitting such a retarded beam through a mesh or a small aperture (to maintain the precise electrostatic field) for more complex surface interaction studies.

An additional complication in any attempt to generalise about particle accelerators arises from the very wide range of possible ion beam requirements. Since any single machine can only cover a small fraction of the overall capability, it is essential to attempt to match the design as closely as possible to the experimental objectives – this again may appear a somewhat trite and obvious conclusion – but it is in fact surprising to note the number of laboratories which have had to restrict the scope of their research programmes because of a failure to undertake such a definition in sufficient depth. Examples include the failure to appreciate the often dramatic loss in performance in certain accelerators as the energy is reduced. This may result in a mediocre or inadequate performance over the most important energy range in order to have an infrequently used higher voltage capability. Equally common is the failure to appreciate that as an almost inevitable consequence of losses due to beam transmission and to the presence in the ion beam of unwanted impurity ions, the available high voltage power at the ion source must be significantly in excess of the required beam on target (in some cases as much as a factor of ten).

To illustrate these points further, we begin below by considering some of the general qualities of an ion beam before proceeding to a more detailed account of the practical considerations which govern the production and extraction of the required flux of charged particles.

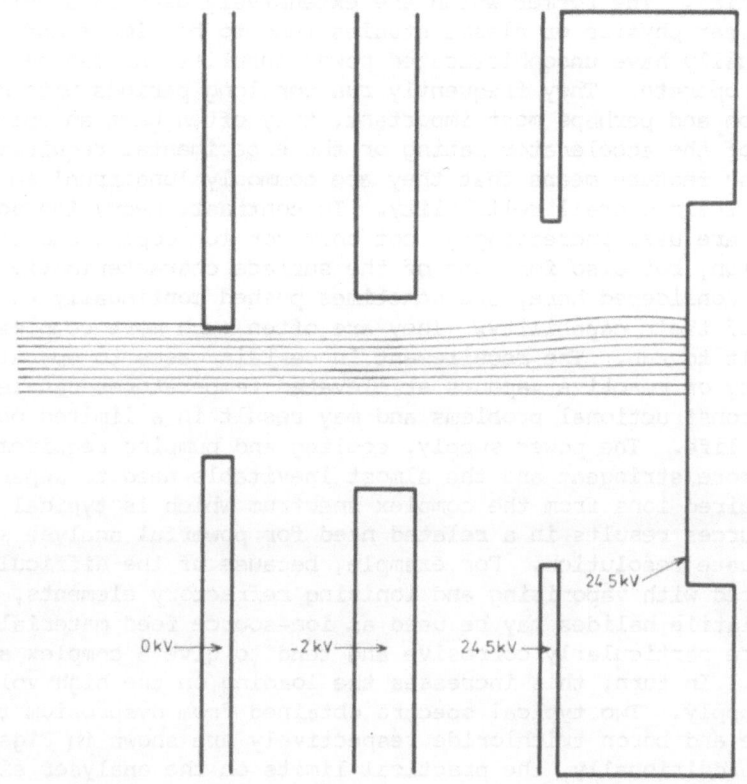

Fig. 3: Retardation lens geometry showing computed ion-beam
        trajectories. Dimensions in arbitrary units.   Initial
        ion beam energy = 25 keV.

2.1 Ion-beam mass

In this instance, there is a fairly clear distinction between
ion sources restricted to hydrogen and helium (with possibly a few
other simple gaseous elements) and those which are specifically
aimed at the production of heavier ions. The differences can be
substantial. The former which are extensively used in accelerators
for nuclear physics or plasma studies tend to be simple and robust.
They usually have unsophisticated power supplies and can be very
easy to operate. They frequently run for long periods with minimal
attention and perhaps most important, they often have an output in
excess of the accelerator rating or the experimental requirements.
This last feature means that they are commonly 'underrun' and this
adds to their overall reliability. In contrast, heavy ion sources,
such as are used increasingly, not only for ion doping and isotope
separation, but also for some of the surface characterisation
studies considered here, are sometimes pushed continually to the
limits of their capability. They are often much more complex and
difficult to run. The requirement in certain cases to operate with
a variety of reactive vapours at elevated temperatures can lead to
severe constructional problems and may result in a limited operational life. The power supply, cooling and pumping requirements
may be more stringent and the almost inevitable need to separate
the required ions from the complex spectrum which is typical of
such sources results in a related need for powerful analysers with
an adequate resolution. For example, because of the difficulties
associated with vaporising and ionising refractory elements, the
more volatile halides may be used as ion-source feed materials.
These are particularly corrosive and tend to give a complex spectrum
of ions. In turn, this increases the loading on the high voltage
power supply. Two typical spectra obtained from dysprosium trichloride and boron trichloride respectively are shown in Figs. 4
and 5. Additionally, the practical limits on the analyser size
commonly set a limit on the achievable energy or mass on target
for a particular accelerator configuration. For similar reasons,
the excellent magnetic focussing lenses which have more significant
aberrations (and which are subject to space-charge defocussing when
used with intense heavy-ion beams) have to be used.

2.2 Ion-beam intensity

In the previous section we identified the important distinction
between the problems associated with the use of 'light' and 'heavy'
ions. A similar very clear distinction exists in the case of ion-
beam intensity. In this instance, the choice resolves largely into
applications requiring currents of some tens of microamperes or
less and those involving beams of hundreds of microamperes or above.
The distinction is, of course, by no means absolute and it is also
dependent upon the ion mass and energy. It is clear, too, that the

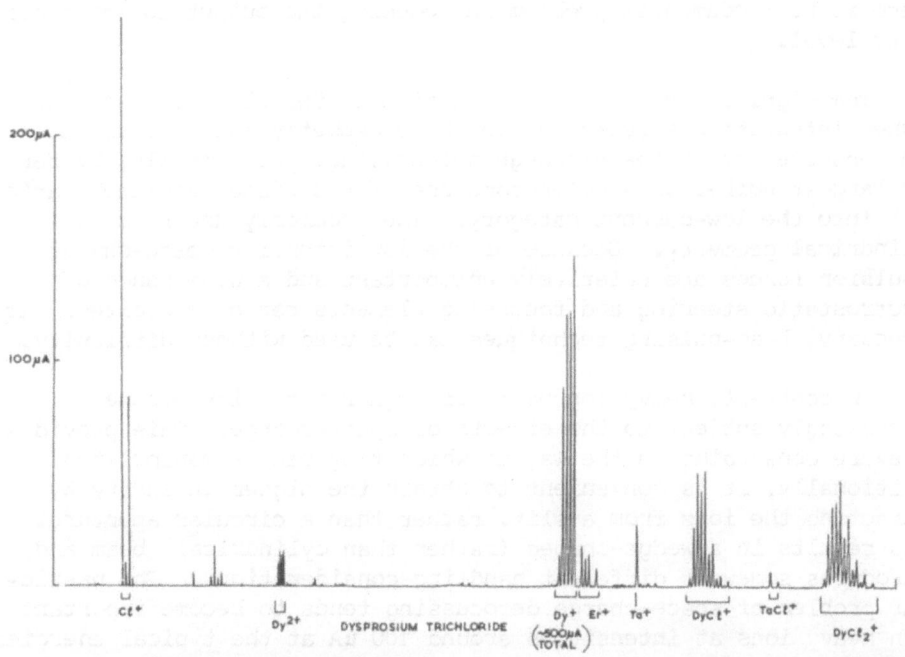

Fig. 4: Dysprosium trichloride spectrum

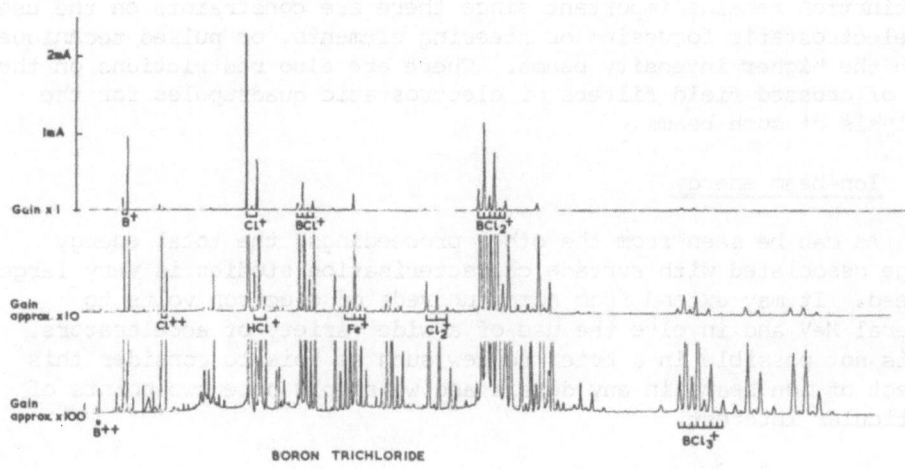

Fig. 5: Boron trichloride spectrum

distinction refers to the maximum current since on any accelerator
there is no fundamental problem in reducing the output to any required
lower level.

The significance of the distinction is twofold. The move to
higher intensities affects the ion-beam geometry and it introduces
the consequences of space-charge defocussing. Fortunately, by far
the largest number of accelerators used for surface characterisation
fall into the low-current category. They commonly use beams of
cylindrical geometry. Because of the low intensity space-charge
repulsion forces are relatively unimportant and a wide range of
electrostatic steering and focussing elements can be exploited. If
necessary, beam-pulsing techniques can be used without difficulty.

In contrast, heavy-ion beams of higher intensity become
increasingly subject to the effects of space-charge. This provides
a severe constraint on the way in which they can be manipulated.
Additionally, it is convenient to obtain the higher intensity by
extracting the ions from a slit, rather than a circular aperture.
This results in a wedge-shaped (rather than cylindrical) beam and
introduces somewhat different handling considerations. The particular problem of space-charge defocussing tends to become important
with heavy ions at intensities around 100 µA at the typical energies
of some tens of kilovolts which are used for a range of ion beam
surface studies. In practice such effects may be minimised by
careful design and operation of the accelerator (for a more detailed
consideration see ref.1), and it is possible to obtain a very high
resolution indeed with intense beams. Nevertheless, this intensity
distinction remains important since there are constraints on the use
of electrostatic focussing or steering elements, or pulsed techniques,
with the higher intensity beams. There are also restrictions on the
use of crossed field filters or electrostatic quadrupoles for the
analysis of such beams.

2.3   Ion-beam energy

As can be seen from the other proceedings, the total energy
range associated with surface characterisation studies is very large
indeed. It may extend from some hundreds of electron volts to
several MeV and involve the use of a wide variety of accelerators.
It is not possible in a brief review such as this to consider this
aspect of ion beams in any detail and we simply note two points of
particular interest.

The first concerns the acceleration of the ion beam, where, as
in the previous sections on ion-beam qualities, we can identify an
important operational distinction. At energies below approximately
100 keV, it is possible to accelerate the ions across a single gap
electrode system. This may incorporate a lens system to focus the
ion beam. A typical simple assembly of this nature is shown in

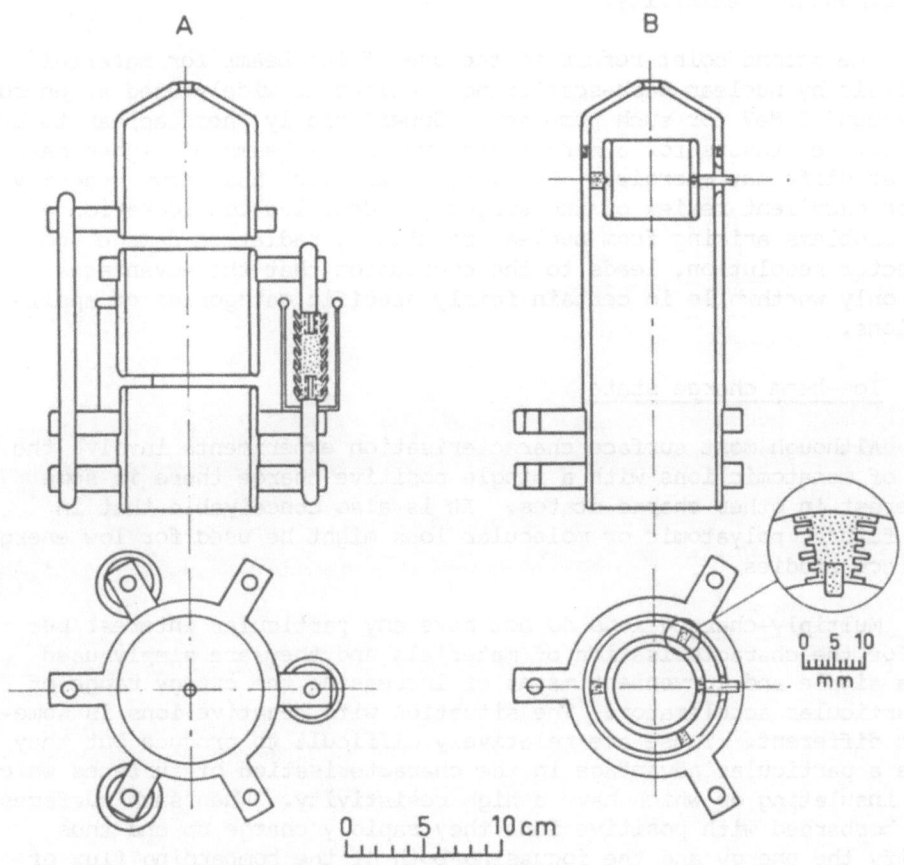

Fig. 6: Extraction and focussing lenses for circular aperture
source (after Alvager and Uhler)

Fig. 6. As the energy is raised, it becomes increasingly necessary
to employ some form of graded acceleration tube to stand-off the
high potential. Such electrode systems have been extensively developed for nuclear physics accelerators and they can be very reliable
indeed. They are, however, inherently more complex than single gap
lenses and when used with heavy ions at high intensity, they are
much more prone to electrical breakdown and to induce space-charge
defocussing. Thus the choice of energy may provide a constraint on
experimental flexibility.

The second point refers to the use of ion beams for material
analysis by nuclear back-scattering. Helium is widely used at energy
of around 2 MeV for such purposes. Superficially there appear to be
attractive reasons for a more extensive use of beams of higher mass
and at different energies. However, as Ziegler[2] has shown recently
in an excellent review of the subject, a detailed consideration of
the problems arising from nuclear reactions, radiation damage and
detector resolution, leads to the conclusion that the advantages
are only worthwhile in certain fairly specific categories of applications.

2.4 Ion-beam charge state

Although most surface characterisation experiments involve the
use of monatomic ions with a single positive charge there is some
interest in other charge states. It is also conceivable that in
the future, polyatomic or molecular ions might be used for low energy
surface studies.

Multiply-charged ions do not have any particular interest per
se for the characterisation of materials and they are simply used
as a simple and convenient means of increasing the energy range of
a particular accelerator. The situation with negative ions is somewhat different. These are relatively difficult to produce but they
have a particular advantage in the characterisation of surfaces which
are insulating or which have a high resistivity. When such surfaces
are bombarded with positive ions they rapidly charge up and thus
modify the energy and the focussing both of the bombarding flux of
particles and of any secondary charged particles emitted from the
surface. This may have important consequences in characterisation
studies. When negative ions are used, the surface charging effect
is minimised by the compensating emission of secondary electrons
and the bombardment can proceed relatively unperturbed.

This last phenomenon raises the interesting question of why
neutral beams are not used much more extensively. The undesirable
formation of fast directed fluxes of neutral particles by charge-
exchange reactions of the ion beam with residual gas in accelerator
transport lines is now widely recognised and precautions are

routinely taken in ion implantation to steer the ions away from the
neutrals. At the energies of interest for many surface characterisation studies (1 - 500 keV), the cross-sections for neutralisation
with air are typically in the range 10^{-15} - 10^{-16} cm^2 and it is
relatively simple to neutralise a significant fraction of an ion
beam.

Thus, for example, with a cross-section of say, 2×10^{-16} cm^2
it would be possible to neutralise around 10% of an ion beam in a
simple charge-exchange chamber 10 cm long at a pressure of 10^{-3} Torr
($f = 3.3 \times 10^{16}$ $plσ$, when p is the pressure in Torr, l the path length
in cm and $σ$ the cross-section). Since the degree of scattering in
such a reaction is quite modest and since it is relatively simple
to deflect the remaining ions out of the beam of fast directed
neutrals, there is therefore no problem in principle - apart from
the difficulty of flux measurement - in using such beams for surface
studies.

2.5 Other ion-beam qualities

In addition to the principal qualities of mass, intensity,
energy and charge-state discussed above, the ion beam is also
characterised by a number of other parameters. Since the importance
of these depends upon the particular application under consideration
and varies considerably from experiment to experiment, we shall only
give them cursory attention in this brief review.

They include the degree of space-charge neutralisation, brightness, emittance, energy spread, stability, quality of focus and
resolution, dispersion and purity.

We can illustrate the overall complexity of the situation by a
very brief description of the factors which influence only one of
these qualities - the ion-beam purity. Let us consider, for example,
an application involving the use of doubly-charged zinc ions obtained
by the ionisation of the convenient volatile zinc sulphide. The
principal zinc beam at mass 32 ($^{64}Zn^{++}$) will inevitably be contaminated by other zinc isotopes (isotopic contamination); it will probably
be contaminated by molecular oxygen, O_2^+, (isobaric contamination);
the beam will contain neutral zinc particles from charge-exchange
(autocontamination) and it may contain charged particles of the
correct mass, but incorrect energy, produced by charge-exchange
before analysis (Aston Bands). In this instance, the reaction,
$Zn_2^+ + X \to Zn^+$ + products, could result in such an impurity beam.
Yet other types of contamination may arise from beam interactions
involving scattering, sputtering or reflection. A careful consideration of these various factors indicates that high beam purity -
above say, 98% - can only be obtained with the most rigorous precautions. Much lower purity levels are commonplace and a high
resolution is not in itself a guarantee of high beam purity. For

many experiments, an impurity level of a few per cent is quite
irrelevant but in others it may lead to quite major errors of
interpretation. An important example of such an effect in surface
characterisation is given later.

2.6 Ion beams – some general considerations

It should be apparent from the above that the behaviour of the
ion beam is complicated. It is in part this complexity which explains
the earlier statement that the quality of the beam is a limiting
factor in many applications and the assertion that the subject merits
attention because of the potential for improvement which exists in
many cases. Although fairly detailed accounts of different aspects
of ion optics and the problems of heavy ion beams have been given by
a number of authors (for example, refs. 3 – 5) there is no general
solution to the problems of accelerator design and operation. At
one extreme, for example, the single problem of ion–beam emittance
has merited a book to itself[6], at the other, in a recent instructive
and entertaining overview of charged particle beams, Lawson[7] describes
a somewhat idealised ion beam with emittance and space–charge in the
single relatively simple equation (following Lawson's nomenclature):

$$a'' + a'\frac{\phi'}{2\phi} + \frac{\phi''}{4\phi}.a - \frac{\varepsilon^2}{a^3} - \frac{(1-f)\,I\,m^{1/2}}{4\sqrt{2\pi}\,\varepsilon_0\,\phi^{3/2}\,q^{1/2}}.\frac{1}{a} = 0$$

where the first three terms represent the paraxial ray equation.
The fourth term is a measure of the beam emittance (or 'entropy')
and characterises the lack of measurable order in the beam. As
Lawson points out the entropy is a rather more subtle measure of
beam quality than the emittance and would be low for a structured
beam containing information which can be focussed to form an image,
as in an ion or electron microscope. Aberrations cause increase of
entropy (or emittance) and hence loss of information. This is manifest as blurring of the image. The final term again using Lawson's
nomenclature, is the space–charge effect where (1–f) denotes the
fractional neutralisation of the ion beam by electrons.

Thus, the first three terms define the essentially linear
aspects of beam behaviour. This is analogous to Gaussian light
optics in respect of the rules relating to the formation and location
of images. The first and last terms of the equation represent the
well-known 'beam-spreading' curve whilst terms 1, 4 and 5 represent
beam spreading with finite emittance. Terms 3 and 5 give the conditions for an accelerated beam with space–charge to exhibit parallel
flow (this yields Child's Law) and finally the relative importance
of emittance or space–charge can be found by comparing terms 4 and 5.

3. THE ION SOURCE

Although a variety of methods of ionisation have been used to produce ion beams, by far the most important is the confined electrical discharge or arc, which is wholly or partially sustained by the vapour of the material to be ionised. Very intense ion beams can be extracted from suitably designed discharge sources and their particular merit lies in the fact that they are universal. All elements can be ionised in a discharge. In contrast, for example, surface ionisation is very restricted in the range of beams of adequate intensity which can be produced. The only limitation to the versatility of discharge sources lies in the technical difficulty of operating over a sufficiently wide temperature range to volatilise the required range of elements. In certain circumstances, this problem is circumvented by the use of volatile compounds, or by the use of sputtering techniques.

Ionising discharges can be produced in a variety of ways (for example, hot cathode, radiofrequency, Penning, laser) and even a brief consideration of the wide variety of discharge sources which are used is quite beyond the scope of this review. We shall simply note below some of the more important features of the discharge to illustrate the role it plays in determining the performance of the accelerator.

Fig. 7 illustrates two versions of a typical discharge source. In general layout, they resemble the familiar mass-spectrometry electron bombardment source. Electrons, extracted and accelerated from a heated cathode, are collimated by a magnetic field (and where appropriate defining slits). Ions produced in the chamber are extracted through the aperture and accelerated. Gas or vapour is introduced into the ionisation chamber through a controlled leak from an external reservoir, or, in the case of low vapour pressure materials, it may be admitted from a furnace.

In the mass spectrometry source, the pressure is normally low and ionisation is largely effected by single collisions between the gas molecules and the monoenergetic electrons. The essential and important, difference in the discharge source is that the pressure is raised (typically to 10^{-4} - 10^{-2} Torr) until a stable discharge is struck between the filament and the anode (in this case, the walls of the arc chamber). In the process, the precise and selective ionisation characteristics of the sampling electron-impact source are lost but the increased possibility of ionisation in the intense arc leads to higher efficiencies and to a very large increase in the resulting ion-beam intensity. In spite of the geometric differences between the circular and the slit extraction apertures illustrated in Fig. 7, the general pattern of ion source behaviour is quite similar. The slit system, which leads to a wedge-shaped rather than

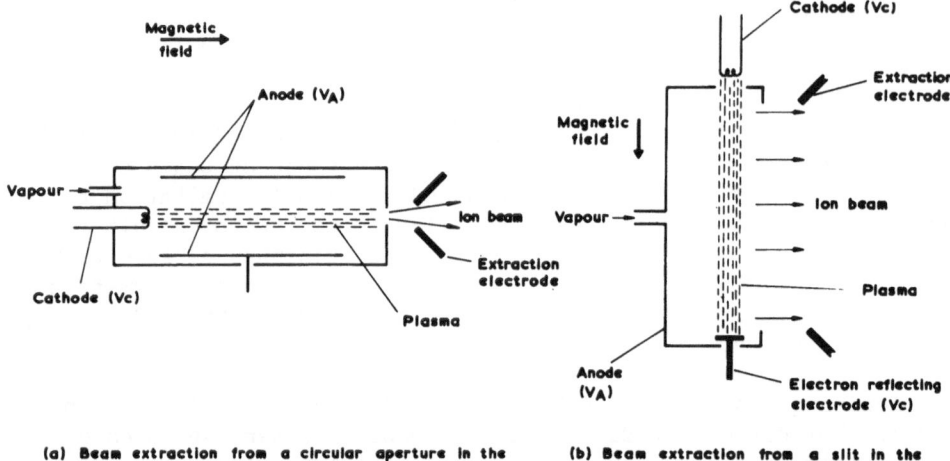

(a) Beam extraction from a circular aperture in the
end of the discharge chamber.

(b) Beam extraction from a slit in the
side of the discharge chamber.

Fig. 7:   Oscillating electron discharge sources

conical beam, provides a simple and convenient means of obtaining
a higher intensity.

A number of the elementary processes which occur in a discharge
are shown diagrammatically in Fig. 8. Even in its simplest form the
gas discharge is a complex phenomenon; the situation is complicated
in heavy ion sources by the geometry of the discharge chamber, the
use of magnetic fields and the wide variety of constructional
materials and source-feed vapours. Although a number of detailed
studies of ion-source discharges have been made, they have unavoid-
ably been restrictive because of the large number of variable parameters. In particular, they have largely concentrated on the
relatively simple case of source operation with pure inert gases.
They thus provide little or no information on the likelihood of
extracting negative, or compound, or polyatomic ions, from an ion
source. They have nevertheless confirmed certain fundamental
postulates and at a general level, the work has led to a considerable
measure of optimisation and understanding of source behaviour.

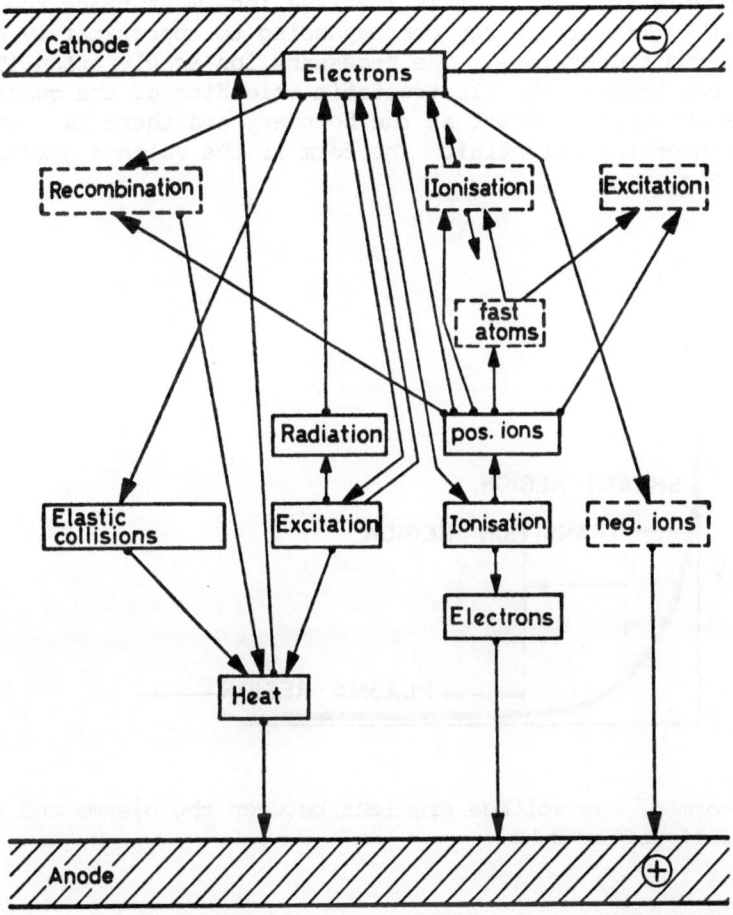

Fig. 8: Elementary processes in a gas discharge. Each process or
transisiton is symbolized by an arrow, the black spot rep-
resenting the start of the process or transition and the
arrow-head it termination. The names of processes and
particles that play a particularly important part in inert
gasses are enclosed by a black line, the others by a broken
one. Liberation of electrons from the cathode arrows ter-
minate at "electrons" on the cathode. (after Penning).

Thus, in an early study of the detailed nature of the ion-source discharge, Bohm[8] showed that a necessary condition for the formation of a stable plasma is that the ions must reach the sheath region with a kinetic energy corresponding to approximately half the mean electron temperature. The necessary ion acceleration ($T^{+} \approx Te/2$) occurs because the electrostatic shielding of the quasi-neutral plasma is not quite perfect at the boundary and there is a small penetration of the potential. The form of the voltage gradient is shown in Fig. 9.

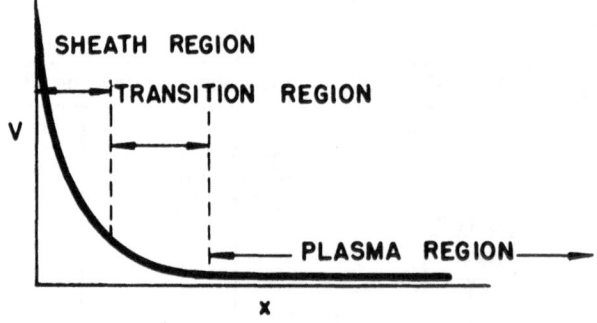

Fig. 9: Form of the voltage gradient between the plasma and the wall (after Bohm)

This very simple conclusion has a number of important conse-quences (for a more detailed account, see ref.4). It follows, for example, that the flow of ions is directed normally to the surface and is determined by the electron temperature and not the ion, or neutral gas, temperature. This explains the otherwise rather puzzling feature that the efficiency of an ion source (the fraction of the input gas extracted as ions) can be considerably in excess of the degree of ionisation - the ions are extracted while the neutral gas can only escape by drifting thermally to the aperture.

The current density of singly-charged ions, j+, through the
plasma boundary is thus given approximately by:

$$\dot{j}_+ = n_+ \vec{v}_+ \simeq n_+ \sqrt{\frac{k \, T_e}{M_+}} \qquad \text{-----} \qquad (1)$$

(Bohm, 1949[8])

where k is Boltzmann's constant, n+ is the density of charged
particles in the plasma, $\vec{v}_+$ is the mean value of the velocities of
the positive particles normal to the plasma boundary and Te is the
electron temperature.

It can readily be shown that the flow of ions at the boundary
of a plasma must also satisfy the familiar "space-charge limited"
form of expression.

$$\dot{j}_+ = K \frac{V^{3/2}}{d^2 \sqrt{M_+}} \qquad (2)$$

where K is a constant, d is the thickness of the plasma sheath and V
is the voltage drop across the sheath.

The mechanism of formation of an ion beam from a discharge is
simply an extension of the general case of ion flow across a plasma
boundary and is illustrated in Fig. 10. This shows how the boundary
of a plasma protruding through an aperture can be progressively
repelled by increasing the voltage between the electrode and the
discharge. The final sketch shows the formation of a beam through
an aperture in the negatively-biassed extraction electrode. At
each stage, the flow of ions across the plasma boundary must simultaneously satisfy the relationships (1) and (2) above. The flow is
determined primarily by the degree of ionisation in the plasma and
the electron temperature. If, for example, the plasma has a uniform density, these must remain approximately constant and it
follows that variations in the extraction conditions must in turn
modify the geometry of the plasma boundary until the two relationships which govern the beam current are satisfied. Therefore, in
the simplest case, as the extraction voltage is increased, the beam
current remains constant while the distance, d, (the sheath thickness, or the gap between the plasma boundary surface and the
extraction electrode) increases accordingly. In a complete analysis
the shape of the boundary must also be considered.

This aspect of the ion source plasma behaviour has been considered at some length because of the sensitivity of accelerator
behaviour to the precise position and shape of this mobile boundary.

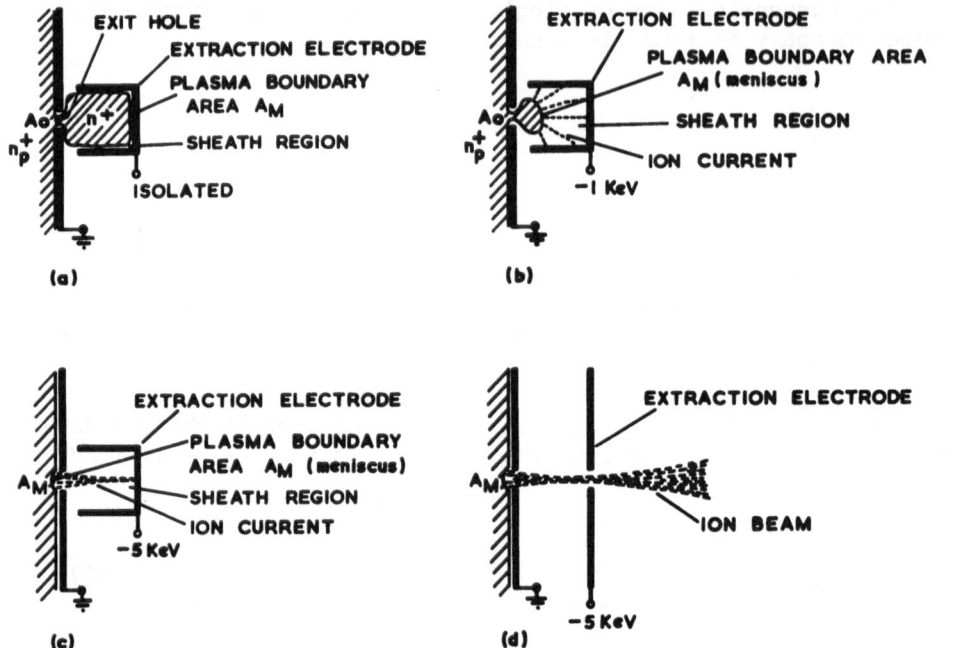

Fig. 10: Mechanism of ion extraction from a discharge (after Walcher)

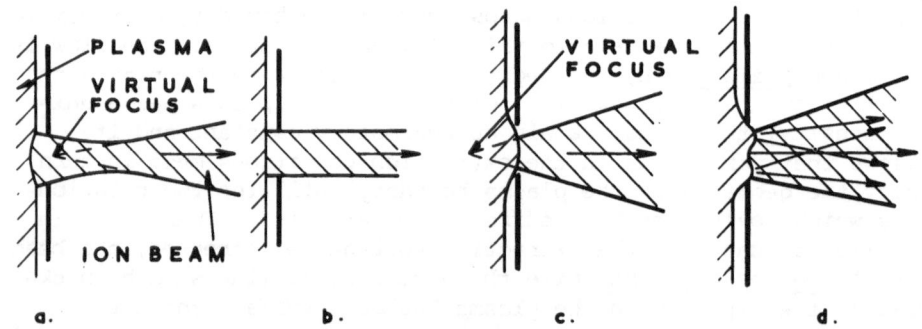

Fig. 11: Simplified representation of typical extracted beam configurations from discharge sources (a) concave plasma
meniscus, (b) flat plasma meniscus, (c) convex plasma
meniscus, (d) distorted plasma meniscus)

Fig. 11 illustrates how very small changes in the shaping of the plasma surface can completely change the focal behaviour of the extracted beam. Although this factor plays a dominant role in ion-source performance, its significance is not as broadly appreciated as it should be. All too frequently the behaviour of ion sources is described simply (and often incorrectly) on the basis of an elementary "space-charge limited" current relationship.

4. SPACE-CHARGE NEUTRALISATION

The final aspect of ion beam formation and manipulation that we consider in this limited review, is that of space-charge compensation. Although such compensation tends to occur with all ion beams in any region of an accelerator which is free of electrostatic fields, the effect is generally only of importance at relatively high intensities. It is thus usually insignificant in most present applications of ion beams to surface analysis but we include a brief description for completeness since it will become of increasing importance as the requirements for beam quality becomes more exacting, or in experiments involving either very low ion energies or an even finer degree of focus than can currently be achieved.

In the beam formation and acceleration region of the ion source considered above, the repulsive forces between the ions result in some degree of beam broadening. This measure of space-charge repulsion cannot be avoided, although the focussing consequences can be minimised by careful design of the extraction optics. The situation is, however, very different in the field-free regions beyond the final extraction electrode. In this volume, any slow electrons, or negative ions, produced by collisions of the primary beam with residual gas molecules are trapped in the positive space-charge potential well of the beam. Positive low-energy secondary ions which are also produced in such collisions are simultaneously accelerated out of the beam volume. If the ion current is stable, the build-up of electrons continues until the electron density fully compensates the positive field. In this state, the beam volume becomes a plasma with equal densities of positive and negative charged particles. It is this fortuitous neutralising effect which permits the sharp focussing of beams of heavy ions of milliampere intensity and at kilovolt energies, but it must be emphasised that the effect occurs automatically, even in very high vacuum conditions, for any ion beam which is not pulsed and which is travelling through a region free of electrostatic fields. The compensation occurs very rapidly ($\sim 10^{-3}$ seconds) at the typical operating pressures of most accelerators. With intense beams, care must be taken to maintain the space-charge compensation over as much of the ion trajectory as possible. This seriously restricts the use of certain types of electrostatic steering and focussing elements (for a more detailed

account, see ref.1) and even with lower intensity beams the
inevitable presence of the neutralising electrons cannot be
ignored in certain aspects of accelerator operation – for example,
they can cause serious errors in electrical measurements of the beam
current (e.g. ref. 4).

5. SOME EXAMPLES OF ION BEAM BEHAVIOUR

In this section, we conclude this brief review by considering
four somewhat unusual examples of recent ion-beam studies which
illustrate on the one hand the complexity of ion beams and on the
other, the scope for further improvements in accelerator performance.

5.1 High brightness ion beams

In recent years, the development of field emission electron
sources of very high brightness has led to significant advances in
electron microscopy. It. now seems possible that a recent analogous
development with ions could have an equally marked effect in ion
microprobe and microscope applications.

The successful generation of intense beams of a variety of
metal ions from thin liquid metal films on the tips of both solid
wire and hollow needles has been reported by several groups of
workers[11-13]. Such ion sources, which are extremely simple, are
apparently based on electrohydrodynamic instabilities induced in
the liquid metal films under high electrical stress. According to
Clampitt et al.[11] direct currents of 50 – 700 uA can be readily
produced from single needles subject to quite modest voltages
(3 – 10 kV). Fig. 12 shows a photograph of an emitting needle.
Examination of the source has indicated that the ion emitting zone
is a stable plasma ball of radius $< 10^4$ Å. This corresponds to ion
current densities over 10^4 A/cm^2. These are far in excess of those
achievable from conventional ion sources.

5.2 Resolution limit in ion-beam experiments

It is commonly assumed that slit scattering provides the
dominant limiting effect in experiments involving the very fine
collimation or precise angular resolution of ion beams. A variety
of experimental studies of the effect have been conducted at MeV
energies. At lower energies, such finely focussed beams are, of
course, of particular relevance to ion microprobe studies of
surfaces. Recently Daglish and Kelly[14] have shown that at energies
up to 200 keV, and at the pressures conventionally used in such
accelerators, the resolution limit is due to background gas scatter
and not to slit scattering. This result is interesting and it
indicates that to improve focussing quality it will be essential
to design improved vacuum systems. This is a fairly practicable

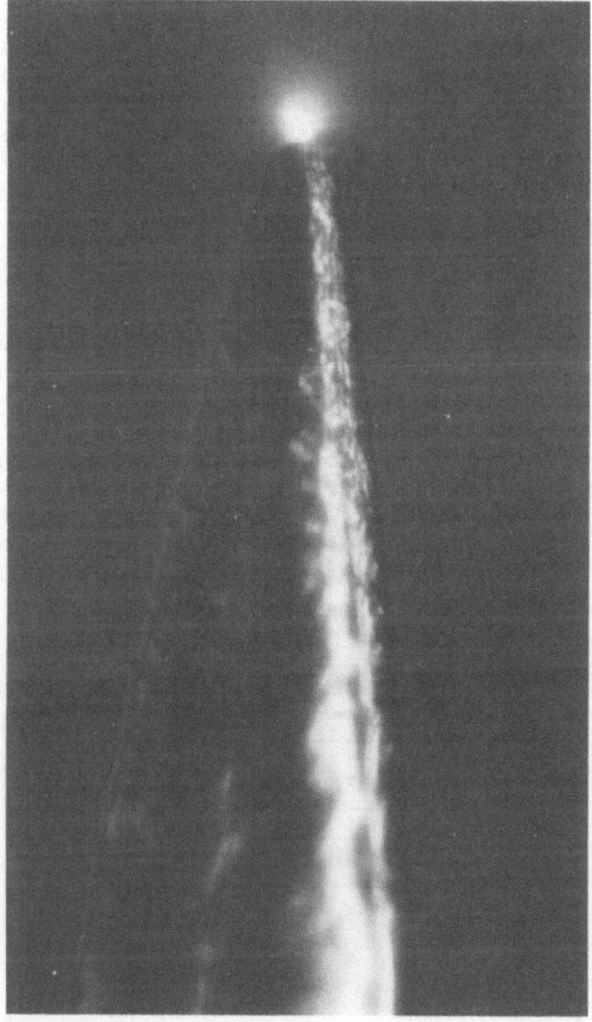

Fig. 12: Photograph of single needle ion source (after Clampitt)

measure in certain types of instrument as a consequence of the small dimensions of the finely collimated beam. A rough estimate of the angular collimation limit can be made from the expression

$$\phi_{min} \approx \left(4\times10^{-6}\right)\frac{Z_2}{Z_1}\frac{P}{E^{1/2}}$$

where Z_2 and Z_1 are the target and projectile atomic number, P is the background pressure in Torr, E the energy in keV and ϕ_{min} the full-width half-maximum angular spread (in degrees).

5.3   Ion-beam purity

We mentioned earlier the several causes of contamination in ion beams.   We included contamination due to Aston Bands – that is, ions arriving at an incorrect position in the mass spectrum as a result of collisions before analysis which result in changes in the charge state or mass of the ion.   One such effect, of particular consequence in surface characterisation using Rutherford backscattering, has been reported by Picraux et al. [15] and by Hemment et al. [16]   This arises from the almost inevitable presence of oxygen as a trace impurity in the helium beam extracted from the ion source.   Some of the singly-charged oxygen ions (mass 16) will lose an electron after acceleration as a result of charge-exchange collisions with residual gas in the accelerator flight tube before mass analysis.   It can readily be shown that these ions will be focussed with the primary helium beam at the mass 4 position.   Even very low levels of such impurity ions will lead to significant errors.   A detailed analysis of the problem is given by Hemment et al. and Fig. 13 shows the spurious peak due to such an oxygen beam in the analysis of gold films on carbon backings.   The solution to this problem (Picraux et al. [15]) is to eliminate the doubly-charged oxygen impurity by electrostatic analysis of the beams after the magnetic analysis.

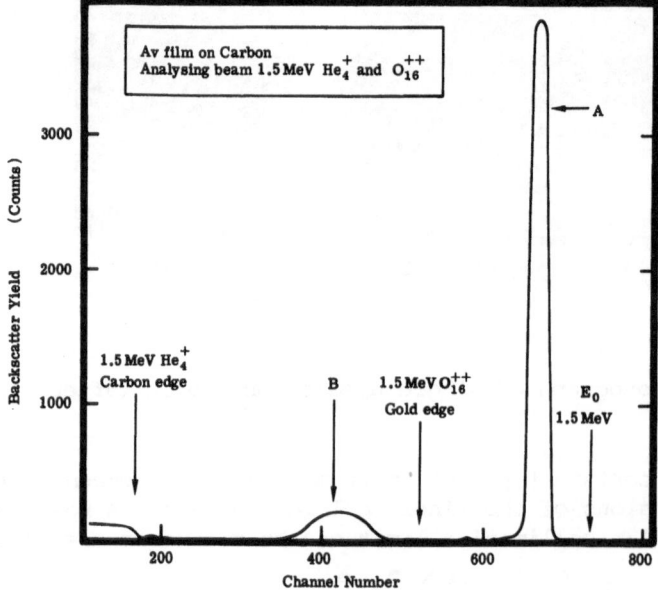

Fig. 13: R.B.S. analysis of gold film on carbon backing showing
         spurious impurity peak (B) due to oxygen contamination of
         ion beam

Fig. 14: Illustration of intense ion-beam steering in an electrostatic mirror (after Beanland et al.)

5.4   The ion mirror

Fig. 14 shows an intense ion beam (~ mA) being steered in an
electrostatic mirror lens[17].   It illustrates some of the points
made earlier about the focussing and manipulation of such beams.
Firstly, the degree of space-charge defocussing is small and the
steering is successful because the dimensions of the lens are small
and care has been taken to minimise penetration of the mirror
electrical field outside the lens space.   In spite of these precautions, it was found that as a consequence of small defocussing
effects within the mirror space, the output beam divergence was
sensitive to the current and care had to be taken to optimise the
operation.   The photograph clearly shows some additional points of
interest.   It can be seen that the flux of neutral particles accom-
panying the ion beam is undeflected in the mirror and strikes the
back reflector plate which is at high voltage.   This bombardment
results in the emission of secondary ions from the surface.   These
are accelerated to the earthed plate opposite, where they produce
a second glowing spot.   In turn the secondary electrons from this
energetic ion bombardment are accelerated back to the high voltage
mirror plate.   Although, in this instance, these effects did not
perturb the operation of the electrostatic ion mirror it is easy
to see how they could lead to the kind of malfunctioning due to
discharges, X-rays, sparking, excessive high voltage drain etc.
which can be a troublesome feature of ion-beam production and
manipulation.   They illustrate the distinction, which we made at
the beginning of this review, between the problems of handling
electrons and ions.

Fortunately, with care and with an understanding and appreciation
of the complexities of ion sources and accelerators, such effects can
generally be eliminated, or reduced to insignificant levels.

REFERENCES

1. J. H. Freeman, W. Temple, D. Beanland and G. A. Gard, Nucl. Inst. & Methods 135, 1, 1976

2. J. F. Ziegler, New Uses of Ion Accelerators (Plenum Press, N. York) 1976

3. A. Septier, Focussing of Charged Particles Vols. I and II (Academic Press, N. York) 1967

4. G. Dearnaley, J. H. Freeman, R. S. Nelson and J. Stephen, Ion Implantation (North Holland, Amsterdam) 1973

5. R. G. Wilson and G. R. Brewer, Ion Beams (Wiley Interscience) 1963.

6. A. J. Lichtenberg, Phase-space dynamics of particles (J. Wiley, N. York) 1969

7. J. D. Lawson, Proc. Int. Conf. on Isotope Separators, Israel, 1976 - in press.

8. D. Bohm, The characteristics of electrical discharges in magnetic fields, edited Guthrie and Wakerling, N.N.E.S. 1 - 5 (McGraw-Hill, N. York) 1949

9. O. Almen and K. O. Nielson, Nucl. Inst. & Methods 1, 302, 1957

10. I. Chavet, Thesis, University of Paris, 1965, also Nucl. Inst. & Methods 44, 77 and 51, 77, 1967

11. R. Clampitt, K. L. Aitken and D. K. Jefferies, J. Vac. Sci. Technol. 12, 1208, 1975

12. V. E. Krohn and G. R. Ringo, App. Phys. Letters 27, 479, 1975

13. J. H. Orloff and L. W. Swanson, J. Vac. Sci. Technol. 12, 1209, 1975

14. R. L. Dalglish and J. C. Kelly, J. Phys. D: Appl. Phys. 9, 581, 1976

15. S. T. Picraux, J. A. Borders and R. A. Langley, Thin Solid Films, 19, 371, 1973

16. P. F. Hemment, J. F. Singleton and K. G. Stephens, Thin Solid Films 28, 1, 1975

17. D. Beanland, J. H. Freeman, D. C. Chivers and G. A. Gard, Nucl. Inst. & Methods - in press, 1976

REFERENCES

Part II

SURFACE STUDIES: keV RANGE IONS

APPLICATIONS OF LOW-ENERGY ION SCATTERING

H.H. Brongersma, L.C.M. Beirens, G.C.J. van der Ligt

Philips Research Laboratories

Eindhoven, The Netherlands

INTRODUCTION

Low-energy (0-10 keV) ion scattering (LEIS, ISS or NIRMS) is used for obtaining information on the atomic composition and structure of surfaces. The extreme sensitivity of the technique to the outermost atomic layer gives it unique properties as compared with other analytical techniques.

The use of lower-energy ion scattering for surface analysis dates back to the early sixties. Walther and Hintenberger[1] were among the first to correlate the energy loss of the scattered ions with the atomic masses of the surface atoms. Later on it was especially D.P. Smith who realised the importance of the application of low-energy (< 10 keV) ion scattering to surface analysis. Since his first paper [2] in 1967 this pioneering work has had a stimulating influence on research in this field. During the last five years many contributions have been made by Taglauer and Heiland [3-5], Brongersma et al.[6-8] and Honig and Harrington [9] in the lower energy range (0-2 keV), while the groups of Boers [10,11] and of Buck[12,13] focussed their attention on somewhat higher energies. Most of these studies were aimed at the fundamental aspects of ion scattering. Special topics were surface sensitivity, multiple scattering, ion neutralisation, surface structure and the influence of atomic vibration on the spectra. In order to compare and contrast the merits of LEIS with other methods, several authors have combined LEIS with low-energy electron diffraction (LEED), Auger spectroscopy and secondary ion mass spectroscopy (SIMS). Although much insight has been gained into the underlying principles of LEIS, many problems remain to be solved. This is particularly true for the mechanisms involved in the neutralisation of ions near the surface.

As compared to the high-energy ($\approx$ 1 MeV) ion backscattering LEIS has the advantage that the production of the ion beam is much cheaper and simpler. The great sensitivity to the outermost layer often makes it necessary, however, to study the target under ultra-high vacuum conditions.

Notwithstanding the remaining problems, LEIS has become a new and powerful tool in surface research. The ease with which the spectra can be interpreted makes the technique a useful tool for applied research as well. The recent development of NODUS[7] increased the sensitivity of ion scattering by more then two orders of magnitude. This development seems to be of great importance to areas such as catalysis, where in the past such research was precluded by the damage due to the sputtering action of the ion beam.

PRINCIPLES OF LEIS

In LEIS a beam of monoenergetic ions (energy E_i, mass M_1) strikes a surface. The energy distribution of the ions scattered in a given direction is measured.

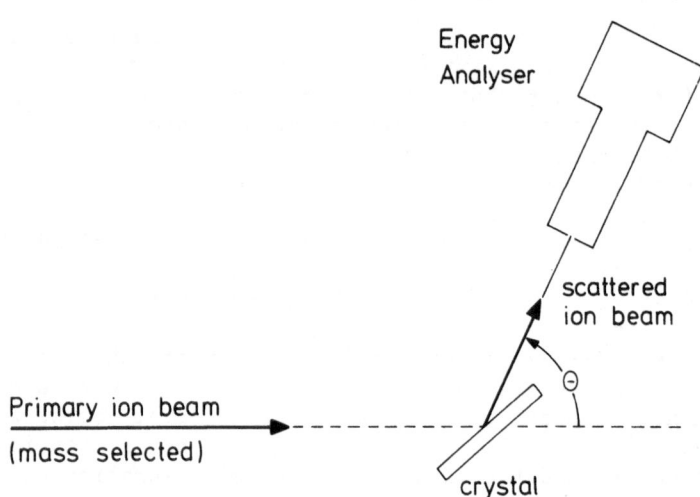

Fig. 1. Set-up of an ion-scattering experiment.

When low-energy noble-gas ions, such as 1 keV He^+ ions, are used it is found that the signal of the backscattered ions results essentially from ions that have made only one collision with a surface atom (mass M_2). Under these simple conditions the energy

E_f of the backscattered ions is given by

$$E_f = k^2 . E_i \qquad\qquad , \quad (1)$$

where

$$k = \frac{\cos\Theta_1 + (r^2 - \sin^2\Theta_1)^{\frac{1}{2}}}{1+r} \qquad ,$$

$r = M_2/M_1$ and Θ_1 = scattering angle in the laboratory.

This expression can be derived by solving the equations for conservation of energy and momentum on the assumption that $M_2 \gg M_1$.
Since the parameters E_i, M_1 and Θ_1 are kept constant in a given
experiment, the energy of the scattered ion is determined by the
mass of the surface atom with which the ion collides. Therefore,
an energy spectrum of the scattered ions is equivalent to a mass
spectrum of the surface atoms. An example is given in fig. 2.

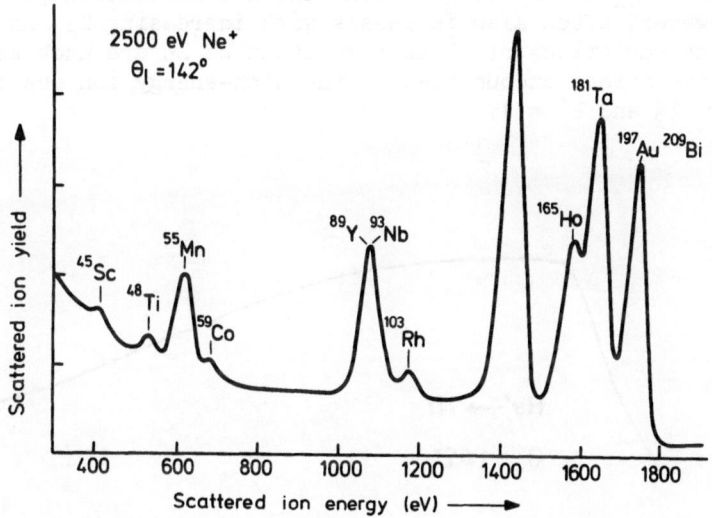

Fig. 2. Energy spectrum of Ne$^+$ ions scattered by a multicomponent
target.

The mass resolution in such spectra decreases with increasing mass
of the surface atom. When heavier ions (A$^+$, Kr$^+$, Xe$^+$) are chosen
as a probe, the mass resolution is improved considerably.

In principle all elements can be detected. The detection of
hydrogen atoms is especially difficult, however, since there are
no noble-gas ions that are lighter than hydrogen. Conservation of
momentum during a collision implies that only forward scattering

exists for the heavier He^+ ions. It was found possible, for instance, to detect deuterium atoms on a polycrystalline tungsten ribbon when studying the scattering He^+ ions over an angle of 24° 14).

The scattered ion yield (y^+) depends both on the scattering process itself and on the neutralisation of the ions. To a good approximation these processes can be regarded as independent. Thus

$$y^+ \sim \sigma^+ = \sigma.P^+ \qquad\qquad (2)$$

where

σ^+ = cross-section for scattering of ions,
σ = cross-section for scattering of particles (independent of their charge state),
P^+ = ion fraction of scattered particles.

The scattering cross-section σ increases with increasing atomic number Z_2 of the target atom. The neutralisation probability, however, often also increases with increasing Z_2. As a result, the variations of σ^+ as a function of Z_2 are much smaller than the variations encountered in the high-energy ion scattering where $\sigma \sim Z_2^2$ and $P^+ = 1$.

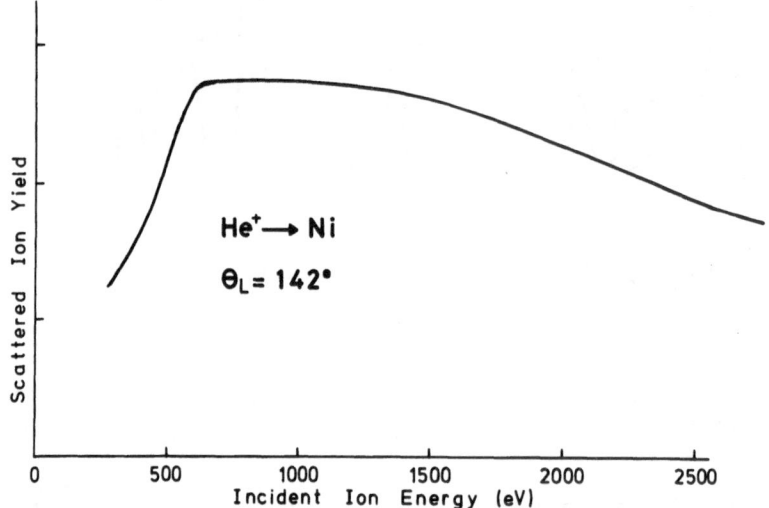

Fig. 3. Characteristic example of the energy dependence of the scattered ion yield.

Fig. 3 shows a fairly typical example of the dependence of the scattered ion yield on the energy of the ions. Recently

Erickson and Smith[15] and Brongersma and Buck[8] found some ion-target combinations which exhibit very pronounced oscillations. This is especially true for the scattering of He^+ ions from Pb or Bi (fig. 4).

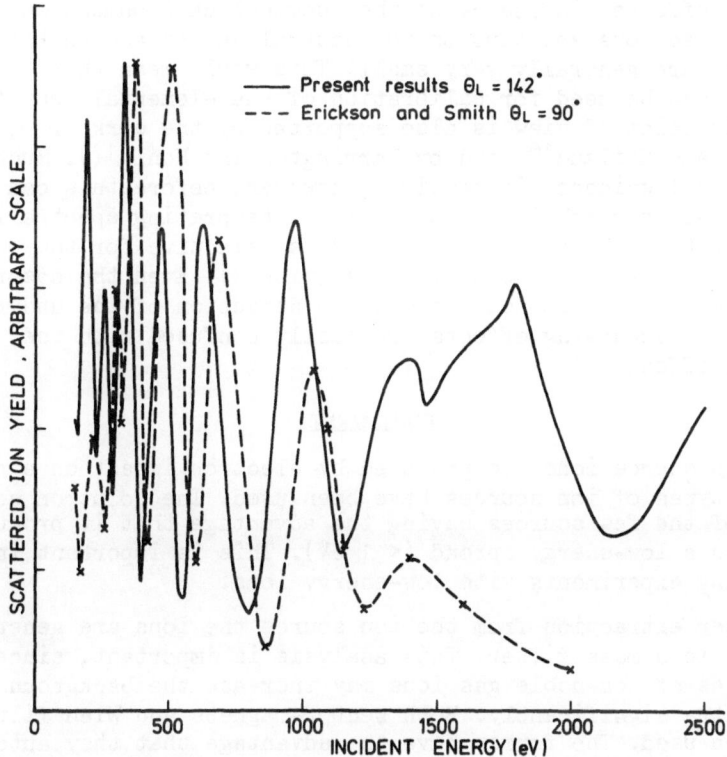

Fig. 4. Strong oscillations are observed for the energy dependence of the scattered ion yield in the case of $He^+ \rightarrow Pb$. The results of Erickson and Smith[15] are given for comparison.

The observed oscillations have been interpreted in terms of a resonant charge transfer between ion and target atom. Since the chemical environment of the target atom may influence the electronic energy levels of the atom and thus the quality of the resonance, spectra such as fig. 4 provide some chemical information about the surface. Of course, in addition to the normal smooth dependence of y^+ on energy and to the pronounced oscillatory dependence of y^+ many intermediate examples exist. When using noble-gas ions that are heavier than helium, one or two broad oscillations are often observed.

The quantitativeness of LEIS has been discussed by many authors.
There is no existing theory that can predict the sensitivity of
the method to the various elements. It seems that the sensitivity
to an element is mainly determined by experimental conditions such
as type and energy of the ion and the angles of the incident and
scattered ion beam with the target. It is believed that pure
chemical effects (influence of the chemical environment on the
valence electrons and thus on the neutralisation and on the ion
yield y^+) are generally very small. This would mean that the pure
elements can be used for calibration of the elemental sensitivities. This point of view is also supported by the work of
Taglauer and Heiland[16] and by Harrington and Honig[17]. Much more
experimental evidence is required, however, before this can be
regarded as an established rule. When interpreting spectra it is
important to realise that the method is selective for the outermost atomic layer of the surface (section 4). Even the adsorption
of hydrogen can sometimes prevent the detection of the underlying
atoms. Such shadowing effects are easily confused with the
chemical effect.

EQUIPMENT

In an ion source ions are produced by electron impact on a gas.
Various types of ion sources have been used. The Colutron source[18]
is one of the few sources having the advantage that it produces
ions with a low-energy spread (< 1 eV). This is important in ion
scattering experiments with low-energy ions.

   After extraction from the ion source the ions are generally
analysed in a mass filter. This analysis is important, since small
quantities of non-noble gas ions may increase the background in
the spectra significantly. Both sector-magnets and Wien filters
have been used. The former have the advantage that they automatically block the line of sight from the ion source to the
target and thus reduce possible contamination of the target. In a
Wien filter the path of ions with the correct momentum is
unaffected. This straight-through geometry facilitates the lining-
up of the various components of the equipment.
The design of the target manipulator depends strongly on the
experiment. In more fundamental studies heatable and/or coolable
target-holders are used which can be put into practically any
position with respect to the ion beam. In more applied work
multitarget holders are employed.
The energy analyser, used to determine the energy distribution of
the scattered ions, is generally of the electrostatic type.
Grundner et al.[19] have successfully added a mass filter to such
an analyser in order to discriminate against high-energy recoil
ions. For surfaces where a large fraction of the sputtered
particles is in an ionised state, such a mass selection can be
very advantageous.

It follows from eq. 1 that an accurate determination of the
scattering angle is essential for a good mass analysis of the
target atoms. Normally this is achieved by limiting both the
angular spread of the incident ion beam and the angular spread
of the scattered ions accepted by the analyser to 1°-2°.

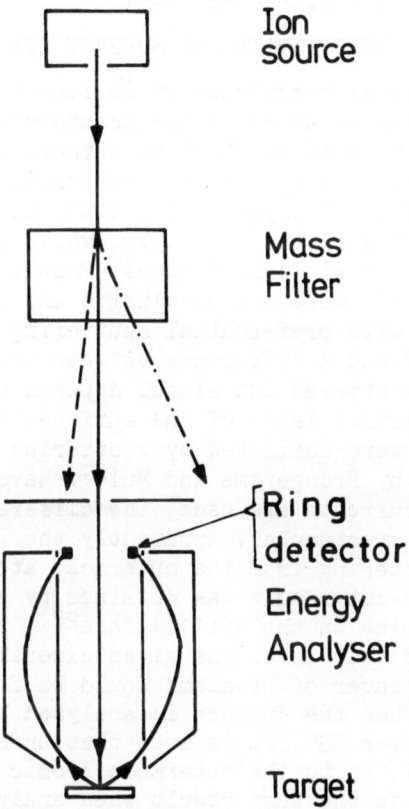

Fig. 5. Schematic diagram of the NODUS equipment. A kind of
    cylindrical mirror analyser is used to analyse the energy
    of the backscattered ions.

Fig. 5 illustrates an experimental set-up of a different design :
the analyser accepts a full cone of the backscattered ions[7]. The
geometry is such that the scattering angle is the same for all
ions within the cone. A kind of cylindrical mirror analyser is
used to energy select the scattered ions. After this selection the
ions are detected by a ringshaped multiplier which surrounds the
incident ion beam. Compared with the other experiments, a great
advantage is the much larger acceptance angle, which leads to a
gain of a factor of 100-1000 in the scattered-ion yield. Moreover,

the large scattering angle improves the mass resolution (eq. 1).

Another important trend is the combination of LEIS with other
techniques. LEIS has been combined, for instance, with Auger
Electron Spectroscopy[20,21] to compare and contrast the relative
merits of the techniques. In studies of the structure of surfaces,
LEIS has also been combined with LEED[22,23], and in other studies
LEIS has been combined with SIMS[19,24].

SURFACE SENSITIVITY AND SURFACE STRUCTURE

For surface analytical techniques it is important to know how
contributions of target atoms to the scattered-ion signal depend
on the depths of these atoms. Various authors have tried to
establish this surface sensitivity. One possibility was to study
the polar faces of a CdS crystal. The (111) face is expected to
contain only Cd atoms in the top layer, while the top layer of the
($\bar{1}\bar{1}\bar{1}$) face should contain only S atoms. Results reported by
Efremenkova et al.[25] were not conclusive and indicated that one
should be careful with preferential sputtering and heating of the
target. Smith[26] found a difference between the two faces, thus
showing that the scattered ion signal depends to a considerable
extent on the outermost layer of the surface. However, for both
faces the spectra were dominated by scattering from Cd atoms.
Later experiments by Brongersma and Mul[27] have shown that when
low incident-ion currents are used, the difference between the
surfaces is quite spectacular. Apparently the LEIS signal is due
mainly to ion scattering from the outermost atomic layer.
An even more clear-cut result was obtained by studying the
adsorption of bromine on the Si(111) face[28]. For this system it
was to be expected that under the given experimental conditions
precisely one monolayer of Br-atoms would be formed on top of the
silicon lattice. When the surface is analysed by scattering
1000 eV Ne$^+$ ions over 33°, it is seen that under these conditions
LEIS is only sensitive to the outermost atomic layer. Harrington
and Honig[17] came to the same result when analysing the (100) and
(111) surfaces of a silicon crystal. They found that the
scattered ion intensities for these faces reflected the difference
in packing density of the top layers.

At low energies this surface sensitivity seems to be quite
general. As will be demonstrated in section 5, this feature opens
up many new possibilities for applications to surface analysis.
A very fundamental consequence of this feature, that the top atoms
can mask the presence of deeper-lying atoms, has led to the use
of LEIS in studies of the structure of surfaces. In such work the
masking of the deeper-lying atoms by the surface atom is studied
as a function of e.g. the angle of incidence or the azimuthal
angle. By careful measurements it is possible to determine the
exact location of surface atoms with respect to the underlying
lattice. The first examples to demonstrate this potentiality were

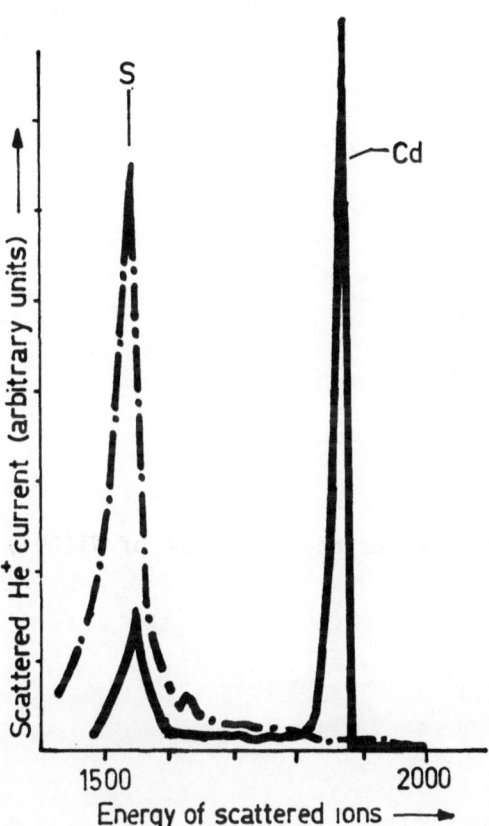

Fig. 6. Energy spectra for He$^+$ ions scattered over 90° from the
Cd (—) and the S(— . —) faces of a Cd S crystal.
The incident energy is 2000 eV.

reported by Heiland and Taglauer[29] for Ni(110) + O, by Heiland
et al.[30] for Ag(110) + O and by Brongersma and Theeten[23] for
Ni(100) + O.

In another study of Theeten and Brongersma[31] qualitative
information was obtained on the Ni(100) + S surface. This example
gives a good illustration of the simplicity of the technique. In
this case the surface contained half a monolayer of sulphur atoms
giving rise to a C(2x2)LEED pattern.
Fig. 7 shows a possible model for the surface structure, which is
in agreement with the LEED pattern. For Ne$^+$ ion scattering
(E_i = 700 eV; Θ_1 = 60°) it is found that the signal for scattering
from sulphur atoms is almost independent of the azimuthat angle ϕ.

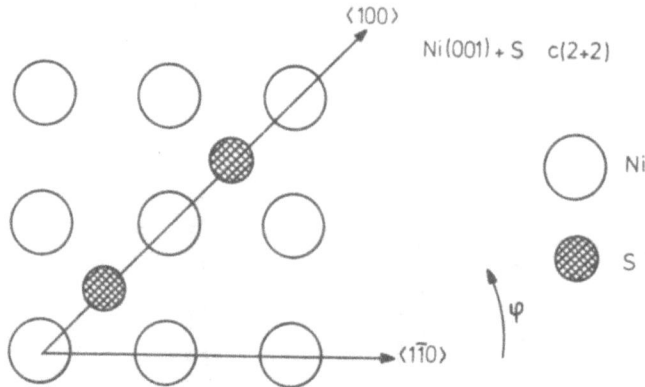

Fig. 7. Model for the surface structure of Ni(001) + S C(2x2).

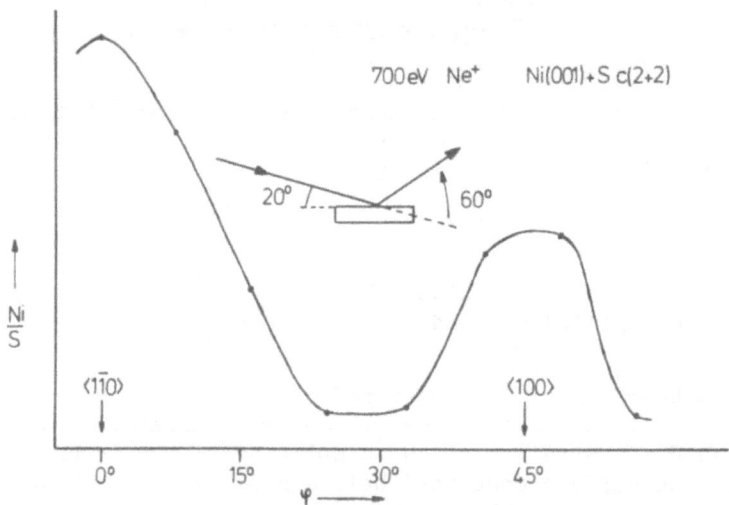

Fig. 8. Dependence of the ratio of the number of Ne[+] ions
        scattered from Ni and S atoms on the aximuth[31]. The
        incident energy is 700 eV and the scattering angle 60°.

As fig. 8 illustrates, this is not the case for the nickel atoms. Apparently the presence of the sulphur atoms cannot be masked and thus the sulphur atoms must be located above the nickel atoms. The maximum intensity ratio for Ni/S is obtained for ion scattering in the [100] azimuth, while half that ratio is found for the [110] azimuth. This is in agreement with the model in fig. 7 : In the [100] azimuth sulphur atoms cannot mask the presence of nickel, while one half of the nickel atoms is masked for the [110] azimuth. For intermediate azimuths one sulphur atom can mask the presence of several nickel atoms, thus reducing the Ni/S signal ratio to very low values. From these measurements other possible surface structures could be rejected. At the same time it shows that one must be very careful with masking effects when interpreting signals quantitatively.

Since LEIS can be used to determine the location of atoms in the surface, this technique is also suited for studies of surface relaxation phenomena. Such studies have been carried out for the GaP(111) surface and the first results were published some years ago[6]. Nowadays, elegant experiments are carried out at higher energies[32,33]. It is expected that the use of high-energy ions will lead to more accurate determinations of the location of surface atoms. However, it should also be realised that LEIS is not restricted to studies for near-channelling directions. For complex surfaces this greater flexibility may be very important.

In all these studies ion impact will cause damage to the surface. In studies of the structure of surfaces great care must be taken to minimise this damage. Taglauer and Heiland[21] have correlated such damage to observations in LEED.

APPLICATIONS

The high sensitivity of LEIS to the outermost atomic layer makes the method an almost ideal tool in studies where this property is important. Examples are chemical reactions of gases with solid surfaces (catalysis, adhesion) and electron emission from solids (cathodes). The ease with which insulators can be studied has also stimulated much LEIS research on compounds such as glasses, plastics and alumina.

It is found that surface segregation plays an important role in many surface problems. Surface segregation means that the concentrations of one or more types of atoms or molecules are larger at the surface than in the bulk, precisely because this situation is thermodynamically more stable. In the case of Cu-Ni alloys, for example, it has been demonstrated with LEIS that upon annealing a strong enrichment in copper occurs at the surface[34]. In general it will be the element or compound with the lowest boiling point that will segregate.
Similar effects have been observed while dispenser cathodes were

activated[35]).

In cooperation with the Department of Inorganic Chemistry of the Technological University in Eindhoven some Bi_2MoO_6 catalysts have been studied. Since these are insulating powders, an electron gun was used to avoid charging of the sample. It was known[36] that the addition of a small amount of MoO_3 to the Bi_2MoO_6 strongly enhances its catalytic activity for oxidation of propene to acrolein, for example, whereas a small addition of Bi_2O_3 has the opposite effect.

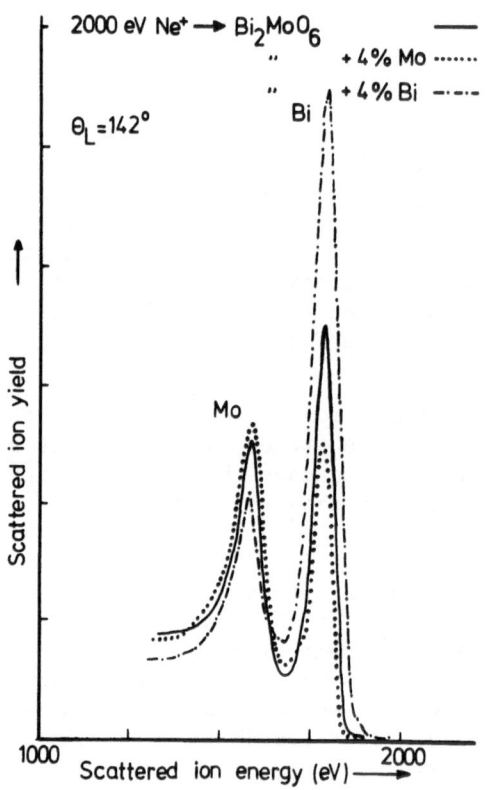

Fig. 9. Analysis of bismuth-molybdate catalysts having small
differences in the bulk compositions.

In fig. 9 some results are presented for He^+ ion scattering from Bi_2MoO_6, Bi_3MoO_6 + 4 at. % Mo (oxidised) and Bi_2MoO_6 + 4 at. % Bi (oxidised). It appears that a small change in the bulk concentration of Bi or Mo has a pronounced effect on the surface compositions of these catalysts. If one assumes that more than one surface atom is involved in the catalytic process, the strong dependence of the catalytic activity on small changes in the bulk composition is easily understood.

A nice application to spinel catalysts has been given by Shelef et al.[37]. A problem encountered in research on practical surfaces is due to the relatively high backgrounds found in the energy spectra for oxidised surfaces. This background is mainly due to the large number of secondary (recoil) ions from such surfaces. In this respect LEIS and SIMS (Secondary Ion Mass Spectrometry) are complementary : when the background in LEIS is high, this means that SIMS is very sensitive. The problem is particularly important when using heavy incident ions. Taglauer and Heiland [16] have shown in the case of Ar^+ ion scattering from an oxidised Cu-Nb surface that the background from the secondary ions is so large that the scattering from Cu or Nb cannot be identified. They also show that this problem can be resolved by mass selection of the scattered ions, or by the use of lighter primary ions.

Sometimes the background is also due to multiply scattered primary ions. In such a case the background cannot be removed by mass analysis of the scattered ions. Sometimes this background provides important information. For example, a comparison of powders of Fe_2O_3 having the α- and the γ-modification shows that the spectra are quite different.

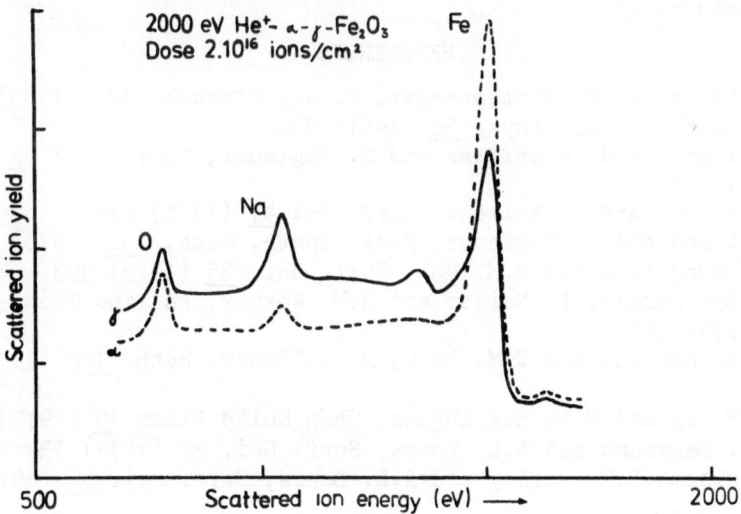

Fig. 10. Comparison of α- and γ-Fe_2O_3 by 2000 eV He^+ scattering.

The spectra were taken after the same dose of ions, since the spectra change in time. After removing a few atomic layers, the

differences become very small. For the α-modification the background is always less than for the γ-modification. For a very
small ion dose, there is no well-defined Fe peak in γ-Fe_2O_3
spectrum.
However, a sharp increase in the background is already observed
at the energy for scattering from Fe atoms. It seems likely,
therefore, that this step is related to scattering from Fe. In the
case of the α-modification, where there is still a pronounced
peak around 1650 eV, the He^+ ions can scatter directly from Fe
atoms. In the γ-modification such scattering is strongly perturbed.
Apparently, the Fe atoms are not located in the top layer but they
must be very close below it in order to account for the maximum
energy of the background. A model assuming the top layer to
consist of oxygen and hydrogen atoms seems plausible. Thus it is
seen that, although the bulk compositions of α-Fe_2O_3 and γ-Fe_2O_3
are the same, chemical effects may strongly influence the spectra.

CONCLUSIONS

During the last decade much insight has been obtained in the
fundamental properties of low-energy ion scattering. Although no
quantitative theory exists, accurate information can be obtained
by calibration against standards.
The technique has a number of features that make it attractive
for use in application research. Of special importance are its
sensitivity to the outermost atomic layer and its applicability
to insulators.

REFERENCES

1. V. Walther and H. Hintenberger, Z. Naturforsch. 18a (1963) 843.
2. D.P. Smith J. Appl.Phys. 38 (1967) 340.
3. W. Heiland, H.G. Schäffler and E. Taglauer, Surf. Sci. 35
   (1973) 381.
4. E. Taglauer and W. Heiland, Surf. Sci.47 (1975) 234.
5. W. Heiland and E. Taglauer, Nucl. Instr. Meth. 132 (1976) 535.
6. H.H. Brongersma and P.M. Mul, Surf. Sci. 35 (1973) 393.
7. H.H. Brongersma, F. Meijer and H.W. Werner, Philips Tech. Rev.
   34 (1975) 362.
8. H.H. Brongersma and T.M. Buck, Nucl. Instr. Meth. 132 (1976)
   559.
9. R.E. Honig and W.L. Harrington, Thin Solid Films 19 (1973) 43.
10.S.H.A. Begemann and A.L. Boers, Surf. Sci. 30 (1972) 134.
11.B. Poelsema L.K. Verhey and A.L. Boers, Surf. Sci. 56 (1976)
   445.
12.T.M. Buck, W. van der Weg, Y.S. Chen and G.H. Wheatley, Surf.
   Sci. 47 (1975) 244.
13.T.M. Buck in Methods of Surface Analysis, ed. A.W. Czanderna,
   Elsevier Sci. Publ. Co.(1975).
14.H.H. Brongersma, 2nd Int. Conf. Solid Surfaces, Kyoto (1974).

15. R.L. Erickson and D.P. Smith, Phys. Rev. Lett. 34 (1975) 297.
16. E. Taglauer and W. Heiland, Appl. Phys. 9 (1976) 261.
17. W.L. Harrington and R.E. Honig, 22nd Annual ASMS conference, Philadelphia (May 1974).
18. M. Menzinger and L. Wahlin, Rev. Sci. Instr. 40 (1969) 102.
19. M. Grundner, W. Heiland and E. Taglauer, Appl. Phys. 4 (1974) 243.
20. H. Niehus and E. Bauer, Surf. Sci. 47 (1975) 222.
21. W. Heiland and E. Taglauer, Radiation Effects 19 (1973) 1.
22. E. Taglauer, M. Melchior, F. Schuster and W. Heiland, J. Phys. E 8 (1975) 768.
23. H.H. Brongersma and J.B. Theeten, Surf. Sci. 54 (1976) 519.
24. W.L. Baun, J. Adhesion 7 (1976) 261.
25. V.M. Efremenkova, I.G. Bunin, D.S. Karpuzov, A.A. Pavlychenko and V.E. Yurasova, Bull. Acad. Sci. USSR Phys. 35 (1971) 375.
26. D.P. Smith, Surf. Sci. 25 (1971) 171.
27. H.H. Brongersma and P.M. Mul, Chem. Phys. Lett. 19 (1973) 217.
28. H.H. Brongersma and P.M. Mul, Chem. Phys. Lett. 14 (1972) 380.
29. W. Heiland and E. Taglauer, J. Vac. Sci. Technol. 9 (1972) 620.
30. W. Heiland, F. Iberl and E. Taglauer, Surf. Sci. 53 (1975) 383.
31. J.B. Theeten and H.H. Brongersma, Revue Phys. Appl. 11 (1976) 57.
32. J.A. Davies, this volume.
33. W.C. Turkenburg, W. Soszka, F.W. Saris, H.H. Kersten and B.G. Colenbrander, Nucl. Instr. Meth. 132 (1976) 587.
34. H.H. Brongersma and T.M. Buck, Surf. Sci. 53 (1975) 649.
35. H.H. Brongersma and W.J. Schouten, Acta Electronica 18 (1975) 47.
36. Private communication, Ph. A. Batist, J.H. van Hooff, I. Matsuura and G.C.A. Schuit.

ION BEAM INDUCED LIGHT EMISSION:

MECHANISMS AND ANALYTICAL APPLICATIONS

W. F. van der Weg

Philips Research Laboratories, Dept. Amsterdam

Oosterringdijk 18, Amsterdam, The Netherlands

1. INTRODUCTION

The interaction of a beam of ions in the energy region from a few hundreds of eV up to a few hundreds of keV with a solid target quite generally results in the emission of atomic particles, electrons and photons. It is natural to expect such emission phenomena, since in the considered energy range the kinetic energy of the beam ions is large compared to both the binding energy of atoms in the solid and to the potential energy of outer shell electrons in the collision partners. The light emission in the visible region of the spectrum can be rather intense and can often be observed by eye alone, even at moderate beam current densities (e.g. 10 $\mu A/cm^2$). It seems therefore surprising that the photon emission is far less documented than beam induced particle or electron emission. A possible reason for this phenomenon could be the extreme sensitivity of the emission intensity on the chemical nature or cleanliness of the bombarded surface, which makes it difficult to arrive at reproducible results. In recent years, however, with the advent of more sophisticated vacuum techniques and target cleaning procedures, an increasing amount of data has become available and it seems therefore appropriate to classify and review the information obtained so far.

Most of the experiments have been performed in the visible and near UV spectral region with a relatively simple set up consisting of a scattering chamber, target manipulator and either a monochromator (usually with resolution of the order of 1-10 $\mathring{A}$) with photomultiplier or a spectrograph with photographic recording. The optical spectra emitted during ion bombardment of solid material always consist of a series of sharp lines, in some cases

superimposed on a broad continuum. The line spectrum can easily be
identified as emission from excited beam particles and excited
target atoms, ions or molecular species. These spectra will be
systematically described in the following section. One important
conclusion, however, can already be made from the fact that a
line spectrum is emitted. This fact means that this emission
results from the deexcitation of both target and projectile
particles outside the target, at such a distance from the surface
that the electronic energy levels of the emitting particle are
not perturbed by the presence of the solid. Obviously, the
observation that the atoms present in or at the target emit
characteristic radiation during ion bombardment, makes it
attractive to investigate the possibility of elemental analysis
of solids using ion beam excitation. In the following we will not
attempt to give an exhaustive review of beam induced light emission,
but rather concentrate on those observation which provide a basis
for understanding of the analytical capability of this technique.

2. ORIGINS OF BEAM INDUCED LIGHT EMISSION

2.1 Excitation of target particles

When one bombards a solid target with a heavy ion beam, it is
well known that a considerable sputtering action takes place, during
which a large amount of target particles are ejected from the
surface. Because of the energy transfers involved it is to be
expected that a fraction of these sputtered particles are excited.
Indeed, deexcitation of sputtered particles is responsible for most
of the line emission observed when bombarding solids with heavy
ions. The relation between optical emission and sputtering was
clearly shown by Kistemaker and coworkers [1-2]. In these studies
a close correlation between light emission intensity and sputtering
yield using different beam ions and a variation of beam incidence
angles was found. Since then, many authors have reported rich
spectra of target atoms and ions, excited by heavy beam ions [3-10].
Also, radiation from sputtered molecular fragments, such as metal
oxides and hydrides has been identified [11]. As an example of the
type of optical spectra obtained when bombarding solids with
40 keV Ar^+ ions, we show in fig. 1 three spectra measured with Si,
Pd and W as target material. Apart from atomic and ionic lines of
sputtered material one also observes a continuum emission in the
W case (see sec. 2.3 for a discussion of continuum radiation).

Some more detailed information about the velocity distribution
of the emitting target atoms was obtained by a measurement of the
luminous region in front of a Cu target bombarded with 60 keV Ar^+
ions [12]. The size of this region was found to be of the orders
of a few mm, when the decay of the Cu I resonance line ($\lambda = 3247$ Å)
in front of the target was measured. Taking into account the

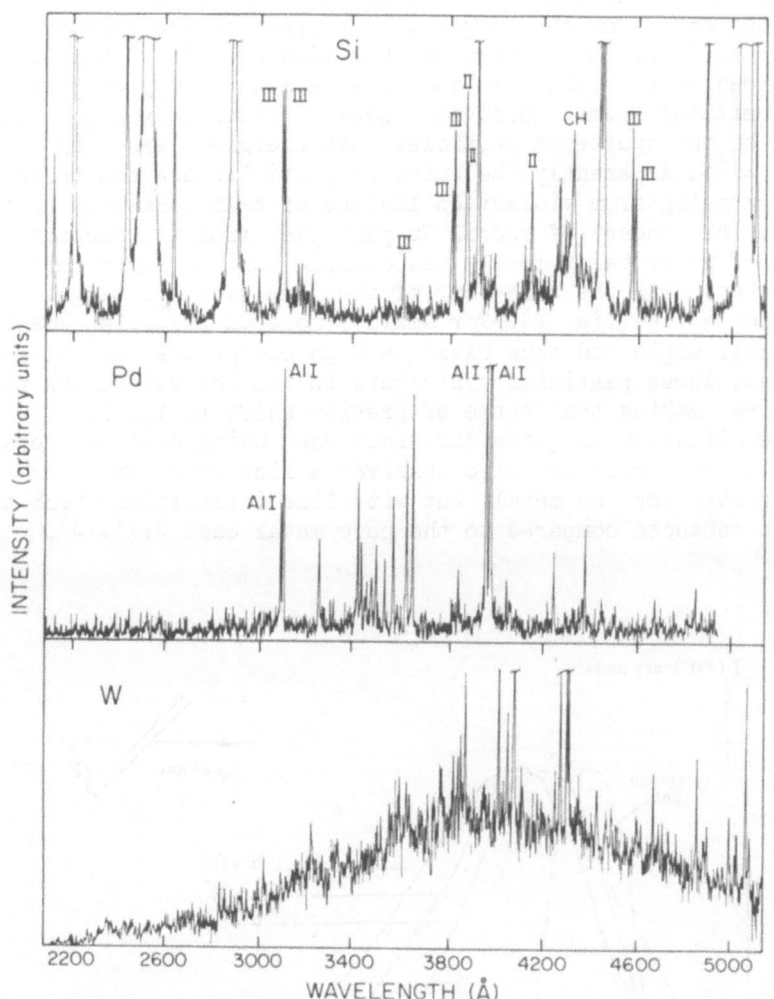

Fig. 1. Spectra obtained for 40 keV Ar$^+$ bombardment of Si, Pd and W targets. The unmarked lines are atomic lines from the target material. In the Si spectrum also lines from single charged (marked II) and doubly charged (III) ions are present (from ref. 8).

lifetime of the considered level (5 nsec), the authors concluded
the mean velocity of the emitting Cu atoms to be 6×10^6 cm/s (the
corresponding energy is 1 keV). This conclusion was substantiated
by a measurement of the Doppler shape and shift of this spectral
line [13]. Fig. 2 shows the profile of this Cu I line ($\lambda = 3247$ Å)
for various angles of the target with respect to the primary ion
beam. The shift of the maximum of this line over 0.6 Å relative
to the expected position also indicates a mean energy of 1 keV
for the emitting atoms. This value seems unexpectedly high, since
the bulk of the sputtered particles have energies less than
100 eV [14,15]. Apparently the emitting particles are scattering
products arising from violent collisions of beam ions with surface
Cu atoms. The concept of recoil Cu particles from interactions
at the surface as being mostly responsible for the light emission
is consistent with the behaviour of the line width vs target
orientation α (see fig. 2). For small α more Cu particles scattered
over a small angle and thus having a high energy are able to leave
the target. These particles contribute to the low wavelength part
of the line, making the centre of gravity shift to low λ.

The situation is quite different when using oxidized metal
targets. In this case one also observes a line spectrum
characteristic for the metal, but with line intensities which are
very much enhanced compared to the pure metal case [5,11,13].

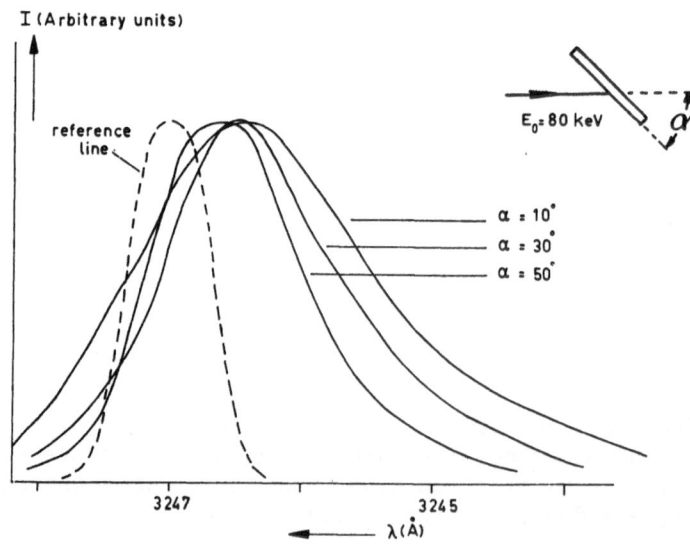

Fig. 2 Line shapes of the 3247 Å Cu I resonance line for
different target orientations α. The intensities are normalized
(from ref. 12).

A further observation is that the Doppler width of spectral lines is considerably reduced when one uses an oxidized target [16]. This phenomenon is illustrated in fig. 3 for 80 keV Ar^+ bombardment of a Si and a SiO_2 target. These chemical effects will be discussed more extensively in sec. 3.3, presently we only note that in the oxidized target case, apparently also slower particles contribute to the radiation.

2.2 Excitation of reflected beam particles

In cases where one uses light projectiles (H or He) on heavier targets, the sputtering of target material becomes less important and the probability for backscattering of beam particles is increased, compared to the heavy beam ion case. This also becomes clear from the optical spectra, which mostly show lines of excited

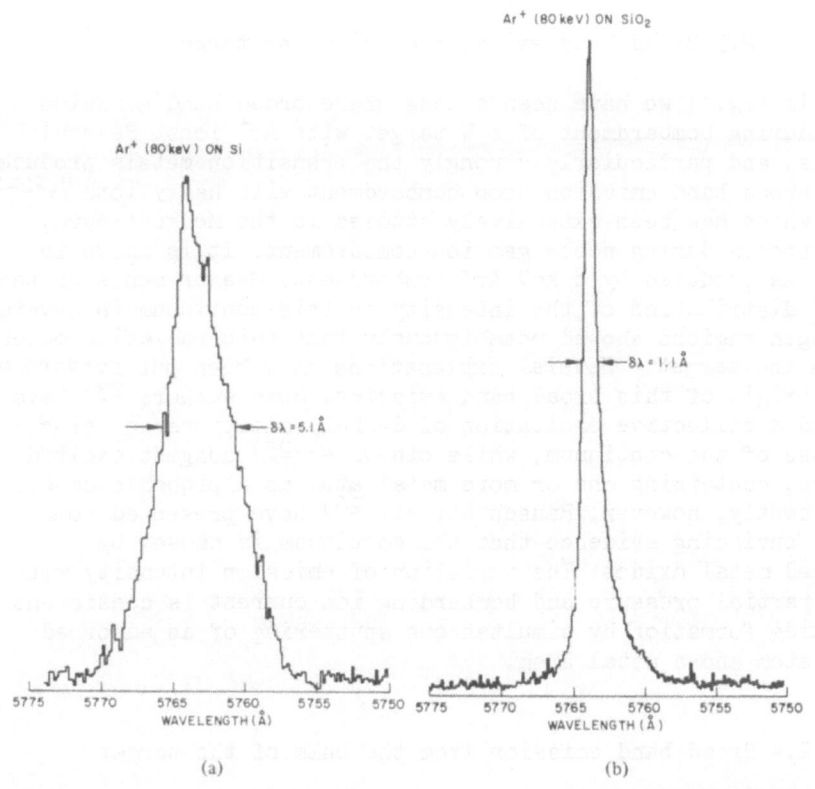

Fig. 3 Measured Si I 2882 Å optical emission line profiles from SiO_2 and silicon. Linewidths measurements were made in second order with an instrumental resolution of $\sim$ 1 Å (from ref. 16).

backscattered particles. During the bombardment of Ta and Cu
targets by 20 keV protons lines of the Balmer series of hydrogen
were detected [17] and in the vacuum UV region also Lyman α
radiation was measured during bombardment of Mo with 12-33 keV
H+ [18]. Bombardment of a Cu target with 2-10 keV He+ results in
several lines of the He I spectrum [19] and at higher primary
energies also lines of the He II spectrum are observed [20].
Analogous to the sputtered atom case (for pure metals) it appears
that the emitting H or He atoms have velocities comparable to the
incident ion velocity. This follows on one hand from direct visual
inspection of the luminous region in front of the target [17] and
from an analysis of the Doppler shift or the spectral lines [18-21].
These authors all ascribe the emission as originating from excited
particles, backscattered from the target with high velocity. These
particles therefore must originate from scattering events close to
the surface. It has not been reported so far what the influence of
oxidation of the target surface is on the line intensities of
emission from backscattered particles.

2.3 Broad band emission outside the target

    In fig. 1 we have seen a case where broad band emission
occurs during bombardment of a W target with Ar+ ions. Several
elements, and particularly strongly the transition metals produce
strong broad band emission upon bombardment with heavy ions [8,22-25].
A case which has been extensively studied is the Mo continuum,
which appears during noble gas ion bombardment. It is shown in
fig. 4, As produced by 8 keV Ar+ bombardment. Measurements of the
spatial distribution of the intensity in this continuum in several
wavelength regions showed unambiguously that this radiation occurs
outside the target. Several explanations have been put forward as
to the origin of this broad band emission. Some authors [23] have
favoured a collective excitation of d-shell electrons as being
the cause of the continuum, while others [24-25] suggest excited
clusters, containing one or more metal atom as a probable cause.
Very recently, however, Rausch et. al. [26] have presented some
rather convincing evidence that the continuum is caused by
sputtered metal oxides. The variation of emission intensity with
oxygen partial pressure and bombarding ion current is consistent
with oxide formation by simultaneous sputtering of an adsorbed
oxygen atom and a metal atom.

2.4 Broad band emission from the bulk of the target

    Emission from the target itself has been observed during
bombardment of various insulators and also metals with 5-30 keV
H+ or He+ ions (or neutralized beams of H or He). Tolk et. al. [6]

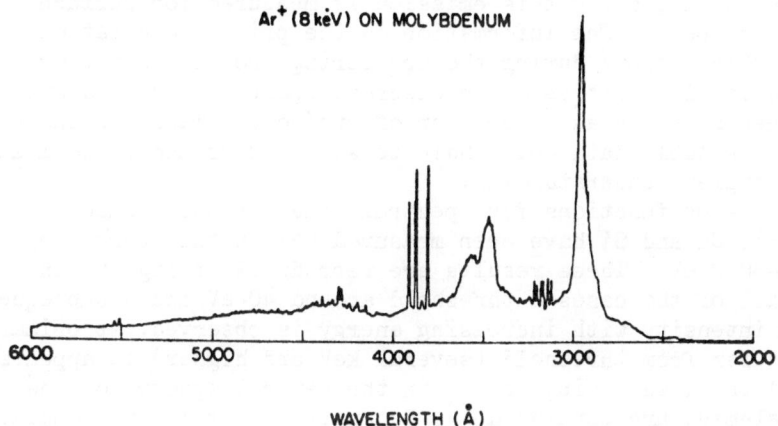

Fig. 4   Continuum emission observed during 8 keV Ar[+] bombardment of Mo. The sharp lines near 3200 and 3800 Å are atomic Mo lines (from ref. 24).

report a strong continuum located between 2000 and 4000 Å upon impact of 5 keV He on CaF_2. They describe this as due to radiative recombination of electron-hole pairs, produced during the interaction of the primary beam with the target. Likewise, Zivitz and Thomas [27] attribute a broad band emission between 4000 and 6000 Å from Al, excited by 25 keV H^+, to recombination of electrons and holes and support this conclusion by a calculation of the emission spectrum on the basis of computed band structures and electron state densities for Al.

Finally, other types of broad band emission, like Bremsstrahlung and transition radiation could be envisaged to occur during ion bombardment of solids. In the energy region considered here, however, the radiation intensity expected from these mechanisms is very small. These processes have been reviewed by Parilis [28].

3. FACTORS INFLUENCING THE EMISSION INTENSITY

3.1 Excitation efficiency

In the following we will mainly concentrate on the emission by target particles under the influence of heavy

ion bombardment, since this emission is measured for surface
analysis purposes. The information on the primary excitation
process which occurs during the sputtering process is rather scarce.
The emission intensities in the discrete spectral lines have in some
cases been measured as a function of projectile mass and energy and
the few available data only indicate some trends and by no means
give a complete understanding.

Emission functions for spectral lines of the atomic spectra
of Sr, Ni, Cu and Si have been measured [29] in the energy region
from 30-3000 eV. These results are reproduced in fig. 5; in
almost all of the cases a threshold around 40 eV and a subsequent
rise in intensity with increasing energy is observed. At primary
energies far from threshold (several keV and higher) it appears
that all the lower lying levels in the neutral spectra of the
target element are populated [7]. Recently, the relative population
of the levels of several metals under noble and reactive gas
bombardment has been studied [30]. These authors found that
especially transition metals like Fe, Cr, Ni, Cu and Zn emit
spectra during ion bombardment which closely resemble an arc
discharge, as far as the relative line intensities are concerned.
A similar conclusion holds for the case of Ar bombardment of Si.
Line intensities of the Si I spectrum obtained by ion bombardment [8]
and from an arc discharge are presented in table I.

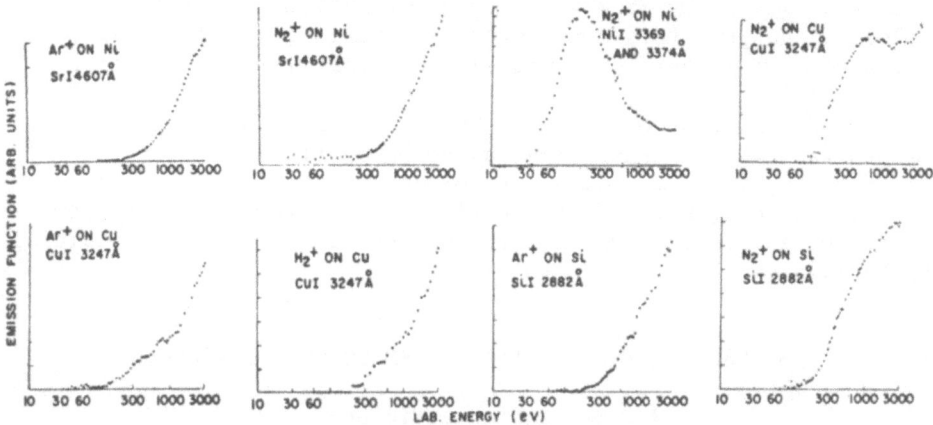

Fig. 5 Emission functions for optical lines produced in ion-
surface collisions. The collision combination and an observed line
are indicated with each emission function. Each graph is plotted
independently in arbitrary units (from ref. 29).

TABLE I

Line intensities in Si I spectrum
(arbitrary units)

λ (Å)	Arc discharge	40 keV Ar$^+$ → Si	$I_{\text{ion bomb.}}/I_{\text{arc}}$
2124	100	600	6.0
2208	110	600	5.4
2211	115	600	5.2
2212	110	600	5.4
2217	120	800	6.6
2218	120	800	6.6
2435	300	2000	6.6
2507	425	2100	5.0
2514 2516 2519 }	1225	7500	6.1
2524	425	2100	5.0
2528	450	2400	5.4
2631	190	950	5.0
2882	1000	600	6.0
2987	150	20	1.3
3905	300	180	6.0

Finally, at high primary energies (> 100 keV) it has been noted
that lines from ionized excited sputtering particles grow relative
to the lines in the neutral excited spectrum [7]. This behaviour
is in accordance with the previously (sec. 2.1) discussed model
for the origin of excited sputtered atoms, in which mainly the
fast recoil target particles are responsible for the light emission.
For these particles it is known that the mean charge state
increases with increasing primary energy [31], in agreement with
the observed trend in the optical spectra. If the total number of

sputtered particles would be responsible for light emission, one
would only note a general decrease of light intensity with
increasing energy along with the sputtering ratio [32]. It is
obvious, however, that many more systematic measurements are needed
to obtain a detailed understanding of the excitation mechanism.

3.2 Survival probability of excited particles
inside and close to a solid

Once the interaction between the beam particle and a target
atom has taken place with sufficient energy transfer, this atom
will move through the solid and eventually be ejected into the
vacuum. It is pertinent here to ask the question whether the
final state of excitation of the target atom is determined during
the primary interaction with the beam particle or whether the state
of excitation is mainly determined at the point of exit from the
surface. The most realistic point of view probably is a combination
of these two mechanisms, in which a target particle first gets
excited by the collision with the beam particle, subsequently
travels through the solid, during which it may lose or capture
electrons and finally emerges from the solid. When the emitted
particle is still close to the surface (< 50 Å distance) electron
exchange with the surface may take place; these eventually cease
to be of importance and the excited state decays at larger
distances by radiation.
There is some evidence that particles in low excited states
can maintain their state of excitation while moving through the
solid. This can be inferred from channeling measurements in which
the intensity of several atomic and ionic spectral lines of Cu
(excited by 100 keV Ar^+) was measured as a function of the angle
of incidence of the beam on the target [20]. Angular scans obtained
from these measurements are shown in fig. 6. The fact that a
reduction in yield is observed for beam incidence along low index
crystal directions already indicates that particles coming from
deeper than one atomic layer contribute to the light emission.
Furthermore, the observation that different spectral lines show
different values of minimum yields in the <110> direction is strong
evidence for the fact that the different excited states originate
from different depths. If excitation would occur upon exit of the
surface, the yield curves for the different lines, if normalized
at one angle, would coincide for all angles, since in this case
the yield curves would simply reflect the number of emerging
particles as a function of angle. A closer analysis [20] makes it
likely that excited ions originate from very close to the surface,
while low excited states in the neutral atom may come from a
depth of a few atomic layers. Again, there is a need for more
measurements and also theoretical calculations to get more
information about the interesting question of the lifetime of

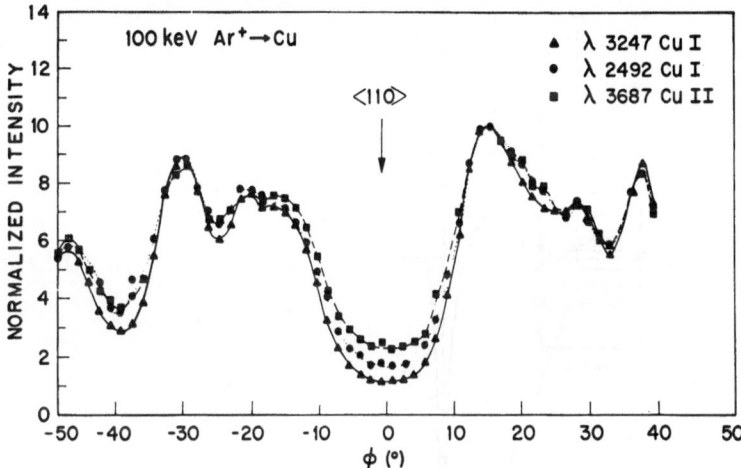

Fig. 6   Intensity of two atomic lines and one ionic line in the
copper emission spectrum vs angle Ø of rotation of the target
around its normal. The three curves have been normalized at Ø = 15°
(from reference 20).

excited atoms in solids. This problem is related to the theory
of ion information in secondary ion mass spectrometry (SIMS).
In fact, one theoretical model [33] calculates the survival
probability for excited particles inside the metal. These then
outside the metal deexcite to form secondary ions by means of
autoionization.

We will now consider the interactions between an excited
particle that has left the solid, and the surface. From the work
of Hagstrum on potential secondary electron emission [34] it has
become very clear that electron transitions between slow particles
and a surface may occur with very high probability, provided the
atom-surface separation is small enough. These transitions can
result in Auger-deexcitation or -neutralization or resonance
neutralization or ionization. A potential energy diagram is
schematically indicated in fig. 7. It has been proposed [35] that
these radiationless deexcitation processes influence the beam
induced light emission since they compete with deexcitation by
photon emission. The transition rate R(s) as a function of
metal-atom separation s for resonant deexcitation is often
approximated by

$$R(s) = A \exp (-as),$$
                                                                                          (1)

A and a being constants. A reasonable value [34] of a is 2 $\overset{\circ}{A}{}^{-1}$ and

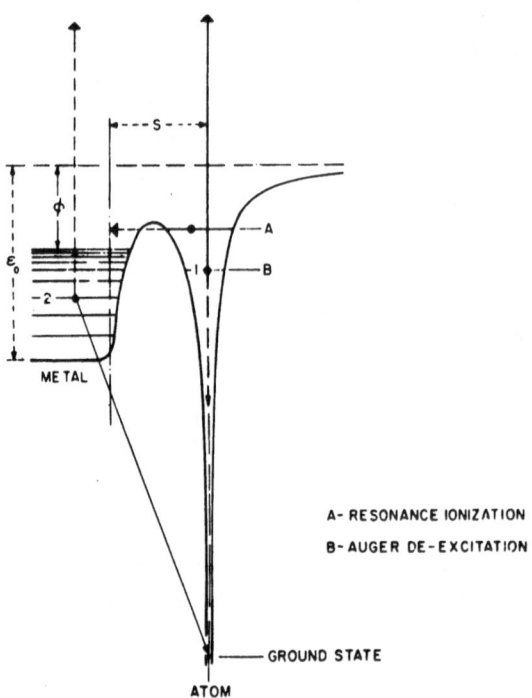

Fig. 7 Schematic representation of the ion-metal energy level
system showing two possible radiationless deexcitation processes,
Auger deexcitation and resonance ionization (from ref. 6).

A is of the order of $10^{14} - 10^{16}$ sec.$^{-1}$. These numbers show that
the transition rate for this process is very large for distances
s of the order of a few $\overset{\circ}{A}$ and several orders of magnitude larger
than the decay rate for optical emission, which is of the order
of 10^8 sec^{-1}. It can be shown [35] that the probability P for a
particle with velocity v to escape from a surface without having
undergone a resonance transition is given by

$$P = \exp\left(-\frac{A}{av}\right) \tag{2}$$

The value of $\frac{A}{a}$, sometimes called the survival parameter, is of the
order of 10^7 cm/sec with the numerical values of A and a quoted
above. With this value of $\frac{A}{a}$, eq (2) shows that low energy particles
have a small survival probability. As an example, the survival
probability for 10 eV excited Cu atom near a Cu surface would be
of the order of 1% [13]. The competition of this type of velocity
dependent radiationless deexcitation with photon emission,

therefore presents a natural explanation for the observation that mainly the most energetic of the sputtered particles contribute to the radiation. In recent years, considerable effort has been invested in a numerical determination of the A/a parameter by means of comparison of measured and calculated Doppler line shapes [13,19,21,36]. The values obtained in this way, however, should be viewed with caution, because of a number of crude assumptions which have to be made about excitation probability and velocity distribution of sputtered or backscattered particles. A severe problem is also posed by the shape of the transition rate for resonance processes for distances s approaching zero. The shape as in eq (1) has been derived for slow ions approaching a surface, but it is not clear what the rate is for sputtered particles, which actually do cross the surface plane. Nevertheless, the occurrence of radiationless deexcitation processes easily explains qualitatively the dominance of relatively fast particles among the emitting sputtered and reflected atoms in the case of bombardment of pure metal surfaces.

3.3 Influence of surface oxidation and other chemical effects

It has been observed that the line intensities are very much enhanced when one bombards an oxidized metal, as compared to the clean metal case [5,13,16]. Also, the admission of oxygen in the scattering chamber during ion bombardment causes a considerable increase of light intensity from sputtered atoms [8,11,37,38]. There has been some speculation and also controversy about the nature of this chemical enhancement effect. Initially it was proposed [35] that the enhancement is caused by the change in band structure of the solid upon oxidation. Because of the for-bidden band gap in the oxide many resonance transitions of electrons from excited sputtered atoms into the solid would then not ben possible any more. This inhibition of the radiation-less deexcitation processes caused the number of excited particles to drastically increase upon oxidation of a surface. It is also clear because of the strong velocity dependence of the survival probability (eq (2)), that especially low velocity particles would have a possibility to radiate, when close to an oxide. This model satisfactorily explained therefore not only the enhancement effect, but also the observed change in Doppler profile (fig. 3) when oxidizing a metal surface. In fact, a measurement of the velocity distributions of particles responsible for different lines in the optical spectrum might be used to obtain information about the electron state density near the solid surface [39].

One example of a systematic study of the influence of the band structure on the light emission probability can be found in

the work on photon emission from clean and oxidized Mg surfaces
by Kerkdijk and Kelly [38]. The optical spectrum emitted from Mg
during Ne+ bombardment shows many Mg I and Mg II lines, some of
which are enhanced, some remain constant and others decrease in
intensity upon oxigenation of the surface. A schematic energy
level diagram of Mg, MgO and the upper levels of the measured
lines is given in fig. 8. A few representative examples of line
intensity behaviour can be indicated. Lines from levels 2 and 4
do not change intensity when going from Mg to MgO, lines 14 a,b,c
increase when going to MgO, while lines 7 a,b decrease. In terms
of the resonance tunneling model, the explanation is quite simple.
Levels 2 and 4 for both metal and oxide are opposite emptly states,
therefore show no oxygen dependence, level 14 can be emptied to
the metal, but not to the oxide (positive oxygen dependence), and
level 7 can be populated from the metal and not from the oxide
(negative oxygen dependence). In this manner it was possible to
explain the oxygen dependence of 16 lines in the spectrum. There
were 5 lines however which did not follow the behaviour expected
from the electron tunneling model. These are all ionic lines, so
it seems likely that other mechanisms are operating in excited ion

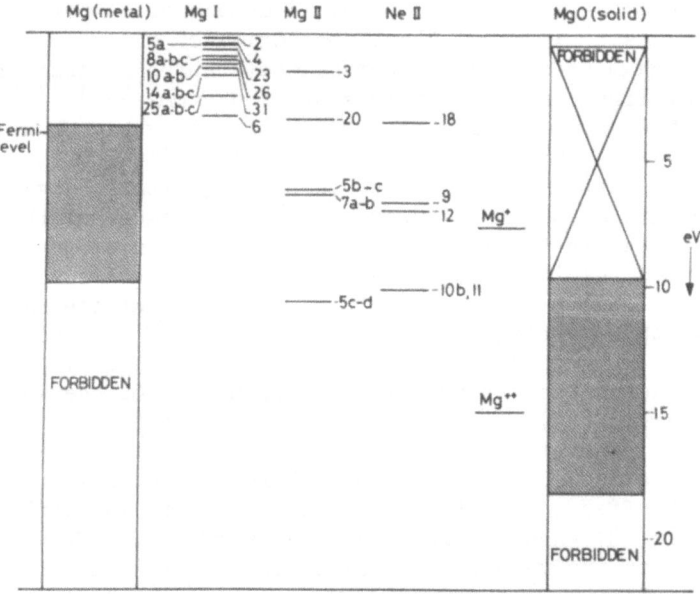

Fig. 8 The energies of excited atoms, ions, and excited ions
observed when Mg is bombarded with 10 keV Ne+. The band structures
of Mg and MgO are also shown. The scale on the right hand side
is in eV (from ref. 38).

formation. It has been suggested that excited ions are dominantly
formed by dissociative excitation of sputtered molecules [40].

Recently, some more doubt about the validity of the resonant
electron transfer model was raised by a rather sophisticated
experiment by G. E. Thomas et. al. [41]. In this work the light
emission from sputtered Cu and Al atoms (using 10 keV Kr^+) was
measured during adsorption of Cs, sometimes with coadsorption of
O_2. The Cs coverage changes the work function of the Al or Cu
metals and this decrease of 2 eV was simultaneously monitored by
laser induced photoelectric emission. In the electron tunneling
model, a change in work function should drastically influence
the probability for resonance tunneling for those lines which are
no longer opposite empty electron states in the solid after a
decrease of work function (compare fig. 7). In the Cs experiment,
actually this enhancement effect was expected for the Cu I
λ = 3247 Å and Al I 3092 Å lines, but surprisingly enough, no such
effect was found. In contrast, O_2 adsorption caused a marked increase
of these lines. These results led the authors [41] to conclude that
the enhancement effect when oxygen is present is directly related
to the formation of substrate atom-oxygen bonds. When these atom-
oxygen molecules are sputtered, a high probability of dissociative
excitation would occur, at distances where electron tunneling
processes would no longer operate [42].

In summary, it is not very clear at present whether the intensity changes of lines from sputtered atoms upon oxidation are related
to a change in the overall band structure, or whether the details
of the metal-oxygen bond formation and bond breaking explain the
observed enhancement effects. For the clean metal case, the
tunneling model describes the velocity distributions (Doppler
shapes) of emitting particles. The change in this velocity
distribution when bombarding an oxide is also naturally explained
by inhibition of tunneling processes, but evidently in the oxide
case, also details of the chemical bonds play a role. An analogous
conclusion was also reached by Mériaux et. al. in their study on
the influence of chemisorption on light emission [43]. It might
also be remarked here that the combination of SIMS and beam
induced light emission could be a powerful technique to identify
the chemical effects [10].

4. SURFACE ANALYSIS WITH ION INDUCED
PHOTON EMISSION

4.1 Elemental analysis

White et. al. were among the first to realize the analytical
capability of light emission by sputtered atoms [44,45]. In their
technique (SCANIIR, Surface Composition by Analysis of Neutral and
Ion Impact Radiation) low energy (< 4 keV) ions and neutral atoms

are used to bombard samples of unknown constituency. Currents are
usually in the range of 0.1 - 1 μA on a spot of 2 mm diameter. It
is clear from the spectra that multi-element analysis is very easy
with this technique, because of the characteristic spectra for each
element. In the initial paper [44], for instance, in the wavelength
scan taken from a tektite, 8 elements show their emission lines.
These are elements present at the surface of the material, since
the sampling depth is only of the order of a few monolayers (as
can be inferred from the discussion in sec. 2 and 3). The analysis
of this work showed detection sensitivities (defined as the volume
concentration of an element in a sample, at which the signal to
noise ratio in the optical emission line is unity) ranging from
10^{-4} to 10^{-6}. It is hard to make general predictions of detection
sensitivities for any desired element, since this number not only
depends on the line strength of the particular transition, but
also on the chemical nature of the surface, especially when oxygen
is present. An illustration of the effect of line strength can be
seen in fig. 1, where trace Al impurities present on a Pd sample
give spectral lines, much stronger than those from the bulk. A
remarkable conclusion from the SCANIIR paper [44] is that in the
spectra obtained with Al_2O_3, LiF and SiO_2 no oxygen or fluorine
lines are present. This could be a result of the fact that these
elements have their strongest lines in the vacuum UV region,
but also could result from the fact that breaking of the strongly
ionic bonds in these compounds during the sputtering would result
in negative O and F ions.

The beam induced light emission technique has also been used
in an analysis of feldspars. In this work [46], referred to as
IBSCA (Ion Beam Spectrochemical Analaysis) several feldspars have
been analysed with the help of a known reference sample. These
authors list several conclusions, as to the analytical capability
of the technique. Some of the more important points are:
1) The necessary instrumentation is cheap and simple, 2) Multi
element analysis is easy, 3) All elements, down to hydrogen, can
be detected, 4) The ion beam used for excitation, also permits
the recording of depth profiles of elements in the sample, because
of the sputtering action. This possibility of studying profiles
by ion luminescence has been applied very fruitfully in the study
of profiles of several elements, like Li and Na in glasses [47,48].
Along the same lines, profiles and profile broadening by recoil
implantation have been measured for Al implanted into Ag and Na
implanted into Si [49].

4.2 Prospects for quantitative analysis

As far as the possibility of performing quantitative analysis
with ion beam luminescence is concerned, the situation is at
present analogous to the one encountered in the SIMS technique. In

the field of secondary ion emission there exists no detailed
theory to quantitatively describe the ion yields. Instead a
phenomenological model is often used, in which the ions are thought
to originate from a plasma in local thermal equilibrium [50,51].
The ionization probabilities are then described with the Saha-
Eggert equation. Usually, the plasma temperature necessary to
describe the experimental ion yields is unrealistically high.
Recently, it was pointed out by Martin and MacDonald that such
a thermodynamic approach may give a good possibility to also
describe the spectral line intensities in a quantitative fashion [30].
The observation that in beam induced emission the relative line
intensities resemble those in an arc discharge (see sec. 3.1)
already points in that direction. If local thermal equilibrium
in a source of photons is assumed, the intensities of spectral
lines are proportional to a Boltzmann factor $\exp(-E_{exc}/kT)$,
where E_{exc} is the upper level of the considered optical transition.
The approach of the authors [30] is then to determine an effective
temperature T, by plotting the line intensities <u>vs</u> E_{exc}. Once
this temperature has been established for a given beam ion-target
combination, it can be used to determine concentrations of trace
elements in the target from their spectral line intensities. As an
example, a temperature of 3800° K was determined for 50 keV Ar^{+}
on stainless steel from the Fe lines in the spectrum. With this
value of the temperature the concentration of Cr in the sample
could be correctly calculated from the Cr spectral lines.
The introduction of oxygen in the scattering chamber, however,
caused a spectral line intensification, which was different for
Fe, as compared to Cr. This clearly complicates the analysis.
In conclusion, it can be said that although the effective arc
temperature seems to depend on the oxygen pressure, the ion
bombardment causes the excited states in the sputtered atoms to
be populated according to a Maxwell-Boltzmann distribution.
This obviously is a promising step towards a quantitative understanding of ion induced light emission.

5. SUMMARY

Although it is clear from the preceding discussion that the
field of beam induced luminescence is still in its infancy, several
conclusions can already be drawn from the available data.
The observed line spectra can be classified as either resulting
from excited sputtered or excited reflected atoms and ions.
Continua originate from sputtered molecules when they are emitted
exterior to the target, while electron-hole recombination gives
rise to continuous radiation from the interior. In the clean
metal case, the dominance of fast particles in the emitting
particle velocity distribution is explained on the basis of the
occurrence of radiationless deexcitation processes. The chemical

enhancement effects probably have a more complicated origin,
involving molecular bond structures. Finally, it is indicated
that a systematic study of relative level populations in the
spectra may lead to a more quantitative understanding.

ACKNOWLEDGEMENTS

The permission of drs. P. J. Martin, R. J. MacDonald,
E. W. Thomas, G. E. Thomas and C. B. Kerkdijk to use parts of
their unpublished material, is gratefully acknowledged.

REFERENCES

1) J. Kistemaker and C. Snoek, Ion Bombardment - Theory and
   Applications, Gordon & Breach, New York, 1964.
2) J. M. Fluit, L. Friedman, J. van Eck, C. Snoek and J. Kistemaker,
   Proc. 5th Inf. Conf. on Ionization Phenomena in Gases, Munich,
   1961, North-Holland Publ., Amsterdam, 1962.
3) I. Terzic and B. Perovic, Surf. Sci. 21 (1970) 86.
4) I. S. T. Tsong, Phys. Stat. Solidi (a), 7 (1971) 451.
5) H. Kerkow, Phys. Stat. Solidi (a), 10 (1972) 501.
6) N. H. Tolk, D. L. Simms, E. B. Foley and C. W. White, Rad.
   Eff. 18 (1973) 221.
7) M. Braun, B. Emmoth and I. Martinson, Physica Scripta 10
   (1974) 133.
8) W. F. van der Weg and E. Lugujjo, Atomic Collisions in Solids,
   Ed. S. Datz et. al., Plenum, New York, 1975, p. 511.
9) T. S. Kiyan, V. V. Gritsyna and Ya. M. Fogel', Nucl. Instr. &
   Meth. 132 (1976) 435.
10) A. R. Bayly, P. J. Martin and R. J. MacDonald, Nucl. Instr. &
    Meth. 132 (1976) 459.
11) G. E. Thomas and E. E. de Kluizenaar, Int. Journal of Mass
    Spectr. and Ion Phys. 15 (1974) 165.
    This paper also contains an interesting historical introduction
    to the subject of beam induced light emission.
12) C. Snoek, W. F. van der Weg and P. K. Rol, Physica 30 (1964) 341.
13) W. F. van der Weg and D. J. Bierman, Physica 44 (1969) 206.
14) K. Kopitzki and H. Stier, Z. Naturforsch. 17a (1962) 346.
15) G. E. Chapman, B. W. Farmery, M. W. Thompson and I. H. Wilson,
    Atomic Collisions in Solids IV, ed. S. Andersen et. al.,
    Gordon & Breach, London, 1972, p. 339.
16) C. W. White, D. L. Simms, N. H. Tolk and D. V. McCaughan,
    Surf. Sci. 49 (1975) 657.
17) V. V. Gritsyna, T. S. Kiyan, A. G. Koval' and Ya M. Fogel',
    Physics Lett. 27A (1968) 292.
18) G. M. McCracken and S. K. Erents, Physics Lett. 31A (1970) 429.
19) C. Kerkdijk and E. W. Thomas, Physica 63 (1973) 577.

20) W. F. van der Weg, N. H. Tolk, C. W. White and J. M. Kraus, Nucl. Instr. & Meth. 132 (1976) 405.
21) W. A. Baird, M. Zivitz and E. W. Thomas, Phys. Rev. A 12 (1975) 876.
22) T. S. Kiyan, V. V. Gritsyna, Yu. E. Logachev and Ya M. Fogel', JETP Lett. 21 (1975) 35.
23) T. S. Kiyan, V. V. Gritsyna and Ya M. Fogel', Nucl. Instr. & Meth. 132 (1976) 415.
24) C. W. White, N. H. Tolk, J. Kraus and W. F. van der Weg, Nucl. Instr. & Meth. 132 (1976) 419.
25) C. B. Kerkdijk, K. H. Schartner, R. Kelly and F. W. Saris, Nucl. Instr. & Meth. 132 (1976) 427.
26) E. O. Rausch, A. I. Bazhin and E. W. Thomas, to be published.
27) M. Zivitz and E. W. Thomas, Phys. Rev. B, 13 (1976) 2747.
28) E. S. Parilis, Atomic Collision Phenomena in Solids, ed. D. W. Palmer et. al., North-Holland Publ., Amsterdam, 1970, p. 513.
29) C. W. White and N. H. Tolk, Phys. Rev. Letters 26 (1971) 486.
30) P. J. Martin and R. J. MacDonald, to be published.
31) W. F. van der Weg and D. J. Bierman, Physica 44 (1969) 177.
32) P. Sigmund, Phys. Rev. 184 (1969) 383.
33) P. Joyes, J. Phys. C, 5 (1972) 2192.
34) H. D. Hagstrum, Phys. Rev. 96 (1954) 336.
35) W. F. van der Weg and P. K. Rol, Nucl. Instr. & Meth. 38 (1966) 274.
36) R. Hippler, W. Krüger, A. Scharmann and K.-H. Schartner, Nucl. Instr. & Meth. 132 (1976) 439.
37) R. Kelly and C. B. Kerkdijk, Surf. Sci. 46 (1974) 537.
38) C. B. Kerkdijk, Thesis, Leiden, 1975, Ch. IV.
39) V. V. Gritsyna, T. S. Kiyan, A. G. Koval' and Ya. M. Fogel', Rad. Eff. 14 (1972) 77.
40) K. Jensen and E. Veje, to be published, Zeitschr. f. Physik.
41) G. E. Thomas and E. E. de Kluizenaar, Nucl. Instr. & Meth. 132 (1976) 449.
42) G. E. Thomas, to be published.
43) J. P. Mériaux, R. Goutte and C. Guillaud, Appl. Phys. 7 (1975) 313.
44) C. W. White, D. L. Simms and N. H. Tolk, Science 177 (1972) 481.
45) C. W. White, D. L. Simms and N. H. Tolk, Characterization of Solid Surfaces, ed. P. F. Kane and G. R. Larrabee, Plenum, N.Y., 1974, Ch. 23.
46) I. S. T. Tsong and A. C. McLaren, Nature 248 (1974) 43.
47) H. Bach and F. G. K. Baucke, Electrochimica Acta 16 (1971) 1311.
48) H. Bach, Rad. Eff. 22 (1974) 73.
49) M. Braun, B. Emmoth and R. Buchta, Proc. 2nd Int. Conf. on Ion Beam Surface Layer Analysis, Karlsruhe, 1975.
50) C. A. Andersen and J. R. Hinthorne, Anal. Chem. 45 (1973) 1421.
51) A. E. Morgan and H. W. Werner, Anal. Chem. 48 (1976) 699.

COMPLEMENTARY ANALYSIS TECHNIQUES : AES, ESCA

J. Tousset

Institut de Physique Nucléaire, Université Claude

Bernard, Lyon-I, Villeurbanne (France).

I. INTRODUCTION

When considering the possibility of electron emission characteristic of the nature of the investigated sample, two effects play
an important role : Auger and photoelectric effects. Both have
given rise to electron spectroscopies : A.E.S. (Auger Electron
Spectroscopy) and E.S.C.A. (Electron Spectroscopy for Chemical Analysis). Because of the depth from which the analysed electrons can emerge without significant alteration of their initial
energy, these spectroscopies appear essentialy as surface analysis methods. A fast technology improvement, a broad scope of
applications resulting from their inherent characteristics (sensitivity, chemical informations, ...) explain the currently increasing developpment of these techniques /1/.

II. FUNDAMENTALS

Electron probing of the surface can cause the emission of
all four types of emitted particles (electrons, ions, neutrals
and photons). But, of these four types of particles, the analysis
of emitted electrons is most commonly used because of the
relative ease of detecting them.

A typical electron distribution, obtained by bombarding a
solid with primary electrons of a fixed energy, may be divided
broadly into three regions (Fig. 1 (1c)).

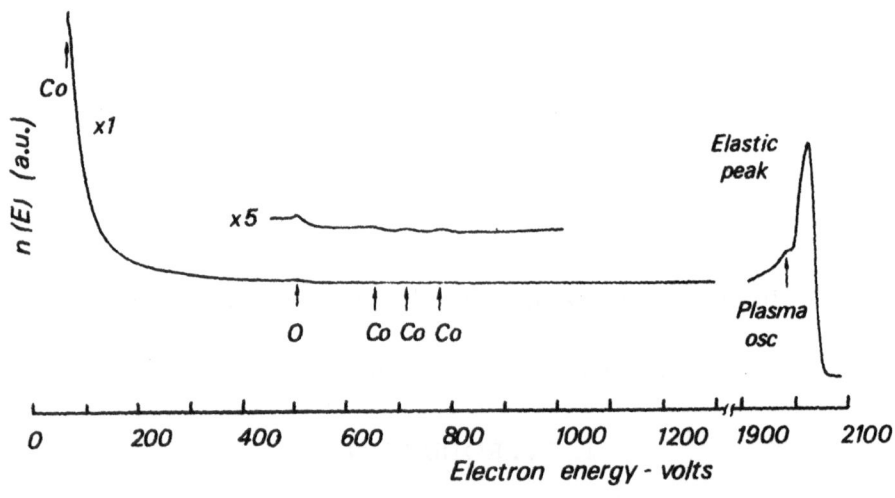

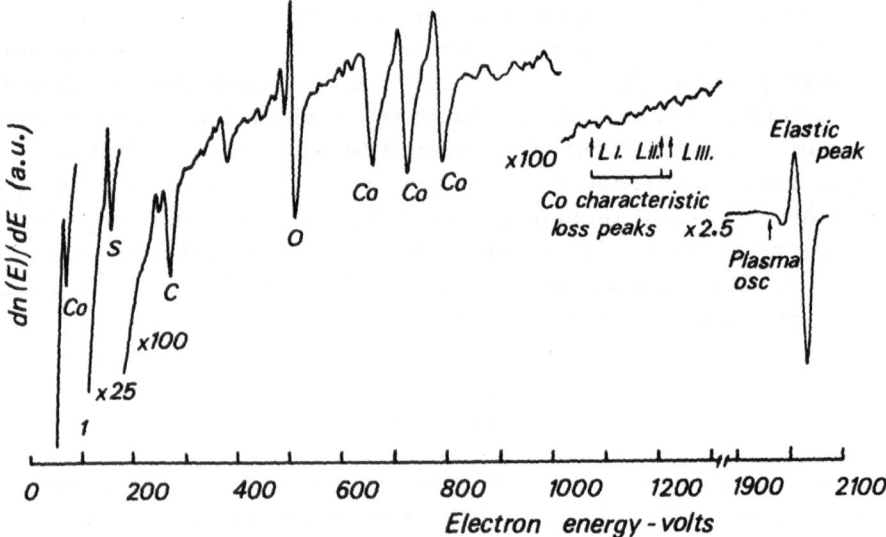

Fig. 1 : Secondary electron energy distribution curves
N(E) and $\dfrac{dN(E)}{dE}$ from a cobalt target (1c).

A large group, at low energies, results from multiple scattering. Another groupe, with energies close to that of the primary beam represents the electrons scattered elastically, that is to say without loss of energy.

Between these extremes, there is a plateau in energy. The Auger electrons typically occur as small bumps on this large background and the technical problem involved in making the Auger electrons useful for chemical analysis is to magnify these small features. That is accomplished by electronic differentiation (Fig. 1) which removes the large background consisting mainly of backscattered primary electrons and inelastically scattered Auger electrons.

What are the Auger electrons ?

When an incident electron with sufficient primary energy ionises a core level, the vacancy is immediately filled by another electron, as depicted for instance by an $L_{II} \rightarrow K$ transition in the Fig. 2.

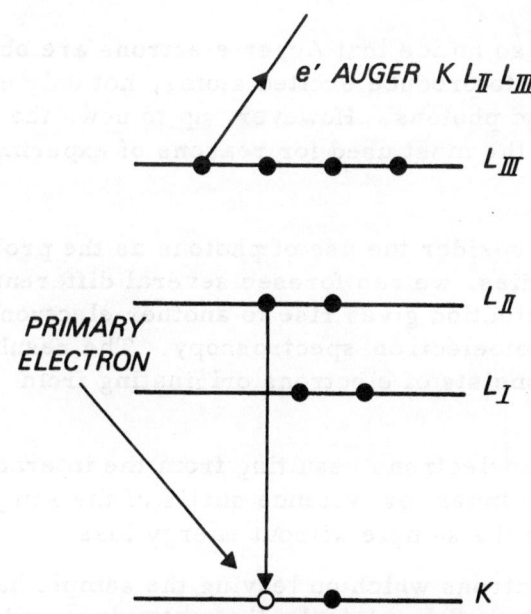

Fig. 2 : Elementary level scheme showing
a typical Auger transition (KLL).

The energy $E_K - E_{L_{II}}$ from this transition can be released in the form of characteristic X-rays (it is the basis for X-ray fluorescence spectroscopy) or be transferred to another electron in the L_{III} level for instance, which is ejected from the atom as an Auger electron.

This Auger electron is called $KL_{II}L_{III}$. Its measured energy is about :

$$E_{KL_{II}L_{III}} = E_K - E_{L_{II}} - E_{L_{III}} - \Phi_A$$

where Φ_A is the work function of the analyser material.

In fact, since the Auger electron actually emerges from a doubly ionized atom, it may be argued that $E_{L_{III}}$ is the energy of the L_{III} shell of a singly ionised atom. Many tables and calculations give energies of Auger electrons (for instance Fig. 3).

The most important fact to notice is that the energy of an Auger electron is characteristic of the target material and independent of the incident beam energy.

We must also notice that Auger electrons are obtained by any probes able to produce excited atoms, not only electrons, but also ions and photons. However, up to now, the electron probe has been the most used for reasons of experimental convenience.

If now we consider the use of photons as the probing beam for surface studies, we can foresee several different techniques. The electron detection gives rise to another electron spectroscopy, called photoelectron-spectroscopy. The resulting photoelectron flux consists of electrons originating from different mechanisms :

- true photoelectrons resulting from the interaction of photons with the inner or valence shells of the sample atoms and which leave the sample without energy loss,

- photoelectrons which on leaving the sample have undergone energy losses through inelastic scattering, and

- Auger electrons resulting from the annihilation of primary holes in inner shells.

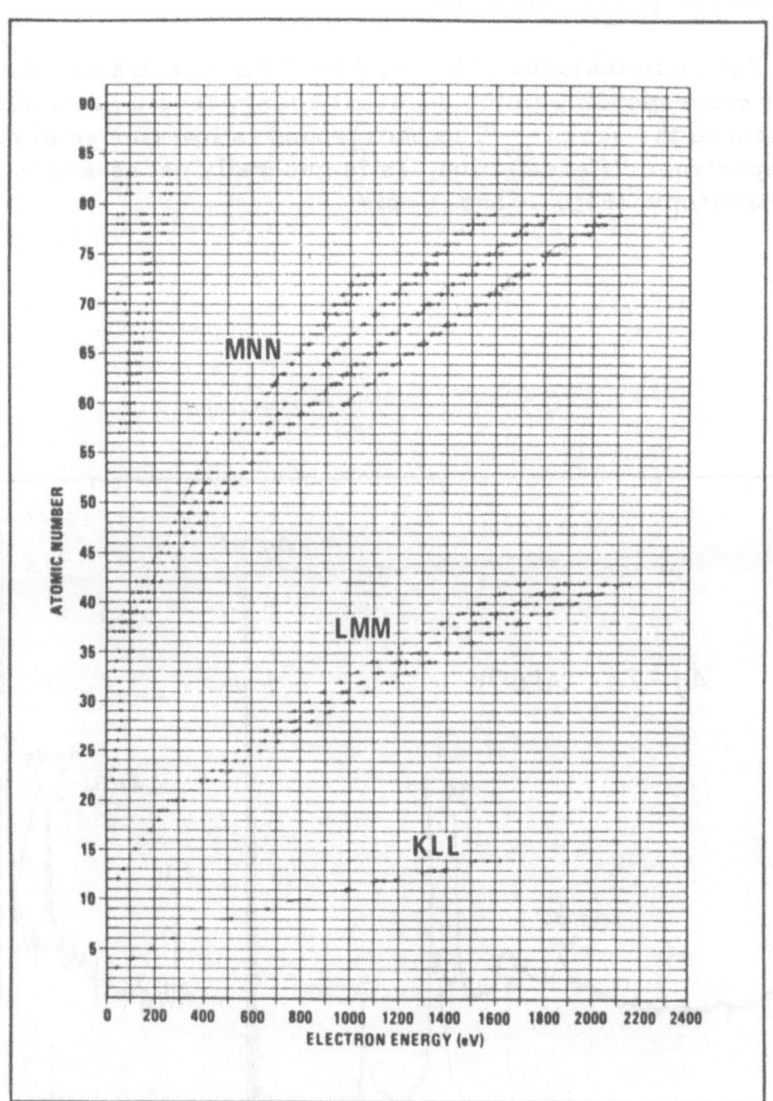

Fig. 3 : Chart of the principal Auger electron energies (26).

Considering the wavelength of the primary source used, it has been customary to divide the field of photoelectron spectroscopy and to distinguish between X-ray (XPS or ESCA) and ultraviolet (UPS) techniques.

The main features of interest in XPS spectra are the photolines corresponding to the different energy levels of the elements present in the sample. The background is low enough to avoid the spectrum differentiation, as is generally necessary in Auger spectroscopy (Fig. 4 and 4 bis).

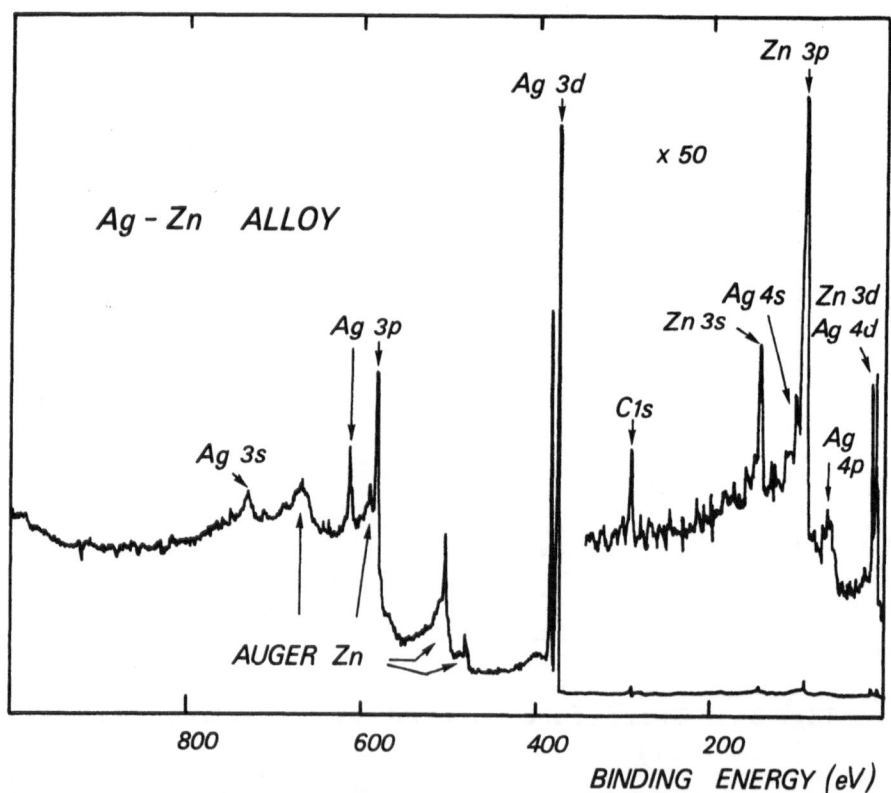

Fig. 4 : ESCA spectrum of an Ag-Zn alloy obtained with monochromatized AlKα X-rays (14).

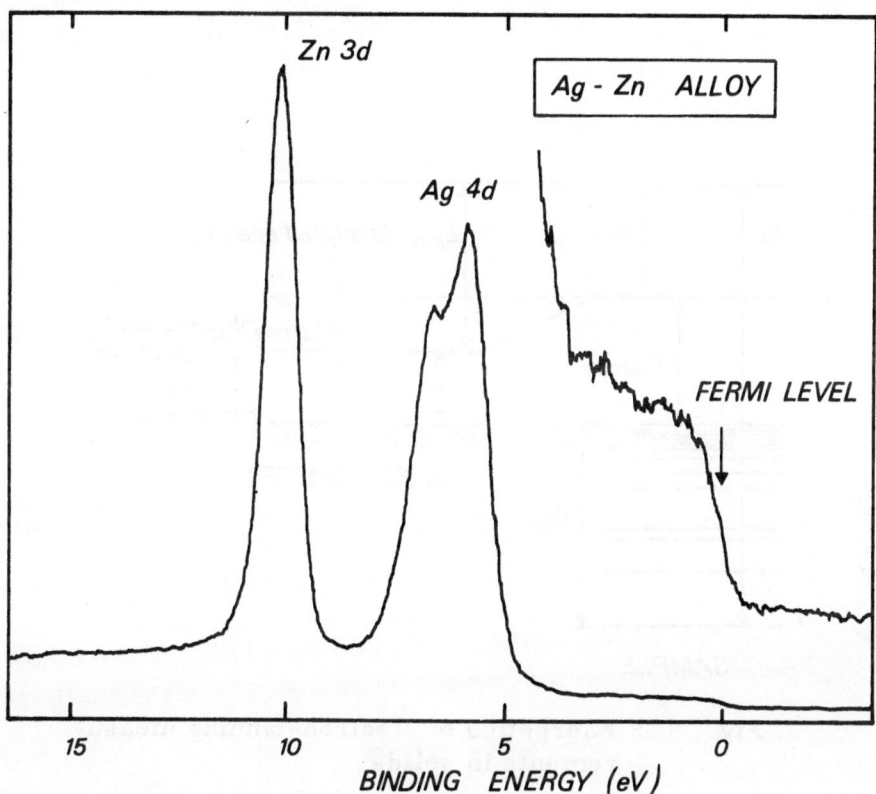

Fig. 4 bis : Valence-band region of Ag-Zn alloy spectrum
of Fig. 4 (14).

The kinetic energy E_{kin} of the analysed photoelectron is linked to its original binding energy E_{bi} by the expression :

$$h\nu = E_{b,i}^{F} + E_{kin} + \phi_{sp} + E_{r}$$

where : $h\nu$: photon energy

$E_{b,i}^{F}$: binding energy, corresponding to orbital i, referred to the Fermi level,

E_{kin} : kinetic energy in the analyser

ϕ_{sp} : spectrometer work function,

E_{r} : recoil energy (normally neglected) (Fig. 5).

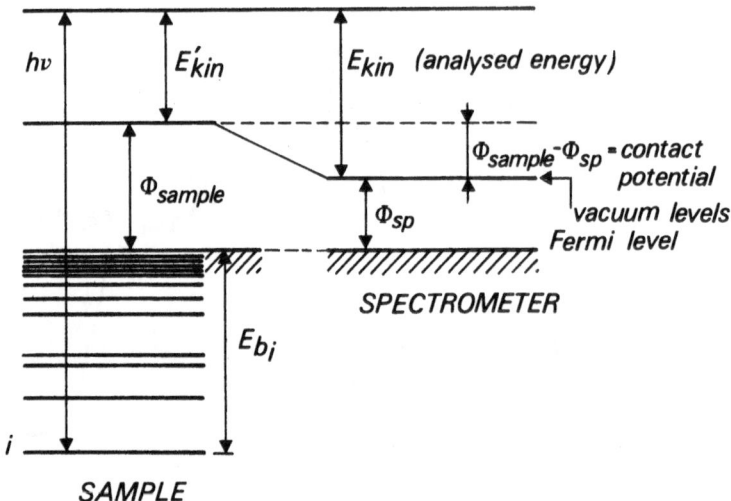

Fig. 5 : Energetics of electron-binding measurements in solids.

It should be noticed that the XPS binding energy, $E_{b,i}^{F}$, is
not necessarily equal to the orbital energy ε_i of the ejected electron. It is only an approximation. As will be discussed later, in
many cases, multiple excitations and relaxation effects may
occur in such a way that :

$$\varepsilon_i = E_{b,i}^{F} + \Delta E$$

In any case, one of the great success of ESCA is the chemical shift of the core lines of an atom, which gives information
on its chemical surroundings and also the valence band spectrum
(Fig. 6), as we shall see later.

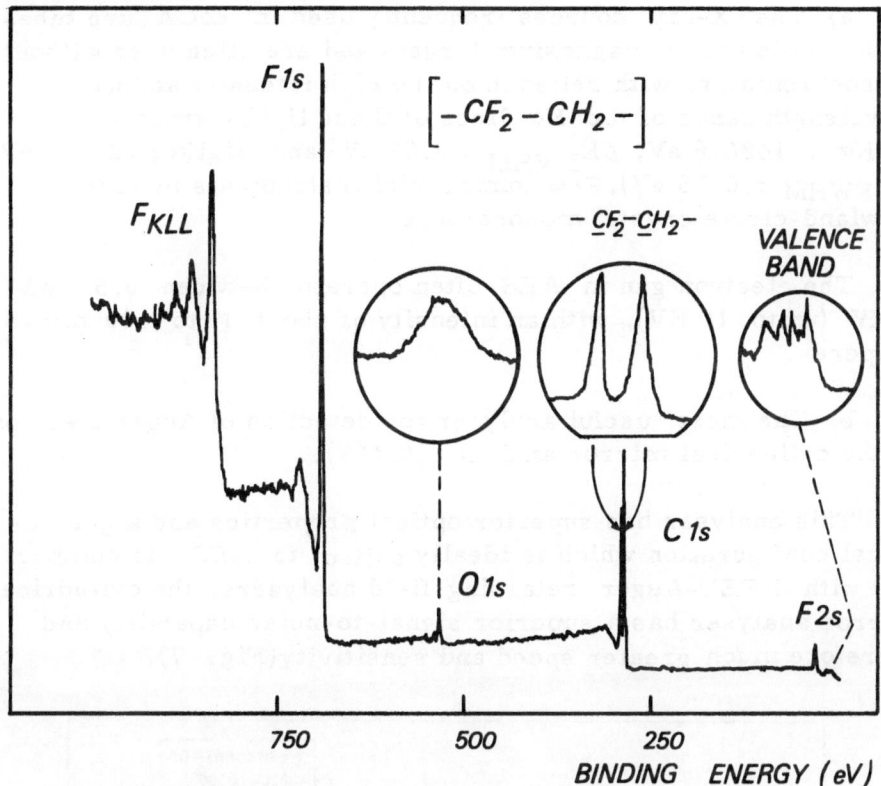

Fig. 6 : Significant informations from an ESCA
spectrum. (14)

III. INSTRUMENTATION

An ESCA or Auger spectrometer must contain the following components :

a) a source of radiation with which to excite the sample (X-rays for ESCA, electrons for AES),

b) an energy analyser

c) an electron detector and

d) a high-vacuum system.

The entire system must be shielded from the earth's magnetic field. In addition, many instruments are provided with an ion-bombardment gun and other accessories.

a) The X-ray sources frequently used in ESCA are tubes with aluminium or magnesium targets and are often used without monochromator, with reliance on the high intensity and narrow wavelength bands of the Kα lines of these light elements (AlKα : 1486.6 eV, ΔE_{FWHM} : 0.95 eV and MgKα : 1253.6 eV, ΔE_{FWHM} : 0.75 eV). Few commercial instruments include a Rowland-circle crystal monochromator.

The electron gun in AES often operates between 0.5 and 3 KV (up to 10 KV), with an intensity of about 1 to 100 micro-amperes.

b) The most useful analyser for detection of Auger electrons is the cylindrical mirror analyser (CMA).

This analyser has superior optical properties and a geometrical configuration which is ideally suited to AES. In comparison with LEED-Auger retarding-field analysers, the cylindrical mirror analyser has a superior signal-to-noise capability and therefore much greater speed and sensitivity (Fig. 7).

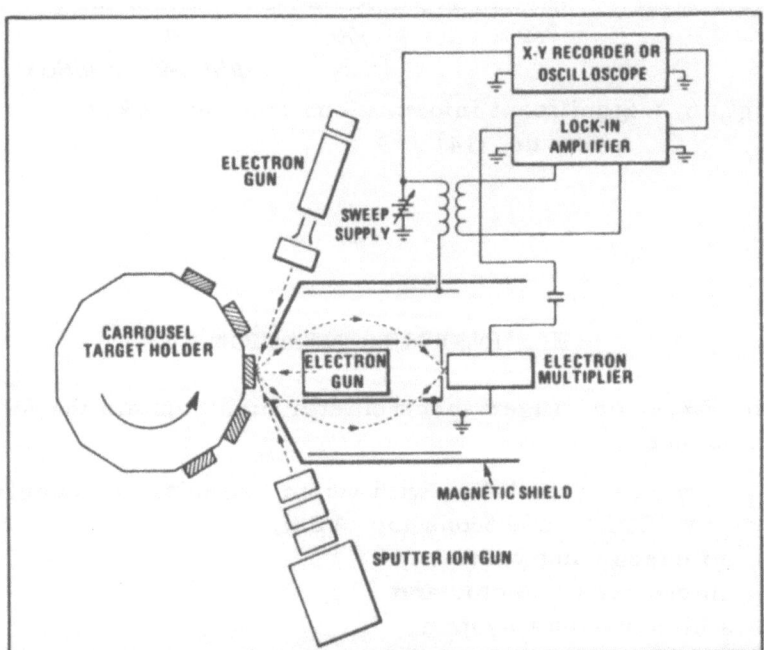

Fig. 7 : Schematic layout of CMA as an Auger spectrometer (26).

In ESCA, three types of analyser geometry are currently
used : the dispersive cylindrical mirror, the dispersive spherical sector capacitor and the non-dispersive energy filter. A
fixed voltage is often applied to the analyser (Fig. 8) and a
retarding field is scanned ; photoelectrons are slowed down
by the retarding field and only those electrons that fit with the
sector voltage will reach the detector. An advantage of this
operation is in particular to keep constant peak widths.

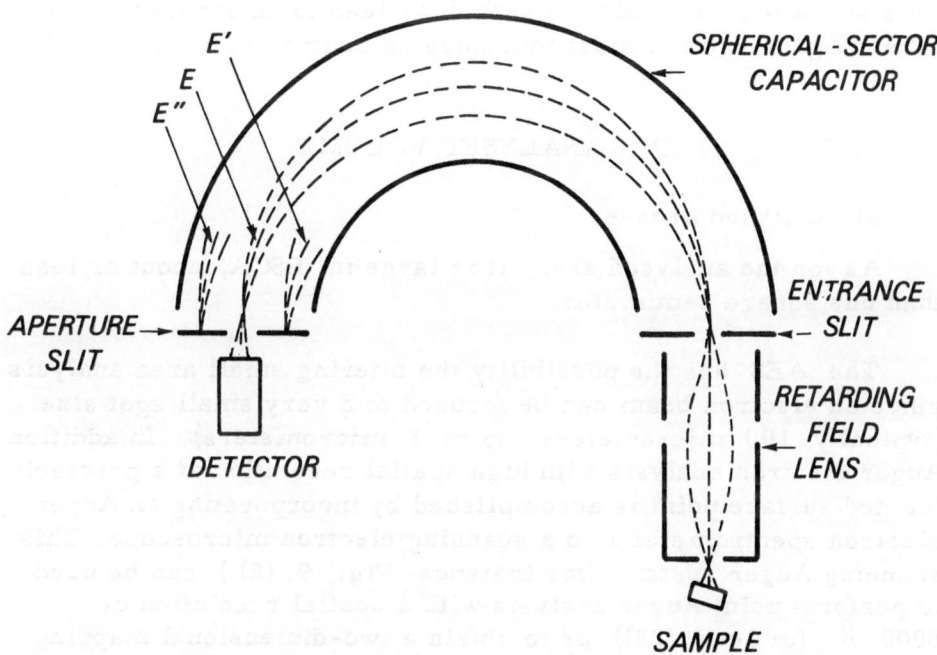

Fig. 8 : Operation of the spherical-sector analyser as
an ESCA spectrometer.

c) The detector commonly used is a channel electron multiplier. An exception is the system with an X-ray monochromator (Hewlett-Packard-5950 A) which employs a large-area detector to compensate for the low X-ray intensity obtained from the
monochromator.

d) The vacuum system must give the analysed electrons a
long mean free path relative to the internal dimensions of the
spectrometer ; it must also reduce the partial pressure of

reactive residual gases. An essential consideration is vacuum composition as well as the ultimate vacuum level. For instance, at a pressure of 10^{-9} Torr, approximately one monolayer adsorbs on the surface in 15 minutes when the sticking coefficient is unity.

The sample may be a solid or a powder and may either be a conductor or non-conductor. Samples that are insulated from the spectrometer can acquire a positive charge at their surface during irradiation because photoelectrons are being ejected. This so-called "charging effect" may lead to an apparent shift in binding energy and must therefore be controlled or cancelled.

IV. ANALYSED VOLUME

a) Analysed area :

As for the analysed area, it is large in ESCA, about or less than one square centimeter.

The AES has the possibility the offering small area analysis since an electron beam can be focused to a very small spot size (typically 100 micrometers, up to 3 micrometers). In addition Auger electron analysis with high spatial resolution at a precisely located surface point is accomplished by incorporating an Auger electron spectrometer into a scanning electron microscope. This scanning Auger system (for instance Fig. 9 (2)) can be used to perform point Auger analysis with a spatial resolution of 5000 Å (or less (25)) or to obtain a two-dimensional mapping of the concentration of a selected surface element (probe current $\sim 10^{-11}$ - 10^{-5} A). To obtain an elemental map, that is to say Auger image, the intensity of the display is controlled by the magnitude of the selected Auger peak. The most negative excursion in the differentiated Auger spectrum is taken as a measure of the Auger current.

b) Analysed depth :

To obtain a peak in an AES or ESCA spectrum, the electron must emerge from the solid without suffering any inelastic scattering. An electron which scatters inelastically with another electron or excites a plasmon loses sufficient energy, so that it no longer appears at the same characteristic energy in the energy

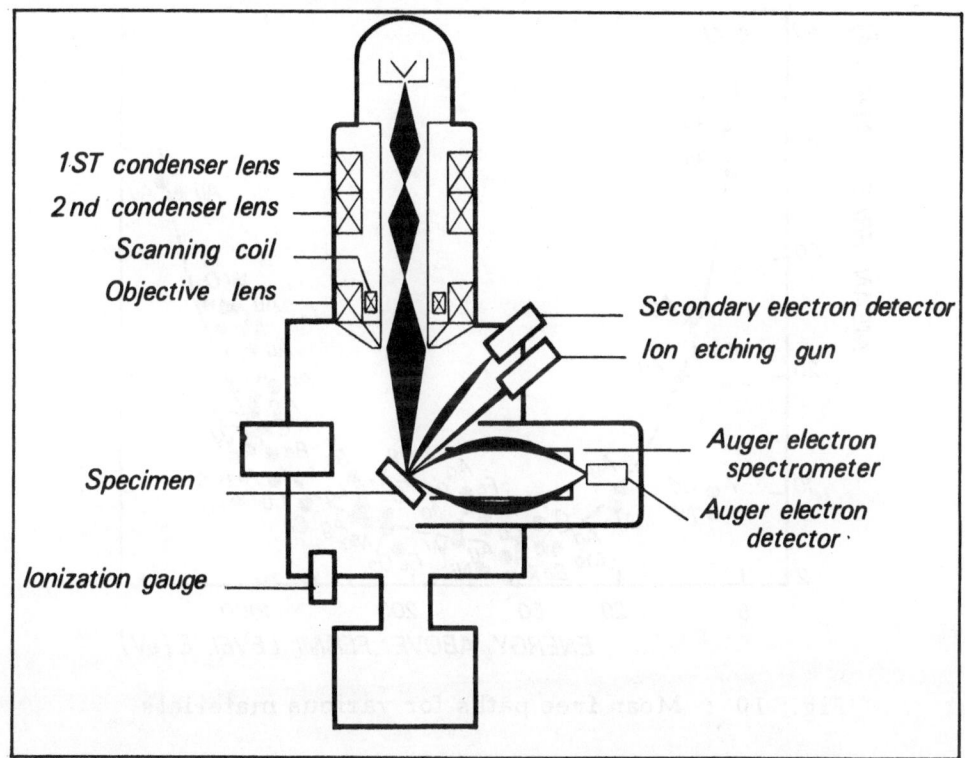

1ST condenser lens
2nd condenser lens
Scanning coil
Objective lens
Specimen
Ionization gauge

Secondary electron detector
Ion etching gun
Auger electron spectrometer
Auger electron detector ·

Fig. 9 : Auger Scanning Electron Microscope (2).

distribution curve, and the information contained about the elec-
electron energy level is lost.

Therefore, in both techniques, it is this escape probability
that will determine the depth of the surface that is being examined.
The mean free path MFP, in ESCA, is always, no matter what
the material is, much larger for a photon of a given energy than
for an electron (~ 100 times larger) and the long penetration
range of the impacting X-rays will be of secondary importance.
This may not be quite true for AES if the primary beam is
near enough in energy to that of the Auger electron for the mean
free path to be comparable.

The mean free path and the way it varies with kinetic energy
of the electron and composition of the matrix is still a subject of
research (Fig. 10).

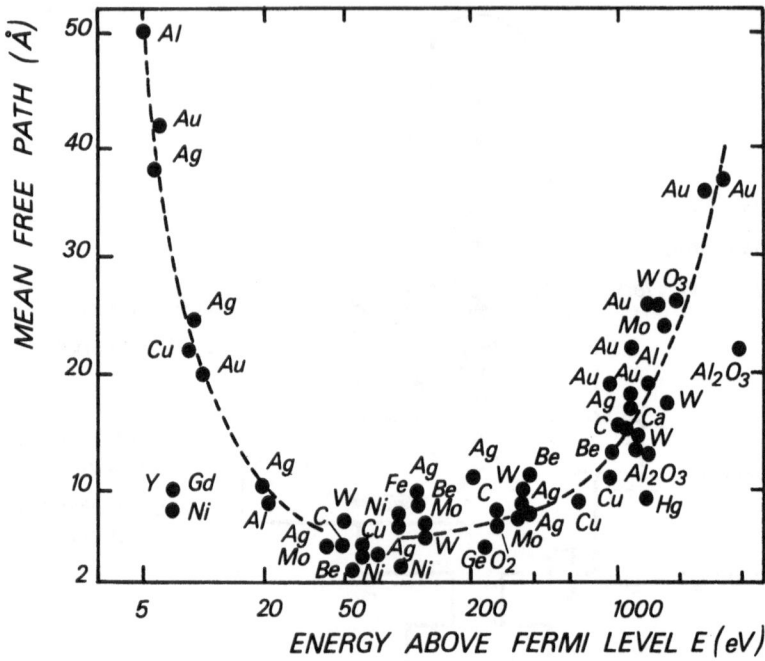

Fig. 10 : Mean free paths for various materials.

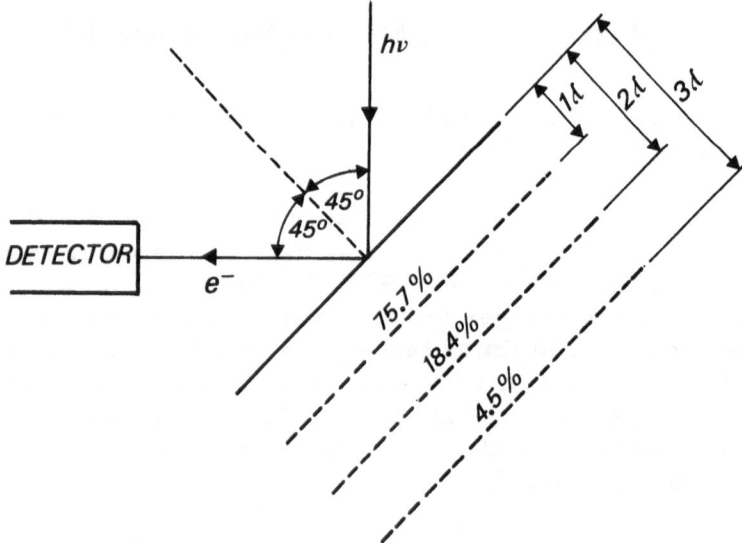

Fig. 11 : Contribution of different regions of the sample to
           the total measured photoelectron signal (assu-
           ming exponential attenuation for measured elec-
           trons and no attenuation of the incident radiation).

If the distribution in direction of excited electrons and the inelastic scattering is isotopic, the probability of a substrate electron traversing an overlayer of thickness x at an angle θ to the surface is proportional to $\exp(-x/\lambda \sin \theta)$.

In the experimental geometry of the Figure 11, the outer-most layer would contribute only to 13% of the total signal assuming a layer thickness of one-tenth the mean free path.

V. QUALITATIVE AND QUANTITATIVE ANALYSIS

a) Qualitative analysis and sensitivity :

Since the atomic structure of each element in the periodic table is distinct from all the others, measurement of the positions of one or more of the electron lines allows the identification of an element present at a sample surface. Therefore, all the elements can be detected, save hydrogen.

However, Auger spectra, especially of the heavier elements, can be quite complicated and some superpositions can occur. Fortunately, even if certain Auger peaks do overlap, there may well be other peaks which do not. Besides, the lighter elements, often important to study, give rise to fairly uncomplicated Auger spectra.

In ESCA, the element adjacent to one another in the periodic chart produce electron lines which are well separated from one another so that no ambiguity exists in identification of adjacent element. We shall return later to the chemical shift.

When one speaks of the sensitivity of AES or ESCA, there are several aspects which must be considered : first, the relative sensitivities for different elements ; second, the sensitivity to bulk constituents and to surface layers and then the comparison between AES and ESCA sensitivities.

The sensitivity of ESCA to different elements does not vary enormously ; ESCA is least sensitive to the light elements and most sensitive to the heavy elements (Fig. 12).

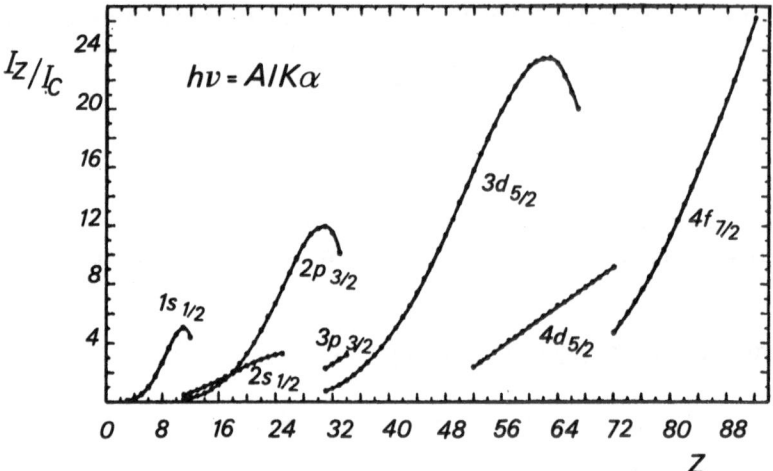

Fig. 12 : Calculated intensities of photoelectron peaks
from X-ray photoelectron spectra of solid.
Results given are relative to the C(1s) peak
for equal atomic concentrations (4).

The most important factor is obviously the cross-section
with which X-rays are absorbed by the electrons of a particular
atomic subshell.

The bulk sensitivity of ESCA is not high ; it is generally
limited to concentrations of approximately 0.1% based on bulk
percentage. This means that ESCA will not be a sensitive
technique for measuring trace materials distributed through the
bulk of the sample. On the other band, as we have seen with the
escape depth of the photoelectrons, ESCA is a sensitive surface
technique. Signals have been detected from as little as 0.2%
of a monolayer of heavy metals which amounts to about 10^{12}
atoms, i.e about in the picogram range.

This surface sensitivity can be enhanced by approximately
one order of magnitude at low angles of electron emission, as
measured from the surface plane /3/5/. We shall return to
this when concentration profiles are discussed. The Figure 13
shows the enhancement of surface sensitivity at a small angle of
emission. This sensitivity should render ESCA a surface
analytical tool with certain capabilities for determining the
composition and electronic structure of both atomically clean
surfaces and such surfaces in interaction with gas-phase species.

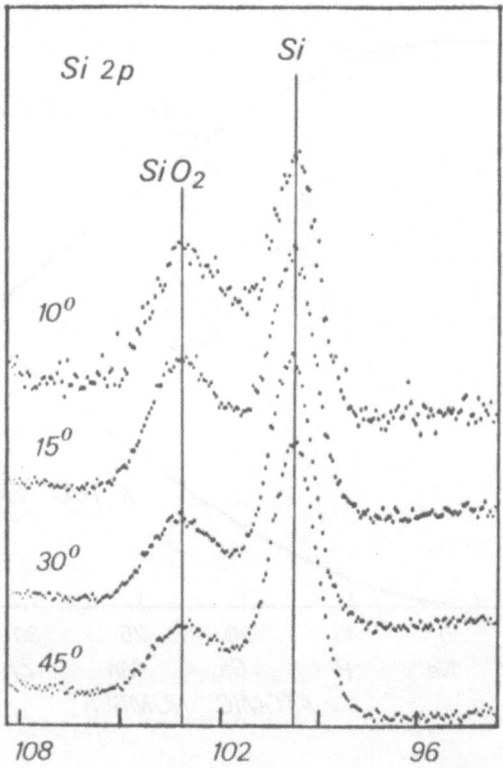

Fig. 13 : Si 2p spectra from an Si specimen with a
surface oxide layer at different angles of
electron emission relative to the surface (27).

We can admit that the AES sensitivity is nearly identical.
However, because of the variations of fluorescence yield (Fig.14)
the sensitivity of AES is very high for the lighter elements.

A comparison between ESCA and AES concerning the
sensitivity can be made on the following points (6) :

- speed :

Under the conventional instrumental condition now used for
each technique, AES is far superior in terms of the speed with
which a signal can be detected (high total count rate, therefore
a given signal-to-noise ratio in a short time). It is possible, for
instance, to obtain a complete spectrum in 0.1 second. This,
in the situation where speed is important, lower detection limits
are likely with AES.

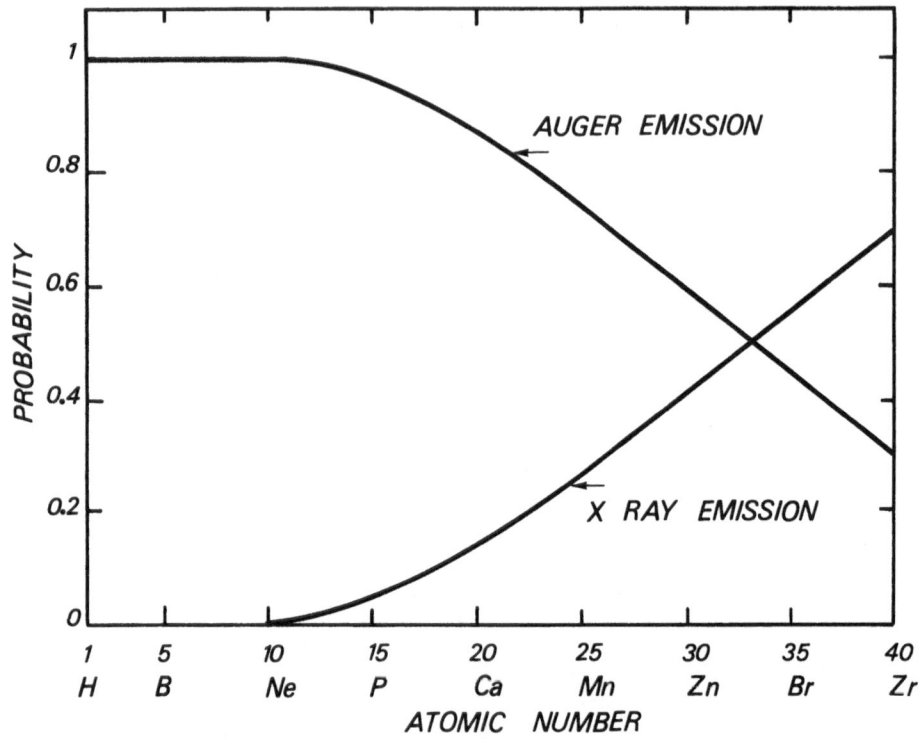

Fig. 14 : Relative X-ray and Auger emissions versus
atomic number.

When considering total signal strengh, two practical factors
play a significant role : electron analyser design and flux of the
primary beam. AES has been developed as a tool for rapid
element identification : thus, electron analysers are generally
set for high sensitivity, but comparatively low resolution (CMA
set ~ 5% transmission). ESCA, however, has been particularly
provided for extracting information form chemical shifts (~ 0.1%
transmission only, but high resolution).

As for the primary beam, a typical XPS AlKα photon flux
is equivalent to about 10^{-8} A, that is to say three orders of
magnitude lower than the typical electron gun current of about
10 μA.

- Analysed area :

We remember that the X-ray beam is diffuse (~ 1 cm^2)

and that fine focusing to a small spot is possible with electrons beams, giving AES, in particular, its fast scanning ability.

- Beam-induced surface effects :

In general, ESCA might be expected to be milder in its effects owing to the lower beam intensity, the unfocused beam and the longer stopping length for X-rays. We shall return to these effects later.

b) Quantitative analysis :

In practical applications of AES and ESCA, semiquantitative determination of the elemental composition is often sufficient. However, it is more and more important to obtain accurate quantitative information.

With a few assumptions, the general expression for the number of photoelectrons or Auger electrons can be written as :

$$dN_i = k I_o \rho_i \tau_{i,j} T \exp\left(-\frac{z}{\lambda}\right) dz$$

where I_o is the excitation flux density

$\tau_{i,j}$ is the ionization cross-section of the core level j

ρ_i is the concentration of the element i at a depth z from the surface

λ is the mean free path for emitted electrons

T is the transmission of the analyser.

In the factor k are included :

- in AES : $(1 - \omega)$, Auger yield after primary ionization and $(1 + R)$, where R is the backscattering factor.

- in ESCA : a factor which takes into account intensity loss due to retardation.

This equation can be integrated in different cases : infinite thickness with constant ρ, thickness t, ... It is difficult to know these most factors to the desired degree of accuracy. Some further complications are surface roughness, diffraction effects, contamination, ... For instance, it is apparent that surface contamination can have a significant effect on measured elemental ratios when the differences in kinetic energies of the measured electrons are large. In addition, the involvement of valence levels in many Auger transitions may impart a strong chemical dependence to Auger intensities instead of just an element concentration one.

At present, several comparative methods are used for quantitative analysis :

- we can compare the Auger spectra from the specimen with that of a standard with a known concentration of the element of interest. The concentration ρ_i^t of element i in the test specimen can be related to that in the standard ρ_i^s by :

$$\frac{\rho_i^t}{\rho_i^s} = \frac{N_i^t \cdot \lambda^s \cdot 1+R^s}{N_i^s \cdot \lambda^t \cdot 1+R^t}$$

in assuming three-dimensional homogeneity of the chemical composition for the region in which the escape probability has significant value.

But when the composition of the standard is not similar to that of the test specimen, careful consideration must be given to the influence of matrix on both the backscattering factor /7/ and the escape depth.

- We can also compare the Auger or ESCA spectrum from an element i in the specimen with that of an other element s of the specimen, choosen as standard. An elemental sensitivity factor can be defined :

$$\frac{N_i}{N_s} = \frac{\rho_i}{\rho_s} K_{s,i}$$

The relative concentration of element i can be expressed as:

$$C_i = \frac{\rho_i}{\sum_n \rho_n} = \frac{\rho_i/\rho_s}{\sum_n \rho_n/\rho_s} = \frac{K_{s,i}^{-1} N_i/N_s}{\sum_n K_{s,n}^{-1} N_n/N_s}$$

But it is evident that these sensitivity factors depend on the nature of the matrix $(R, \lambda, \ldots)$ and hence this method is normally only semi quantitative $(\sim 30\%)$. It is convenient, however, and very useful, for the samples of a same family (Fig. 15 (8)).

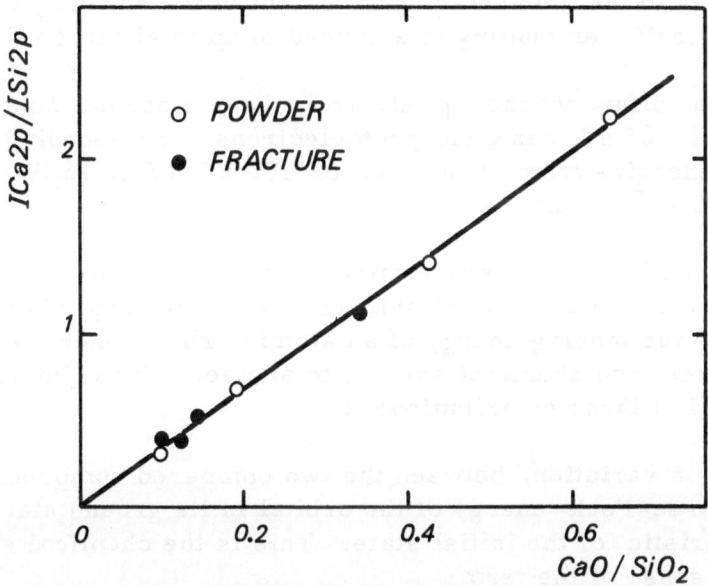

Fig. 15 : Linear correlation between intensity ratio and composition for some glasses (8).

VI. CHEMICAL INFORMATIONS

The greatest success of ESCA and to a lesser degree of AES, has been, in addition to its sensitivity to almost all elements in the periodic table and its ability to study surfaces, its capacity to detect the chemical states by :

a) The chemical shift, that is the changes of binding energies of atomic electrons due to the presence of other atoms when they are chemically bonded together

b) The existence of satellite lines (shake-up, plasmons, ...)

c) The possibility of obtaining a good representation of the valence bands.

a) Chemical shift :

Even though the electrons come from the inner core shells, their binding energies are affected measurably by the state of chemical combination. Hence, the binding energy exhibits a sensitivity to the molecular environment of the emitting atom, a chemical shift, amounting to a spread of up to about 15 eV.

Thus sulfur, whose 2p electrons have a normal binding energy of 165 eV can yield photoelectrons corresponding to binding energies from about 160 to 170 eV (Fig. 16 (9)) and even 175 eV[(9) and (9 bis)].

In practice, the interpretation of chemical shifts is often complicated. The chemical shift in ESCA corresponds to the change in the binding energy of an atomic orbital when the atom passes from one chemical species to another. It is generally composed of three contributions :

- a variation, between the two compared compounds, of the Hartree-Fock energy of the orbital in its ground state $\Delta\varepsilon^{HF}$, characteristic of the initial state. This is the chemical shift in the real sense of the term.

- a relaxation contribution ΔE_R which corresponds to the different response of the solids to the creation of a vacancy.

- the possible displacement ΔE_{ref} of the Fermi level as compared with an absolute energy scale (potential at infinity) in the two materials being compared.

One may then write :

$$\Delta E_B = \Delta\varepsilon^{HF} + \Delta E_R + \Delta E_{ref}$$

In the past, the last two terms were often neglected and the shift measured by ESCA was confused with the chemical shift.

An approximate interpretation gave the simple, useful so called "potential model" :

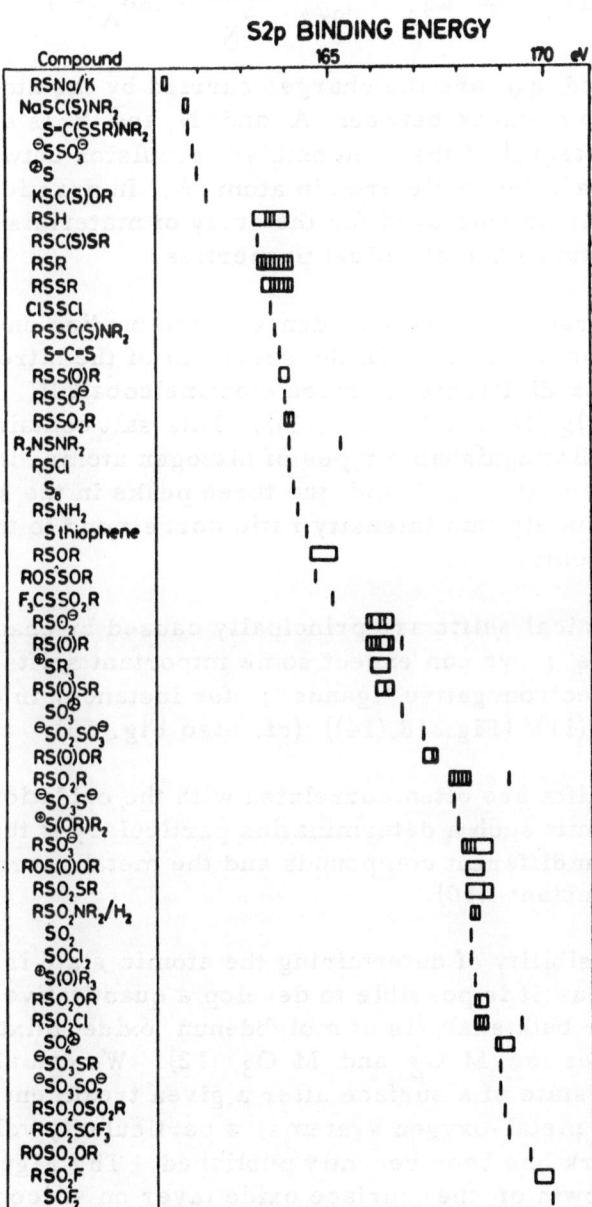

Fig. 16 : S2p electron binding energies for sulfur
compounds. R signifies alkyl or aryl
carbon (9).

$$\Delta E_B \neq \Delta \epsilon^{HF} \neq kq_A + \sum_{N \neq A} \frac{q_N}{r_N} = kq_A + V$$

where q_A and q_N are the charges carried by the atoms A and N, r_N is the distance between A and N and k is a constant equal to an integral of the Coulombian repulsion between a core electron and a valence electron in atom A. In practice, this model can only be employed for the study of materials of closely related structures and chemical properties.

An illustration of the dependence of the binding energy on the chemical structure is seen in the spectrum of the nitrogen 1s levels of trans-dinitrobis (ethylen diamine)cobalt III nitrate $[Co(en)_2(NO_2)_2]NO_3$ (Fig. 17 (13)). This salt contains three structurally distinguishable types of nitrogen atoms, in an abundance ratio of $4/2/1$ and the three peaks in the spectrum with approximately this intensity ratio correspond to these types of nitrogen atoms.

The chemical shifts are principally caused by changes in the atomic charge ; we can expect some important shifts in atoms bound with electronegative ligands ; for instance, in oxides or fluorides (11') (Fig. 18 (14)) (cf. also Fig. 13).

These shifts are often correlated with the oxidation state and often permit such a determination particulary if the ligands are similar in different compounds and the metal coordination number is constant (10).

This possibility of determining the atomic state is of course valuable. Thus it is possible to develop a quantitative analytical procedure for bulk analysis of molybdenun oxide mixtures by using 3d lines for $M O_2$ and $M O_3$ (12). We can also study the chemical state of a surface after a given treatment. For instance, for metal-oxygen systems, a particularly voluminous amount of work has been recently published. The Figure 19 (28) shows the growth of the surface oxide layer on silicon.

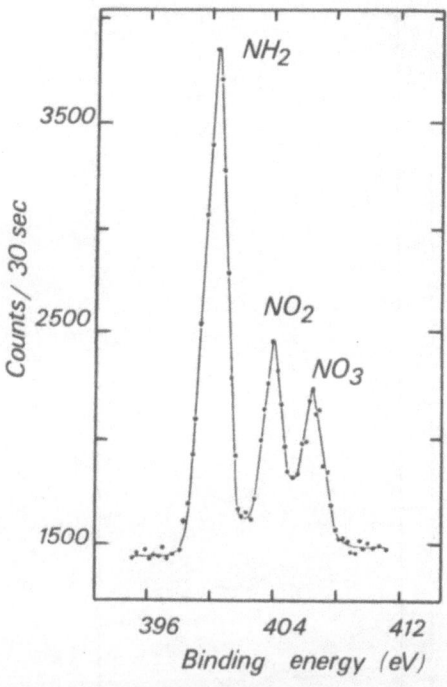

Fig. 17 : Nitrogen 1s
spectrum $[Co(en)_2(NO_2)_2]NO_3$
(13).

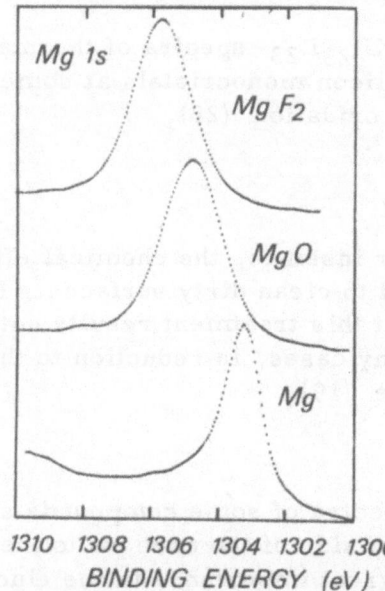

Fig. 18 : Magnesium 1s
spectra of metal, oxide and
fluoride (14).

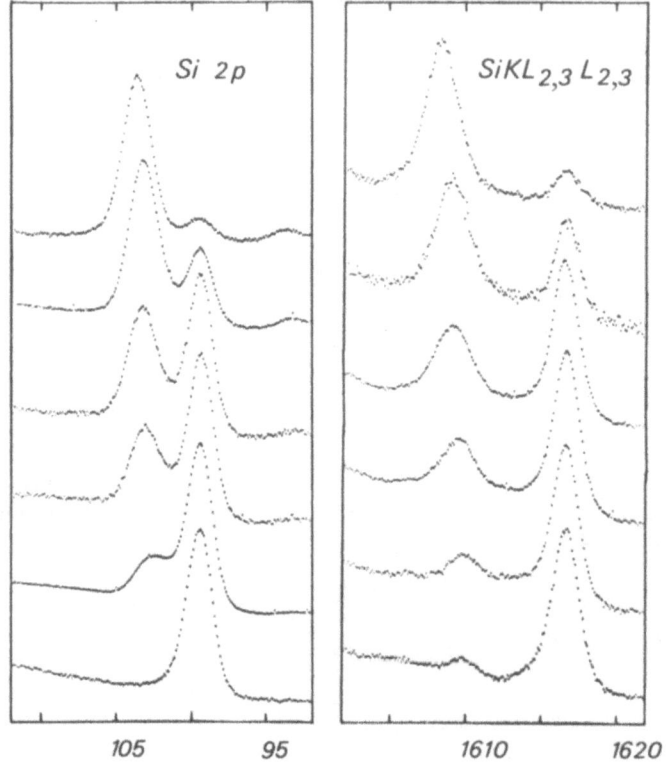

Fig. 19 : Si 2p and Auger $KL_{23}L_{23}$ spectra of thermally
grown oxides on silicon monocristals at some
different stages of oxidation (28).

It is also possible to study, for instance, the chemical effects
of ion bombardment often employed to clean dirty surfaces. For
many oxides, ESCA has shown that this treatment results not
only in sputtering, but also, in many cases, in reduction to the
corresponding metal or lower oxide (19).

b) Satellites lines :

The X-ray photoelectron spectra of some compounds show
satellites on the high binding energy side of the core photolines.
The study of these satellites is of great importance in the elucida-
tion of the nature of chemical bonding.

These satellites have been attributed to two-electron "shake-up" processes in which valence shell excitations take place in parallel with core electron ejection. In addition, if there are unpaired valence electrons in a compound more than one final state may occur because the magnitude of the exchange interaction of the unpaired valence electrons with the unpaired electron in the core shell (after photoelectron ejection) depends on whether or not the latter has its spin up or down. If a core s electron is ejected, two final states are formed and a doublet spectrum is expected (10) (29).

These splittings and shake-up lines are still not well-understood in detail, but one may look forward to their correlation with chemical structure in the near future. An illustration is presented, for instance, in the Figure 20 (14).

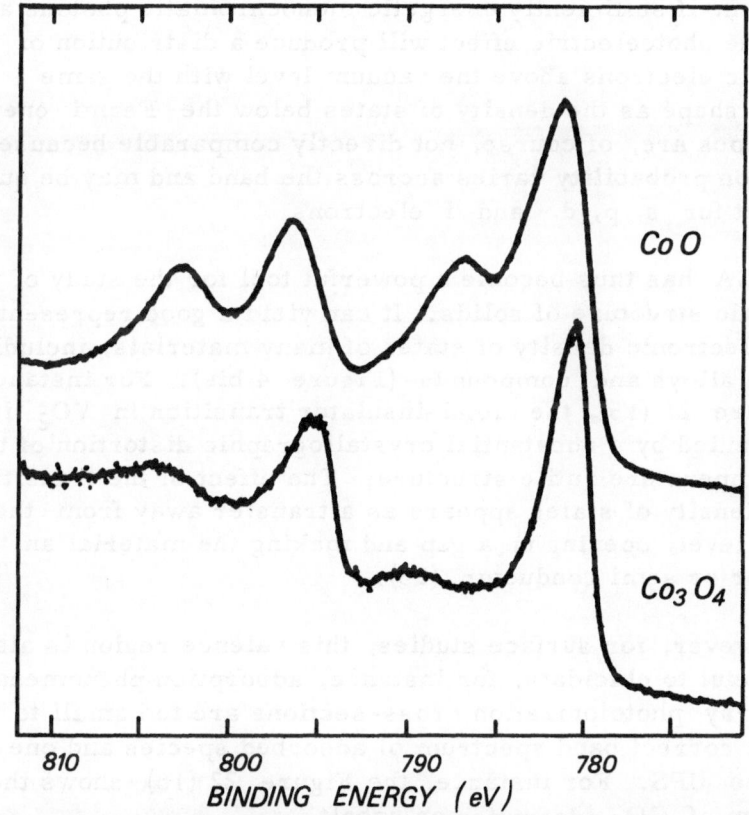

Fig. 20 : Cobalt 2p spectra of CoO and Co$_3$O$_4$ showing presence of shake-up satellite only on CoO (14).

This Figure 20 shows the easy distinction between the two cobalt oxides by the existence of satellites only on CoO peaks.

c) Valence bands :

The lower binding energies of valence-shell orbitals make them accessible to low-energy photons in the ultraviolet region. However, the use of very high energy photons, i.e X-rays, is interesting because the final state energy of the ejected electron is so high that this state can be treated as a continuum state, so that a direct image of the occupied density of states is obtained.

In addition, the range of energetic electrons in solids makes the experiment less surface sensitive and one can also observe deeper electronic states.

Then, if sufficiently energetic monochromatic photons are used, the photoelectric effect will produce a distribution of energetic electrons above the vacuum level with the same general shape as the density of states below the Fermi energy. The shapes are, of course, not directly comparable because the transition probability varies accross the band and may be quite d different for s, p, d, and f electrons.

ESCA has thus become a powerful tool for the study of the electronic structure of solids. It can yield a good representation of the electronic density of states of many materials, including metals, alloys and compounds (Figure 4 bis). For instance, in Figure 21 (15), the metal-insulator transition in VO_2 is accompanied by a substantial crystallographic distortion of the high-temperature rutile structure. The effect of the transition on the density of states appears as a transfer away from the Fermi level, opening up a gap and making the material an insulator or semi conductor (15).

However, for surface studies, this valence region is also very useful to elucidate, for instance, adsorption phenomena. But X-ray photoionization cross-sections are too small to obtain a correct band spectrum of adsorbed species and one must use UPS. For instance, the Figure 22 (16) shows the spectrum of CO_2 adsorption on cobalt.

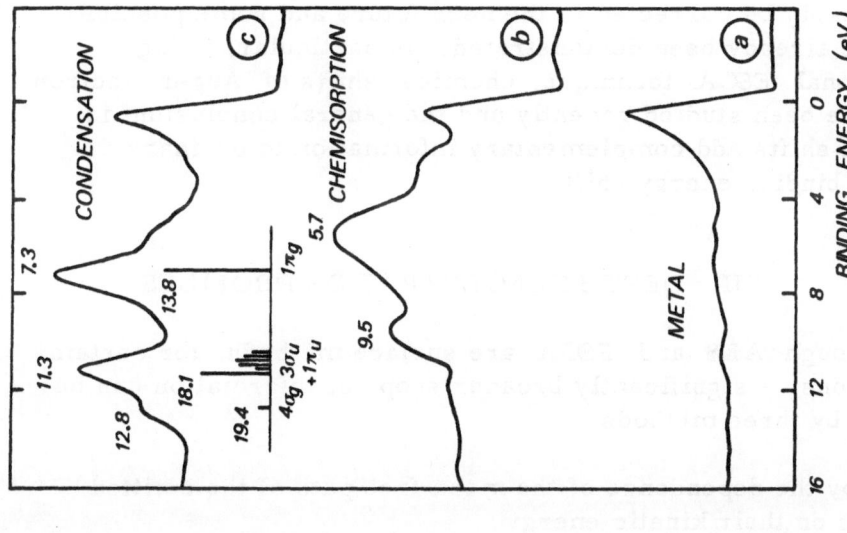

Fig. 22 : UPS spectra (He II) of clean Co (a) and CO₂ adsorption on Co at 300 K (b) and 100 K (c) (16).

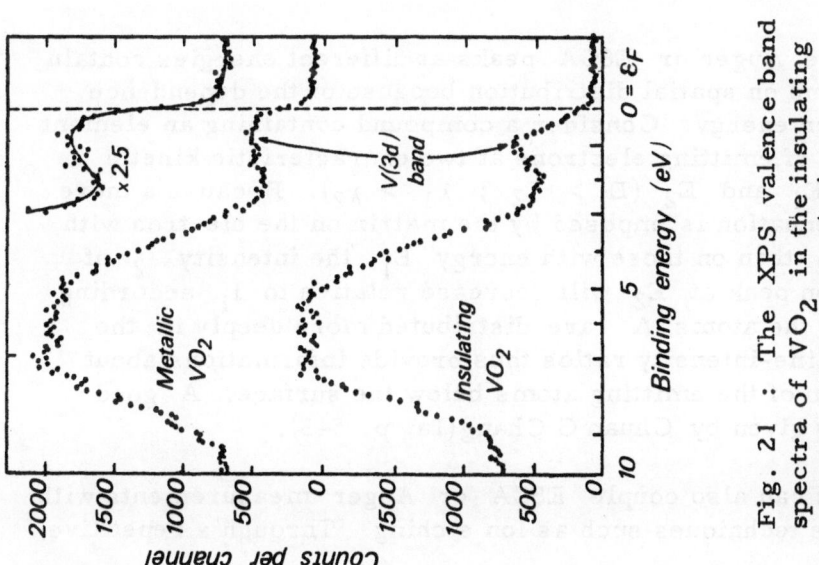

Fig. 21 : The XPS valence band spectra of VO₂ in the insulating and metallic states (15).

In AES, chemical information is often poorer. The chemical shifts in AES are more complex than for ESCA since three energy levels are involved, two of which may be valence. In practice, AES, when applied to surfaces, has been used in the majority of cases purely for atomic identification. This situation will probably be corrected in the near future and some possibilities have already been demonstrated. In particular, using conventional ESCA technique, chemical shifts of Auger electron lines have been studied recently and the general conclusion is that such shifts add complementary information to ordinary electron binding energy shifts.

VII.   DEPTH CONCENTRATION PROFILES

Although AES and ESCA are surface methods, for certain applications, a significantly broader scope of information can be obtained by three methods :

a) by the dependence of the mean free path of the emitted electrons on their kinetic energy

b) by careful sputtering whilst monitoring Auger or ESCA lines

c) by means of angular distribution measurements.

a) The Auger or ESCA peaks at different energies contain informations on spatial distribution because of the dependence of MFP on energy. Consider a compound containing an element A capable of emitting electrons at two characteristic kinetic energies E_1 and E_2 ($E_1 > E_2$; $\lambda_1 > \lambda_2$). Because a more strong attenuation is imposed by the matrix on the electron with energy E_2 than on those with energy E_1, the intensity I_2 of the electron peak at E_2 will decrease relative to I_1 according to whether the atoms A are distributed more deeply in the matrix. Line-intensity ratios thus provide informations about the position of the emitting atoms below the surface. A good example is given by Chuan C Chang (1a, p. 545).

b) We can also couple ESCA or Auger measurements with destructive techniques such as ion etching. Through a repetitive,

sequential process of measurement followed by ion etching, depth
profiling data can be obtained. For instance, the Figure 23 (17)
shows the characterization of passive films formed on stainless
steel (26 % Cr) in an aqueous NaCl solution (3.5 % at p_H 2.5).

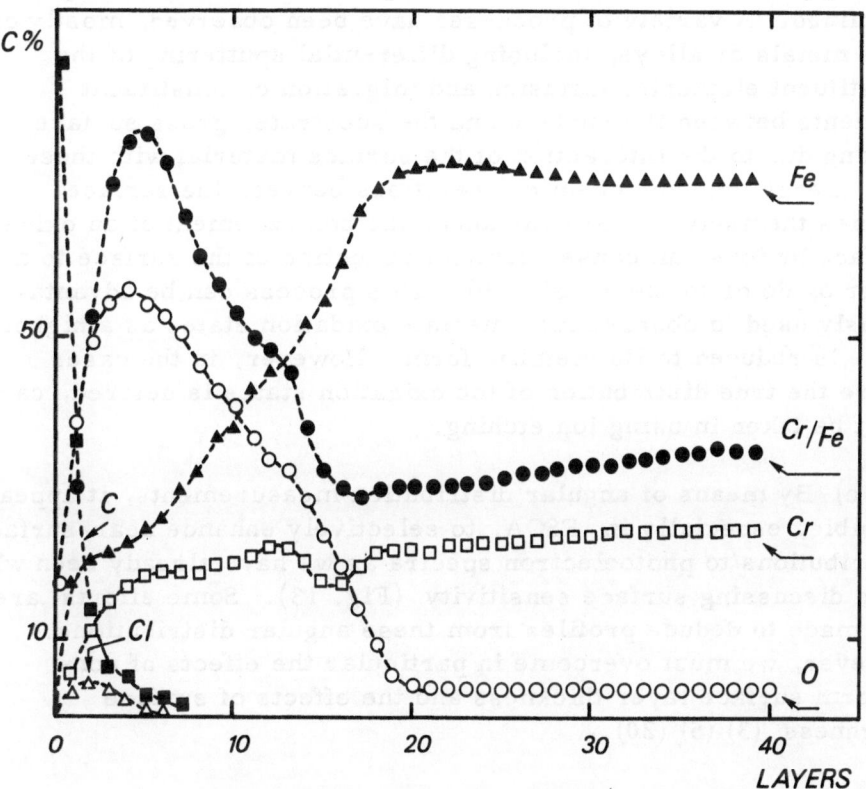

Fig. 23 : Calculated atomic fractions C of Fe, Cr, O,
          C and Cl in the passive film. The peak-to-
          peak height of the Auger lines are treated
          in order to yield the atomic fraction of the
          various elements present in a given subsur-
          face layer (17).

An advantage of ESCA against AES is the greater facility
to obtain chemical information (18). However, in ESCA, rather
large diameter ion beams with uniform current density become
a necessity because of difficulty in focusing X-rays onto small
target areas.

But the use of ion etching is extremely difficult. First, it is difficult to measure precisely the amount of material removed, and, above all, caution should be exercised in chemical interpretation of ESCA data following ion bombardment since the ion etching process itself can cause changes in the surface being examined. A variety of processes have been observed, mostly on pure metals or alloys, including differential sputtering of the constituent elements, diffusion and migration of constituent elements between the surface and the substrate, gross surface heating due to the interaction of the surface material with these high energy ions and chemical reactions between the surface species themselves. For instance, the bombardment of an oxide surface by ions can cause chemical reduction of the surface to a lower oxide or to the metal (19). This process can be advantageously used to observe intermediate oxidation states as a higher oxide is reduced to its metallic form. However, in the cases where the true distribution of the oxidation states is desired, care must be taken in using ion etching.

c) By means of angular distribution measurements, it appears possible, especially in ESCA, to selectively enhance near-surface contributions to photoelectron spectra as we have already seen when when discussing surface sensitivity (Fig. 13). Some efforts are now made to deduce profiles from these angular distributions. However, we must overcome in particular the effects of non-uniform surface layer thickness and the effects of surface roughness (3) (5) (20).

VIII. BEAM-INDUCED EFFECTS

Composition or **bonding** changes arising from the photon or electron bombardment are of concern in both ESCA and AES, like many other techniques (6).

Electron beams (AES) seem to cause more problems than photon beams (ESCA), through the differences may be exaggerated because there are more data available for AES and because it is easier to look for beam effects in this technique since it is possible to scan an electron beam. If ESCA seems to be a more gentle method than AES, this is principaly because the power delivered to the analysis volume in ESCA is much lower than in classical AES (30).

It is possible to distinguish the following effects :

- thermal migration :

Heating will depend on the sample, beam current, etc...

- beam-assisted diffusion :

An example of this effect is the difficulty associated with the measurement of sodium within a glassy silica matrix using AES (21). The Na ions have sufficient mobility at room temperature to migrate readily under the influence of electric fields produced by electron beam irradiation upon the glass surface (Fig. 24) (21).

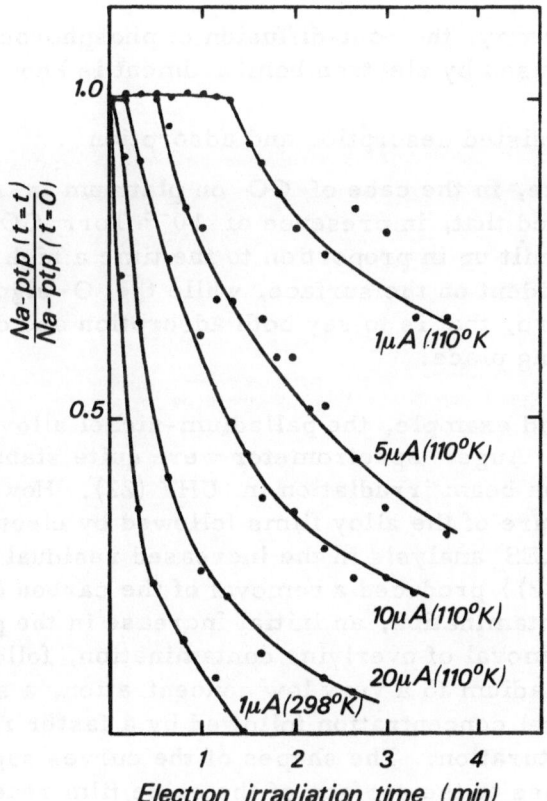

Fig. 24 : The decay of the Na peak-to-peak height, normalized to the peak-to-peak height at zero time
as a function of the time of electron irradiation (21).

The measurements were carried out on soda-lime silicate glasses of commercial composition (SiO_2 74 %, Na_2O 13 %, Ca O 12 %).

At ∼ 1 μA of beam current and with the sample at room te temperature, the Na signal decreased markedly within a few seconds. With the sample mounted on a stage cooled with liquid nitrogen, it was found that the Na peak remained essentialy unchanged and could be repeatedly measured for a period of 80s using a beam current of ∼ 1 μA. Thereafter the signal level decreased slowly, reaching about half its initial level after a number of minutes. At higher current levels, the signal remained constant for a much shorter period and then decreased with time as can be seen in the curves of the Figure 24.

In the same way, the out-diffusion of phosphorous from doped SiO_2, caused by electron bombardment is known.

- beam-assisted desorption and adsorption :

For instance, in the case of CO on platinum in AES, it was demonstrated that, in presence of 10^{-8} Torr CO, thick carbon layers built up in proportion to the time a high beam current was incident on the surface, while the O signal decreased to zero, that is to say both adsorption and decomposition were taking place.

As an second example, the palladium-nickel alloy films deposited in the Auger spectrometer were quite stable under primary electron beam irradiation in UHV (22). However, a small air exposure of the alloy films followed by electron irradiation in the AES analysis in the increased residual gas pressure (Fig. 25 (22)) produced a removal of the carbon due to atmospheric contamination, an initial increase in the palladium signal due to removal of overlying contamination, followed by a decrease in palladium to a very low concentration, a show initial increase in nickel concentration followed by a faster rise which then reaches saturation. The shapes of the curves suggest that nickel ions diffuse to the surface of the oxide film reacting with available oxygen. The reaction is probably accelerated by the formation of active oxygen from adsorbed water under the influence of the electron beam (22).

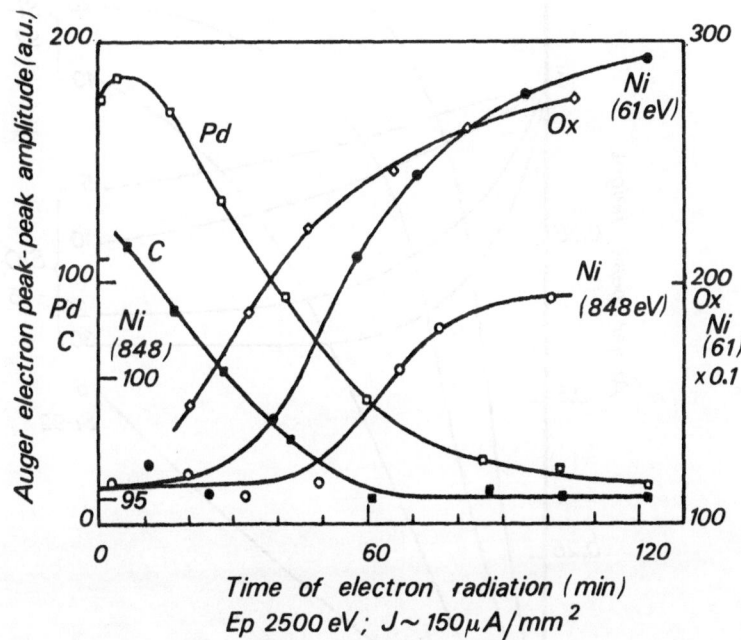

Fig. 25 : Effect of primary electron beam irradiation
          on the surface composition of an air-exposed
          palladium-nickel alloy film (22).

- substrate decomposition :

Dissociation or break-up of molecules under electron beam
bombardment does occur even at surfaces of rather stable insulators such as SiO_2 or Al_2O_3. For instance, the primary effect
of the electron beam on the surface is the dissociation of SiO_2
into elemental silicon and oxygen. Under prolonged electron
bombardment, the oxygen is desorbed from the surface leaving
the surface enriched in elemental silicon (Fig. 26 (23)).

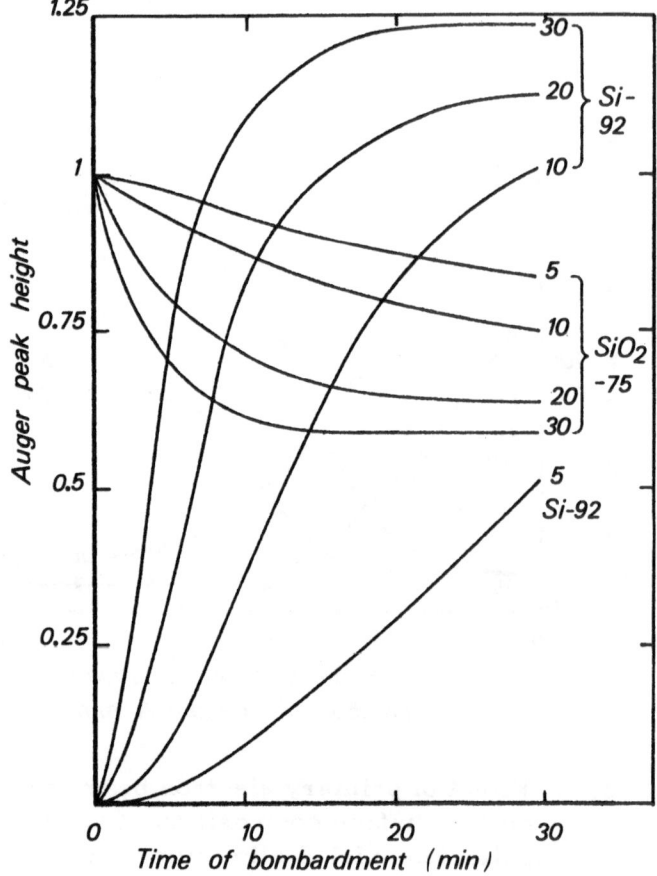

Fig. 26 : Variations in the Si-92 eV and SiO$_2$-75 eV
Auger peak heights with time of bombardment.
The number beside each curve gives the pri-
mary electron current in μA (1 μA ≃ 5mA.cm^{-2})
- Ep = 3 keV (23).

IX. CONCLUSION

A review of applications of both techniques wanders quite
from this paper. In conclusion, in comparing AES and ESCA,
we can observe that, as ESCA has not yet got the spatial resolution of AES, AES with its small area of analysis and ESCA
with its chemical contribution can act as complementary techniques.

A trend of AES, however, is to borrow from ESCA the high resolution spectrometer to exploit the Auger shifts and to realise significant improvements in using low intensity electron beams in order to avoid the previously discussed beam-effects. For instance, it is possible to obtain Auger spectra without differentiation replacing the classical energy modulation by a brightness modulation of primary beam. This modification permits one to lower the electron beam to about 5.10^{-9} A and to obtain a scanning Auger system with a spot diameter of ~ 1500 Å (24).

REFERENCES

1)   a - Chuan C Chang, Analytical Auger Electron Spectroscopy in "Characterization of Solid Surfaces" ed. by P.F.
Kane and G.B. Larrabee, Plenum Press, 20, 509-575, (1974).

b - D. Lichtman, A Comparison of the Methods of Surface
Analysis and their Applications, Chap. 2.

W.M. Riggs and M.J. Parker, Surface Analysis by X-ray
Photoelectron Spectroscopy, Chap. 4.

A. Joshi, L.E. Davis, P.W. Palmberg, Auger Electron
Spectroscopy, Chap. 5.

in "Methods of Surface Analysis" ed. by A.W. Czanderna,
Elsevier Scientific Publishing Company (1975).

c - G. Stupian, Auger Electron Spectrometry, Chap. 3,
Vol. I.

W.G. Proctor, X-ray Photoelectron Spectrometry, Chap. 19,
Vol. II.

in "Systematic Materials Analysis" ed. by J.H. Richardson
and R.V. Peterson, Acad. Press (1974).

d - K. Siegbahn, Jl. of Electron Spectroscopy, 5, 3-97,
(1974).

e - T.A. Carlson, Photoelectron and Auger Spectroscopy,
Plenum Press, (1975).

2)   Auger Scanning Electron Microscope JAMP-3, JEOL.

3)   R.M. Friedman, Silicates Industriéls, 9, 247, (1974).

4)   W.J. Carter, G.K. Sweitzer, T.A. Carlson, Jl. of
Electron Spectroscopy, 5, 827-835, (1974).

5)   W.A. Fraser, J.V. Florio, W.N. Delgass, W.D. Robert-
son, Surface Science, 36, 661-674, (1973).

6)   C.R. Brundle, Jl. Vac. Sci. Technol., 11, 1-212, (1974).

7)   M.L. Tarng, G.K. Wehner, Jl. Appl. Phys., 44, 1534,
(1973).

8) D. Brion, J. Escard, Jl. de Microscopie et de Spectros-
    copie Electroniques, 1 , 2-227, (1976).

9) B.J. Lindberg, K. Hamrin, G. Johansson, U. Gelius,
    A. Fahlman, C. Nordling, K. Siegbahn, Physica Scripta,
    1, 286-298, (1970).

9bis) K. Siegbahn, C. Nordling, A. Fahlman, R. Nordberg,
    K. Hamrin, J. Hedman, G. Johansson, T. Bergmark,
    S.E. Karlsson, I. Lindgren, B. Lindberg, Electron
    Spectroscopy for Chemical Analysis - Atomic, Molecular
    and Solid State Structure Studies by Means of Electron
    Spectroscopy, ed. by Almqvist and Wiksells Boktryekeri
    AB, Stockholm, Sweden, (1967), Nova Acta Regie Soc. Sci.,
    Upsaliensis, Ser. IV, 20 , (1967).

10) W.L. Jolly, Coord. Chem. Rev., 13, 47, (1974).

11) K. Hamrin, G. Johansson, U. Gelius, C. Nordling,
    K. Siegbahn, Physica Scripta, 1, 277-280, (1970).

12) W.E. Swartz Jr., D.M. Hercules, Analytical Chemistry,
    43 , 13-1774, (1971).

13) D.N. Hendrickson, J.M. Hollander, W.L. Jolly, Inorg.
    Chem., 8 , 2642, (1969).

14) G. Hollinger, Y. Jugnet, P. Pertosa, L. Porte, Tran
    Minh Duc, Unpublished results.

15) G.K. Wertheim, Jl. Franklin Inst. (USA), 298 , 4-289,
    (1974).

16) Y. Jugnet, G. Hollinger, P. Pertosa, L. Porte, Tran
    Minh Duc, Jl. de Microscopie et de Spectroscopie Elec-
    troniques, 1, 2-187, (1976).

17) R. Rondot, F. Pons, J. Le Hericy, M. da Cunha Belo,
    J.P. Langeron, Colloque de Physique et Chimie des
    Surfaces, Brest, France, (1975).

18)   C.R. Brundle, Jl. of Electron Spectroscopy, 5 , 291-319,
       (1974).

19)   a - K.S. Kim, W.E. Baitinger, J.W. Amy, N. Winograd,
       Jl. of Electron Spectroscopy, 5 , 351-367, (1974).

       b - K.S. Kim, W.E. Baitinger, N. Winograd, Surface
       Science, 55 , 285-290, (1976).

20)   C.S. Fadley, R.J. Baird, W. Siekhaus, T. Novakov,
       S.Å.L. Bergström, Jl. of Electron Spectroscopy, 4 ,
       93-137, (1974).

21)   C.G. Pantano Jr., D.B. Dove, G.Y. Onoda Jr.,
       Jl. Vac. Sci. Technol., 13 , 1-414,(1976) .

22)   C.T.H. Stoddart, R.L. Moss, D. Pope, Surface Science,
       53 , 241-256, (1975).

23)   Simon Thomas, Jl. Applied Physics, 45, 1-161, (1974).

24)   a - Cl. Le Gressus, D. Massignon, R. Sopizet, Compte-
       Rendus Acad. Sci. Paris, 280 B, 439, (1975).

       b - Cl. Le Gressus, Personal Communication.

25)   J.A. Venables, A.P. Janssen, C.J. Harland, B.A. Joyce,
       Philosophical Magazine, 34 , 495-500, (1976).

26)   P.W. Palmberg, High Vacuum Report, Analytic Technique,
       KG4 Balzers, (1972).

27)   G. Hollinger, Y. Jugnet, P. Pertosa, L. Porte, Tran Minh
       Duc, Analusis (to be published).

28)   G. Hollinger, Y. Jugnet, P. Pertosa, L. Porte, Tran Minh
       Duc, Jl. de Microscopie et de Spectroscopie Electroniques,
       1 , 335-358, (1976).

29)   J.C. Carver, G.K. Schweitzer, T.A. Carlson, Jl. Chem.
       Phys., 57, 2, 973-982, (1972).

30)   J.P. Coad, M. Gettings, J.C. Rivière, Faraday
       Discussions of the Chemical Society, 60, 269-278, (1975).

Part III

IN-DEPTH ANALYSIS

FUNDAMENTAL ASPECTS OF ION MICROANALYSIS [*]

G. BLAISE

Laboratoire de Physique des Solides
Associé au C.N.R.S.
Bât. 510 - 91405 - ORSAY - FRANCE

INTRODUCTION

The interaction of ions of several keV energy with a solid results in the ejection of various secondary particles such as photons
(U.V., X Rays), electrons, neutral or ionized atoms and molecules.
In so far as the nature and the energy of these particles are characteristic of the chemical composition of the irradiated solid, they
can be used for analytical purposes. Together with this analytical
interest, secondary emissions, give rise to a number of fundamental
problems on inelastic atomic collisions and atom-surface interactions which constitute a very active field of research at the present
time.

Our purpose is to present here the secondary ion emission phenomenon and the analytical method associated with it.

Secondary ion emission is composed of particles ejected in an
ionized state during the sputtering of a solid. It gave rise, about
fifteen years ago, to the Solid Ion Microanalysis method |1-3|. As
this method requires the use of mass spectrometric techniques it was
further popularized under the name of Secondary Ion Mass Spectrometry
(SIMS).

By penetrating into the crystal the primary ion suffers collisions with lattice atoms located in the first atomic layers and sets
them into motion. In turn, these atoms initiate collisions cascades
which can end up by the ejection of several particles. This is the

---

[*] Lectures presented at the Summer School on "Material Characterization using ion beams", Aléria, Corsica, August 29th September
12th, 1976. Sponsored by Nato.

sputtering phenomenon. The fraction of sputtered particles which leave the surface in an ionized state, composes the ion emission.
Therefore, ion emission results from a sputtering process associated
with an ionization process. In so far as there is not a mutual dependence of these two processes, the intensity of any secondary ion species, atom or molecule, can be formally written as :

ion intensity = Sputtering process × Ionization process

This formal expression must be considered as the first approximation
to the theoretical approach of the ion emission phenomenon. It will
be the basis of these lecture series dealing with the fundamental aspects of ion microanalysis. This means that emphasis will be laid on
the study of sputtering and ionization phenomena, in turn in two adjacent sections. However, in order to present the matter in a logical
way, these two major sections will be preceded by a brief review of
the basic instrumental concepts and a detailed survey of ion emission.
To conclude this presentation some applications directly related to the
formalism which has been developped will be given in a fifth section.

Initially G. Slodzian |1-3| defined   Ion Microanalysis as a
method of local chemical and isotopic analysis of solids. This definition expresses the capability of providing compositional information
on microareas ($\sim 1\mu m^2$) but, owing to the erosion of the solid, this
information was considered as   bulk information. In fact, under the
bombarding conditions usually used in Ion Microanalysis, the greater
part of atoms involved in the sputtering process come from the first
3 or 4 top layers. Regarding the ionization processes it is generally agreed that they take place outside the solid, within the first
few Å. Therefore Ion Microanalysis really appears as a method capable
of providing surface analysis. This broad capability for both surface and bulk analysis means that the method occupies a unique position
among the new methods of solid analysis.

I. EXPERIMENTAL CONDITIONS FOR SECONDARY ION MICROANALYSIS

For a detailed description of the various instruments used in secondary ion microanalysis and their respective advantages and limita-
tions the reader is referred to a number of recent reviews |1-17|.

Our purpose here is confined to recalling the analytical capaci-
lities of the method in relation to the basic instrument concepts.
Any secondary ion mass analyser is composed of a primary ion gun, a
secondary ion collection and focalisation system, a mass spectrometer for mass / charge analysis and an ion detection system. Two different instrumental conceptions are given in fig.1.

I-1 Bombarding conditions :

The bombarding conditions must provide a sufficient sputtering
of the material for analysis. In this respect the most important factor is the energy of the bombarding ions rather than their nature.
Below 1 keV the sputtering yield of many materials may become very

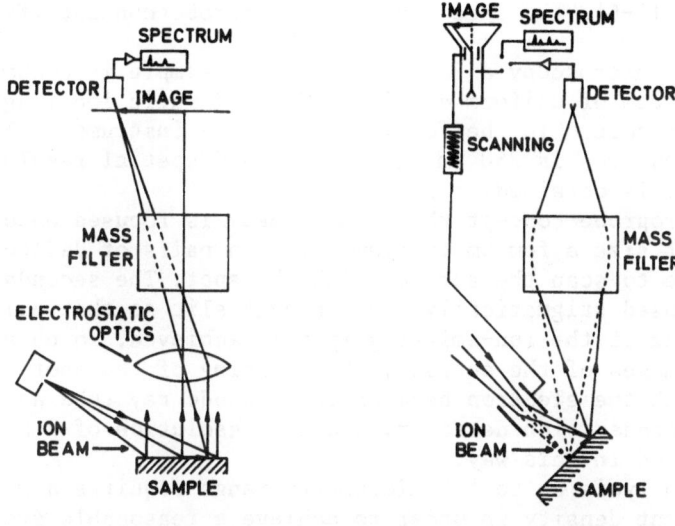

Fig.1 - Instrument concepts utilized in SIMS analysis |9|.
         a- emission microscopy |1-6|
         b- microprobe |7-9|

low, whereas on the contrary there is no advantage to bombarding beyond ∿ 20 keV energy since the sputtering yield changes very little
in this energy range and even tends to decrease at higher energy
|96| . Most of the ion guns in these instruments operate within
the interval 1 to 20 keV.
     The incidence of the bombarding beam on the target is of little
importance provided it does not exceed a critical angle θ ∿ 70° beyond which the sputtering yield falls off (see sect. III.1.2).
     Any ion species can in principle be used for sputtering. How-
ever it is not advisable to use very light ions such as H⁺ or even
He⁺ because they produce a very low sputtering effect. We must distinguish between two types of bombardment:
   - bombardment by ions which do not alter the sample chemically.
These ions are exclusively positive rare gas ions (except for the
possibility of He⁻) and argon is the gas most commonly used.
   - bombardment by ions which alter the sample chemically on the
surface and in depth owing to implantation. Oxygen is the typical
example of this kind of bombardment. It is used as positive ions
(O⁺, O₂⁺) or negative (O⁻).

I-2 Conditions for spatial and in-depth analysis :

     There are two different ways to accomplish spatial analysis with
SIMS instruments |9,15| : the first is based on the concept of emis-

sion microscopy |1-6|, the second on the microprobe concept (fig.1)
|7-10|.

In emission microscopy a large area of the sample is bombarded
with a primary beam of uniform density and a stigmatic ion image of
the surface is produced in the focal plane of the instrument. Typically, the imaged area is 250 µm in diameter and spatial resolution
better than 1 µm is obtained.

In the microprobe concept the primary beam is focused onto a
sample to a spot size a few µm in diameter. Two pairs of deflection
plates allow one to scan the surface with the spot. The secondary
ion beam is focused stigmatically onto an exit slit so that point-
to – point imaging of the ion-emitting spot is achieved. To obtain
secondary ion images of the surface, the scanning of the spot is
synchronized with the electron beam of the cathode ray tube associated with the secondary ion detector. Spatial resolution of 2 µm or
larger is obtained in this way.

In-depth analysis up to the micrometre range requires a primary
beam of sufficient density in order to achieve a reasonable erosion
time : densities within the interval 10 µA/cm^2 to 10^3 µA/cm^2 are u-
sually used. With a mean density of 100 µA/cm^2 and primary ions of
a few keV energy, the erosion rate of the solid is typically of the
order of one atomic layer/second (table I). The removal of a certain
amount of solid depends only on the total primary ion dose being received on the surface and not on the way in which this dose is distributed as a function of time. This means that in-depth analysis in the
monolayer range might also be achieved with the primary densities indicated above in spite of the rapid erosion rate. For providing such
analysis a first method consists of bombarding the sample for very
short periods of time with an intense ion beam |18|. During each
bombardment, all the secondary ions are simultaneously detected with
a Mattauch – Herzog mass spectrometer. Another method would consist
of counting secondary ions during very short periods of time ($\sim$10^{-3}s
for instance) and directing them successively into multichannel plates and storing them in memory. In fact, the most popular method of
surface analysis in the monolayer range consists of bombarding the
sample with very low ion densities $\sim$10^{-3} to 10^{-1}µA/cm^2 (table I) so
that the material consumption can be less than 10^{-4} monolayer per se-
cond |19|. As the recording time of a complete spectrum is about 400s,
a material consumption equivalent to about 1% of a monolayer is removed
during each recording time. Therefore, one can consider that each
spectrum is characteristic of a given surface composition. This is
the reason why the method is termed "static" SIMS. For a better sensitivity of the procedure, a large area $\sim$ 0.1cm^2 is bombarded. Spatial
analysis is not achieved by this method for the moment.

Although the basic phenomenon of secondary ion emission is exactly the same in the various instruments one can expect different experimental procedures, particularly, in relation with the primary beam
density used. Two experimental procedures have to be distinguished :
a "dynamic" procedure |6,10| and a "static" procedure |4|. Each of

Table I

	Static conditions|4|	Dynamic conditions|6,10|
residual pressure torr	10^{-10}	10^{-7}
positive or negative ion energy keV	~ 3	6 to 15
range of the primary beam density $\mu A/cm^2$	10^{-3} to 10^{-1}	$10 - 10^3$
analysed area cm^2	10^{-1}	$5\ 10^{-4}$
erosion rate atomic layer/sec	10^{-5} to 10^{-3}	10^{-1} to 10

these corresponds to experimental conditions mentioned in either column of table I. Both procedures will be discussed and compared in
section II.2.1. devoted to ion emission induced by surface reactions.
Historically, these procedures were employed on instruments of different design but, in principle, they can be employed on the same instrument.

I-3 Conditions for chemical analysis :

      To achieve chemical analysis secondary ions emitted from the
sample are first collected then focused by electrostatic optics before entering the mass spectrometer. The mass/charge separation is
accomplished with either a magnetic or a quadrupole analyser where
mass resolution usually does not exceed $M/\Delta M = 300$ |1-12|. However,
owing to the presence of an interfering molecular ion spectrum (see
sect. II.1.3.), one requires the use of a high mass resolution analy-
ser |16, 18, 20|.
      From an analytical standpoint the parameter of major importance
is the sensitivity of the instrument. The sputtering of a given volume of material v which contains $No = dv$ atoms, where d stands for the
number of atoms per unit volume, leads to the production of $N(M^+)$
ions of species M^+. The number of ions M^+ collected at the detector
is $n(M^+)$. Then one defines   the transmission of the instrument η by:

$$\eta = \frac{n(M^+)}{N(M^+)} \qquad (1)$$

If the atomic concentration of element M is C_M the number $N(M)$ of
sputtered atoms M, initially located in the volume v is $N(M)$ =
$C_M\ N_o$ (see sect. III.2.1.). To achieve |6| a precision of $p/10^2$ in a
measurment a minimum number $n(M^+) = 10^4/p^2$ of M^+ ions have to be detected (here we suppose neither background at the detector nor spu-

rious effects of any kind) which requires a sputtered volume:

$$v = \frac{10^4}{p^2} \frac{1}{\eta C_M d\tau^+} \qquad (2)$$

where $\tau(M^+) = N(M^+)/N(M)$ is the absolute ionization ratio of M atoms.*
$\tau(M^+)$ is dependent on the characteristics of the element itself and
on the chemical composition of the matrix in which the element is
present (see sect. II.1.2). The ion transmission efficiency η depends
on the particular SIMS instrument for a given ion species. It is also
very sensitive to the initial energy and angular distributions of the
emitted M^+ ion (see sect. II.1.1.). Therefore it is not at all a constant for the various species emitted by the same sample. In instruments which use a quadrupole mass analyser, secondary ions cannot be
accelerated beyond $\sim$50-100 eV before entering the analyser, and the
entrance slit of the quadrupole is very small so that η is low, of
the order of $\sim10^{-5}$ |19|. On the contrary, in instruments of the other
type a strong accelerating electric field ($\sim$ 1000 V/mm) is applied
at the surface of the sample which favors the collection of ions |6|.
In that case, the transmission efficiency may reach high values up to
$\eta \sim 10^{-2}$ to 10^{-1}.

Using equat.2, let us estimate the minimum volume v_m to be sputtered under the following conditions :

$$p = 10 \text{ (precision 10\%)} \; ; \; \eta\tau^+ \sim 10^{-3} \; ; \; d \sim 10^{23} \; \text{at/cm}^3$$

The calculated value is $v_{min} = 10^{-18}/C_M$ in cm^3.

Supposing a concentration $C_M \sim 1$ ppm, $v_{min} \sim 10^{-12}$ cm^3. If the analysis is performed on the area of the viewing field 5 10^{-4} cm^2, supposing
a uniform concentration C_M of element M, the minimum thickness to be
eroded is ~ 0.2 Å . If M is only contained in an inclusion of 1 μ^2 ,
the minimum thickness of inclusion to be sputtered is now $\sim 10^4$ Å.

These numbers demonstrate the capability of SIMS for chemical
analysis.

II. SURVEY OF ION EMISSION

It is convenient to distinguish two types of ion emission :
- the intrinsic ion emission which is composed of atomic species
really present in the sample. Their exclusive observation implies that
neither the chemical nature of the sample is disturbed by the bombarding ions nor that a contamination of the surface is produced by the
active components of the residual atmosphere which surrounds the sample. The former condition requires bombardment by an inert gas (He,

---

* $\tau_p^+ = \eta\tau^+$ is usually called "practical ion yield" |16|.

Ne, Ar ...), whereas the second demands a residual pressure as low
as possible and a sufficiently high primary beam density (see sect.
II.2.4.).
   - the ion emission induced by surface reactions due to the adsorption of active gas components of the residual atmosphere on the surface. Effects due to the chemical activity of the bombarding ions into the sample (oxygen for instance) will also be included within this
type of ion emission.

II-1 Instrinsic ion emission :

II-1-1 Angular and energy distributions of secondary ions

II-1-1-1 Angular distribution

      In the first approximation the angular distribution positive
singly ions emitted from polycrystalline metals appears to be nearly
a cosine-like distribution |21,22|. Therefore, using the notations of
fig.2 the number of ions dN^+ emitted in a solid angle $d\Omega$ can be expressed by a Lambert's law ; namely

$$\frac{dN^+}{d\Omega} = \frac{N\perp^+}{\cos \theta_i} \cdot \cos \phi \qquad (3)$$

where θ_i is the incidence of the bombarding particles
        ϕ is the emission angle of seconday ions
        $N\perp^+$ is the number of ions emitted per steradian normal to the
        surface, under normal bombarding incidence ($\theta_i = 0$).
The factor $\dfrac{1}{\cos \theta_i}$ represents the dependence of the sputtering
yield on the bombarding incidence θ_i (see sect. III.1.2).
      In fact the actual angular distributions are different from one
metal to another and, furthermore, they depend more or less on the
direction of the bombardment. This led some authors |21| to propo-
se empirical expressions such as :

$$\frac{dN^+}{d\Omega} = \frac{N\perp^+}{\cos \theta_i} \cos \phi \ \frac{1- \lambda \cos\alpha}{1- \lambda} \qquad (4)$$

with $\cos \lambda = \cos \theta_i \cos \phi + \sin \theta_i \sin \phi \cos \psi$
λ is a fitted parameter whose modulus is always < 1.
when $\lambda=0$ equat 4 is equivalent to Lambert's law (eq.3),
       $\lambda<0$ corresponds to overcosine distribution,
       $\lambda>0$ to undercosine.
For example, using primary ions of 8 keV energy, Hennequin |21|
found $\lambda \sim 0.3$ to 0.4 for many transition metals such as Ni and Cu and
$\lambda = 0.2$ for Al.
      In fact the deviation from a cosine-like distribution is mainly
due to the high energy ions |21-22| as shown in fig.3; therefore expres-
sion (3) is a good approximation for low energy ions, typically below
30 eV, and can be used advantageously for studying the optical proper-
ties of instruments |6,23|.

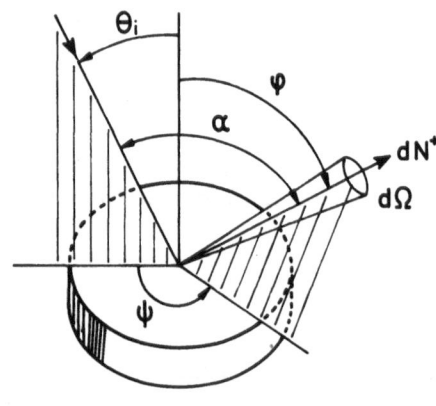

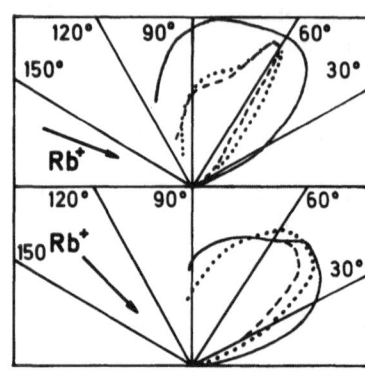

Fig.2   Fig.3

Fig.2 - Angular parameters
Fig.3 - Angular distribution of Cu^+ ions during the bombardment of
    polycrystalline copper targets with 8 keV Rb^+ ions |from
    22|
    ——— 30 eV ; - - - 110 eV ; ..... all ions up to 500 eV.

    Concerning the angular distributions of molecular and multi-
charged ions, it is observed that the former have a cosine-like dis-
tribution rather insensitive to the bombarding incidence |21|, whe-
reas the latter exhibits a strong anisotropy in relation with the
bombarding incidence, the distribution being tilted in the direction
of the primary momentum |21|. This indicates that multicharged ions
are produced by violent collisions of primary ions with target atoms
located in the first atomic layers |24,25|.
    Results on the angular distribution of ions emitted from single
crystals were still very contradictory a few years ago. It is now well
established that they exhibit a strong anisotropy more or less rela-
ted with the spot patterns of neutral atoms |22,26|. For example,
preferred Al^+ ion emission of an (100) Aluminium target bombarded
by 8 keV. K^+ ions is observed along the <100> and <111> crystal axes
(fig.4) |22|. However, the detail of the anisotropy is very complica-
ted : some deviation of the preferred ion emission often exists from
crystals axes (Cu for example |22| and furthermore, the anisotropy
changes with the energy of secondary ions as is illustrated in fig.5.
Bombarding a (100) Ni target with A^+ ions, Bernheim |6| recorded the
Ni^+ ion emission at different energies around a cone at 45° from the
normal to the surface. He observed that maxima corresponding to low
energy ion emission (< 40 eV) in <$\bar{1}$10> like directions became maxima for ions of higher energy (80, 100 eV ...).

NICKEL FACE(100), IONS INCIDENTS A⁺, 9 keV

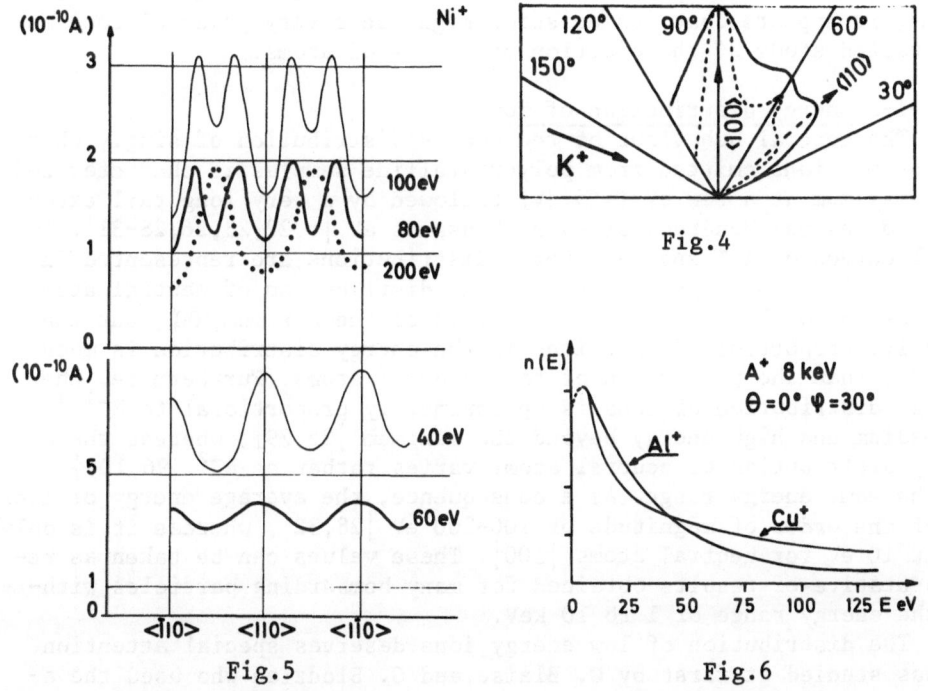

Fig.4

Fig.5   Fig.6

Fig.4 – Angular distribution of Al⁺ ions during the bombardment of
         (100) Aluminium single crystal with 8 keV K⁺ ions |from 22|
         ———— 8 eV ; ------- 100 eV.
Fig.5 – 58 Ni⁺ currents of ions emitted by a (100) face at θ_i = 45°
         and different azimuthal angles and emission energies as re-
         corded with a field free collecting system |from 6|.
Fig.6 – Energy distributions of Al⁺ and Cu⁺ ions from polycrystalli-
         ne samples bombarded with 8 keV A⁺ ions |from 28|.

    The angular distribution of molecular ions is not well known.
However, some preliminary results obtained by Bernheim et al. clear-
ly show the anisotropy is more pronounced than for atomic ions |27|.
    These results show that the angular distribution of ions cannot
be explained just by considering the anisotropy due to focusing collision sequences; elastic scattering processes of the emitted ions
on surface atoms must also be taken into account |22|.
    It will be mentioned in sect. IV.3 that ionization of atoms ta-
kes place outside the surface of a solid. Therefore, the angular distribution of ions is imposed, to a large extent, by collisional pro-

cesses which precede ionization. As it is relatively easier to analyse angular and energy distributions of ions rather than of neutral
atoms, it appears that ion emission might be a very powerful tool for
a detailed study of the ejection processes of atoms.

II-1-1-2 Energy distribution of ions :
     The general behaviour of the energy distribution of singly char-
ged atomic ions emitted from polycrystalline samples is characterized
by a maximum at a few eV ($\sim$ 5 eV) followed by a very long tail exten-
ding to several hundreds or even thousands eV |6,22,23,26,28-31|. Ty-
pical curves of Al$^+$ and Cu$^+$ energy distributions are represented in
fig.6 |28|. An analogy with the energy distribution of neutral atoms
is observed with regard to the position of the maximum|100| , but the
relative proportion of fast ions in the energy distribution is much
greater than the proportion of fast neutral atoms. Furthermore, the
energy distribution of ions is approximately proportional to $E^{-1.3}$
at medium and high energy beyond the maximum |26,29|, whereas the e-
nergy distribution of neutral atoms varies rather as E^{-2} |96,100|
in the same energy range. As a consequence, the average energy of ions
is of the order of magnitude of 100-200 eV |28,32|, whereas it is only
about 10 eV for neutral atoms |100|. These values can be taken as re-
presentative of results obtained for many bombarding particles within the energy range of 1 to 10 keV.
     The distribution of low energy ions deserves special attention.
It was studied at first by G. Blaise and G. Slodzian who used the e-
lectrostatic mirror of the ion-microanalyser |1-6| as an energy fil-
ter |23|*. They obtained energy distributions from zero to 20 eV with
a precision of 5.10^{-2} eV below 1 eV and 0.25 eV from 1 to 20 eV **.
Their study revealed that ion energy spectra are very diversified as
compared to neutral energy spectra. For example, the energy spectra
of neutral atoms emitted by the metals of the first transition row

---

* In that respect it is advisable to point out that it is better to
  ensure a strong acceleration of ions at the surface of the sample
  rather than a field free region. In the latter case the trajectories of ions emitted at low velocity will suffer violent distortion
  when they will be accelerated and focused before entering the mass
  separator. Most of them will not be collected at the detector. On
  the contrary, in the former case, these ions are very well collected but, of course, it is necessary to know the optical properties
  of the extracting system to derive the energy distribution |23|.
** More recently Wittmaack measured energy spectra in the same ener-
  gy range with an energy band pass of 1 eV |31|.

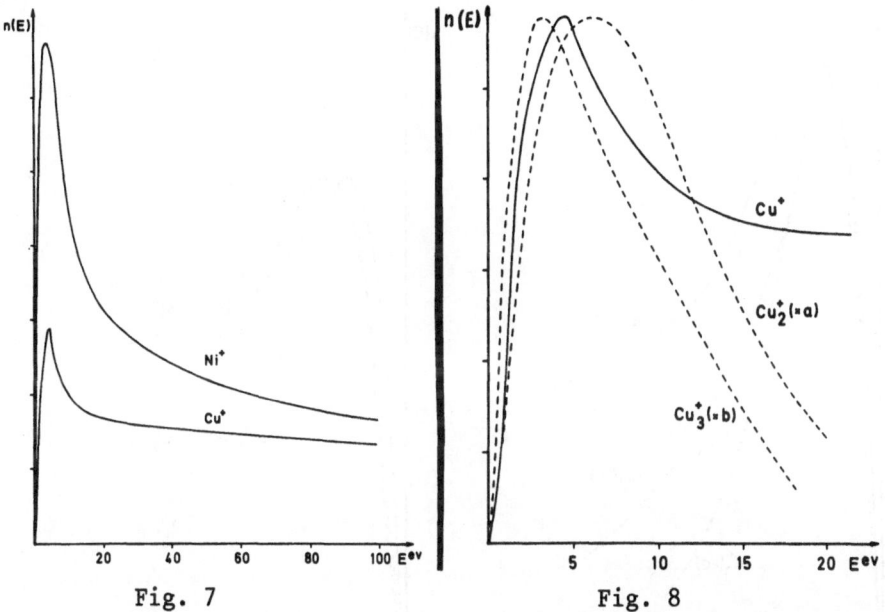

Fig. 7

Fig. 8

Fig.7 – Energy distributions of Cu^+ and Ni^+ ions from polycrystalline
metals bombarded with 6 keV A^+ ions |from 23|.

Fig.8 – Energy distributions of Cu^+, Cu_2^+ (a=3.6) and Cu_3^+ (b=2.1)
ions from polycrystalline copper bombarded with 6 keV A^+ ions
|from 23|.

(Ti to Cu) are very similar |100|. On the contrary, energy spectra of
ions are different. The most typical example is represented in fig.7
for Cu^+ and Ni^+ ions: at energies below 20 eV both energy distributions differ greatly then tend to become identical as energy increases. The relationship found experimentally between these emissions
is:

$$\frac{n(Cu^+)}{n(Ni^+)} = 0.72 \exp - \frac{1700}{v} \qquad (5)$$

where v is the velocity of atoms in m/s.
It is clear that the difference in the ion energy spectra are related with ionization processes since no significant difference exists
in the neutral atom emission of these metals.
   The energy spectra of molecular ions also exhibit a maximum at
a few eV, but the energy spread is always smaller than it is for
atomic ions (fig.8-9-10). The shape of these distribution clearly
shows that the emission process of molecular ions is not thermal. One
rather imagines a kinetic process in which several atoms are simultaneously ejected from adjacent sites on the surface. Under these conditions a cluster cannot receive a large amount of kinetic energy
without being dissociated. This explains the small energy spread of
the distributions. On the other hand, it is quite remarkable that molecular ions of low energy, below 1 eV, are almost nonexistent.

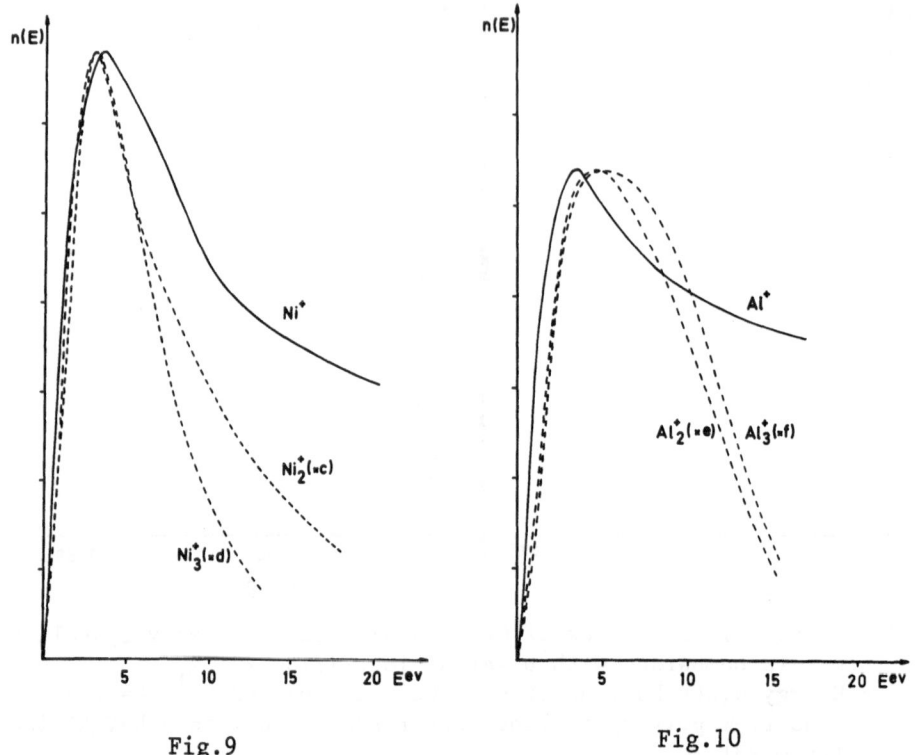

Fig.9   Fig.10

Fig.9 – Energy distributions of Ni^+, Ni_2^+ (c=2.7), Ni_3^+ (d=24.2) ions
   from polycrystalline nickel bombarded with 6 keV A^+ ions
   |from 23|.
Fig.10 – Energy distributions of Al^+, Al_2^+ (e=3), Al_3^+ (f=11.5) ions
   from polycrystalline aluminium bombarded with 6 keV A^+ ions
   |from 23|.

The pecularities of molecular energy spectra also appear in the comparison of Cu_2^+, Ni_2^+ or Cu_3^+, Ni_3^+ energy distributions (fig.8-9-10).
   Energy spectra of multicharged ions are represented in fig.11.
They hardly exhibit a maximum at relatively high energy, $\sim$ 10 eV,
followed by a very slow decrease. This behaviour confirm these ions
are produced by hard collisions.
   All the energy distributions presented here are the most commonly observed. However, some unusual behaviours are also mentioned in
the litterature. For example, when argon is incorporated by bombardment into aluminium at high temperature, $\sim$ 300°C, it tends to preci-
pitate as bubbles |33,34|. This precipitation gives rise to a very
intense emission of molecular secondary ions from the noble gas
(A_2^+, A_3^+,...,A_{18}^+) (see fig.20) whose energy spread does not exceed
1 eV (fig.12). Another spectacular example is observed when bombard-

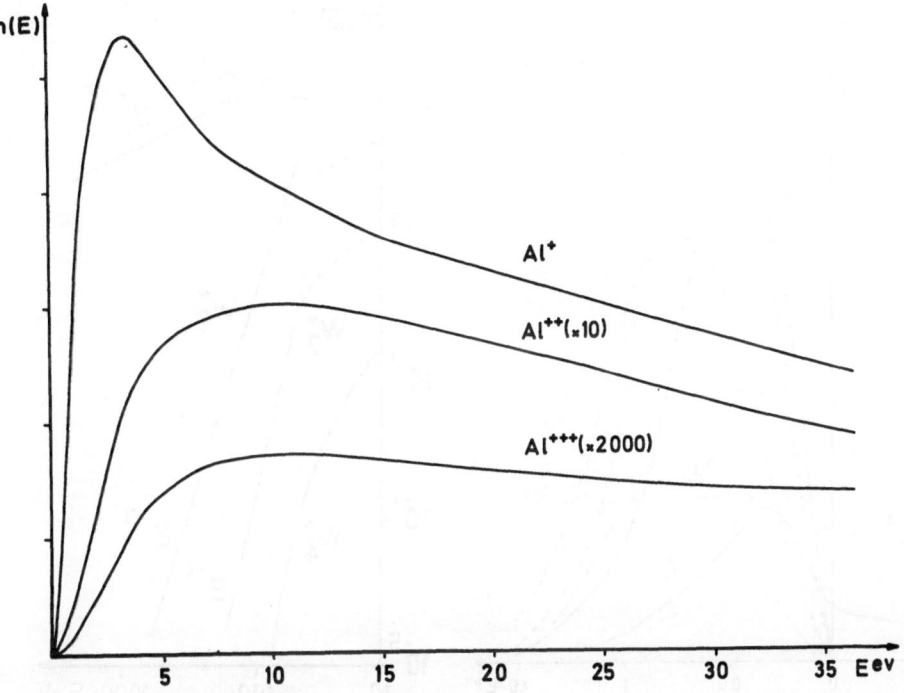

Fig.11 - Energy distributions of Al^{++}, Al^{+++} ions from polycrystalline
   aluminium bombarded with 6 keV A^+ ions |from 23|.

ding tungsten with very high energy ions of 150 keV |30| (fig.13) :
W_2^+ ions of 1000 eV and W_3^+ ions of more than 100 eV are still detected.
However, this example very well illustrates the decrease of the intensity of a cluster with energy, the decrease becoming steeper as the
number of its atomic components increases.
   Energy distributions of ions emitted from single crystals depend on the emerging direction : along preferred directions of ejection corresponding to focusing sequences, the ion distribution has a
much greater low-energy component than the random spectrum |26|.
   Very little work has been reported on energy spectra of negati-
ve ions. Jurela |32| mentions that negative atomic ions have a smal-
ler average energy than do positive atomic ions.

II.1.2 Ion yield and ionization ratio of atoms
   For any positive or negative, atomic or molecular, ion species
$A^\pm$, the ion yield $K(A^\pm)$ is defined as the ratio of the number of emitted ions $n(A^\pm)$ to the number Np of primary incident ions

$$K(A^\pm) = \frac{n(A^\pm)}{Np} \qquad (6)$$

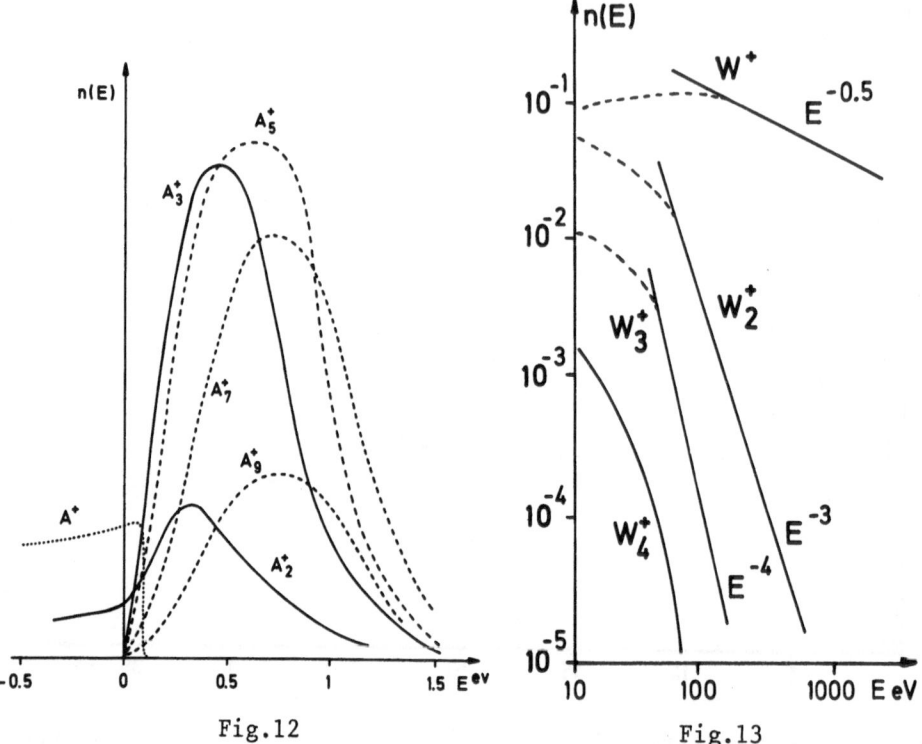

Fig.12

Fig.13

Fig.12 - Energy distributions of A_n^+ ions emitted by argon bubbles produced by argon implantation into aluminium at 300°C |from 33-34|.

Fig.13 - Energy distribution of W_n^+ ions from polycrystalline tungsten bombarded with 150 keV A^+ ions |from 30|.

In principle K is obtained experimentally by measuring the totality of primary and secondary ion currents. However, this is not as easy as it might appear at first sight. On the one hand, it is advisable to point out that measurment of the total primary current delivered to the sample is not always sufficient because the size of the bombarded area can be different from the viewing field of the instrument. On the other hand, only a small part of the emitted ions $A^\pm$ is collected and detected by the instrument.

In practice, most investigators deduced partial ion yields ΔK from the ratio of the secondary intensity $I(A^\pm)$ detected on the collector to the primary intensity Ip

$$\Delta K(A^\pm) = \frac{I(A^\pm)}{Ip} \qquad (7)$$

Then, having estimated the transmission factor η of their instrument (equat.1), some of them calculated absolute ion yields K' using the relation :

$$K' = \frac{\Delta K}{\eta} \qquad (8)$$

However, there is no advantage to providing absolute ion yields in this way, with respect to partial ion yields ΔK, since η has been taken as the same for all ion species, whatever their angular and energy distributions (see sect. II.1). K' is generally different from K and must, in fact, be considered as the partial ion yield ΔK. Consequently, it matters to know exactly in what energy range ions are collected for the determination of K' as well as for ΔK.

It is noted that when operating always under the same bombarding conditions, measurment of Ip is not necessary, since secondary currents $I(A^{\pm})$ will be directly proportional to the partial ion yields.

Some absolute ion yields measurments have been performed by Hennequin |21,35|, who used a special instrument. For that, the ion current emitted from the sample in a well defined solid angle was integrated in angle and energy. This is the only correct procedure for obtaining absolute ion yields in spite of the difficulties.

For any atomic species $M^{\pm}$ one defines the ionization ratio $\tau(M^{\pm})$ as the ratio of secondary ions $n(M^{\pm})$ to the total number $n(M)$ of sputtered atoms M, namely :

$$\tau(M^{\pm}) = \frac{n(M^{\pm})}{n(M)} \qquad (9)$$

This parameter has a real meaning only for atomic species, because molecular ions more often have no equivalent neutral parents ($\tau \to \infty$ for many molecular species).

If $S(M) = \frac{n(M)}{Np}$ is the partial sputtering yield of M atoms (see section III.2.1), by combining (6) and (9) one gets for atomic species only :

$$K(M^{\pm}) = S(M) \ \tau(M^{\pm}) \qquad (10)$$

For the reasons presented previously about absolute measurments of K, it is difficult to obtain absolute values of τ, and more particularly because measurments of S are relatively scarce for compounds (see section II.2). However as, in many cases, S(M) does not exceed a few units, $\tau \sim K$.

Partial ion yields of transition metals bombarded with A^{+} ions are indicated in fig.14 in comparison with copper ion yield |23| : points refer to measurments performed with ions from zero to 15 eV; triangles are for ions collected in an energy bandwidth of 15 eV centred at 50 and 100 eV. Measurments at 50 and 100 eV always coincide for metals of the first transition series. The reason is that the energy distributions are similar above 30 eV. On the contrary when M^{+} and Cu^{+} ions have different energy distributions below 30 eV, points and triangles do not coincide. However, whatever the measurments it is noted that the general trend of ion yields is a decrease from the beginning to the end of each transition series.

Absolute ion yields indicated in Table II have been reported by Hennequin |21|, Benninghoven |36| and Jurela |37,38|.

Table II

Ion Yields of pure elements

Band 1

Ref	Sec. Ions	Mg	Al	Si
(1)	+	$4\ 10^{-3}$	$2.5\ 10^{-2}$	10^{-3}
(2)	+	10^{-2}	$7\ 10^{-3}$	$8.4\ 10^{-3}$
(3)	+	$3\ 10^{-2}$	10^{-7}	$2\ 10^{-3}$
(3)	–	no	10^{-4}	$1.8\ 10^{-4}$

Band 2

Ref	Sec. Ions	Ti	V	Cr	Mn	Fe	Co	Ni	Cu	Ge
(1)	+	$1.6\ 10^{-3}$				10^{-3}		$8\ 10^{-4}$	$6\ 10^{-4}$	
(2)	+	$1.3\ 10^{-3}$	10^{-3}	$1.2\ 10^{-3}$	$6.4\ 10^{-4}$	$1.5\ 10^{-3}$		$6\ 10^{-4}$	$3\ 10^{-4}$	
(3)	+				$5\ 10^{-3}$		$3\ 10^{-3}$	$3.3\ 10^{-3}$	10^{-3}	$4.4\ 10^{-3}$ $2.4\ 10^{-3}$
(3)	–				no		$4\ 10^{-5}$		$4.8\ 10^{-5}$	$3.3\ 10^{-5}$

Band 3

Ref	Sec. Ions	Sr	Nb	Mo	Ag
(1)	+				$<5\ 10^{-5}$
(2)	+	$\sim 10^{-4}$	$6\ 10^{-4}$	$6.5\ 10^{-4}$	
(3)	+			$5\ 10^{-2}$	$8.5\ 10^{-4}$
(3)	–				$1.7\ 10^{-5}$

Band 4

Ref	Sec. Ions	Ba	Ta	W	Pt	Au
(1)	+		$5\ 10^{-4}$			$8.2\ 10^{-4}$
(2)	+	$\sim 10^{-4}$	$7\ 10^{-5}$	$9\ 10^{-5}$		
(3)	+		$5.4\ 10^{-3}$	$5.3\ 10^{-3}$		$5.5\ 10^{-5}$
(3)	–		$5.3\ 10^{-5}$			$5.3\ 10^{-5}$

(1) J.F. Hennequin |21| : A^+ ions 8 keV; beam density 800 μA/cm²; residual pressure 2 10^{-7} torr.

(2) A. Benninghoven |36| : A^+ ions 3 keV; beam density 10^{-3} μA/cm²; residual pressure ~ 10^{-10} torr.

(3) Z. Jurela |37,38| : A^+ ions 40 keV; beam density 60 to 100μA/cm²; residual pressure 2 10^{-6} torr.

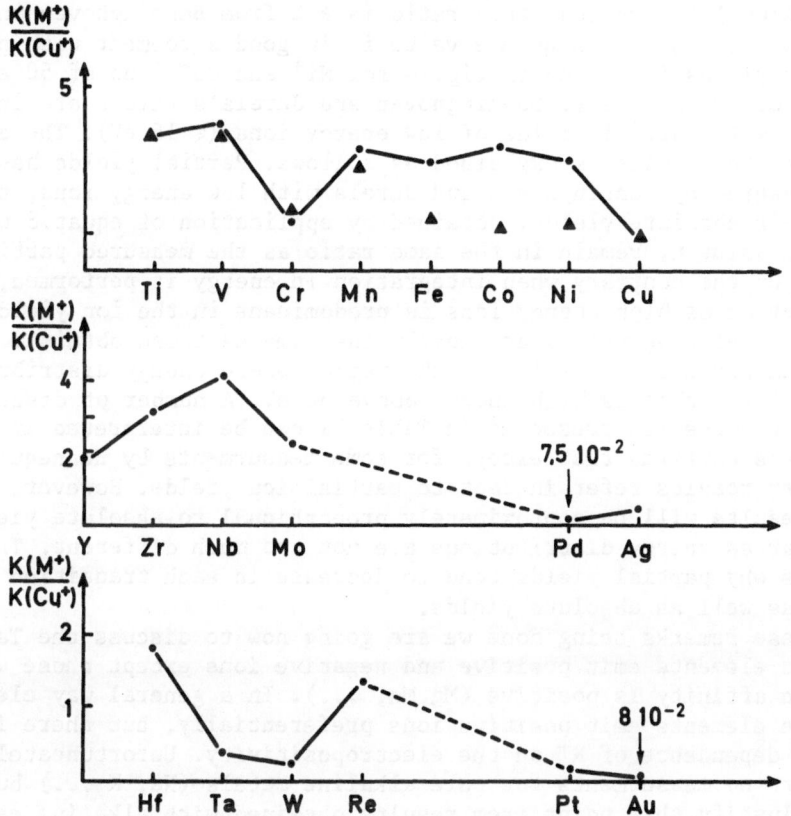

Fig.14 - Partial ion yields of polycrystalline transition metals bombarded by 6 keV A⁺ ions |from 23|.

● K(M⁺)/K(Cu⁺) for 0-15 eV ions ; ▲ K(M⁺)/K(Cu⁺) for ions collected in an energy bandwidth of 15 eV centred at 50 eV and 100 eV.

Hennequin's yields were obtained by the integration method mentioned previously whereas Benninghoven and Jurela's results were obtained by using a constant transmission factor η. It is seen that ion yields also tend to decrease in each transition series as indicated above for partial ion yields, but one notices some disagreement between the various authors. The disagreement can be explained, to some degree, by the two experimental procedures used *. For example from Hennequin

---

* For some readily oxidizable metals large divergences can also be due to the influence of the residual atmosphere (see sect. II.2.4).

$K(Ni^+)/K(Cu^+) \sim 1$ whereas this ratio is $\geqslant 2$ from Benninghoven and Jurela's results. Hennequin's value is in good agreement with partial ion yields indicated in fig.14 for Ni^+ and Cu^+ ions of 50 and 100 eV. On the contrary, Benninghoven and Jurela's values are in accordance with partial yields of low energy ions ($\leqslant 15$ eV). The difference in the results is explained as follows. Partial yields have been measured by Benninghoven and Jurela with low energy ions, therefore their absolute yields, obtained by application of equat.8 with the same value η, remain in the same ratio as the measured partial yields. On the contrary when integration in energy is performed, the contribution of high energy ions is predominant in the ion yield. In that case relative values are nearly the same as those obtained by measuring partial ion yields in the region where energy distributions are similar ; that is high energy above 30 eV. A number of discrepancies in the results presented in Table II can be interpreted in this way[*]. This confirms that except for some measurments by Hennequin all other results refer in fact to partial ion yields. However, these results will be approximately proportional to absolute yields in so far as energy distributions are not too much different. This explains why partial yields tend to decrease in each transition series as well as absolute yields.

These remarks being done we are going now to discuss the Table II. Most elements emit positive and negative ions except those which electron affinity is positive (Mg,Mn, ...). In a general way electropositive elements emit positive ions preferentially, but there is not a clear dependence of K^+ on the electropositivity. Unfortuneately, there are no measurments for pure alkaline metals (Na, K ...) but we can justify this point from results obtained with alkaline earth : ion yields of Mg, Sr, Ba are not higher than those of most other metals.[**]

Except for some indubitable cases of negative ion emission of metals (Si^-, Ag^-, Au^- ...) most other results have to be considered with reservation because a great deal of MH^- ions are emitted which

---

[*] For some readily oxidizable metals large divergences can also be due to the influence of the residual atmosphere (see sect. II.2.4)

[**] It is even expected that alkaline metals (Na, K...) have very low positive ion yields (see sect. IV.3 on ionization processes).

can be mistaken with M^- ions $|39|^*$. There is a lack of measurments
for elements from III B to VII B groups. The first reason is that a
number of them are gaseous and the second that, except for few cases,
most others are poor emitters of positive ions as well as of negative ions.

Some values of the ionization ratio of pure metals are given in
Table III $|21,38|$. In spite of a large dispersion of some measurments,
this table shows that the ionization ratio covers a large range of
values : aluminium is probably the most emissive metal whereas gold
(and also Pd) is one of the least emissive metal.

In fact one judges the real importance of the ion emission phenomenon through the study of complex compounds. This point will be
illustrated by several examples. At first ion emission of alkaline
halides is presented in Table IV $|41,42|$. It appears that alkaline
elements(Li, Na, K ....) have about 100% of ionization ratio whereas
halogen have only 30%. A simple explanation $|6|$ of these numbers is
that sodium sits in the lattice already as a positive ion and therefore a collision may expel sodium directly as Na^+ ; the same is true
for Cl^- but if the collision is too violent, Cl^- may be deionized and
leave the solid as neutral.

Although this explanation is probably a little too naive, it has
the advantage of drawing attention on the role played by the chemical
bond of atoms on their ionization efficiency. This dependence is
clearly confirmed by experiments on oxides. On table V $|43|$ for exam-
ple, $\tau(Al^+)$ appears to be much greater in Al_2O_3 oxide than in pure a-
luminium. Likewise in Table VI $|44|$ the ratio of absolute ion yields
of some metal elements in oxides and in pure metals largely exceed
unity. It is obvious that the exaltation is due to the formation of
metal - oxygen bonding. But the situation is less clear than in the
case of alkaline halides because in oxides the ionic character is often less pronounced - and sometimes even nonexistent. The influence
of the metal-oxygen bonding on the ionization of atoms decreases in
proportion as the velocity of the particles increases : atoms of low
energy are much more affected than those of high energy. Therefore,
when measuring partial ion yields in the low energy range (<20 eV)
frequently one obtains enhancement higher than those indicated in
Table VI for the whole emission. This is for example the case in
Table VII for Ti, Cr and Ni $|45,46|$.

---

* Intense negative ion yields of metals are observed when Cs^+ ions
are used for bombardment$|40|$, but in that case, the secondary ion
emission is influenced by the chemical nature of primary ions. This
is not a case of intrinsic emision $|93,94|$.

Table III

Ionisation ratio

authors see Table II.	Sec. Ions	Al	Si	C	Cu	Ag	Ta	Au
(1)	+	$1.2\ 10^{-2}$	$5.5\ 10^{-4}$		10^{-4}	$\sim 5\ 10^{-6}$	$\sim 5\ 10^{-4}$	
(3)	+	$1.2\ 10^{-1}$	10^{-3}*	$1.6\ 10^{-4}$	$1.4\ 10^{-4}$	$8\ 10^{-5}$	$3.3\ 10^{-3}$	$5.4 10^{-6}$
	−			$1.4\ 10^{-4}$				

* This value has been reestimated by the author using a more realistic sputtering yield of silicon

Table IV

Ion emission from KBr and Na Cl $|41,42|$.

Sputtered particles	KBr		NaCl	
	K	Br	Na	Cl
S	0.3	0.3	0.45	0.45
ions	K^+	Br^-	Na^+	Cl^-
$K^\pm$	0.28	0.08	0.4	0.1
$\tau^\pm$	0.93	0.27	0.89	0.22

Table V

Ion emission from aluminium and alumina bombarded by 8 keV A^+ ions $|43|$.

	pure Al	Al_2O_3
S(Al atoms only)	2	0.5
K^+	0.025	0.12
τ^+	0.01	0.24

Table VI

Ratios of absolute ion yields of pure metals and
oxides $|44|$.

	Mg O	$Al_2 O_3$	$Si O_2$
$\dfrac{K(M)^+ \text{oxide}}{K(M)^+ \text{ pure metal}}$	10	4	40

Table VII

Ratios of partial ion yields (ions with energy
< 20 eV) of pure metals and oxides $|45,46|$.

	TiO_2	Cr_2O_3	NiO	CuO
$\dfrac{\Delta K^+ \text{oxide}}{\Delta K^+ \text{metal}}$	$150\|45\|$	$500\|46\|$	$100\|46\|$	$20\|45\|$

This influence of the metal-oxygen bonding on the ionization
ratio has very important analytical implications. The recent study
of a number of silicates by A. Havette $|47|$ shown that ion yields of
most elements may vary from one silicate to another. This implies
that quantitative analysis requires the use of standards of the same
family as the sample to be analysed.

Independently of the role played by oxygen on ion emission, it
is established that, in a general way, the ionization efficiency is
closely connected with the initial environnement of atoms in the so-
lid $|48,49|$. This connection clearly appears in results presented
in Table VIII for dilute alloys $|50|$. The relative ionization coef-
ficient $\rho(T)$ of a solute element T defined by (see section V.1) :

$$\frac{K_c(T^+)}{K(T^+)\text{pure}} = \rho(T)c$$

depends on the matrix. For example $\rho(Cr)$ is much higher when chromium is diluted into Ni rather than Cu or Fe matrices. For some elements such as Mn or Cu the variations of ρ are small; for others,Ti,
V, Cr, large variations are observed.

The dependence of the ionization efficiency of a solute element
on the matrix is known as matrix effect $|48,51,52|$. It is pointed

Table VIII

Relative ionization coefficient $\rho(T)$ of a
solute element T in various matrices $|50|$. ρ
is measured here for ions collected in an
energy bandwidth of 15 eV centred at 50 eV.

alloy	Fe T	Co T	Ni T	Cu T
solute element T				
Ti			10	27
V			13	5.7
Cr	2.7		46	10
Mn	2.8	3.7	3.1	2.7
Fe	——	0.8	0.7	1.3
Co	0.7	——	0.4	0.2
Ni	1.3	1.1	——	2.8
Cu	1.1	1.7	1.2	——

out   that matrix effect is important when the solute element is on
the left hand of the transition series and the matrix on the right
hand (Examples Cu Ti, Cu V, Cu Cr). On the contrary small matrix effects are observed when both elements are neighbouring atoms (Ni Cu
alloys for example). This rule is very general and can be applied
qualitatively to the three long periods of transition elements.*
    As the environment of an atom depends on the concentration of
the alloy, variations of the ionization coefficient can be observed
as a function of concentrations. For example, in Ni Fe alloys (fig.
15) $|53|$ $\rho_{Ni} = 0.8$ is independent of the concentration whereas ρ_{Fe}
increases when the Fe concentration increases.
    All the results on alloys will be discussed theoretically la-
ter on (sectIV.3). Here we just would like to draw attention on the

---

* This rule is based on the physical properties of dilute alloys
   specific heat and resistivity,related to the electronic structure
   of the solute element (see the auto-ionizing state model in sect.
   IV.3 and ref. 48, 157).

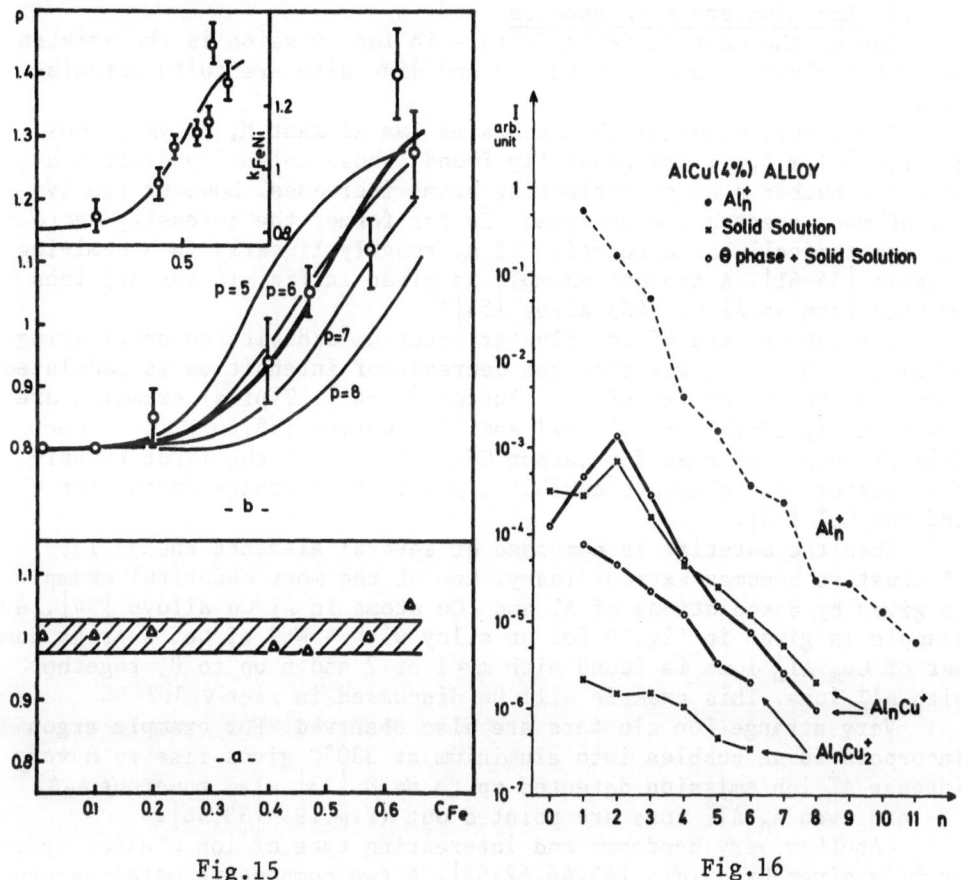

Fig.15   Fig.16

Fig.15 – Ion emission of Ni Fe alloys |from 53| bombarded with 6 keV
         A^+ ions. $\rho(Ni)$ (▲) and $\rho(Fe)$ (○) as a function of iron concen-
         tration. Comparison between theoretical curves (thin lines)
         and experimental data (broad lines). In the top of the left
         corner, concentration dependence of k_{FeNi}.

Fig.16 – Amounts of molecular ions from an Al Cu (4%) alloy bombarded
         with 6 keV A^+ ions |from 54|.

fact that matrix effects and non-linear concentration dependence are
the most common features of ion emission. This explains the difficulties encountered in elementary quantitative analysis but, on the o-
ther hand, it is one of the most exciting aspect of ion emission. We
shall try in sections IV.4 and V.1 to give some enlightenment on
these two points.

II.1.3. Ion clusters mass spectra

One of the most striking feature in ion emission is the existence of ion clusters whose intensity and diversity are quite astonishing.

For a pure material that contains an element M, polymer ions $M_n^{\pm}$ with n=1,2,3,4.. are generally found, whose intensity decreases when the number n of participating atoms increases. However two types of mass spectra are observed. In the former the intensity decreases monotonically as a function of n, roughly linearly in a semi-log diagram |54-61|. A typical example is given in fig. 16 for Al_n^+ ions emitted from an Al Cu (4%) alloy |54|*.

The latter type of ion cluster spectrum exhibits an oscillating behaviour in such a way that the decrease of intensities is modulated according to the parity of the cluster |55-61|. Typical examples are given in fig.17-18 for Cu_n^+ |61| and C_n^- clusters |58,60|: for copper $Cu_{2p+1}^+ > Cu_{2p}$, whereas for carbon $C_{2p}^- > C_{2p+1}^-$. In the first transition series most elements exhibit a monotonic decrease except for Cr_n^+ and Cu_n^+ |55|.

When the material is composed of several elements the variety of clusters becomes extraordinary. One of the most beautiful example is given by associations of Al and Cu atoms in Al Cu alloys |54|. An example is given in fig.16 for an alloy with 4 Wt% of Cu. A great number of $Cu_m Al_n^+$ ions is found with m = 1 or 2 and n up to 8, together with Al_n^+ ions. This example will be discussed in sect V.3.2.

Very strange ion clusters are also observed. For example argon incorporated as bubbles into aluminium at 330°C gives rise to a very intense A_n^+ ion emission detected up to n=18 |33|; also numerous $A_n Al^+$ and even $A_n Al_2^+$ ions are pointed out (fig.19) |33,56|.

Another very handsome and interesting type of ion cluster spectrum is given by oxides |45,46,62,63|. A two components metal-oxygen oxide emits ions of the general composition $M_n O_m^{\pm}$ (n,m = 0, 1, 2 ... M stands for metal, 0 for oxygen). Some examples are given in fig.20 for positive ions. Ion clusters are classified as a function of the number of participating metal atoms : $MO_m^{\pm}$, $M_2O_m^{\pm}$, $M_3O_m^{\pm}$. Although spectra are far from being complete for most oxides, it seems that a steep decrease of the $MO_m^{\pm}$ ions intensities as m increases up to 2 is the general rule. For m > 2 it is likely that intensities are too low to be detected except for Ti O_2 oxide. In the $M_2O_m^+$ series a maximum for M_2O^+ is always observed. It is remarkable that Ti O_2 emits the most intense M O_m^+ ions and the least $M_2O_m^+$ emission. In the $M_3O_m^+$ series the maximum should be for M_3O^+ ions (possibly $M_3O_2^+$ in Cr_2O_3 oxide).

---

* In this alloy the Al_n^+ molecular ion spectrum is about the same as in pure aluminium because the copper concentration is low ($\sim$ 2 at. %).

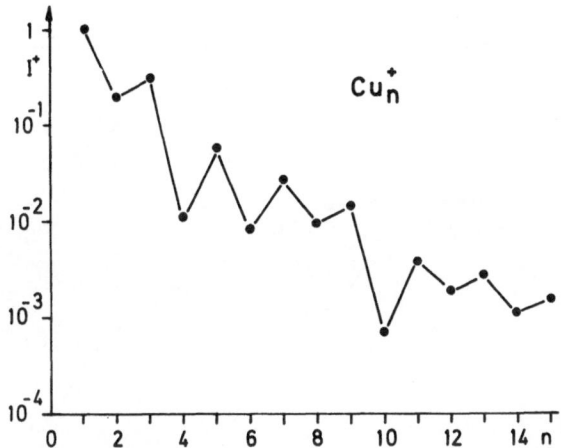

Fig.17 - Cu_n^+ ion spectrum from polycrystalline copper bombarded with 11 keV A^+ ions |from 61|.

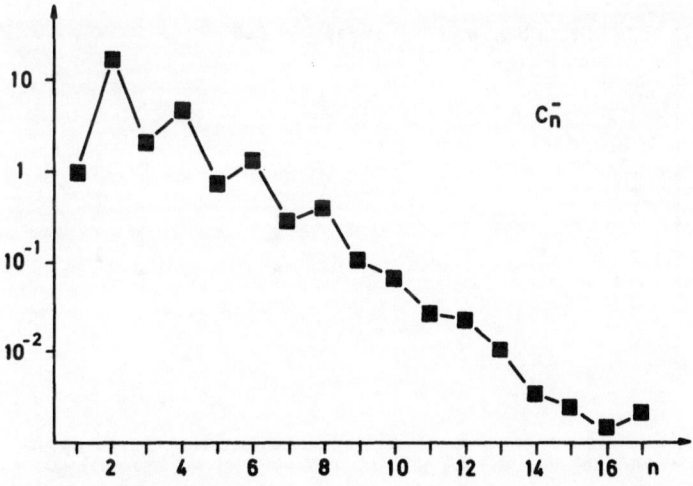

Fig.18 - C_n^- ion spectrum from graphite bombarded with 6.5 keV A^+ ions |from 58,61|.

At last, a part of a very beautiful oxide spectrum given by a complex geological sample is shown in fig.21 | 47 |. Together with two-compo-nents ions (M and O), Aegirine which formula is $(Si\ O_3)_2$ Fe Na gives rise to three-components ions $M\ O_m\ N^+$ involving two different metal

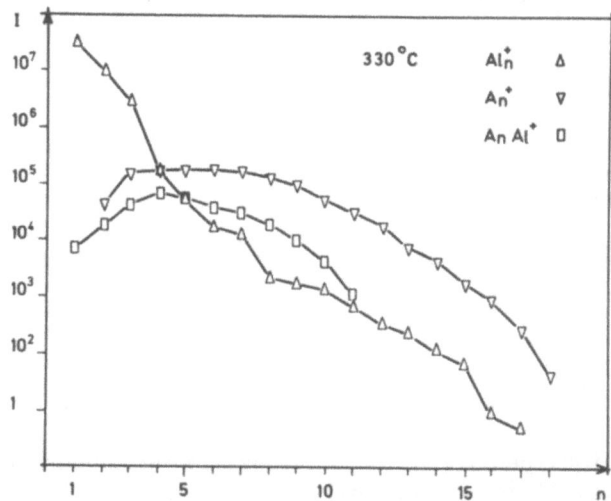

Fig.19 - Molecular ions emitted by an aluminium crystal containing
argon bubbles |from 33|.

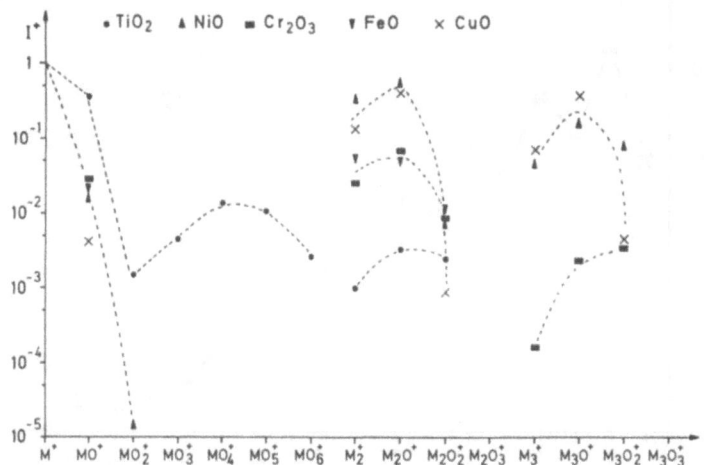

Fig.20 - Ion clusters mass spectra of oxides bombarded by 6.5 keV A^+
ions. Ti O_2 and Cu O |45|; Ni O, Cr_2 O_3, Fe O |46|.

atoms M and N. Associations of three metal atoms with oxygen have
not been detected however.

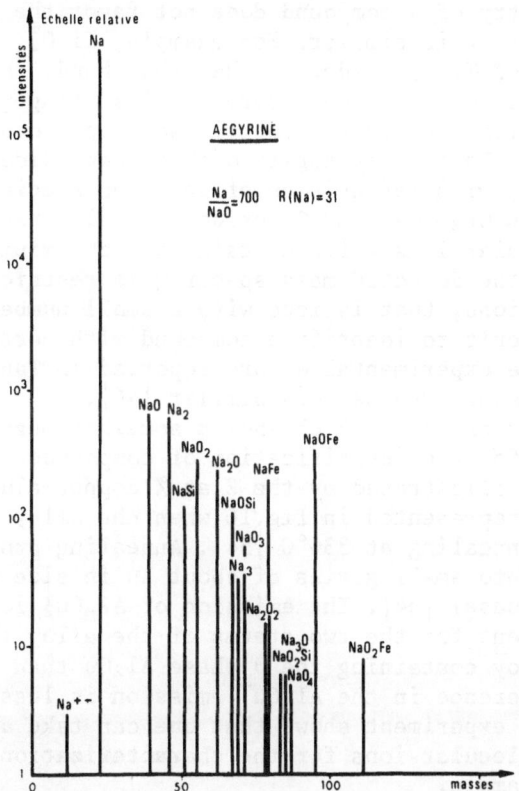

Fig.21 – Molecular ion spectrum of Aegirine bombarded with 6.5 keV
A$^+$ ions |from 47|.

Table IX

Intensities of molecular ions in various iron oxides bombarded with
6.5 keV A$^+$ ions |from 46|: I (molecular ion) / I (Fe$^+$).

oxides	FeO$^+$	Fe$_2^+$	Fe$_2$O$^+$	Fe$_2$O$_2^+$
Fe O	2 10^{-2}	5 10^{-2}	5 10^{-2}	8.5 10^{-3}
Fe$_3$ O$_4$	2.6 10^{-2}	3.7 10^{-2}	5.8 10^{-2}	9.4 10^{-3}
Fe$_2$ O$_3$	2.3 10^{-2}	5 10^{-2}	5.1 10^{-2}	7.4 10^{-3}

All the spectra presented here can be considered as pure ; that
is spectra characteristic of a compound. If the sample is contamined
at the surface or in the bulk a forest of peaks mainly composed of
molecular ions associated with H, C, N, O ... atoms is observed
|18,19|. Such spectra have no proper signification, they just prove
the existence of a pollution.

What is the analytical interest of molecular ions?
At first, from the spectra presented in fig.20 it is clear

that the stechiometry of a compound does not favor the production of
ions whose composition is similar. For example, Ti O_2^+ is one of the
least intense ion of Ti O_2 oxide. On the other hand, the overall ion
spectrum of a compound can be considered as its "fingerprint" and u-
sed to its identification. This principle was applied by Werner |63|
to chromium oxides. In fact it appeared that such oxides were clearly
differentiated only with intensities of some ion species. For example,
distinction between $Cr_2 O_3$ and $Cr O_2$ oxides is only based upon the intensities of molecular ions which contain many chromium atoms : Cr_2^+,
Cr_3^+, Cr_3O^+ ... If the detected mass spectrum is restricted to the most
intense molecular ions, that is ions with a small number of atoms, it
becomes very difficult to identify a compound with certainty. For
example, within the experimental errors reported in Table IX, the
three iron oxides appear to be very similar |46|.

Besides these examples which show a somewhat negative aspect
of molecular ions for the identification of compounds one finds a very positive aspect illustrated by the 2 at % copper-aluminium alloy
whose spectrum is represented in fig.16 when the alloy is a solid so-
lution and after annealing at 330°C |54|. Annealing produces the pre-
cipitation of Cu into small grains of about 1μ in size whose composi-
tion is Al_2Cu (θ phase) |64|. The emission of $Al_nCu_2^+$ ions was found
to be quite different for the two states of the alloy : it is much
greater in the alloy containing the θ phase Al_2Cu than in the solid
solution. The difference in the Al_nCu^+ emission is lesser but significative too. This experiment shows that one can take advantage of
the presence of molecular ions for the characterization of the structural state of a sample.

To conclude this discussion on the interest of molecular ions
without anticipating the problem of their formation, one just says
that molecular ions are very useful for detecting small amount of
phases and less well adapted for their chemical identification. Both
these points will be discussed in sect. V.3.

II-2 Ion emission induced by surface reaction :

It is well known that chemically active gas components adsorb
on most solid surfaces where they form a contaminant layer. An intense parasite ion emission resulting from this contamination is detected
at the very beginning of the bombardment and even permanently if precautions are not taken. This emission is generally considered as a
nuisance. However, the sensitivity of ion emission to the surface
composition may be used as an advantage for gas-solid interface
reactions studies on the condition that parameters which regulate the
adsorption process are well controlled.

II-2-1 general conditions for adsorption studies |65,66| :

For the sake of clarity we consider a metal with n_o equivalent
adsorption sites per unit area, surrounded by a gas at pressure p

and temperature T °K. The rate of collisions ν of the molecules of the gas per unit surface area is :

$$\nu = p(2\pi mkT)^{-1/2} \qquad (11)$$

where m is the mass of the molecules of the gas, and k the Boltzmann constant. The fractional coverage θ is defined as the ratio of the number n_a of occupied adsorption sites to the number n_o of available sites :

$$\theta = \frac{n_a}{n_o} \qquad\qquad 0 \leqslant \theta \leqslant 1 \qquad (12)$$

The rate of molecules sticking to the surface per second is :

$$n_1 = s(\theta)\nu \qquad (13)$$

where $s(\theta)$ is the sticking coefficient at coverage θ.

Adsorbed atoms are also ejected by bombardment. If we assume that they are mainly sputtered by collisional sequences initiated in the metal, the number n_2 of sputtered adsorbed atoms per unit time will be proportional to the number n_a of adsorbed atoms and to the number $N(\theta) = S(\theta) Np$ of metal atoms sputtered per unit time and unit area; namely :

$$n_2 = a \; n_a \; s(\theta) \; Np \qquad (14)$$

where a may be interpreted as the ejection cross section of adsorbed atoms.

Depending on the initial values of n_1 and n_2 , the surface will be cleaned as long as $n_1 < n_2$ or progressively covered by the gas if $n_1 > n_2$. However, in both cases a dynamic equilibrium corresponding to a permanent coverage θ_o given by :

$$n_1(\theta_o) = n_2(\theta_o) \qquad (15)$$

will be always reached. From 11,12,13,14, one obtains :

$$\theta_o = K \; s(\theta_o) \; \frac{p}{S(\theta_o)Np} \qquad (16)$$

$$K = \frac{1}{a \; n_o(2\pi mkT)^{1/2}} \qquad (17)$$

At room temperature and pressure $p \sim 10^{-6}$ torr, $\nu \sim 10^{15}$ collisions/ cm². sec. If we assume that n_o is roughly equal to the surface atom density of the solid N_o, that is $n_o \sim N_o \sim 10^{15}$ atoms/cm², it is seen that the flux of molecules apt to stick on the surface will be equivalent to the deposit of one atomic layer per second.

According to the experimental conditions indicated in Table I, two different procedures are used for adsorption studies.

dynamic procedure :|65| Under dynamic conditions the erosion rate is about 1 layer/sec. This means that the equilibrium condition (15) leading to the permanent coverage θ_o is obtained for pressure of the order of 10^{-6} torr. In practice the gas pressure is varied in the range of 10^{-7} to 10^{-5} torr. Ion intensities are measured at any pressure after the equilibrium coverage is reached and they are plotted as a function of p:

$$I_d^{\pm} = f(p)$$

static procedure : |19| The erosion rate is $\sim 10^{-4}$ layer/sec.,
therefore the equilibrium coverage should be obtained at pressures of
$\sim 10^{-10}$ torr. It should be difficult to use the dynamic procedure under
these conditions. One likes better to keep the pressure constant p_o,
such that $n_1 \gg n_2$, that is with $p_o \gg 10^{-10}$ torr. The surface is then
progressively covered by the gas without undergoing any appreciable
destruction. The variation dn_a of the number of occupied adsorption
sites during a time dt is given by :

$$dn_a = s\ (\theta)\ p_o (2\pi mkT)^{-1/2} dt \qquad (18)$$

If $dL = p_o\ dt$ is the esposure, the variation of θ is

$$d\theta = \frac{s(\theta)}{n_o} \frac{dL}{(2\pi mkT)^{1/2}} \qquad (19)$$

Integration of equat. (19) is possible on the condition that $s(\theta)$ is
known. At very low coverage one can admit that $s(\theta) \sim 1$ |67|, then
equat. (19) yields :

$$\theta = a\ K\ L \qquad (20)$$

where K is given by (17).
Beyond this limiting case expressions of $s(\theta)$ have been derived by
Krueger and Pollack |68| regarding some elementary adsorption proces-
ses but it should be unwise to use them in a general way.
In practice ion intensities are recorded as a function of the exposure L :

$$I_s^{\pm} = f(L)$$

L is expressed in Langmuir, with 1 Langmuir = 10^{-6} torr $\times$ 1 sec.
Both experimental procedures have been used extensively for studies
of oxygen adsorption on metal surfaces |13,18,19,36,62,69-87|. It is
quite remarkable to observe that the evolutions of ion intensities
as a function of the two parameters p and L are qualitatively very similar (see sect. II.2.2). This similarity can be explained to some extent by considering the limiting case $\theta \to 0$. Combining equat. (16) and
(20) one gets :

$$L = \frac{p}{a\ S\ Np} \qquad (21)$$

This means that the same coverage is obtained in the two experimental
procedures for a value of L proportional to the pressure p. Therefore
it is not really surprising that a certain similarity exists in the
two evolutions of ion intensities.

II-2-2 nature of ion emission induced by oxygen adsorption :

We deal with the case of oxygen adsorption on pure metals. In
proportion as the surface is covered with oxygen, ion species of the
same composition $M_n\ O_m^{\pm}$ (n,m = 1,2,3 ...) as those observed with oxides appear and increase in intensity. In the same time metal ion

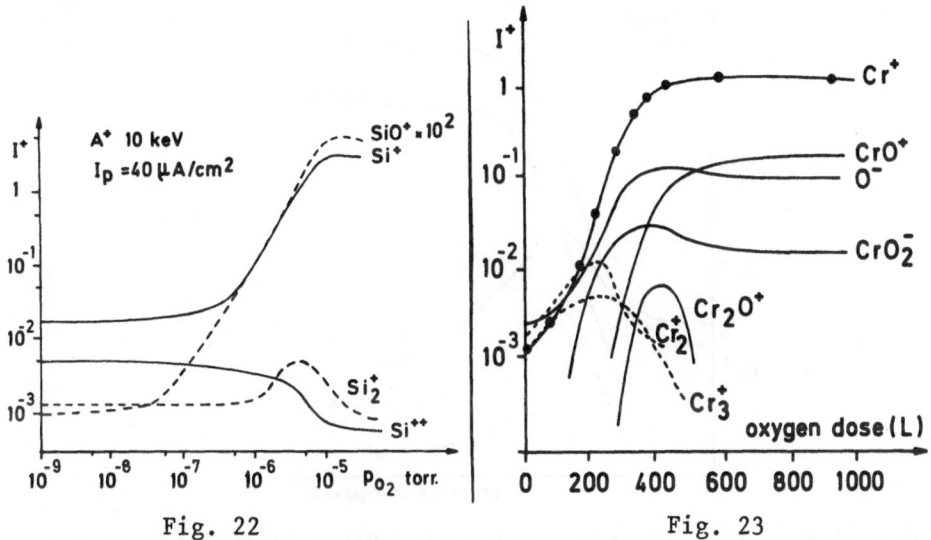

Fig. 22   Fig. 23

Fig.22 - Ion emission from silicon as a function of oxygen pressure:
        dynamic procedure |from 62|.
Fig.23 - Ion emission from chromium as a function of oxygen exposure:
        static procedure |from 69|.

emissions M^+, M_2^+, M_3^+ ... also increase in intensity to various degree
|69-87|. The overall effect of oxygen adsorption on the various ion
species is represented in fig.(22)(23). Results in fig.(22) were obtained by a dynamic procedure and in fig.(23) by a static procedure
|69|. It is seen that ions of the same species exhibit the same evo-
lution in the two procedures, as the coverage proceeds.

One finds 5 types of ion intensity evolution with oxygen - or
exposure - ideally represented in fig.(24).

a - Metal ion M^+ : Under ideal vacuum conditions where the surface of the metal is perfectly clean, only the intrinsic ion emission
exists. This emission is represented in fig.24 by the lower plateau
of the $I(M^+)$ curve, obtained at low pressure or exposure. In practice
this plateau is really observed if the residual atmosphere has a negligible influence on the surface (this problem will be discussed later
on) in section II.2.4. This is not exactly the case in experiments
carried out under static conditions (fig.23). The lower plateau is
followed by a steep increase corresponding to the progressive coverage of the surface by oxygen. At last, an upper plateau is reached
which corresponds to the saturation of the surface with oxygen. From
the lower to the upper plateau intensities usually vary by a factor
100.

b - Metal polymer ions M_m^+ (m⩾2) : as for M^+ ions, the intrinsic
emission of these ions is represented by a plateau. Then their intensities increase as the coverage proceeds and reach a maximum before

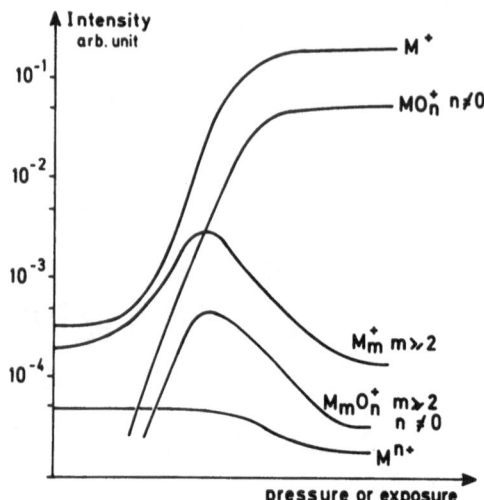

Fig.24 – Different qualitative evolutions of ion intensities as a
function of gas pressure or exposure.

falling to a level below the initial plateau.

c – <u>Molecular ions</u> MO_n^+ ($n \neq 0$) (also $O_n^\pm$) : If no oxygen atom is
adsorbed on the surface these ions must not exist. Therefore, their
intensity is normally zero under ideal vacuum conditions. Oxygen
adsorption must lead to a rapid increase of their intensity without
exhibiting a lower plateau. However, a plateau is observed in Witt-
maack's experiments |62| (fig.22) which probably denotes the presence
of residual oxygen in silicon or at the surface. After the steep increase an upper plateau is reached.

d – <u>Molecular ions</u> $M_m O_n^\pm$ ($m \geqslant 2$; $n \neq 0$) : Their evolution combines evolution of $MO_n^\pm$ and M_m^+ ions ; that is no lower plateau but a
steep increase followed by a maximum and a fall.

e – <u>Multicharged ions</u> M^{n+} : These ions are only detected on
light elements, Mg, Al, Si when the primary ion energy does not exce-
ed few keV |88|. They are rather insensitive to the presence of oxy-
gen. A long plateau corresponding to the intrinsic emission is observed ; it is followed by a decrease when the coverage becomes important and a plateau at oxygen saturation. Multicharged ions are due to
violent collisions which produce inner shell electronic excitations
(excitation of 2p electrons) |24,25|. The behaviour of this emission
as a function of oxygen coverage suggests that the excitation proba-
bility is reduced by oxidizing the surface |89|. In addition a part
of the decrease must be attributed to the reduction of the sputte-
ring yield (sect.III.2.1) |43|.

The evolutions of ion intensities presented here are the most
commonly observed with pure polycrystalline metals. When dealing

with compounds, metal alloys for instance, some differences in the
behaviour are pointed out. For example, in dilute Cu Al alloys the intensity of multicharged aluminium ions increases with oxygen coverage
|85|. In dilute Ni Cr alloys a very pronounced hump is observed on
the I (Cr$^+$) curve, before the surface is saturated with oxygen |65|.
At last, in carbides, most ions associated with carbon atoms decrea-
se very much in intensity when oxygen is adsorbed on the surface |90|.
Most of these particular phenomena are not yet well understood. A
great deal of works has to be done before getting a complete understanding of the ion emission induced by adsorption. Anyway, the sensitivity of ion emission to surface conditions appears to be of a great
interest for adsorption studies.

II-2-3 ion emision of metals at maximum oxygen coverage :

     When the surface is completely covered with oxygen , ion inten-
sities take well defined values characteristic of the nature of the
oxide layer formed. In principle, these values correspond to the saturation levels obtained at high pressure or exposure. The question
to be asked is to know the composition and the structure of the oxide layer formed at saturation. A first attempt to answer this question was given by simultaneously recording the M$^+$ current and the
change $\Delta\phi$ in surface potential against the oxygen partial pressure
|39,65,91|. Two examples are given in fig.25-26 for Mg and W samples.
The sign and the total change in surface potential are in good agreement with results obtained by non-destructive methods. This shows
that the nature of the dipole layer formed by adsorption does not
suffer a great alteration under ion bombardment. Therefore we can
conclude that ion emission at saturation coverage is characteristic
of the metal-oxygen reaction, even under dynamic conditions.
     Incidentaly it is noticed that oxygen has always an enhancing
effect on M$^+$ ion yields, whereas it may cause either a decrease (Mg)
or an increase (W) of the surface potential or even leave it nearly
unchanged (Al) |91|. This shows there is no direct relation between
the ionization of atoms due to oxygen adsorption and surface potential.
     Benninghoven |36| has recorded a great number of mass spectra
characteristic of the oxide layer formed on the surface of metals saturated with oxygen. Yields given in Table X have been obtained by
using a constant transmission factor, therefore, they must be considered, in fact, as partial ion yields (see sect. II.1.2). Elements
at the beginning of each transition series preferably emit positive
molecular ions whereas those which are at the end rather emit negative molecular ions. This explains that similarities in the ion spectra
of elementsof the same column of the periodic table are observed.
Light elements such as Mg, Al emit a great deal of negative molecular
ions. These observations recall the general trend of oxide ion emission (see sect. II.1.3).

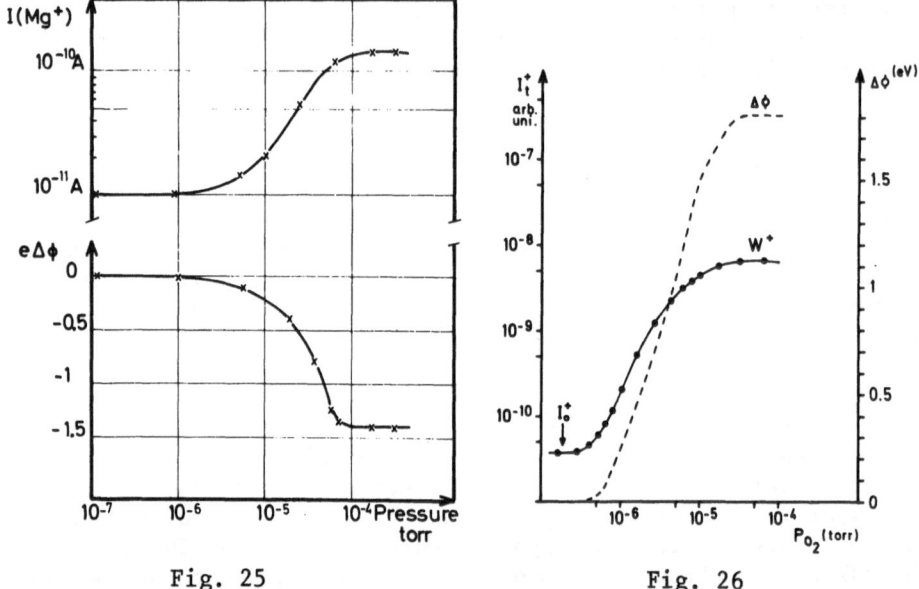

Fig. 25   Fig. 26

Fig.25 - Mg$^+$ ion intensity and surface potentiel $\Delta\phi$ eV variations
          of a polycrystalline magnesium sample, versus oxygen pressu-
          re (bombardment by 6 keV A$^+$ ions; primary ion current densi-
          ty $\sim$ 100 $\mu A/cm^2$). |from 39|.

Fig.26 - W$^+$ ion intensity and surface potential $\Delta\phi$ eV variations of
          a polycrystalline tungsten sample, versus oxygen pressure
          (bombardment by 6 keV A$^+$ ions; primary ion current density
          $\sim$ 100 $\mu A/cm^2$)|from 65|.

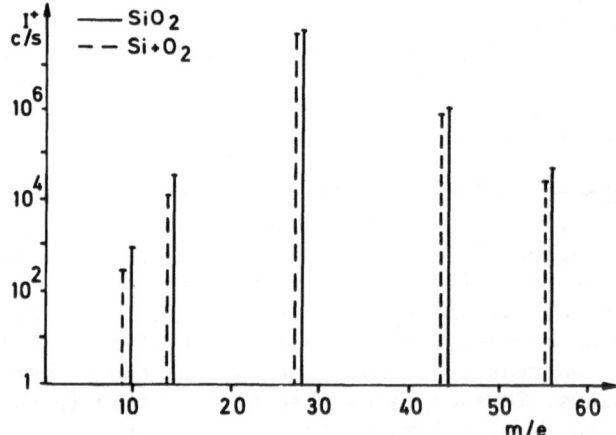

Fig.27 - Intensity of various secondary ions emitted from Si O_2
          and silicon completely covered with oxygen |from 62|.

Table X

Ion yields characteristic of the oxide
layer formed on the surface of metals
saturated with oxygen |from 36|.

Metal	Me^+	MeO^+	MeO_2^+	MeO^-	MeO_2^-	MeO_3^-	MeO_4^-
Mg	0.9	0.0015	–	0.01	0.0025	–	–
Al	0.7	0.0006	–	0.02	0.02	–	–
Ti	0.4	0.5	0.007	–	0.008	0.018	–
V	0.3	0.6	0.01	0.0001	0.002	0.01	0.0001
Cr	1.2	0.2	0.0025	0.00025	0.018	0.07	0.006
Mn	0.3	0.007	–	0.004	0.03	0.004	–
Fe	0.35	0.014	–	0.0007	0.0085	0.0035	–
Ni	0.045	–	–	0.007	0.06	–	–
Cu	0.007	–	–	0.0015	0.015	–	–
Sr	0.16	0.035	–	0.013	0.006	–	–
Nb	0.05	0.3	0.06	–	0.0008	0.02	–
Mo	0.4	0.3	0.017	–	0.0014	0.085	0.014
Ba	0.03	0.017	–	0.0009	0.007	–	–
Ta	0.002	0.02	0.005	–	0.001	0.008	0.0002
W	0.035	0.15	0.012	–	0.0012	0.13	0.01
Si	0.58	0.011	–	0.00038	0.058	0.058	–
Ge	0.02	0.0012	–	0.00081	0.045	0.0081	–

A first step in the identification of the oxide layer formed
on metals exposed to oxygen can be obtained by a direct comparison of
their mass spectra with those of the corresponding bulk metal oxides.
Fig.27 shows that the mass spectrum of Si O_2 is the very same as the
mass spectrum of Si + O_2 saturated|62|. On the contrary for most tran-
sition elements large divergences in mass spectra are observed with
respect to the most common oxides : oxides always emit less $M O_n^+$ and
more $M_2 O_n^+$ ions than the metal covered with oxygen (fig.28).
    The two different situations reported in Table XI can be ex-
plained by the influence of oxygen on the surface of an oxide : the-
re is no oxygen adsorption on the surface of Si O_2, $Al_2 O_3$, MgO |44,
62| , whereas adsorption is still possible on transition metal oxides
such as Ti O_2 , $Cr_2 O_3$, Ni O |46|. It is easy to know by ion emission
whether oxygen adsorption is possible on an oxide . For example, when
Si O_2 is exposed to increasing oxygen pressure, the Si^+ signal remains

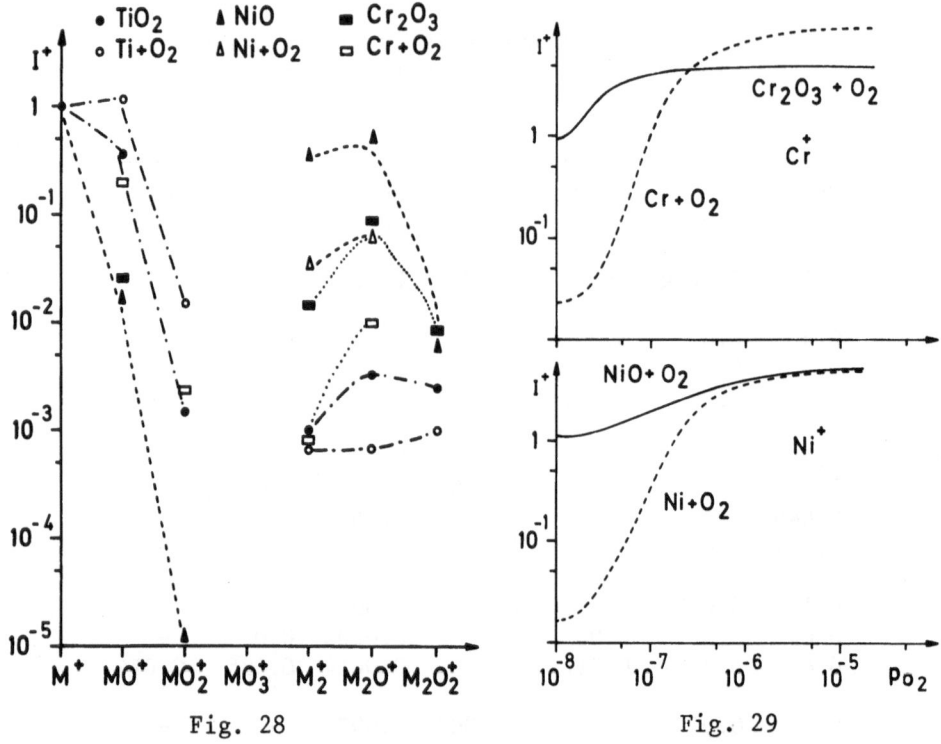

Fig. 28

Fig. 29

Fig.28 – Comparison of ion spectra from metals completely covered
        with oxygen and oxides |from 46|.
Fig.29 – Variations of Cr^+ and Ni^+ ion intensities of Cr and Ni me-
        tals and $Cr_2 O_3$ and Ni O oxides as a function of oxygen
        pressure |46|.

strictly constant |44,92 |. This means that oxygen has no influence
on the surface. On the contrary when the same experiment is done on
a transition metal oxide, one observes an increase of the positive
metal ion emission as a function of pressure (fig.29)|46|. Then we
can conclude that in the former case a Si O_2 layer is formed on silicon exposed to oxygen whereas in the latter case the oxide layer is
different from the bulk oxide.

    The second step in the classification of an oxide layer con-
sists of comparing the mass spectra of metals and oxides under the
same experimental conditions, that is under oxygen saturation. This
situation is illustrated in fig.29 by a comparison, at oxygen saturation,of the M^+ intensities from chromium metal and $Cr_2 O_3$ and from
nickel metal and NiO |46|: at saturation the Ni^+ intensity is the sa-
me for the metal and the oxide, whereas the Cr^+ intensity is twice

<u>Table XI</u>

Comparison of the mass spectra of oxides
and metals saturated with oxygen |from
44,46,62|.

metal + O_2 saturated	oxide (no oxygen adsorption)		metal + O_2 saturated	oxide (oxygen adsorption is still possible)
Mg + O_2	=	MgO	Ti + O_2	$\neq$ Ti O_2
Al + O_2	=	Al$_2$ O$_3$	Cr + O_2	$\neq$ Cr$_2$ O$_3$
Si + O_2	=	Si O$_2$	Ni + O_2	$\neq$ NiO

higher for the metal. Furthermore the mass spectra are the same for
nickel and nickel oxide and different for chromium and chromium oxi-
de |46|.Then we can conclude that the oxide layer

$$Ni + O_2 \text{ saturated } \equiv NiO + O_2 \text{ saturated}$$

$$Cr + O_2 \text{ saturated } \neq Cr_2 O_3 + O_2 \text{ saturated}$$

In other words the oxide layer formed on nickel is certainly
NiO whereas it is sure that chromium oxidation does not lead to the
formation of Cr$_2$ O$_3$ (fig.30).

These few examples show the interest of ion emission for the
identification of oxide layers but, so far, little work has been done in that field.

II-2-5 <u>cleaning of surfaces by ion bombardment</u> |45| :

Ion bombardment is usually used for cleaning the surface of
samples. But, during the cleaning, the sample is surrounded by an
active atmosphere (oxygen essentially) which partial pressure is p.
Therefore, the lowest coverage reached under bombardment will be given by expression (16). The extreme sensitivity of ion emission to
surface composition can provide an excellent cleanness criterion.
It was mentioned (section II.2.2) that the reinforcement of M^+ ion
emission is commonly of a factor 100 at complete oxygen coverage
of the surface. Then, supposing a reinforcement proportional to the
coverage, an enhancement of less than 10% of the intrinsic M^+ emission will correspond to a coverage of $\theta < 10^{-3}$. To estimate the parameter a of exp.(17) we used Wittmaack's experimental conditions

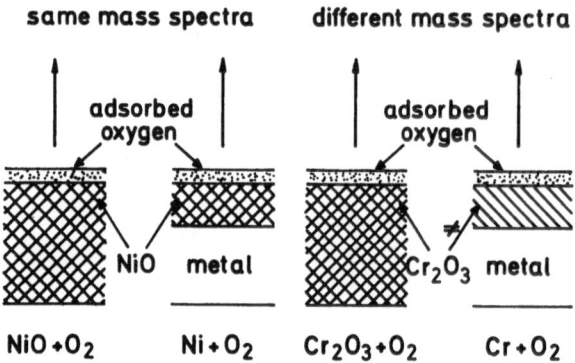

Fig.30 – Comparison of the mass spectra of metals (Cr and Ni) and
          oxides (NiO and Cr₂ O₃) both completely covered with oxy-
          gen |46|.

indicated in fig. 22 |62|. The Si⁺ emission has doubled when the
pressure was ∿ 4x10⁻⁷ torr. This means that θ ∿ 10⁻² at this pressure.
     Under these conditions we found a = 10⁻¹⁸m² assuming s(θ) ∿ 1.
Finally the cleanness criterion may be expressed as

$$\text{Ip} > 6 \; 10^7 \; \frac{p}{S} \qquad\qquad (22)$$

where Ip is the primary beam current in μA/mm² and p the pressure in
torr.
For experiments carried out with primary ions of a few keV energy
directed at an angle θi ∿ 45°, the sputtering yields of most metals
will be of the order of 4 to 10. Supposing a residual pressure of
∿ 10⁻⁷ torr the cleaning of the surface requires a primary density
≳ 1 μA/mm².
     Condition (22) will surely be considered as drastic by many ex-
perimentators. In fact, it is valid for metals the most sensitive to
oxygen. For example every time this condition is not fulfilled the lower plateau of the intrinsic emission of these metals is not clearly
observed. It is the case in Benninghoven's experiment presented in
fig.23 |69| where the chromium emission was recorded as a function
of oxygen exposure with a primary density of 10⁻⁴μA/mm² under a resi-
dual pressure of 10⁻¹⁰ torr. Jurela's |37,38| experimental conditions
are still worse for measuring the intrinsic ion emission since the
primary density is ∿ 1 μA/mm² and the residual pressure ∿ 2 10⁻⁶
torr (Table II). This probably explains the high ion yields measured
by Jurela on metals very sensitive to oxygen (Mg, Al, Ni, Mo, Ta, W
in Table II).

II-2-5 ion emission induced by the implanted gas :

When the primary ions are chemically active their implantation
in the solid will cause a specific ion emission |93,94|. For example,
in the case of oxygen bombardment, ions of the same species as those
given by oxides will be observed. Furthermore, these chemically active ions have an influence on secondary ion yields. Bombarding with
electronegative species (e.g oxygen), positive secondary ion yields
are enhanced |93,94| , but the enhancement is much smaller than in
the case of oxides. When electropositive ions (e.g Cs^+) are used,
an enhancement of negative secondary ion yields is observed |40,93,
94|.

The surface concentration $c = \alpha/S$ of implanted ions is proportional to the probability α for an atom to be implanted and inversely
proportional to the sputtering yield S. Therefore, the effect of implantation on ion emission will greatly depend on the material to
be sputtered : the smaller the sputtering yield, the higher the implanted dose. In particular, in the case of polycrystalline samples
the effect of implantation will be different from one grain to ano-
ther |95|.

On the other hand a part of the primary beam is scattered by
the surface and neutralized in the sample chamber. Then, the residual
atmosphere is enriched in oxygen which adsorbs on the surface. This
influence of adsorbed oxygen on ion emission has to be added to the
influence of implanted oxygen. Both effects have not been clearly
decoupled but, anyway, it seems that they cause very selective per-
turbation on the various atomic species which compose the sample|95|.

III – SPUTTERING

III-1 Sputtering yield of a monoatomic solid :

III-1-2 amorphous solids :

The sputtering of a solid results in the ejection of atomic
particles under the impact of energetic ions or atoms. The theory
of sputtering was developped by Sigmund |96,97| who used the Bolt-
zmann transport equation to describe the random collision cascades in
a solid. This theory applies to random monoatomic targets that is,
in practice, to amorphous solids or – in a rather good approximation
– to polycrystalline solids.

For ions of mass M_1, atomic number Z_1 and energy E_o bombarding
at normal incidence the surface of a target composed of atoms of mass
M_2 and atomic Z_2 , the sputtering yield S_o, i.e. the average number
of target atoms ejected from the solid per incident ion, is expressed by :

$$S_o = 0.042 \ \alpha(M_2/ \ M_1) \ \frac{Sn(E_o)}{U} \qquad (23)$$

where – U is the binding energy of surface atoms
       – $Sn(E_o)$ the stopping cross section of the incident particle

in the solid.
- $\alpha(M_2/M_1)$ has been calculated as a function of the mass ratio M_2/M_1 (fig.31) $|97,98|$. It shows only a slight variation with the bombarding energy E_o for heavy projectiles,
- 0.042 is calculated in $\overset{\circ}{A}{}^{-2}$

Expression (23) can be written more conveniently by introducing a dimensionless variable $|99|$.

$$\varepsilon = 6.95 \ 10^{-2} \ \frac{a}{Z_1 Z_2} \ \frac{M_2}{M_1 + M_2} \ E_o \qquad (24)$$

$$\text{with} \quad a = 0.47 \ \left[z_1^{2/3} + z_2^{2/3}\right]^{-1/2}$$

In these expressions E_o and U are in eV and a in $\overset{\circ}{A}$.
One gets then :

$$S_o = 4.2 \ 10^{-2} \ a^2 \ \alpha \ \frac{Tm}{U} \ \frac{sn(\varepsilon)}{\varepsilon} \qquad (25)$$

$$\text{with} \quad Tm = \frac{4M_1 M_2}{(M_1 + M_2)^2} \ E_o \qquad (26)$$

The reduced nuclear stopping cross section $s_n(\varepsilon)$ is represented in fig.33 $|96|$.

Theoretical sputtering yields predicted by (26) have been compared rather extensively with experimental data $|96|$. At bombarding energies in the keV range a good quantitative agreement between theory and measurments was found for light or intermediate mass of the projectile and the target atoms. Systematic deviations are observed for heavy bombarding particles such as X_e^+ $|100|$.

The influence of the bombarding incidence θ_i on the sputtering yield is represented in fig.32 for high and low energy bombardment, 20 keV and 1 keV respectively $|98,100|$. At first, when θ_i is small the sputtering yield always increases with θ_i then it reaches a maximum at about $\sim 70°$ and drops steeply when θ_i approaches 90°. This behaviour is not completely explained by the theory. However at high bombarding energies ($\sim$ 20 keV) Sigmund's theory predicts the dependence $|97|$:

$$S(\theta_i) = S_o \ | \cos.\theta_i|^{-5/3} \qquad (27)$$

which is in good agreement with most experimental data for $\theta_i < 70°$. At low bombarding energy there is not a very clear θ_i dependence for all metals $|100|$. For $\theta_i < 60°$ the best approximation seems to be :

$$S(\theta_i) = S_o |\cos.\theta_i|^{-1} \qquad (28)$$

III-1-2 crystalline solids :

When the first few atomic layers (5 or 6) of the surface of a solid remain crystallized under ion bombardment, an influence of the arrangment of atoms on sputtering is observed. The lattice structure has an effect on the penetration of primary ions and ejection of se-

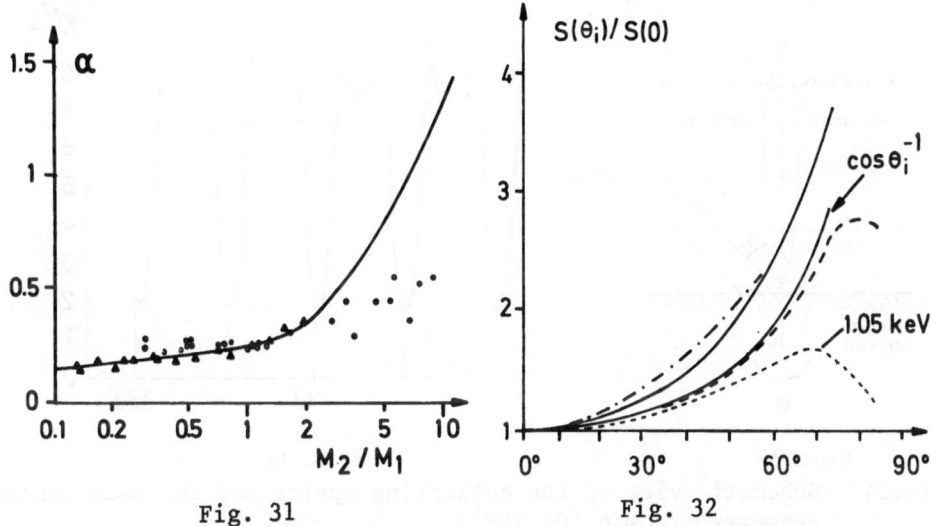

Fig. 31   Fig. 32

Fig.31 - Experimental and theoretical dependence of α on M_2/M_1 |from
            98|. ——— Sigmund's theory |96|; ● ▲ ○experiments |98|.

Fig.32 - Experimental and theoretical dependence of the sputtering
            yield of polycrystalline copper, on the bombarding angle
            θ_i |from 100|.
            Bombardment by A^+ ions :
            - with energy ≥ 20 keV : ——— Sigmund's theory |96|;
            ---- V.A.Molchanov et al. |101| ; —·—·— P.K. Rol et al.
            |102|.
            - with energy = 1.05 keV : ----- H. Oechsner |103|.

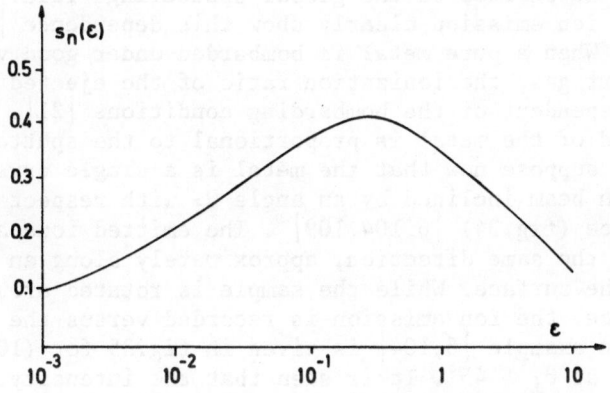

Fig.33 - Reduced nuclear stopping cross section |99|.

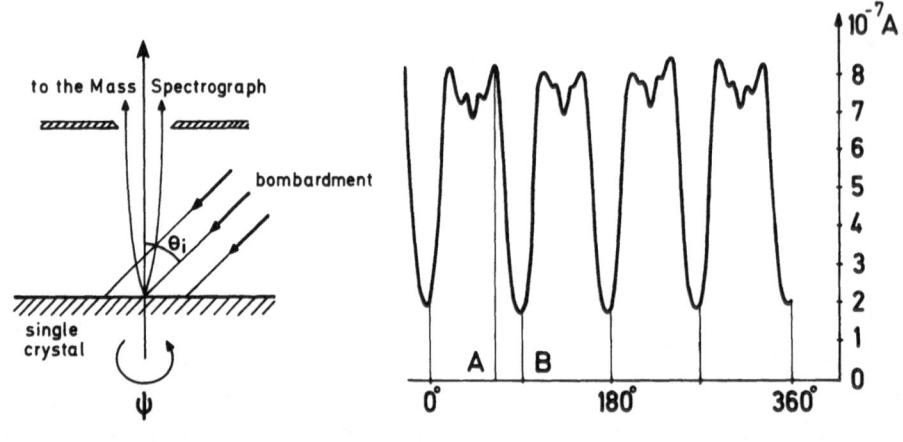

Fig. 34   Fig. 35

Fig.34 – Schematic view of the collecting optics and the beam target
       geometry |from 6,104,109|.

Fig.35 – Variation of Al$^+$ ion intensity of a(100) Aluminium
       single – crystal as a function of the azimuthal angle ψ
       (see fig.34). The primary argon beam is nearly along <110>
       direction when $\psi = \psi_B$ ($\theta_i \sim 45°$) |from 6,104|.

condary particles.

     The anisotropic angular distribution of atoms reveals the in-
fluence of the lattice on the ejection direction of the particles.
The main features described for ions in sect.III-1-1-1 are also va-
lid for neutral atoms |105-108|.

     The effect of the lattice structure on the penetration of the
primary ions is related to the global sputtering yield. Numerous ex-
periments on ion emission clearly show this dependence |6,34,86,87,
95,104,109|. When a pure metal is bombarded under good vacuum condi-
tions by inert gas, the ionization ratio of the ejected atoms is cons-
tant and independent of the bombarding conditions |21|. Therefore,
the ion yield of the metal is proportional to the sputtering yield.

     Let us suppose now that the metal is a single crystal bombar-
ded by an ion beam inclined by an angle θ_i with respect to the normal
to the surface (fig.34) |6,104,109| . The emitted ions are always
collected in the same direction, approximately along an axis perpendicular to the surface. While the sample is rotated around the normal
to the surface, the ion emission is recorded versus the angle of ro-
tation ψ. An example |6,104| is given in fig.35 for (100) Al surfa-
ce bombarded at $\theta_i \sim 45°$. It is seen that the intensity of Al$^+$ ions
varies strongly with the azimuthal angle ψ : deep minima are observed
when the beam is directed along <110> directions ($\psi = 0, 90°, 180°$

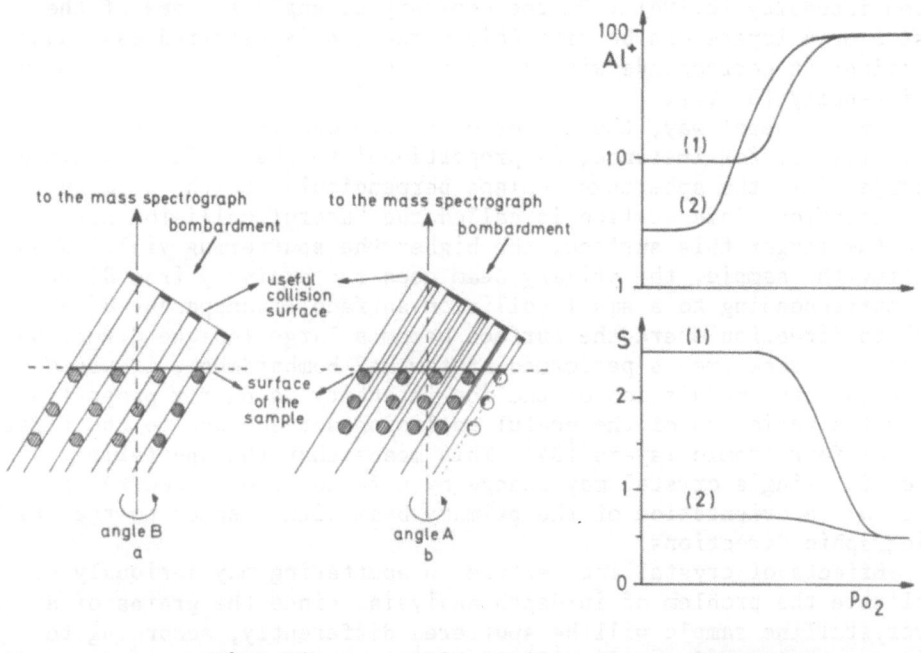

to the mass spectrograph

bombardment

to the mass spectrograph

bombardment

useful
collision
surface

surface
of the
sample

angle B
a

angle A
b

Fig. 36   Fig. 37

Fig.36 - Useful collision surface corresponding to a transparent
         direction (angle B) and to an opaque direction (angle A)
         of bombardment |from 66|.
Fig.37 - Qualitative variations of the Al^+ ion emission and sputte-
         ring yield S as a function of oxygen pressure |from 104|.
         (1) opaque direction of bombardment ; (2) transparent di-
         rection of bombardment.

... in fig.35). As $\tau(Al^+)$ is constant in the experiment, the variations of the intensity are related with the ψ dependence of S. This
dependence can be understood if one agrees that the sputtering yield
depends mainly on the collision cascades initiated by primary ions
in the first 3 or 4 atomic layers for bombardment at a few keV ener-
gy |34,108|.
        In the hard sphere collision model each atom is represented by
a sphere which radius is smaller than the interatomic distance(fig.36)
|108,110,111|. A collision occurs with a lattice atom when the tra-
jectory of the primary particle intercepts a sphere. When the bombardment is directed along a <110> direction (angle B in fig.35-36) only
atoms of the first layer can be hit by the primary ions since other
atoms are masked. This direction is referred as transparent direction.
It corresponds to a low sputtering yield and therefore a minimum

of ion intensity $|6,104|$. On the contrary at angle A atoms of the
first 3 or 4 layers can be hit. This direction is referred as opaque
direction; it corresponds with high sputtering yield and maximum in
ion intensity $|6,104|$.

In a general way, the number of collisions in the first three
atomic layers, for instance, is proportional to the surface obtained
by projecting the spheres on a plane perpendicular to the primary
beam direction. This surface is called the "useful collision surface"
$|34|$; the larger this surface, the higher the sputtering yield. When
rotating the sample, the primary beam goes successively from direction corresponding to a small collision surface (transparent direction) to direction where the surface becomes large (opaque direction).
For all the experiments performed at few keV bombardment, it was observed that the modulation of the ion intensity is nearly proportional to the variation of the useful collision surface due to the first
three or four atomic layers $|34|$. This means that the sputtering
yield of a single crystal may change by a factor 3 or 4 according to
the relative orientation of the primary beam with respect to the crystallographic directions.

Effects of crystalline texture on sputtering may seriously
complicate the problem of in-depth analysis, since the grains of a
polycrystalline sample will be sputtered differently, according to
their respective orientation $|86,95,109|$.

Even, the sputtering yield of a single crystal may change during erosion if the lattice structure is affected by the bombardment.
The typical example is given by single crystals of covalent elements
Si, Ge which become amorphous under bombardment at room temperature
$|112,115|$. This means that during erosion of the first 100-200 Å the
sputtering yield will increase in proportion as the crystal structure is becoming amorphous. Proceeding at high temperature (250 to
500°C) these materials remain as single crystal under bombardment
$|112|$.

III-2 <u>Sputtering of compounds</u> :

No theory accounts for the sputtering yield of compounds composed of several atomic species. However, the global composition of
the sputtered products is exactly the same as that of the initial
solid, whatever the nature of this latter *. If the amount of atoms
A, B, C ... is n_A, n_B, n_C ... in the sputtered products, after bombarding with N_p primary ions, the global sputtering yield is expres-

---

* Transient regime will be discussed later on.

sed as :

$$S_T = \frac{n_A + n_B + n_C + \ldots}{N_p} \qquad (29)$$

The atomic concentration of any species - A for example - in the sputtered products is then given by :

$$\left(C_A\right)_{sputtered} = \frac{n_A}{n_A + n_B + n_C \ldots} = \left(C_A\right)_{solid} \qquad (30)$$

But in a complex sample, the various atomic species are ejected dif-
ferently according to their mass and their binding energy |95, 100,
116-119|. It results that the permanent sputtering regime is reached
after a certain time of bombardment to the detriment of a change in
the surface composition |117-122|.

III-2-1 <u>sputtering of a one-phased system</u> :

    One defines the partial sputtering yield for any element - A
for example - of the sample by :

$$S_A = \frac{n_A}{Np} = C_A \ S_T \qquad (31)$$

Measurments of the total sputtering yield of compounds |116| are very
scarce, therefore, approximations have to be used more often. For e-
xample, in binary alloys AB where B is the solute element at low
concentration , one can admit that the total sputtering yield of the
alloy $S_T(AB)$ is nearly equal to the sputtering yield S(A) of the pu-
re metal A |50|

$$S(AB) \sim S(A)$$

When A and B are neighbouring atoms it is also frequently assumed
that the sputtering yield of the alloy is constant and independent
on the respective concentrations C_A and C_B |100|.
    The lack of informations concerning the sputtering of compounds
is an handicap for analytical applications. However this difficulty
can be overcome by employing convenient experimental procedure (see
section V.1).
    The adsorption of oxygen on the surface of a metal also has an
effect on the sputtering yield |43,86,104| . For example, the average
sputtering yield S(Al) of a polycrystalline Al sample, bombarded with
8 keV A^+ ions, varies from $\sim$ 2 when the surface is perfectly clean
to $\sim$ 0.5 when the surface is completely covered with oxygen |43|.
When bombarding an Al single crystal the sputtering yield varies ac-
cording to the initial transparency of the metal (fig.37) |86, 104|.
But, whatever the initial transparency, the ion intensity Al^+ as well
as the sputtering yield always reach the same value at complete coverage. This means that lattice effects disappear completely when the
surface is saturated with oxygen |86|. Furthermore, it is noticed

that the final value S(Al) = 0.5 corresponds to the sputtering yield
of aluminium atoms in amorphous Al_2O_3|43|.This confirms that alumina
is obtained after exposing Al to oxygen (see sect. II.2.3). The same
conclusion applies to Si and probably Mg |43|. For many transition
metals and alloys the sputtering yield is reduced by oxygen adsorp-
tion but some lattice effects always persist |6,87,104|. It appears
that oxygen adsorption always decreases the sputtering yield of a
material, but the decrease is important at the end of adsorption
|104|.

Analysis of bombarded surfaces by Auger Spectroscopy revealed
that sputtering induced changes in the surface composition of com-
pounds |117,122|. It is out of the scope of this lecture to give a
thorough explanation of this phenomenon, but, we would like to show
the interest of ion emission in this field. In the following we will
use a very crude model |45| for describing the change in surface com-
position, based on Sigmund's equat.23.

In equat.23,α is a function of the mass ratio M_2/M_1 and Sn
characterizes the slowing down of the incident particles in the solid.
Let us consider an alloy AB where C_A,C_B and $U_A(AB)$, $U_B(AB)$ are respectively the bulk atomic concentrations and the binding energies of
atoms A and B in the alloy. Regarding the slowing down of incident
particles in the alloy we can introduce a stopping cross section
$S_n(AB)$ which will be function of concentrations. If A and B are
neighbouring mass atoms one can admit that $\alpha(AB)$ is roughly constant
with respect to the concentrations (fig.31). If the solid was only
composed of atoms A, its sputtering yield will be :

$$S_A^\circ = 0.042 \; \alpha(AB) \; \frac{S_n(AB)}{U_A(AB)} \qquad (32)$$

As the surface concentration of A atoms is really C_A^S in layers concerned by sputtering, the partial sputtering yield is

$$S_A = S_A^\circ \; C_A^S \qquad (33)$$

Likewise for B atoms one obtains

$$S_B = S_B^\circ \; C_B^S \qquad (33')$$

The compositional conservation law of the sputtered products (equat.
30) can be written :

$$\frac{S_A}{S_A + S_B} = C_A \qquad \text{idem for B atoms (34)}$$

Finally by combining exp.32-33-33'-34 one obtains

$$\frac{C_A^S}{C_B^S} = \frac{U_A(AB)}{U_B(AB)} \; \frac{C_A}{C_B} \qquad (35)$$

Approximations used to derived exp.35 are very crude. They suppose, in particular, that all atomic species of the alloy have the same energy distribution *. However the surface composition of Cu Ni alloys was found to be in good agreement with expression (35)** |118|.

Let us suppose now that the composition of the alloy is initially the same at the surface and in the bulk, that is C_A and C_B. A change of composition from C_A to C_A^S and C_B to C_B^S occurs progressively in time, as the bombardment proceeds. We consider a surface unity of the sample whose atomic surface density is N_0 at/cm^2. If $n_A(t)$ and $n_B(t)$ are the number of surface atoms A and B at time t, the surface concentrations are given by :

$$C_A^S(t) = \frac{n_A(t)}{N_0} \qquad \text{idem for B} \qquad (36)$$

The number $\Delta'n_A$ and $\Delta'n_B$ of atoms disappearing from the surface during a time Δt is :

$$\Delta'n_A = S_A(t) \, Np \, \Delta t \qquad (37)$$

$$\Delta'n_B = S_B(t) \, Np \, \Delta t$$

where $S_A(t)$ and $S_B(t)$ are the sputtering yields given by (33-33') at time t.

When the bombarding energy is $\lesssim 1$ keV most ejected particles come from the topmost layer. Therefore, we can assume that the com-

---

* From Sigmund's theory |96| the energy distribution is given by $\frac{dN}{dE} = \frac{Sn}{E^2}$. It depends only on the stopping cross section of the solid. Therefore according to this model, the energy distributions of A and B atoms will be proportional to the stopping cross section $S_n(AB)$ of the alloy, namely :

$$\frac{1}{C_A} \frac{dN_A}{dE} = \frac{1}{C_B} \frac{dN_B}{dE} = \frac{S_n(AB)}{E^2}$$

As a consequence it appears that both energy distributions are similar. This result can be understood if a great number of collisions occurs before ejection, because it implies the equipartition of energy between the various atomic species of the alloy.

** Using a different approach Shimizu et al. obtained an expression nearly similar to (35) |118|.

position is mainly changed in this layer. The sputtering of $\Delta'n_A + \Delta'n_B$ atoms from the surface produces the appearance of :

$$\Delta n_A = C_A (\Delta'n_A + \Delta'n_B) \tag{38}$$

atoms A from the second layer.

The variation of the surface concentration of A atoms per unit time is then given by :

$$\frac{dn_A(t)}{dt} = N_0 \qquad \frac{dC_A^s(t)}{dt} = \lim_{\Delta t \to o} \frac{\Delta n_A - \Delta'n_A}{\Delta t}$$

namely :

$$\frac{dC_A^s(t)}{dt} = \frac{Np}{N_0} \left[C_A S_B^o - (C_A S_B^o + C_B S_A^o) C_A^s \right] \tag{39}$$

Expression (40) shows that C_A^s will vary exponentially in time with an average time :

$$T = \frac{N_0}{Np} \frac{1}{C_A S_B^o + C_B S_A^o} \tag{40}$$

To estimate T we suppose :

$$C_A \sim C_B \sim 0.5 \quad , \quad S_A^o \sim S_B^o \sim 3 \quad \text{and} \quad N_0 \sim 10^{15} \text{ atoms/cm}^2$$

For a primary density $Ip \sim 1 \ \mu A/cm^2$ which corresponds to quasi-static conditions one obtains : $T \sim 20^s$.

III-2-2 sputtering of a multi-phased system :

We consider a multi-phased system formed by juxtaposed grains which size is large as compare to the average lengh of the collision cascades responsible for the sputtering. This means that the major part of the collision cascades initiated in a grain will end up by ejection of atoms of the same grain, except a few of them generated at the fringe of each grain. Under these conditions the sputtering yields of the different phases will be considered as independent each other, since the sputtering of one phase by collisions initiated in another phase will be negligible.

Let S_ϕ be the sputtering yield of a given phase ϕ. If we consider the unit surface of the system where Σ_ϕ is the area occupied by the phase ϕ, the number of atoms ejected from ϕ per unit time is :

$$\Sigma_\phi S_\phi Np \tag{41}$$

If C_ϕ^i is the atomic concentration of an element i in the phase ϕ the partial sputtering yield of i atoms is :

$$S_\phi^i = C_\phi^i \Sigma_\phi S_\phi \tag{42}$$

The knowledge of S_ϕ^i requires the determination of Σ_ϕ in particular. We shall distinguish between two cases.

III-2-2-1 <u>sputtering of a small grains phase</u> : |123| .

For the sake of clarity we deal with a two-phase system in which a phase ϕ is formed of small precipitates disperced into a matrix.

The surface Σ_ϕ occupied by the precipitates is expressed as:

$$\Sigma_\phi = n_s \, \sigma \tag{43}$$

where σ and n_s are respectively the mean surface and the surface density of the precipitates.

The surface density n_s results from a balance between
- the number $\Delta'n_s$ of precipitates disappearing from the surface by their sputtering, during time Δt,
- the number Δn_s of precipitates appearing at the surface by sputtering of the matrix during Δt.

If N_ϕ is the average number of atoms (all species) in a precipitate, one obtains :

$$\Delta'n_s = \frac{\Sigma_\phi \, S_\phi \, N_p \, \Delta t}{N_\phi} \tag{44}$$

During Δt the number of atoms ejected from the matrix is :

$$(1-\Sigma_\phi) \, S_m \, N_p \, \Delta t \tag{45}$$

If $\bar{v}_m$ is the average volume of a matrix atom, the volume sputtered from the matrix is :

$$\Omega_m = \bar{v}_m \, (1-\Sigma_\phi) \, S_m \, N_p \, \Delta t \tag{46}$$

The sputtering of the matrix has extracted a volume Ω_ϕ of precipitates such as :

$$\frac{\Omega_\phi}{\Omega_m + \Omega_\phi} = \bar{V}_\phi \, n_v \tag{47}$$

where n_v is the number of precipitates per unit volume and $\bar{V}_\phi$ the mean volume of a precipitate.
Therefore, the number Δn_s of precipitates appearing at the surface is :

$$\Delta n_s = \frac{\Omega_\phi}{\bar{V}_\phi} \tag{48}$$

namely :

$$\Delta n_s = \frac{n_v}{1-\bar{V}_\phi \, n_v} \, \bar{v}_m \, (1-\Sigma_\phi) \, S_m \, N_p \, \Delta t \tag{49}$$

Expressions 44-49 can be written more conveniently by introducing D_m and D_ϕ, respectively the number of matrix atoms and precipitate atoms per unit volume.

$$D_m = \frac{1-\bar{V}_\phi \, n_v}{\bar{v}_m} \tag{50}$$

$$D_\phi = n_v \, N_\phi$$

At equilibrium, when $\Delta n_s = \Delta' n_s$ the surface density of precipitates n_s^e is given by :

$$n_s^e = \frac{1}{(1+ \dfrac{S_\phi}{S_m} \dfrac{D_m}{D_\phi})} \tag{51}$$

If n_s^o is the initial surface density of precipitates, after polishing the sample for example, the time evolution of $n_s(t)$ towards the equilibrium value n_s^e is given by :

$$\frac{dn_s}{dt} = \frac{\Delta n_s - \Delta' n_s}{\Delta t} \qquad \Delta t \to 0$$

that is :

$$n_s(t) = n_s^e + (n_s^o - n_s^e) \ e^{-\frac{t}{T}} \tag{52}$$

with

$$T = \frac{N_\phi}{\sigma S_\phi} \ \frac{1}{(1+ \dfrac{S_m}{S_\phi} \dfrac{D_\phi}{D_m}) \ N_p} \tag{53}$$

In practice most of the two-phased systems encountered are composed of a phase very dilute into a matrix, therefore $\dfrac{D_m}{D_\phi} \gg 1$. Then, expressions 51-53 can be approximated by :

$$n_s^e \simeq \frac{1}{\sigma} \ \frac{D_\phi}{D_m} \ \frac{S_m}{S_\phi} \tag{54}$$

$$T \simeq \frac{N_\phi}{\sigma S_\phi N_p} \tag{55}$$

To estimate the average time T which characterizes the time evolution of the surface density we consider the case of spherical predipitates of radius R :

$$\bar{V}_\phi = \frac{4}{3} \ \pi R^3 \quad , \quad \sigma = \frac{2}{3} \ \pi R^2 \qquad N_\phi = \frac{\bar{V}_\phi}{v_\phi}$$

Then

$$T = \frac{2R}{N_p \ S_\phi \ v_\phi} \tag{56}$$

With the following numerical values :

$$\bar{v}_\phi \simeq 7 \overset{\circ}{A}^3 \quad , \quad S_\phi \sim 3 \quad , \quad I_p \sim 10^2 \ \mu A/cm^2$$
$$\text{and } 2R \sim 50 \overset{\circ}{A}$$

one finds $\qquad T \sim 30s$

This example shows that the equilibrium surface density n_s^e will rapidly be reached when operating under dynamic conditions on samples which contain precipitates of a few tens of $\overset{\circ}{A}$ in diameter.
Inserting (54) in S_ϕ^i one obtains under equilibrium conditions

$$S_\phi^i \neq C_\phi^i \; \frac{D_\phi}{D_m} \; S_m \qquad (57)$$

This expression shows that the sputtering of precipitates is regulated by the sputtering S_m of the matrix. In other words there is a coupling in the sputtering of the two phases. An example of this type will be presented in section V.4.2.

III-2-2-2 sputtering of a large grains phase : |90|

Under the same conditions as above, but for precipitates of large diameter, $2R \sim 10 \; \mu$ for example, one obtains $T \sim 2 \; 10^4$ s !!!

The change in the surface density will be quite undiscernible during an experiment. The surface composition is then different from the equilibrium composition given by 51. The problem to be solved now is the knowledge of the area Σ_ϕ covered by the phase ϕ.

A simple method consist of obtaining an image of the surface which reveals the presence of the various phases either by classical metallurgical methods or by using the ion emission itself |1-6,16 90,124 |. The proportion of the phase ϕ will be determined by using quantimetric methods. This is the only way of obtaining Σ_ϕ for inclusions (oxide grains for example) fortuitously embodied in the sample during its preparation, or if the phases have been selectively attacked during the preparation.

When the decomposition into two phases results from a thermodynamic equilibrium, the phase diagram can be used to determine the respective areas Σ_ϕ and $\Sigma_{\phi'}$ of the two phases (fig.38) |90|. With the notations of fig.38, the ratio of the mass abundances m_ϕ and $m_{\phi'}$ of the two phases is given by :

$$\frac{m_\phi}{m_{\phi'}} = \frac{C_{\phi'} - C}{C - C_\phi} = \frac{d_\phi \; V_\phi}{d_{\phi'} \; V_{\phi'}} \qquad (58)$$

where $d_\phi, d_{\phi'}$ are the densities of the two phases.
V_ϕ, $V_{\phi'}$ are the volumes occupied by the phases.
If the surface composition also results from the thermodynamic equilibrium one obtains from Delesse's theorem :|125|

$$\frac{\Sigma_\phi}{\Sigma_{\phi'}} = \frac{V_\phi}{V_{\phi'}} \qquad (59)$$

As $\qquad \Sigma_\phi + \Sigma_{\phi'} = 1$ one gets :

$$\Sigma_\phi = \frac{V_\phi}{V_\phi + V_{\phi'}} \qquad (60)$$

The partial sputtering yield of the phase ϕ is then :

$$S_\phi^i = C_\phi^i \; S_\phi \; \frac{1}{1 + \dfrac{d_\phi}{d_{\phi'}} \; \dfrac{C - C_{\phi'}}{C_{\phi'} - C}} \qquad (61)$$

In that case the phases are sputtered independently each other.

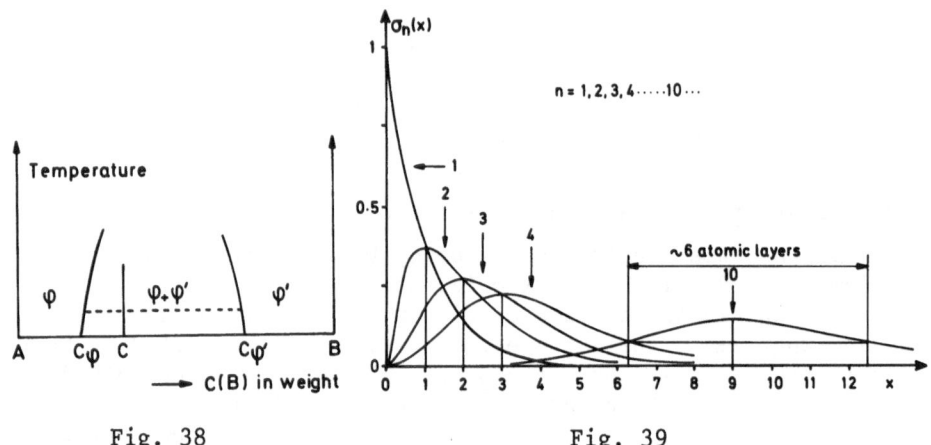

Fig. 38   Fig. 39

Fig.38 - Phase diagram of a binary alloy AB including a two- phased
          region $\phi + \phi'$.
Fig.39 - $\sigma_n(x)$ function (eq.65) corresponding to the erosion of the
          successive atomic layers |from 126|.

III-3 Erosion of a solid :

       It is easily understood that ion bombardment produces a relief
on the surface of a sample which sputtering yield is not uniform :
polycrystalline or multi-phased system for example. But, on the other
hand, it is not easy to find an answer to the problem of the evolution under bombardment of a perfectly homogenous surface such as that
of a single crystal. In fact this problem has not yet received a sa-
tisfying solution.
       A very simple erosion model was developped by Benninghoven
|126| and applied to surface analysis in the monolayer range. It con-
sists of taking a "frozen" solid with a perfectly smooth initial surface and supposing that atoms are always ejected from the topmost layer. If the latter assumption is quite reasonable with bombarding
ions of about $\sim$ 1 keV energy, the former is, on the contrary, not at
all realistic. Therefore, this model must be used carefully because
it can lead to erroneous interpretations |65,66|.
       As the bombardment removes the first atomic layer, the second
monolayer is uncovered and eroded in its turn. Then, the 3rd and
4th atomic layer will be involved in the sputtering process. After
a while, one obtains a step surface.
       Let $s_1(t)$, $s_2(t)$ ..., $s_n(t)$ be the remaining surfaces of 1st,
2nd, ... nth monolayer at time t after the beginning of the bombardment. If the bombardment surface area is unity, we have at any time:

$$1 = \sum_{n=0}^{\infty} s_n(t) \tag{62}$$

During a time dt, the variation of the surface ds_n of any atomic layer is :

$$ds_n = -s_n \frac{dt}{\tau} - ds_{n-1} \qquad (63)$$

with $\tau = \dfrac{N_o}{NpS_o}$ and N_o = the surface atom density of the solid.

On the right side of equation (63), the first term represents the surface of the nth,plane being sputtered during time dt and the second term, the variation of the surface of the $(n-1)^{th}$ layer. One finds by integration :

$$s_n(t) = \frac{1}{(n-1)!} \left(\frac{t}{\tau}\right)^{n-1} \exp - \frac{t}{\tau} \qquad (64)$$

If we introduce a dimensionless variable $x = \dfrac{t}{\tau}$, equat.64 becomes:

$$\sigma_n(x) = \frac{1}{(n-1)!} x^{n-1} \exp - x \qquad (65)$$

$\sigma_n(x)$ are represented in fig.39. One notices these functions are maximum at integer $x = n-1$ values.

According to equat.65 the surface $s_1(t)$ of the first atomic layer decreases exponentially as a function of t. Therefore, the sputtering of an element M which would only be present in the first atomic layer of the surface, would decrease exponentially as a function of time.

From (42) one obtains :

$$s^i = c^i s_T e^{-\frac{t}{\tau}} \qquad (66)$$

If M is present at concentration c^i in the first two atomic layers, one obtains :

$$s^i = c^i s_T (1 + \frac{t}{\tau}) e^{-\frac{t}{\tau}} \qquad (67)$$

In a general way if the concentration of the element M is c_1^i, c_2^i ... in the successive atomic layers :

$$s^i = s_T \sum_{n=1}^{\infty} c_n^i s_n(t) \qquad (68)$$

Let us suppose now that the element M is only present in the nth atomic layer : the 2nd, 3rd ... or another layer.

The sputtering of this monolayer is represented by $\sigma_n(x)$. According to eq. 65, the width of $\sigma_n(x)$ increases continuously as a function of n. For instance, one finds that the Δx interval which corresponds to 50% variation of $\sigma_n(x)$ with respect to the maximum, is about 6 at n = 10 (fig.39) and about 75 at n = 1000 |66|.Consequently, in-depth localization of the atomic layer which contains the M element becomes less and less definite as n increases. Finally we can conclude that, except in the first two or three atomic layer range, in-depth localization rapidly becomes impossible. If we characterize in-depth resolution by the width Δx previously defined, one

can say in-depth resolution is at least 6 atomic layers when the maximum of the surface of the 10th atomic layer is reached and about
75 atomic layers in the neighbourhood of the 1000th atomic layers.
In other words, the erosion model of a frozen solid leads to the conclusion that any method using ion bombardment is unsuited for in-
depth analysis on a large scale. On the contrary, it has been demonstrated that diffusion profiles of several microns can be determined with an excellent in-depth resolution of about 50Å (that is
~ 15 atomic layers) by using ion erosion |6,15,92| In fact the model
used to describe the erosion is not realistic. It leads to a considerable increase of the free surface of the solid which is inconsistent with the stability of this surface. It is sure that surface
tension will reduce the roughness of the relief in proportion to the
progress of erosion. It will result in a softening of the surface
which probably explains the good experimental in-depth resolution,
but which is also responsible for the mixing of the structures of
several atomic layers. On the other hand, this model completely i-
gnores the surface diffusion which, in fact, may have a very impor-
tant role in the surface structure, even at room temperature |127|.
Neglecting surface diffusion has led to erroneous interpretation of
the thickness of the atomic layers formed on metals exposed to oxy-
gen (sect. V.1) |65|.

III-4 Ejection of clusters :

        The existence of ion clusters does not necessarily implies that
of neutral clusters. However, before ionization an ejection process
takes place which must be propitious to the gathering of atoms. It
is this point that will be examine here.
        The formation of a molecule resulting from the ejection of seve-
ral atoms requires three conditions as a preliminary:
- the mutual distances of the ejected atoms must be of the order of
the interatomic distances in molecules. The most favorable case corresponds to the ejection of atoms from adjacent sites on the surface.
- all the participating atoms must be ejected simultaneously. We denote f_1 , f_2 ... f_n the probabilities for the simultaneous ejection
of 2,3 ... n atoms consecutive to a primary impact |60,128|. As it is
obvious that f_n decreases as n increases, the intensities of molecular species will fall off rapidly as the number of particles in combination becomes larger and larger.
- the magnitude and the direction of the velocity must be about the
same for the various atomic components.
        These conditions lead to conclude that polyatomic clusters ha-
ve initially a plane structure even if arrangement takes place fur-
therly during their flight |60,129|.
        Let us consider a solid formed by a two-isotopes element A
and B of respective natural abundances C_A and C_B. It is always ob-

served that the partial intensities of the various ion clusters
$A_m B_n^+$ including a constant number n+m of atoms were proportional to
the isotopic   abundance of the cluster $A_m B_n$:

$$I(A_m B_n^+) \propto \frac{(n+m)!}{n!m!} \quad C_A^n \; C_B^m \qquad (69)$$

This law just implies a random distribution of atoms without
taking the ejection process into account. In fact, one can justifies
its validity in a context which agrees with the conditions imposed to
the formation of molecular ions in sputtering |45|.

Let us suppose that an atom A preferably associates with its
nearest neighbours. If A and B atoms are distributed at random, the
probability of finding A and B atoms around A are respectively C_A
and C_B. Then, the probability of finding AA and AB associations will
respectively be proportional to C_A and C_B. As C_A also is the bulk
concentration of A, one obtains C_A^2 AA associations and $C_A C_B$ AB
associations. In the same way for B atoms one obtains $C_B C_A$ BA associations and C_B^2 BB associations. Finally the total number of AA,
AB and BB associations is respectively C_A^2, $2C_A C_B$, C_B^2. This is exactly the result given by the binomial exp 69, but it has been obtained
in accordance with the condition that cluster atoms are located at the
surface on adjacent sites.

This formation of clusters can be easily extended to the case
of 3, 4 ... atoms by supposing that each atom of the chain associates
with its nearest neighbours.

This description of the clustering of atoms is quite general,
whatever the nature of atoms (isotopes or different atomic species).
But the binomial law given in 69 is only valid if atoms are distributed at random so that the bulk concentrations C_A, C_B are the same as
the "local concentrations" around each atom. In the case of ordered
structures both concentrations may be different and the reasoning has
to be slightly modified. Instead of C_A and C_B we consider the "local
concentrations" around A and B, deduced from the atomic arrangement.
For example, if n_A and n_B are the number of A and B atoms nearest
neighbours of an atom A, the local concentrations are defined as :

$$a = \frac{n_A}{n_A + n_B} \quad \text{for A atoms} \quad ; \quad b = \frac{n_B}{n_A + n_B} \quad \text{for B atoms.}$$

Likewise around a B atom the concentrations will be α and β. According to the diagram in Table XII, the association probabilities are
respectively :

$$
\begin{aligned}
&C_A a & &\text{for AA} \\
&C_A b + C_B \alpha & &\text{for AB} \\
&C_B \beta & &\text{for B}
\end{aligned}
\qquad (70)
$$

The binomial law is no longer valid in that case and the calculation
becomes very intricate in proportion as the cluster increase.

Table XII

	local concentrations	cluster	probability

bulk concentrations

$A \\ C_A$ $\begin{cases} a \longrightarrow A \\ b \longrightarrow B \end{cases}$ AA   AB $C_A a$ $C_A b$

$B \\ C_B$ $\begin{cases} \alpha \longrightarrow A \\ \beta \longrightarrow B \end{cases}$ BA   BB $C_B \alpha$ $C_B \beta$

An example of such a calculation based on the local concentration model will be given in sect. V.3.2 for copper-aluminium clusters of a
θ phase (Al_2Cu).
     Whatever the situation, atoms at random or in an organized struc-
ture, the number of clusters will be proportional to the number of
ejected atoms, that is, in fact, to the total sputtering yield S_T.
This leads to define a "pseudo sputtering yield" for clusters by :

$$S_{A_m B_n} = f_{m+n} \, F_{m,n} \, (C_A, C_B, a, b, \ldots) \, S_T \qquad (71)$$

where f_{m+n} is the probability for the simultaneous ejection of m+n
atoms consecutive to a primary impact and $F_{m,n}$ is the association
probability of A and B atoms into a cluster $A_m B_n$. $F_{m,n}$ is connected
with the concentrations C_A, C_B of   elements and the structure of
the solid (a, b, α, β). For this reason, it will be called "strutural
factor". According as the solid is disordered or organized, $F_{m,n}$ is
given either by the binomial law (exp 69) or by expressions derived
from the diagram represented in Table XII (exp.70).
     We have just examined here the "spatial aspect" of the formation
of clusters. The "energy aspect" will be developed with the ionization process in sect.V

IV.   IONIZATION PROCESSES OF SPUTTERED ATOMS :

IV-1 General remark :

     Historically V.I. Veksler and M.B. Ben'Iaminovich |130| exa-

mined the first the problem of the positive ion emission from metals
on the basis of the Saha-Langmuir surface ionization equation |131|.
They concluded it was impossible to explain their experimental re -
sults by including the thermodynamic temperature and the work function of the metal, in this equation.

When a great deal of experimental results were gathered later
on, one noticed that the ionization efficiency of atoms was approxi-
mately inversely proportional to their ionization energy I |52,55,
93|. This dependence is illustrated in fig.40 |55|for transition ele-
ments. Indeed, a parallel exists between low ion yields and high ionization energies of elements at the end of each transition series
(Cu,Ag,Pt,Pd,Au) but numerous anomalies are also found (Cr,Fe,Co,Ni,
Ta,W ...). Therefore, the ionization energy dependence of ion yields
must be considered as a general trend rather than as a physical experimental law. Nevertheless it is interesting to look for a phenomenological relation between the ionization probability P^+ of an atom
and its ionization energy I. Noticing that variations of I are very
small (a few eV) as compared to ion yields which may vary by several
orders of magnitude (Table II), the only acceptable relation is, in
first approximation, exponential :

$$K^+ \propto \exp- \frac{I}{\alpha_o} \qquad\qquad (72)$$

where α_o is a coefficient.

Plotting the ion yields as a function of I in a semi-log dia-
gram, a roughly linear dependence must be obtained |10,17|. To the
question of the physical meaning of the coefficient α_o several answers were given, all based on thermal considerations. It resulted
that α_o was assimilated to an absolute temperature T.

Essentially two ionization models were developped on thermal
concepts. One is known as plasma model under Local Thermodynamic
Equilibrium (LTE) |10,17,93,94,132| , the other as Thermodynamic non-
equilibrium model of Surface Ionization |37,38|. In both cases the
temperature is deduced from experimental results. It varies usually
from 5000 to 15000°K depending on the sample. Furthermore instead of
I a reduced ionization energy I'=I-ΔE was always introduced in the
equations. In the plasma model the reduction ΔE is attributed to the
Coulomb interaction :

$$\Delta E = 2.09 \ 10^{-8} \times 0.68 \times (\frac{n_e}{T})^{1/2}$$

where n_e is the electron density in the plasma. In the surface ionization model ΔE represents the reduction in energy due to the image for-
ce : |131|

$$\Delta E = \frac{e^2}{4 \ x_c}$$

where x_c is a critical distance $\sim$ 1/2 the interatomic distance.

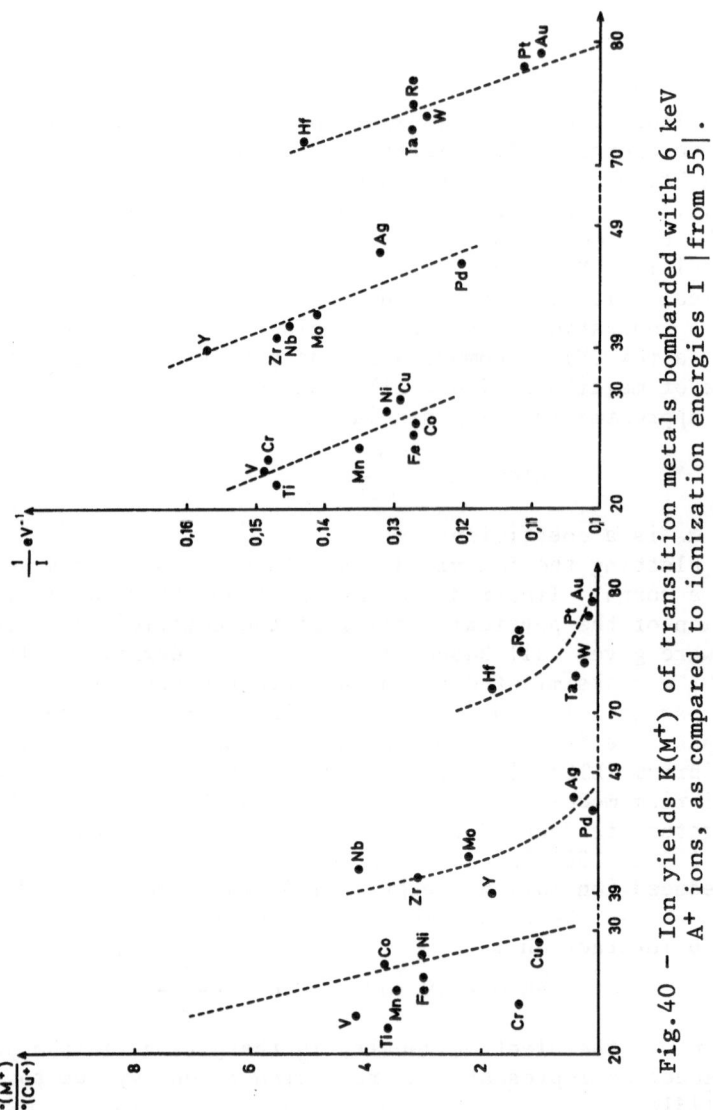

Fig. 40 – Ion yields K(M⁺) of transition metals bombarded with 6 keV A⁺ ions, as compared to ionization energies I |from 55|.

IV-2 Thermal ionization models :

In their LTE model Andersen and Hinthorne postulated that the impact of a primary ion gave rise to a local plasma at temperature T. Ionization results in an equilibrium reaction such as that given below for positive ions :

$$M \rightleftharpoons M^+ + e^-$$

According to this reaction the positive ionization ratio τ_i^+ of any atomic species i is given at thermal equilibrium by the Saha-Eggert equation : |133|

$$\tau_i^+ = \frac{2Q_i^+}{Q_i^o} \frac{(2\pi M_e kT)^{3/2}}{n_e h^3} \exp - \frac{I_i - \Delta E}{kT}$$

(73)

where Q_i^o and Q_i^+ are the partition functions of ionized and neutral atoms ; h and k are respectively Boltztmann and Planck's constants ; M_e is the electron mass and n_e the electron density in the plasma.

From the reaction :

$$M^- \rightleftharpoons M^o + e^-$$

one derives the negative ionization ratio :

$$\tau_i^- = \frac{Q_i^-}{2Q_i^o} \frac{n_e h^3}{(2\pi M_e kT)^{3/2}} \exp \frac{A_i}{kT}$$

(74)

where Ai is the atomic electron affinity.

For any another ion species, multicharged or molecular ions, the ionization degree can, in principle, be deduced from the corresponding equilibrium reaction.

The surface ionization model proposed by Jurela is based on the surface ionization theory developped by Dobretsov to account for thermoionic emission from hot metals |131|. However, in thermoionic emission ejection and ionization of atoms proceed from a thermal process whereas it is not the case in ion emission for ejection of particles at least. In that meaning, ion emission appears to be a "non-equilibrium" application of the surface ionization theory. It results that application of Dobretsov's treatment to secondary ion production is restricted to the ionization process. Therefore, instead of Saha-Langmuir equation this treatment leads to the following expressions for positive and negative ionization ratio :

$$\tau_i^+ = \frac{Q_i^+}{Q_o} \exp \frac{\phi - I_i + \Delta E}{kT}$$

(75)

$$\tau_i^- = \frac{Q_i^-}{Q_o} \exp \frac{A_i - \phi}{kT}$$

where ϕ is the work function of the solid.

Independently of contradictions that might exist between these two thermal models, it may be asked whether, regarding the experimen-

tal results, ion emission really exhibits a pronounced thermal character. It was mentioned in sect.II.1.1 that many aspects of ion emission, angular anisotropy, energy distribution, crystalline effects,
were clearly connected to sputtering and it is well known that sput-
tering produced by ions of medium energy has no thermal aspect |96|.
Let us examine the problem in a different way: should the temperature control ionization without playing a role in sputtering so that
most of the effects mentioned above would be understandable in a
thermal context? According to Andersen's plasma model it is not at
all possible since all particles, neutral atoms, ions, electrons arise from the same hot center produced by the impinging primary ion.
On the contrary it is possible in Jurela's surface ionisation model
where sputtering keeps its entity. But, on the other hand, how from
Jurela's model to explain the influence of an adsorbed gas on ion
emission since the enhancement is independent of the variation of the
work function! (see sect. II.2.3). Moreover, matrix effects observed
in alloys (sect. II.2.1) |50-52| cannot be explained in this model.
Finally two of the most remarkable experimental features of ion emission seem to be inconsistent with the surface ionization model. On
the other hand, the plasma model is not much more probative regarding the experimental results. As it is in principle valid for any
ion species it should have to account for the whole molecular and
multicharged ion spectra. In fact, it seems that the only attempt to
justify the presence of molecular ions is based on three MO^+ ions
(CaO^+, AlO^+ and SiO^+) emitted from a silicate. A more general con-
frontation of the model with molecular ion spectra such as those presented in fig.16 to 21 has never been considered, even qualitatively.
Anyway, the model does not seem in accordance with experiments regarding multi-charged ions, since their intensities are expected to
be higher the lower the ionization energy. In fact, bombarding with
primary ions of a few keV energy, double charge ions, for example,
are only observed with light elements (Mg, Al, Si ...) and not with
transition elements * |88,134| for which the second ionization ener-
gy is nevertheless smaller. The plasma model also has very strange
consequences which leave one perplexed regarding its physical basis.
For example, to justify the values of the two parameters T ∿ 11000°K
and n_e ∿ 10^{19} electrons/cm included in eq. 73 to account for ion
emission from silicates, a work function ϕ ∿ 11.25 eV has to be assumed. Likewise the metal work function would be ϕ ∿ 8.5 eV. Such high
values are quite surprising.

---

*   double-charge ions observed with transition elements are due to
     charge exchange between the primary beam and sputtered atoms.
     They are produced in front of the surface of the sample (pre-
     pic). They are not relevant to the secondary ion emission process
     discussed here |88,134|.

Moreover, they have not been confirmed by work function measurments directly carried out under ion bombardment |39,91|. At last, with such high values the secondary kinetic electron emission produced at the impact of ions should be quasi inexistent since it requires low work function. In particular, Parilis's theory |135,136|, which accounts very well for the kinetic electron emission, should be unvalid under the conditions imposed by the plasma model.*

Although thermal models raise a number of insurmountable physical problems, it cannot be denied that, in practice, they have been rather successful, in particular the plasma model, in providing elementary quantitative analysis |10,17,132|. This appears to be somewhat disconcerting regarding the criticism above. In fact, the two thermal models propose an exponential dependence of ion emission as a function of atomic ionization energy so, it is not surprising that they are well adapted for quantitative analysis since this exponential dependence corresponds to the general trend observed experimentally (eq.72). As there are two fitting parameters, T and n_e, in eq.73 one can imagine to fit these parameters at best for a certain type of compounds, silicate for example |17| in order to get a better quantitative analysis. The same success should be obtained by considering eq.72 with two fitting parameters without any reference to a thermal mechanism.

As a conclusion, eqs.72 to 76 relative to thermal processes certainly correspond to the best empirical approach to the problem of quantitative analysis by ion emission, but these equations must be considered as phenomenological laws including a certain number of fitting parameters. The attempts to justify them by thermodynamic considerations are not at all convincing because they are in opposition to the best established experimental features *.

IV-3 <u>Quantum models of secondary ion emission</u> :

From a quantum point of view the ionization of a particle coming out of a metal results in the time evolution of its electronic struture from an initial state located in the metal to a final state distributed on the energy levels of the free atom ; the ionization probability, positive or negative, being determined by the difference in the electronic population of the final state as compared to the normal population of the neutral atom. This is a problem of strong

---

* Recently a new approach to the plasma model was proposed by J.N. Coles in a letter |137|. The plasma would be formed outside the surface by all the particles leaving the solid. The temperature involved is that associated with secondary electron emission only. More details would probably be given in a further paper.

time dependent perturbation which cannot be treated by linear response theory. A perturbative many-body treatment was first proposed by
|138,139| but is only valid at the approach of the surface. Recently
a rigorous treatment was developped by A.B. Blandin, A. Nourtier,
D.W. Hone |140,141| on the basis of the Keldish |142| formalism for
irreversible processes. However this treatment was applied to ion
emission in the approximation of a one-body time dependent Hamiltonian.

        Let us consider the time dependent coupling $V_{dk}(t)$ of a d ato-
mic orbital of energy $\varepsilon_d(t)$ with the k   states of the conduction
band of the metal. A simple expression for the metal-atom Hamiltonian
is provided by the Anderson Hamiltonian |143| where time dependent
parameters $\varepsilon_d(t)$ and $V_{dk}(t)$ are included. Using this one-electron
approximation, the d - level occupancy $n_d(t)$ was calculated as a function of time.

        When $t \to \infty$, $\varepsilon_d(t)$ tends towards the energy level $\varepsilon_d(\infty)$ of the
free atom and the d-level occupancy tends to $n_d(\infty)$. The positive io-
nization probability is then given by :

$$P^+ = 1 - n_d(\infty) \qquad\qquad (77)$$

For a slow variation of the parameters the system will follow adiabatically the perturbation and the final result will be 0 or 1 depen-
ding on the position of $\varepsilon_d(\infty)$ with respect to the Fermi level ε_f
(fig.41).

$$n_d(\infty) = 0 \qquad\qquad\qquad \varepsilon_f < \varepsilon_d(\infty)$$

$$= 1 \qquad\qquad\qquad \varepsilon_f > \varepsilon_d(\infty)$$

If the variation of the parameters is no longer slow, $n_d(\infty)$ may be
different from 0 or 1. The general result is that the d-level occupancy at time t exhibits important retardation effects involving the
instantaneous position and width of the level at all times before   t.
A calculation has been performed for the most reasonable physical situation of an exponential time decrease of the width of the d level:
|141|

$$\Delta = \Delta_o \exp - avt \qquad\qquad (78)$$

where v is the velocity of the particle
        Δ_o is the width of the d level in the metal
        a^{-1} characterizes the decrease of Δ as a function of x (fig.41).
Assuming ε_d constant, the obtained ionization probability is :

$$P^+ = \frac{2}{\pi} \exp - \frac{\pi|\phi - E_d|}{h\,a\,v} \qquad\qquad (79)$$

where E_d is the d-level energy with respect to the vacuum-level(fig.41).
The ionization probability estimated from 79 by including reasonable
values of the parameters appears to be much lower than that given by
experiment (TableII). The reason of this discrepancy holds in the va-

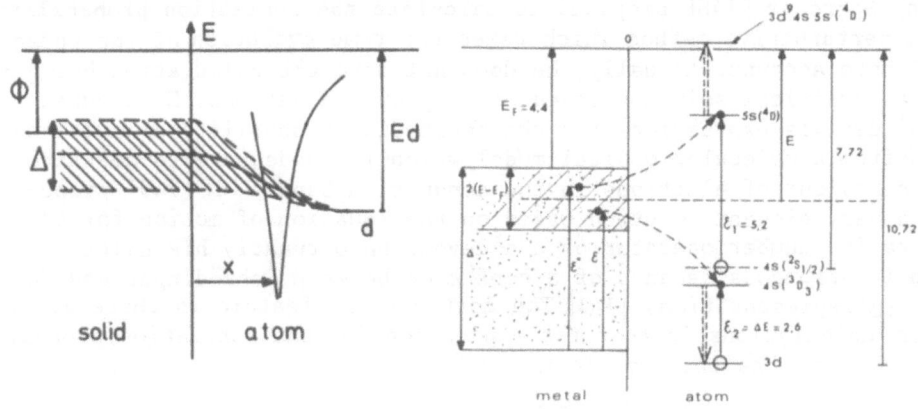

Fig. 41   Fig. 42

Fig.41 – Qualitative variation of the width Δ of the d level of an
        atom as a function of the distance x from the surface of the
        solid.
Fig.42 – Electronic structure of the 3d 4s5s (4D) auto-ionizing sta-
        te of copper atom as compared to the electronic structure
        of copper metal |from 49,144|.

lidity of the one-body treatment used by Blandin et al. |140, 141|.
These authors have in fact demonstrated that the problem of the ionization of an atom near a metal surface is in principle exactly sol-
vable whenever the Hamiltonian can be written in a one-body form. In
that meaning exp.79 must be considered as exact. However, it is obvious that a coherent treatment of the problem would require a many-
body approach. In spite of a recent attempt to do this, much work has
still to be done before solving the problem. After discussing a two-
many interaction A. Nourtier |141| led to the conclusion that positi-
ve or negative ionization probability could be expressed in a general
way by :

$$P^+ \propto \exp\left(-\frac{v_o}{v}\right) \qquad (80)$$

where v_o is a coefficient related to an energy.
    It is worth noticing that an expression of this type was obtai-
ned experimentally by Mac Donald |145| for the positive ionization
of copper, with v_o = 200 m/s (see also exp.5).
    In the recent past years other expressions for the ionization
probability have been derived from more or less whimsical quantum
models |146| – some of them even are completely devoid of meaning
|147|- by sometimes following a confused mathematical route |148,149|.
Assuming that atom leaves the metal as neutral (adiabatic approxima-

tion) Schroeer |148| proposed to calculate the ionization probability by a perturbative method which takes the time evolution of the potential into account. Actually, he does not make the calculation but gives an empirical solution based on hazy considerations. Z. Sroubek |146| derives expressions for the ionization probability of metal atoms from a molecular orbital model which is inadequate to describe the behaviour of electrons in the conduction band. Cini |149| proposed a very elegant solution based on the equation of motion for the occupation number operator of a d level. Unfortunatly his calculation is erroneous because of a confusion between Schrödinger and Heisenberg representations |141|. The most strange feature in these dissimilar calculations is that the expression for the ionization probability always   has the same form:

$$P^+ = r \; \frac{(\hbar \, a \, v)^n}{(E_a - \phi)^m} \tag{81}$$

with n and m $\sim$ 1, 2, 3

    r is a numerical coefficient

Depending on the metal and the atomic level to be considered E_a is associated either with the d-level |149|- as in exp.79 - or to the ionization energy I |148|. Expression 81 is in complete discordance with exp.79.

    Owing to the slight variations of $E_a - \phi$ from one system to another, exp.81 leads to small variations of the ionization probability from one atomic species to another. This is not in accordance with experiments |48,50| which reveal large variations of the ionization probability, in particular in the case of alloys (Table VIII).

    The quantum models discussed previously have the advantage to account for the velocity dependence of ionization probabilities. But, on the other hand, even in the most sophisticated model |140,141|, the atomic structure is reduced to a single orbital. In fact, the presence of photon emission |150,153| together with ion emission testifies to the necessity of considering a more realistic atomic structure from which it should be possible to look to the formation of excited states. This remark, a few years ago, led G. Blaise and G. Slodzian |144| to propose an ionization process based on the formation of highly excited states called auto-ionizing states |154|. Owing to the difficulty of dealing with the problem, these authors postulated the adiabaticity for the particles leaving the metal. This assumption is justified, to some degree, since the velocity of the ejected atoms is always much smaller than the velocity of the conduction electrons near the Fermi level*. Therefore it is supposed that conduction electrons are able to follow atoms in their displacement inside the metal and to maintain them as neutral outside. But the perturbation acting upon the outer electronic shells of an atom, while it crosses the metal-

---

* This really corresponds to near-adiabatic conditions.

vacuum interface, produces excited states. When the excitation energy E^* exceeds the ionization energy I of the atom, one deals with
auto-ionizing states ; that is states capable of leading to a sponta-
neous ionization of the atom in the vacuo, through an auto-ionizing
process. It results from this process that the ionization probability
of an atom $P(M^+)$ is determined by the probability to form auto-ionizing states $P(M^+)$.

An auto-ionizing state results from a multi-electronic excitation of the atom. . A typical example is the $3d^9 4s5s(4D)$ of copper
the energy of which is 1/10 eV above the ionization energy. This state is represented in fig.42 in one electron model, beside the elec-
tronic structure of copper metal $|49,144|$. A description of the au-
to-ionizing states of elements of the first transition series has
been proposed on the basis of a one-electron picture of the atomic
structure $|144,155|$. For these elements there are two different ba-
sic configurations $3d^\nu 4s$ and $3d^{\nu-1} 4s$ where ν and $\nu-1$ are the numbers
of d electrons. From these configurations 3 types of auto-ionizing
states have been considered:

$$\text{I} - \quad 3d^{\nu-1} n1n'1' \rightarrow (3d^\nu)^+ + e$$

$$\text{II} - \quad 3d^{\nu-2} n1n'1'n''1'' \rightarrow (3d^{\nu-1}4s)^+ + e$$

$$\text{III} - \quad 3d^{\nu-2}4sn1n'1' \rightarrow (3d^{\nu-1}4s)^+ + e$$

In a one-electron picture the excitation energy is given by the energy required to promote electrons from levels 4s and 3d to levels n1,
n'1', n''1''. Furthermore, all auto-ionizing states with the same mean
energy level E are equivalent and cannot be distinguished. In the
example given in fig.42 the mean energy level E of the state is E =
5.32 eV and the excitation energy $E^* = \varepsilon_1 + \varepsilon_2 = 7.8$ eV. It exceeds
the ionization energy I = 7.72 eV by 0.08 eV. If E_m (fig.43) is the
mean energy level of the auto-ionizing state,which energy justs equal
the ionization energy associated with the final state of the ion,
$(3d^\nu)^+$, $(3d^{\nu-1}4s)^+$, all the auto-ionizing states will be characterized by a mean energy level $E < E_m$.

The formation of auto-ionizing states has been considered on
the basis of the approximations usually used in Dobretsov's surface
ionization theory $|131|$. According to this theory, it is supposed that
the atom is included to the metal as long as its distance x from the
surface is smaller than a critical distance x_c. On the other hand,
the atom is considered as completely free when $x > x_c$ (fig.43). The
formation of auto-ionizing states is then described as a transition
in x_c which keeps constant the mean energy of electrons. The condition
for the process to be possible is that the mean energy level $E_m-\delta$ of
the first auto-ionizing state * is below the Fermi level :

* δ is the energy of the ion in the final state. Both δ and $E_m-\delta$ refer to the vacuum energy level (fig.43).

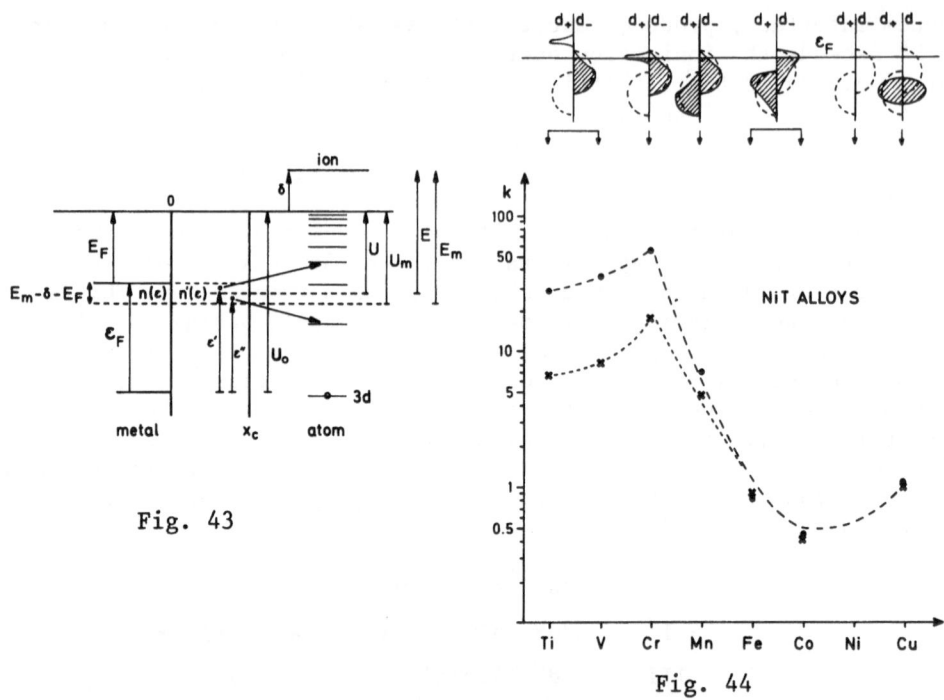

Fig. 43

Fig. 44

Fig.43 - Formation of auto-ionizing states |from 155|.
Fig.44 - Upper part : electronic structure of transition elements
in nickel based alloys.
Experimental (●) and theoretical (×) k values deduced from
exp   94   and   95 |from 48|.

$$E_m - \delta - E_F > 0$$

This condition implies that there exists in the initial state pairs
of electrons (processes I and III) - or triplets according to the
process II - whose mean energy level corresponds to that of an auto-
ionizing state. According to this mechanism only auto-ionizing states
with a mean energy $E - \delta$ included within the interval $E_m - \delta$ and E_F
will be occupied.
    The calculation of the probability of formation of auto-ioni-
zing states has been done assuming an electron density $n'(\varepsilon)$ in the
initial state equal to the electron density $n(\varepsilon)$ of the metal and
supposing an uniform density of states $g(E)$ on the free atom. This
probability is written as :

$$P^*(E_m-\delta-E_F) = \frac{\int_{E_F}^{E_m-\delta} C(E-\delta) \, dE}{\frac{1}{2} N^2} \qquad (82)$$

for processes I and III involving 2 electrons.

and $\qquad P^*(E_m-\delta-E_F) = \dfrac{\int_{E_F}^{E_m-\delta} C(E-\delta) \, dE}{\frac{1}{3} N^3} \qquad$ (83) for process II involving 3 electrons.

N is the number of d+s electrons for transition elements. $C(E-\delta)$ is a convolution function which represents the number of pairs - of electrons - or triplets according to the process - in the conduction band whose mean energy is $E-\delta$. $C(E-\delta)$ depends on the energy distribution $n(\varepsilon)$ of electrons in the conduction band of the solid.

All the auto-ionization processes considered in I,II,III involve the excitation of a d electron with respect to the basic atomic configuration. In other words, the d level must be partially empty at distance x_c. On the other hand, auto-ionizing states can still be destroyed beyond x_C by electronic exchange with the metal |155,156| . Therefore, we shall introduce two other functions in the calculation of the ionization probability: the probability $F_d(v)$ for the deficiency of an electron on the d level of an atom at x_c and the survival probability $P_o(v)$ of an auto-ionizing state beyond x_c. Finally, the ionization probability is expressed as :

$$P^+ = F_d(v) \, P_o(v) \, P^*_{x_c}(E_m-\delta-E_F) \qquad (84)$$

Estimations of P^+ have been performed assuming $F_d(v)$ is the same for transition elements of the same series |48| . The survival probability $P_o(v)$ depends on the configuration of auto-ionizing states |155| , but it always tends towards unity when the velocity of ions increases. In other words, the high energy ion emission (typically > 50 eV - see sect. II.1.2) of transition elements must appear to be proportional to P^*. In practice, numerical calculations of P^* require the knowledge of $n(\varepsilon)$ in a small bandwidth $A_m=E_m-\delta-E_F \sim 0.5$ to 1 eV. In some very simple but physically acceptable cases, P^* is easily obtained. For example, when $n(\varepsilon)=\alpha$ is approximately constant near the Fermi level, |144| one obtains :

$$P^* = \frac{\alpha^2 \ A_m^2}{1/2 \ N^2} \qquad \text{for processes I and III}$$

$$\qquad (85)$$

$$P^* = \frac{3/2 \ \alpha^3 \ A_m^3}{1/3 \ N^3} \qquad \text{for process II}$$

Now, when the d electrons are localized near the Fermi level in a virtual bound state |157| , all they can participate to the formation of auto-ionizing states and one obtains for process II :

$$P^{*} \sim \frac{1/3 \ N_d^3}{1/3 \ N^3} \tag{86}$$

where N_d is the number of d electrons of the atom.

This model explains the general trends of the ion emission of pure transition metals and alloys |48,53|. In particular, it explains rather well the matrix effects observed in dilute alloys (Table VIII). An example is given in fig.44 for NiT alloys : for each alloy, k_{NiT} was measured according to the method indicated in sect.V.1 and compared with theoretical values. It is seen that agreement between calculations and experiments is rather good.

It is noticed that this model established a strong correlation between the ionization probability of atoms and their electronic density near the Fermi level. This has the result that the ionization efficiency of atoms must be related to other physical properties of solids such as specific heat or resistivity as is actually observed experimentally |48,157|. On the other hand, exp. 85-86 show that a small change in the electronic density of an atom, near the Fermi level, may have a big effect on its ionization efficiency. As a consequence, ion emission appears as a tool capable of providing informations on the electronic properties of solids. Interesting results have already been obtained in this field |53|(see sect. V.1.2.1).

An auto-ionization state results from a multielectronic process which requires 2 valence electrons, at least. Therefore, the model can, in principle, be applied to most metal elements except for alkaline elements which are monovalent. So, it is expected that pure alkaline metals exhibit very low ion yields, but there is not yet experimental evidence of this.

At last it is worth noticing that the present model is not incompatible with the rigourous theoretical treatment proposed by Blandin et al. |140,141|. Both approaches are even very complementary. For example, exp.79 could be interpreted as the probability $F_d(v)$ of finding a hole on the d level of an atom. It is clear that a better description of the phenomenon would result in a close synthesis of the two point of view.

IV-4 Formation of ion clusters :

The theoretical interest in ion clusters is related to the recent development of the physics of small aggregates. Most of the earlier works concern neutral clusters but, owing to the great diversity of ion clusters revealed by secondary ion emission (see sect. III.1.3) one can hope to see a growing interest in the study of ionized aggregates. P. Joyes and M. Leleyter are among the pioners in this field |57,58,60,158-160|.

We have described in sect. III.4 the departure of atoms from the surface; the problem to be handled now is the evolution of the initial cluster towards a stable ionized molecular configuration. At the outset, the cluster atoms form a pseudo molecule with very high

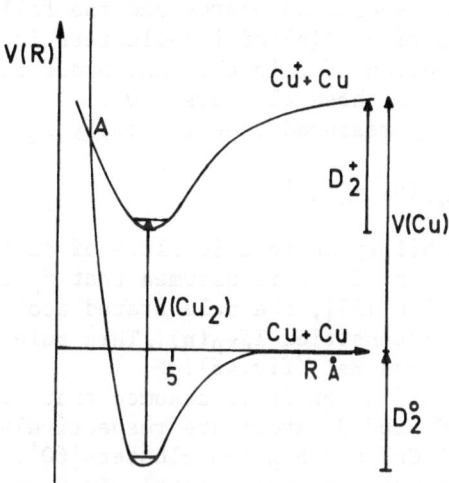

Fig.45 – Schematic representation of energies of the $Cu_2(^1\Sigma g^+)$ and $Cu_2^+(^2\Sigma g^+)$ states |from 158|.

vibrational energy. Let us take the example of the Cu_2 molecule |158|. The two potential energy curves V(R) corresponding to the fundamental configurations of Cu_2 and Cu_2^+ are represented in fig.45. When the distance R of the two atoms reach about three atomic units ($\sim$ 1.5Å) a crossing point A occurs. Imagine now that two Cu atoms leave the surface from adjacent sites with slight different ejection conditions. If their distance becomes < 3 a.u. during their flight the molecule has a certain probability of passing to the configuration of Cu_2^+, that is to be ionized with emission of an electron. Depending on whether the potential energy of the system exceeds the dissociation energy into Cu^+ + Cu or not, the molecular ion will dissociate into Cu^+ + Cu or will form a stable Cu_2^+ ion. It is seen that the process leads to the transformation of kinetic energy into electronic energy so that the system may be stabilized by ionization. This mechanism requires a crossing point – or at least the closest approach of the potential energy curves – below the ionization energy of the dissociated system V(Cu). As the energy level of the crossing A is likely more or less related to the ionization energy $V(Cu_2)$ of the neutral molecule, it also clearly appears that the higher the probability of formation of an ion cluster, the smaller $V(Cu_2)$ and the higher the dissociation energy D_2^+ of the ion cluster. In other words, the formation of an ion cluster depends on : |60|
- the ionization energy of the neutral cluster $V(A_nB_m)$
- the dissociation energy of the ion cluster $D^+(A_nB_m)$.
From a great number of experiments |60|, it appeared that the predo-

minant term was the dissociation energy and the following rule was stated * : the intensities $I^+(n)$ of ion clusters involving n atoms are classified as function of n in the same order as the dissociation energies per electron of these clusters $|60|$.

The ion intensity measured experimentally $I^+_{exp}(n)$ is related to $I^+(n)$ by :

$$I^+_{exp}(n) = f_n \, I^+(n)$$

where f_n is the probability defined in III.4 of setting n atoms simultaneously into motion. If it is assumed that f_n is a monotonic decreasing function of n $|59|$, the rule stated above also is valid for the measured ion intensities $I^+_{exp}(n)$. This rule is illustrated by two examples : Cu^+_n and Be^+_n (fig.46).

To a first approximation it is assumed that electrons of the 4s and 2s levels of Cu and Be atoms are respectively responsible for the binding energy of Cu^+_n and Be^+_n ion clusters $|60|$. Let E_a be the energy of the corresponding s atomic level. In Huckel's $|161,162|$ approximation ** a n-atom cluster has n energy levels given by :

$$\epsilon_j = E_a + \beta h_j \qquad\qquad j = 1,2,3 \ldots n \qquad (87)$$

where $\beta < 0$ is the transfert integral supposed constant.

For a linear chain $h_j = 2 \cos. \dfrac{j\,\pi}{n+1}$

The electron energy of a n-atom ionized cluster is given by the energy of n-1 electrons, namely :

$$E^+_e(n) = \sum_{i=1}^{n-1} \epsilon_j(i) = (n-1)\, E_a + \beta \sum_{i=1}^{n-1} h_j(i) \qquad (88)$$

The summation in 88 must be carried out by setting 2 electrons i on each j level. The binding energy is given by :

$$D^+_e(n) = E^+_e(n) - n\, E_a$$
$$= -E_a + \beta \sum_{i=1}^{n-1} hj(i) \qquad (89)$$

The energy levels of Cu^+_n and Be^+_n are represented in fig.46. On each diagram the binding energy per electron is represented by das-

---

* For negative ion clusters the rule is based upon electron affinity $|60|$.
** This approximation is less good for Be than for Cu because Be atom has two 2s electrons. A more refined treatment must take hybridation into account for Be $|60|$.

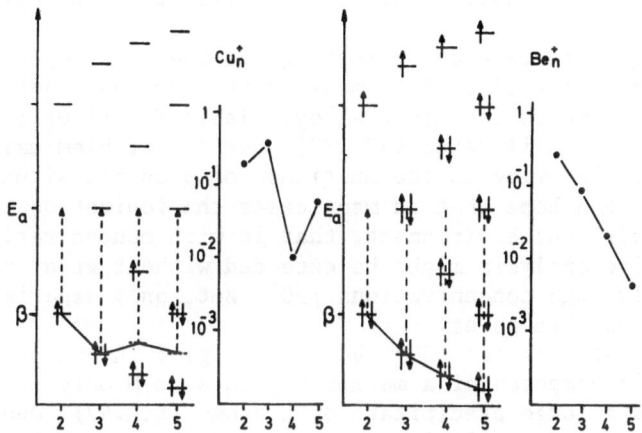

Fig.46 – Energy levels for Cu_n^+(a) and Be_n^+(b) in Hückel approximation.
Binding energies are represented by dashed arrows and compa-
red with intensities of ion clusters |from 60|.

hed arrows. It is seen that the binding energies of Cu_n^+ clusters exhibit an oscillating behaviour as the intensities, whereas the binding energies of Be_n^+ varies linearly as the intensities of the ion
clusters.

IV-4 <u>An approach to quantitative analysis</u> : |45|

    It is less likely that one succeeds in calculating accurately
the ionization probabilities in a near future. Therefore, it is essentially by experimental means that the problem of quantitative analysis
might be solved for the moment. However, most quantum theories put
emphasis, to various degree, on the fact that ionization probability
is connected to the electronic structure of the atom in the solid (see
in particular the auto-ionizing state model in sect.IV.3). This allows
us to understand why the ionization probability of an atom depends experimentally on the nature of the solid in which it is incorporated
(see Table VIII).

    In fact, the electronic structure of an atom is imposed by its
atomic environment and this influence can, in most cases, be restric-
ted to the nearest neighbours |163,164|. Consequently, the problem
of ionization of atoms can be convert into a problem of environment
and two very simple rules are proposed :
  - The ionization ratio of an element remains constant as long as its
    environment does not change as a function of its concentration.
  - In a general way the ionization ratio of an element in a solid

is a linear combination of the ionization ratios related to the various types of environment of this element in the solid under question.

Finally quantitative analysis requires the knowledge of a great number of ionization ratios. Some measurments have been determined for transition elements in dilute alloys (Table VIII) |50| ; others for many elements in silicates |47,132|, but the problem may appear to be insurmontable owing to the multitude of possible situations. Fortunately one can hope that in many cases the ionization ratio will change slowly with the environment, that is with concentration, so that quantitative analysis might be extended without great difficulty to relatively high concentrations |90|. But, only experiment can give an answer on this point.

To illustrate the second rule we shall give an example. Let us consider a solid composed of a matrix M', an atom M only surrounded by M' atoms and a large precipitate of M atoms (fig.47). Due to the different environment the isolated atom M and the atom M located at the center of the precipitate (black square in fig.47) will have different electronic structures and consequently different ionization probabilities. This is true as well as for the intrinsic emission (clean surface) than under oxygen adsorption, since in this latter case, the metal-oxygen bonding MO will not be the same in both situations. Let $P_o(M)$ and $P_M(M)$ - or $P_o^g(M)$ and $P_M^g(M)$ - be the respective ionization probabilities in both cases *. If we imagine now a solid solution $M_cM'_{c'}(c+c'=1)$ where atoms M and M' are distributed at random, all the intermediate environments between the two extreme cases seen above will be statistically possible with well defined probabilities.

Supposing the electronic structure is determined by the environment of the nearest neighbours |163,164|, the probability $\mathscr{P}(p)$ of finding p atoms (M or M') among the n nearest neighbours of an atom is given by :

$$\mathscr{P}(p) = C_n^p \, c^p \, c'^{n-p} \qquad\qquad (90)$$

with

$$C_n^p = \frac{n!}{p!(n-p)!}$$

It results that each atom M will be ionized with a probability $P_o(M)$, $P_1(M)$, $P_2(M)$ ... $P_M(M)$ according as the number of nearest neighbours M atoms is 0, 1, 2, 3 ... . Naturally the situation is completely symmetric for M' atoms. Finally the mean ionization probability $\overline{P(M)}$

---

* Subscript o indicates that M is isolated into M'; subscript M indicates that M is completely surrounded by M atoms; superscript g is related to gas adsorption (fig.47).

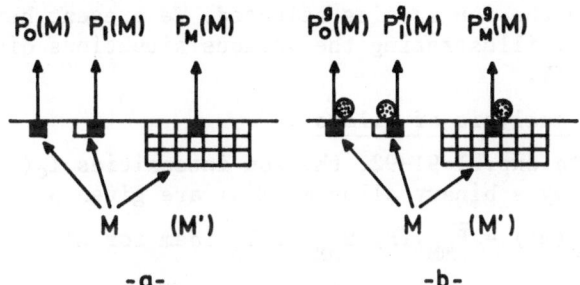

Fig.47 - Various ionization probabilities of M atoms in an alloy
$M_c M'_{c'}$ according to their environment. Only M atoms are represented by squares □ or ■ .
a - Intrinsic emission.
b - Emission induced by oxygen adsorption (⊛ adsorbed
oxygen atom).

will be :

$$\overline{P(M)} = \sum_{p=0}^{n} \binom{}{}(p)\ P_p(M) \tag{91}$$

where p is the number of M atoms around each atom M.
An expression similar to 91 is also obtained for M' atoms. Expression 91 corresponds to the general case of disordered systems. In
principle the knowledge of $\overline{P(M)}$ requires the measurment of P_p for all
possible environments. Fortunately one can expect simplification of
exp.91 in many cases (see sect. V . 1).

V - APPLICATIONS

Expression 10 can be used for any secondary ion species $A^{\pm}$, in
any system, on the condition that the ionization ratio τ, defined by
exp.9, is replaced by the more general notion of ionization probability $P(A^{\pm})$ and the definition of the sputtering yield S is extended to
any ion species as indicated in sect.III. Furthermore, for practical
reason, it is more convenient to introduce in eq.10 the primary and
secondary ion intensities I_p and $I(A^{\pm})$. Finally, the general expression for the intensity of any ion species can be written as :

$$I(A^{\pm}) = P(A^{\pm})\ S(A)\ I_p \tag{92}$$

In this expression, $S(A)$ and $P(A^{\pm})$ will be expressed respectively by
expressions indicated in sections III and IV, according to the nature

of the sample and the ion species selected. We present here some applications of exp.92, illustrating the various situations discussed in
sections III and IV.

V-1 Quantitative analysis of alloys :

        According to exp.31-91-92, the ion intensities $I_c(M^+)$ and
$I_{c'}(M'^+)$ emitted by a binary alloy $M_c M'_{c'}$ are given by :

$$I_c(M^+) = \bar{P}_{MM'}(M)\, S_{MM'}\, c\, I_p \quad \text{idem for M'} \tag{93}$$

The ratio of the two intensities is : $|49, 50|$

$$\frac{I_{c'}(M'^+)}{I_c(M^+)} = k_{MM'}\, \frac{c'}{c} \tag{94}$$

with:

$$k_{MM'} = \frac{\bar{P}_{MM'}(M'^+)}{\bar{P}_{MM'}(M^+)} \tag{95}$$

It is also interesting to compare the ion emission of an element of
the alloy to that of the corresponding pure metal. For a pure metal
M, the M^+ ion intensity is :

$$I(M^+) = P_M(M^+)\, S_M\, I_p \tag{96}$$

then, the ratio of the two intensities is :

$$\frac{I_c(M^+)}{I(M^+)} = \rho_{MM'}(M)\, \frac{S_{MM'}}{S_M}\, c \qquad \text{idem for M'} \tag{97}$$

with :

$$\rho_{MM'}(M) = \frac{\bar{P}_{MM'}(M^+)}{P_M(M^+)} \tag{98}$$

$\rho_{MM'}(M)$ is called relative ionization coefficient.

V-1-1 Application to dilute solid solutions : $c' \ll 1$

        In that case we can admit that $S_{MM'} \sim S_M$. As $c' \ll 1$, the most im-
portant term in exp. 90-91, written for the dilute element M', corresponds to p=0. This means that we can consider that M' atoms have only
one type of environment formed by M atoms (fig.48). It results that
their ionization probability is constant with respect to the concen-
tration c' and given by : $\bar{P}_{MM'}(M') \approx P_o(M')$. The coefficient $\rho_{MM'}(M')$
(eq.98) is also constant and the measured ion intensity $I_{c'}(M'^+)$ is
proportional to c'(eq.93). This linear concentration dependence of the
ion intensity of the dilute element is observed as well as for the in-
trinsic emission of alloys $|23|$ than for emission induced by gas ad-
sorption (oxygen). Some measurments of $\rho_{MT}(T)$ have been given in Ta-
ble VIII for the intrinsic emission of transition element $|50|$.

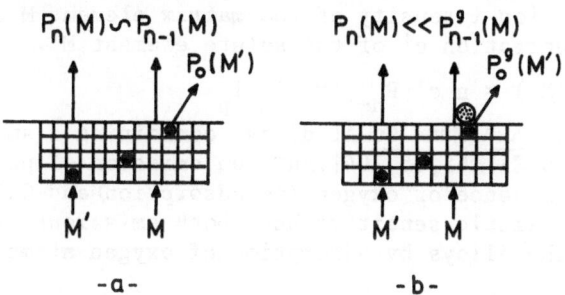

Fig.48 — Ion emission from a dilute solid solution $M_c\,M'_{c'}$ ($c' \ll 1$).
□ M atoms; ● M' atoms; ⊛ adsorbed atoms.
a) Intrinsic ion emission : $P_n(M)$ and $P_{n-1}(M)$ are the ionization probabilities of M atoms corresponding to the two types
of environment. $P_o(M')$ is the ionization probability of M'
atoms.
b) Ion emission induced by gas adsorption : it is supposed
here that adsorption only occurs on M' atoms and not on the
matrix M (this is the case of oxygen adsorption on Au Al alloys).

There are two types of environment for M matrix atoms (fig.48).
One type for M atoms immersed in the matrix, that is completely surrounded by n other atoms M. Their ionization probability is $P_n(M)$. There
is another type of environment for M atoms located near an atom M'.
These are surrounded by n-1 atoms M and their ionization probability
is $P_{n-1}(M)$. According to 90-91, the average ionization probability is:

$$\bar{P}_{MM'}(M) = c^n\,P_n(M) + n\,c^{n-1}\,c'\,P_{n-1}(M)$$

$$\text{as } c \neq 1 \quad \bar{P}_{MM'}(M) \neq P_n(M) + n\,c'\,P_{n-1}(M) \tag{99}$$

Two cases may occur :
 — if the ionization probability of M atoms is not very sensitive
to the presence of M' atoms, that is $P_{n-1}(M) \lesssim P_n(M)$, one obtains
$\bar{P}_{MM'}(M) \backsim P_n(M)$ (fig.48 a). The ion intensity of matrix atoms is practically constant and independent of the concentration of the dilute element ($\rho_{MM'}(M) = 1$). This is generally the case for the intrinsic emis-
sion of alloys |49,50|.
 — under oxygen coverage the situation can be completely different
because it may happen that the ionization probability of matrix atoms
(M) located near a solute atom (M') is considerably increased by the
presence of an adsorbed oxygen atom whereas the matrix itself is not
at all. - or very little - sensitive to oxygen (fig.48 b). In that
case $P_{n-1}^g(M) \gg P_n^g(M)$ (superscript g refers to gas adsorption), so that:

$$\bar{P}_{MM'}^g(M) \backsim n\,c'\,P_{n-1}^g(M)$$

It results that the ion intensity of the matrix element M depends linearly on the concentration c' of the solute element M'.

$$I_c(M^+) \simeq n\, c'\, P_{n-1}(M)\, S_M\, I_p \tag{100}$$

Typical examples are AuAl and Cu Al alloys containing a small amount of aluminium ($\leqslant$a few %) |65,85,104|.Au+ ion emission of pure gold is unaffected by the presence of oxygen (no adsorption) and Cu+ emission of pure copper is a little sensitive but, both emissions are considerably enhanced in the alloys by adsorption of oxygen atoms near aluminium atoms.

V-1-2 Concentrated solid solutions :

Here we present two examples which illustrate the general expression (91) for the ionization probability in disordered systems. The first application is concerned with the intrinsic emission of Ni Fe alloys |53|, the second with the emission of Ni Cr alloys under oxygen coverage |165|.

If $\rho_p(M) = \dfrac{P_p(M)}{P_M(M)}$ represents the relative ionization coefficient (eq.98) of an atom M surrounded by p other atoms M, according to (91) the ionization coefficient of the alloy $\rho_{alloy}(M)$ is given by:

$$\rho_{alloy}(M) = \sum_{p=0}^{n} \mathcal{C}(p)\ \rho_p(M) \tag{101}$$

V-1-2-1 intrinsic emission of Ni Fe alloys :

$\rho(Ni)$ and $\rho(Fe)$ are represented as a function of iron concentration C_{Fe} in fig.15|53|.One can see that $\rho(Ni)=1$ over the whole concentration range. This means that the ionization probability of Ni atoms is the same in the alloy as in the pure metal. In other words the electronic structure of nickel atoms does not change appreciably with the alloy composition.

On the contrary $\rho(Fe)$ increases with C_{Fe}. This behaviour was explained in relation with the magnetic properties of these alloys which suggest the existence of two possible magnetic states, that is two types of electronic structures of iron atoms, depending on their environment |166,167|. Consequently, we assume the existence of two kinds of iron atoms having two different ionization probabilities:
- iron atoms of type a, with concentration $c_a(Fe)$ and ionization coefficient ρ_a.
- iron atoms of type b, with concentration $c_b(Fe)$ and ionization coefficient ρ_b :

$$\rho_{alloy}(Fe) = c_a\, \rho_a + c_b\, \rho_b \tag{102}$$

with : $c_a + c_b = 1$

The simplest model of environment-dependent magnetic states |168|

consists of assuming that the magnetic state of an iron atom changes
if surrounded by at least p iron atoms among its n nearest neighbours:
- When n<p iron atoms are assumed to be of type a, that is

$$\rho_a = \rho_1 = \rho_2 = \ldots = \rho_{p-1}.$$

- When n≥p iron atoms are assumed to be of type b, that is

$$\rho_b = \rho_p = \rho_{p+1} = \ldots = \rho_n$$

Using this environment model in a f.c.c. lattice with n=12 one finds:

$$c_a(c_{Fe}) = \sum_{q=0}^{p-1} C_{12}^q c_{Fe}^q (1-c_{Fe})^{12-q}$$

$$c_b(c_{Fe}) = \sum_{q=p}^{n} C_{12}^q c_{Fe}^q (1-c_{Fe})^{12-q}$$

c_a and c_b were computed for p=5,6,7,8. Clearly, only the value p=6
agrees very well with experimental data in fig.15. Therefore the concentration dependence of the Fe^+ ion emission can be expressed numerically by :

$$\frac{I_{c_{Fe}}(Fe^+)}{I(Fe^+)} = \left[0.8 + 0.5\; c_b(c_{Fe})\right] c_{Fe} \tag{103}$$

with

$$c_b = \sum_{q=6}^{12} C_{12}^q c_{Fe}^q (1-c_{Fe})^{12-q}$$

V-1-2-2 <u>ion emission of Ni Cr alloys completely covered with oxygen</u>
<u>(θ=1)</u>.
    ρ(Ni) and ρ(Cr) are represented in fig.49 as a function of the
chromium concentration c_{Cr} |165|. Both ionization coefficients increa-
se linearly with c_{Cr}. It results from eq.93 that $I(Ni^+)$ and $I(Cr^+)$
vary parabolically with the chromium concentration.
    To explain this behaviour, we assume that the ionization proba-
bility of an atom M in an alloy $M_c\,M'_{c'}$ is proportional to the number
p of nearest neighbours M atoms, namely:

$$P_p(M) = P_o(M) + p\, \Delta\, P(M) \tag{104}$$

where $\Delta P(M)$ is an increment independent of p.
Using 91 the ionization probability can be written :

$$\overline{P(M)} = P_M(M) - \overline{(n-p)}\, \Delta P(M) \tag{105}$$

with

$$\sum_{p=0}^{n} \mathcal{C}(p) = 1$$

and

$$\sum_{p=0}^{n} \mathcal{C}(p)\, (n-p) = \overline{n-p} = n(1-c).$$

    Then , the expression of the ionization coefficient of an atom
M is :

$$\rho_{all.}(M) = 1 - n(1-c)\, \Delta P(M) \tag{106}$$

A symetric expression is obtained for M' atoms.

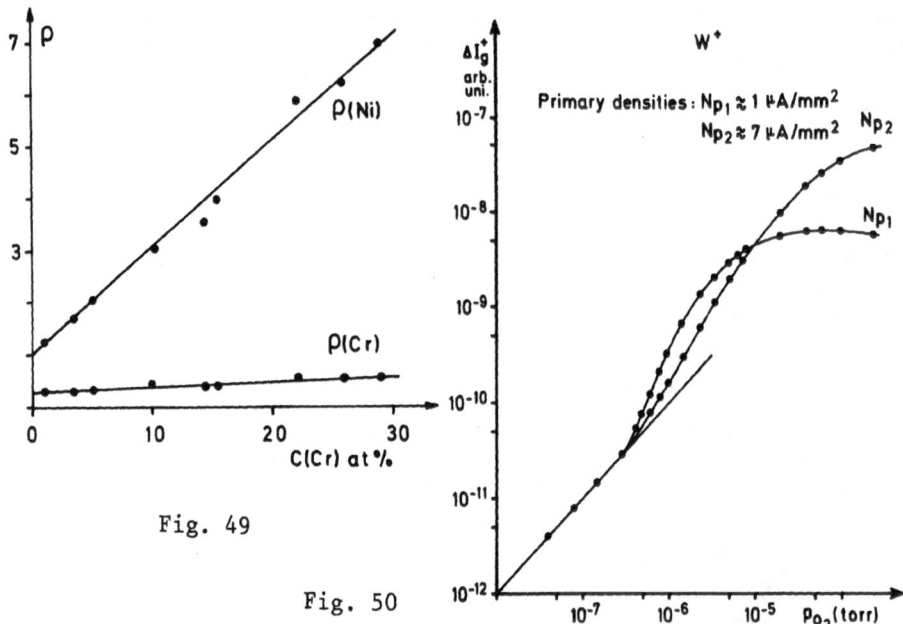

Fig. 49

Fig. 50

Fig.49 - Variations of $\rho(Ni)$ and $\rho(Cr)$ in Ni Cr alloys under complete oxygen coverage, as a function of chromium concentration |from 165|.

Fig.50 - ΔI_g^+ intensity of W^+ ion emission of a polycrystalline tungsten sample as a function of oxygen pressure and at different primary ion densities Np. Bombardment by A^+ ions of 6 keV |from 65|.

It appears that the ionization coefficient of M atoms at concentration c is proportional to the concentration $c'=1-c$ of M' atoms. In other words, there is a coupling between M^+ and M'^+ ion emissions: increasing the concentration of M' - that is decreasing the concentration of M - will change the ionization probability of M in a way which depends on the sign of ΔP.

In Ni Cr alloys one obtains:

$$\rho_{Ni\ Cr}(Cr) = 1 - n\ (1-c_{Cr})\ \Delta P(Cr)$$

$$\rho_{Ni\ Cr}(Ni) = 1 - n\ c_{Cr}\ \Delta P(Ni) \qquad\qquad (107)$$

These expressions are in good agreement with the linear dependence of $\rho(Ni)$ and $\rho(Cr)$ observed experimentally, with $\Delta P(Ni)<0$ and $\Delta P(Cr)>0$ |165|.

V-2 <u>Adsorption of gases studied by ion emission</u>:

V-2-1 <u>characteristics of positive metal ion emission induced by ad-</u>
        <u>sorption</u> |65|:

The intensity I_o^+ of M^+ ions emitted from a clean metal M bombarded under good vacuum conditions is given by:

$$I_o^+ = P^+ \, S_o \, I_p$$

where S_o is the sputtering yield of the clean metal.
When an atom of a reactive gas is adsorbed on the surface, the ionization probability of the neighbouring metal atoms becomes P_g^+. At a
coverage θ the measured ion intensity $I_t^+(\theta)$ consists of the contribution $I_o^+(\theta)$ of metal atoms not involved in the adsorption process (ionization probability P^+), plus the contribution $I_g^+(\theta)$ of metal atoms
involved in the adsorption process (ionization probability $P_g^+(\theta)$:

$$I_t^+(\theta) = I_o^+(\theta) + I_g^+(\theta) \tag{108}$$

If $N(\theta)$ metal atoms are sputtered, a fraction $C(\theta) \, N(\theta)$ are involved
in the adsorption process; therefore, I_o^+ and I_g^+ are expressed as :

$$I_o^+(\theta) = e \, P^+ \, [1-C(\theta)] \, N(\theta)$$

$$I_g^+(\theta) = e \, P_g^+ \, C(\theta) \, N(\theta) \tag{109}$$

e is the electron charge.
Let us write the difference in ion intensities:

$$\Delta I_g^+(\theta) = I_t^+(\theta) - I_o^+(\theta)$$

$$= P^+ [S(\theta)-S_o] I_p + [P_g^+(\theta)-P^+] C(\theta) \, N(\theta) \, I_p \tag{110}$$

The first term of this expression is negative because $S(\theta)$ decreases
as a function of θ(fig.37), but $|S(\theta)-S_o|$ is usually small. Therefo-
re, this term is always negligible with respect to the second term
since $P_g^+ \gg P^+$. As a consequence, to a good approximation:

$$\Delta I_g^+(\theta) \simeq P_g^+(\theta) \, C(\theta) \, S(\theta) \, I_p \tag{111}$$

represents the ion intensity relative to metal atoms involved in the
adsorption process. In practice, ΔI_g^+ is obtained by substracting to
the measured ion intensity I_t^+, the ion intensity I_o^+ of the clean
metal. The dependence of $\Delta I_g(W^+)$ of a tungsten sample is represented
in fig.50 as a function of oxygen pressure (dynamic conditions) for
various primary ion densities.

We consider three steps of coverage to interpret the oxygen pressure dependence of ΔI_g^+.

<u>small coverages</u>:$\theta \ll 1$.

In the low pressure range θ is small (equat.16). Therefore, the
average distance between adsorbed atoms is large as compared to the
interatomic distances of metal atoms. It results that the influence
of adsorbed atoms on metal atoms is proportional to the coverage. This
means that $C(\theta)=\alpha\theta$ where α is the number of metal atoms influenced
by each adsorbed atom |169|.

Taking into account eq.16 and assuming $S(\theta) \approx S_o$ at small θ, eq.111 becomes: $\Delta I_g^+(\theta) = e \, K \, \alpha \, P_g^+(0) \, s(\theta) \, p$ (112)

where $P_g^+(0)$ is the ionization probability at $\theta \to 0$.

As long as $s(\theta)$ does not change very much with θ (at small θ, $s(\theta) \approx 1$) $|170|$ ΔI_g^+ will be proportional to the pressure and independent of S and Ip. This analysis is well supported by experimental results presented in fig.50.

high coverages: $(\theta \approx 1)$

At high coverage all the metal atoms are influenced by adsorbed atoms. $C(\theta)$ reaches unity and the ion intensity is given by:

$$\Delta I_g^+(1) = P_g^+(1) \, S(1) \, Ip$$ (113)

It is proportional to Ip and S(1) as it appears in fig.50 $|65|$.

intermediate coverages:

The dependence of ΔI_g^+ on oxygen pressure is very difficult to explain at intermediate coverages because all the parameters included in eq.111 may vary with θ. However, whatever the metal one observes a systematic deviation from linearity, which suggests an increase of the ionization probability as a function of θ $|65|$.
A series expansion is proposed:

$$P_g^+(\theta) = P_g^+(0) \left[1 + \beta\theta + \gamma\theta^2 + \ldots\right]$$

By using the first two terms of this series expansion and assuming $S(\theta) \approx S_o$ we obtain the following expression:

$$\Delta I_g^+ = e \, K \, \alpha \, P_g^+(0) \left[1 + \beta K \frac{s(\theta)p}{S_o \, Np}\right] s(\theta)p$$ (114)

Expression 113 gives a good qualitative representation of the experimental curves at intermediate coverages (fig.50).

This description of the ion emission induced by gas adsorption is well supported by a great number of experiments on metals and alloys covered with oxygen $|65,66|$. It appears that the dependence of the M^+ ion intensity on oxygen coverage is rather complicated except for the first stage of the adsorption process (small θ) where the ionization probability $P_g^+(0)$ is constant (eq.112) and could be interpreted as a characteristic property of the metal-oxygen bonding $|65,66|$. This is the most striking feature of the ion emission phenomenon induced by adsorption.

V-2-2 oxygen adsorption on (100) nickel single crystal $|65|$:

The effect of oxygen on Ni^+ ion emission and work function of (100) surface of nickel is represented in fig.51. At low oxygen pressures $\Delta I_g(Ni^+)$ increases linearly with p, but it seems to tend rapidly towards a saturation value until a critical pressure p_c is reached at which it jumps suddenly to higher values (this effect is not observed on polycrystalline sample). Beyond p_c the current tends slowly toward the saturation. The kinetics of the oxidation process is not the same on both sides of the critical coverage θ_c which corresponds to the pressure p_c. At $\theta < \theta_c$ the ion intensity instantaneously follows the variation of the pressure, whereas a few minutes are necessary before the equilibrium (eq.16) is reached at $\theta > \theta_c$. It is obvious that above θ_c a reaction occurs on the surface. This is confirmed by the variation of work function (fig.51). The potential of the surface increases rapidly with the coverage and remains high until the reaction begins at θ_c. At $\theta > \theta_c$ the work function shows a dip and rises again up to the saturation coverage. Such a behaviour was also found by Park and Farnsworth |171|. These authors showed that the decrease in work function is associated with the formation of NiO. This decrease is explained by the penetration of oxygen into sublayers of nickel. Therefore, it is clear from ion emission results that the nucleation of NiO starts at the critical coverage θ_c. The behaviour of the Ni^+ ion intensity is explained as follows.

At low coverage, oxygen atoms are adsorbed at random and progressively form p(2×2) and c(2×2) organized structures |172|. This implies that the sticking coefficient decreases rapidly (fig.51) |170|, and becomes zero theoretically at complete arrangment, before coverage unity is reached, if no further transformation can take place. For (100) nickel surface the complete arrangement of the c(2×2) structure corresponds to a coverage of 0.5 |172|. But, between 0.3 and 0.4 the nucleation process of NiO begins |170|. Then oxygen incorporation releases new adsorption sites, leading to an increase of the sticking coefficient. The singularity in the ion intensity evolution represented in fig.51 is probably associated with the critical coverage θ_c of about 0.3 to 0.4 at which the formation of NiO starts.

V-3 Molecular ions:

According to eq.71, the general expression for the intensity of a two atomic species cluster is:

$$I(A_m B_n^{\pm}) = f_{m+n} \; P(A_m B_n^{\pm}) \; F_{m,n}(c_A, c_B, \alpha, \beta) \; S_T \; I_p \tag{115}$$

V-3-1 random distribution of atoms:

V-3-1-1 case of oxides: A number of oxides become amorphous under ion bombardment, therefore, to a first approximation, their molecular ion emission can be discussed by assuming a random distribution of atoms. It was mentioned in sect.II.1.3 that $Cr_2 O_3$ and $Cr O_2$ |63| oxides were differentiated by ion clusters which contained many Cr atoms. From eq.115 the ratio of $M_m O_n^{\pm}$ ion intensities in the two oxides is given by the ratio of the structural factors, that is:

$$\frac{F_{m,n}(Cr_2 O_3)}{F_{m,n}(Cr O_2)} = \frac{(2/5)^m (3/5)^n}{(1/3)^m (2/3)^n} = 1.2^m \; 0.9^n$$

It clearly appears that the difference in intensities will be large for high m and small n values. The best choice is for Cr_m^+ or $Cr_m O^{\pm}$ ions with m as high as possible.

Likewise, it is seen that structural factors calculated in Table XIII for three iron oxides are of a little difference. This

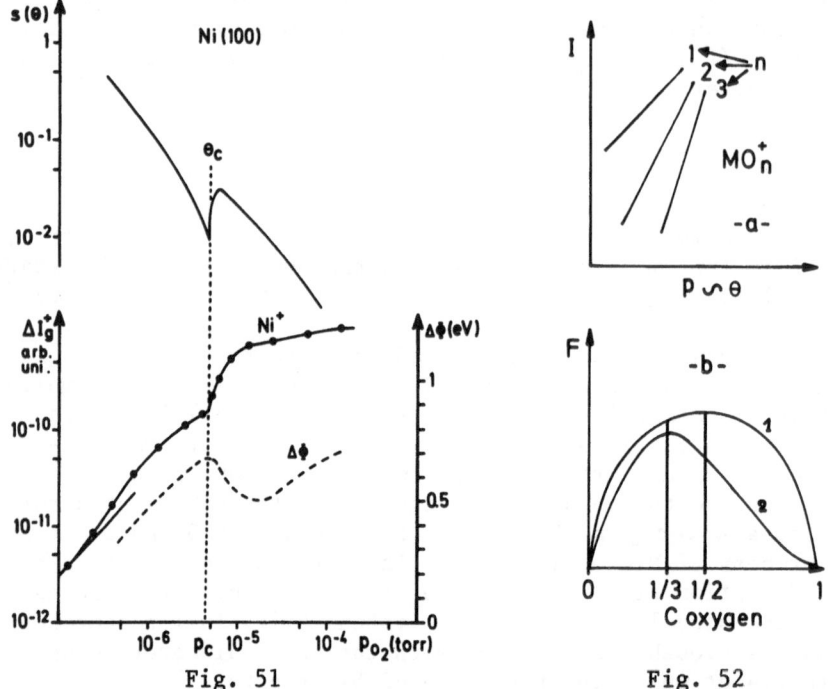

Fig. 51   Fig. 52

Fig.51 – ΔI_g^+ intensity of Ni$^+$ ion emission of (100) nickel single
        crystal and change in work function $\Delta\phi$ as a function of oxy-
        gen pressure |from 65|. Bombardment by A$^+$ ions of 6 keV and
        primary density $\sim$ 1 μA/mm . The upper curve represents the
        variation of the sticking coefficient |from 170|.
Fig.52 – a – Qualitative variation of $MO_n^\pm$ ion intensity at low oxygen
        coverage. Intensity I and pressure p are in log coordinates.
        b – Variation of the structural factor F for $MO^\pm$ ions (cur-
        ve 1) and $M_2O^\pm$ ions (curve 2) as a function of oxygen surface
        concentration.

explains why it is difficult to identify these oxides (see Table IX).

V-3-1-2 case of oxygen adsorption: We give there a qualitative inter-
pretation |45| of the behaviour of molecular ions as a function of o-
xygen coverage (fig.22–23), assuming that atoms are distributed at
random in the surface oxide layer. The evolution of the molecular ion
intensities $M_mO_n^\pm$ represented in fig.22–23 result from the change of
the surface composition as the coverage θ proceeds. From eq.115 the
three parameters $P(M_mO_n^\pm)$, S_T and $F_{m,n}$ may depend on θ. However, the
first two vary slowly |65,104| as compared to the structural factor.
Therefore, we can consider that the behaviour of molecular ions is
controlled, to a large extent, by the variation of the structural factor.

Table XIII

	MO^+	M_2^+	M_2O^+	$M_2O_2^+$
Fe O	0.25	0.25	0.125	0.062
$Fe_3 O_4$	0.245	0.185	0.104	0.06
$Fe_2 O_3$	0.24	0.16	0.096	0.058

If $c(M)$ and $c(O)$ are respectively the surface concentration of metal and oxygen atoms, the structural factor for $M_mO_n^\pm$ ions is :

$$F_{m,n} = \frac{(n+m)!}{n!m!}\ c(M)^m\ c(O)^n = \frac{(n+m)!}{n!m!}\ [1-c(O)]^m\ c(O)^n \qquad (116)$$

At low coverage $c(M) \sim 1$ and $c(O) \propto \theta$, and, as under dynamic conditions θ is proportional to the pressure p (eq.16), one obtains:

$$F_{m,n} \propto c(O)^n \propto \theta^n \propto p^n$$

Variations of $MO_n^\pm$ ion intensities are qualitatively represented in fig.52-a as a function of p (log-log diagram) for various n: the higher n, the steeper is the slope of the straight line.

For a large variation of coverage, structural factors corresponding to $MO^\pm$ and $M_2O^\pm$ ions are represented in fig.52-b. A comparison of these curves with the experimental curves in fig.24 suggests that the maximum of $MO^\pm$ ion intensity, obtained when the surface is saturated in oxygen, corresponds approximately to the maximum of F(fig.52-b). It is concluded that the highest oxygen surface concentration is generally ~ 0.5 or, eventually, a little bit more because, in some cases, a slight decrease of intensity is observed after the maximum (case of Si in fig.22). In the same way, the maximum of $M_2O^\pm$ ion intensity is related to the maximum of F which corresponds to 1/3 of oxygen concentration. It results that the strong decrease of the intensity after the maximum is explained by the decrease of F beyond 1/3 of oxygen concentration.

According to the structural factor, the maximum of $M O_2^\pm$ ion intensity should be reached at 2/3 of oxygen concentration. In most cases, the experimental evolution of $MO_2^\pm$ ions is very similar to that of $MO^\pm$ ions. This behaviour is consistent with the fact that the maximum of the surface oxygen concentration probably never reaches 2/3 on most metals.

The structural factor of M_n^+ ions decreases continuously as oxygen coverage proceeds. So it is expected the same for M_n^+ ion emissions, but, as oxygen adsorption enhances these emissions, one observes the behaviour represented in fig.24. However, as n increases, the enhancement is lesser and lesser and the influence of the structural factor becomes predominant. That explains why Cr_3^+ ion intensity only exhibits a slight maximum and drops more rapidly than Cr_2^+ ion intensity.

This succint description of the behaviour of ion clusters intensities as a function of oxygen coverage agrees rather well with experiments. However, the detail of the behaviour is not always clearly
understood since several parameters may vary in the same time; sputtering yield and ionization efficiency among others.

V-3-2 organized structures : Al_2 Cu θ phase |54| :

If we take the parameter of the lattice of the θ phase Al_2 Cu
as unity one finds around each copper atom |64| :

    2 Cu atoms at $\sqrt{9}/6$
    8 Al atoms at $\sqrt{7.25}/6$

Likewise, around each Al atom one finds:

    4 Cu atoms at $\sqrt{7.25}/6$
    5 Al atoms at $\sqrt{8}/6$

As the Cu-Cu distance ($\sqrt{9}/6$) is relatively large as compared to others,
we will consider that the environment of a Cu atom is only composed
of Al atoms. On the contrary Al-Cu and Al-Al distances are nearly
equal so that we will consider both types of atoms as equivalent regarding the environment. Aluminium and copper concentrations are respectively 2/3 and 1/3 in the θ phase and the parameters a, b,α,β
used in the calculation of structural factors are :

environment atoms	Al	Cu
Cu	a=1	b=0
Al	α=0.55	β=0.45

Some structural factors calculated as indicated in section III.4 are
given below:

$$F_{CuAl^+} = 1/3(a+2\beta)$$

$$F_{CuAl_2^+} = 1/3(a\alpha+2\alpha\beta+2\beta a)$$

$$F_{CuAl_3^+} = 1/3(a\alpha^2+2\alpha^2\beta+4\alpha\beta a)$$

$$F_{Cu\ Al^+_2} = 1/3(a\beta+ba+\beta b)$$

$$F_{Cu_2\ Al_2^+} = 1/3(a\alpha\beta+a^2\beta+ba\alpha+2\alpha\beta b+2\beta^2 a+2\beta ba)$$

(117)

The intensities of ion clusters emitted by a pure θ phase have been
compared to those emitted by a AlCu solid solution containing 98 at.%
of Al and 2 at.% of Cu. Assuming a random distribution of Cu atoms in
the solid solution, the structural factors have been calculated according to exp.69 with the following parameters:

Environment	Al	Cu
atoms		
Cu	a=0.98	b=0.02
Al	α=0.98	β=0.02

From eq.115 the ratio of the ion intensities of the same cluster, emitted by the pure θ phase and the solid solution is:

$$\frac{I_\theta(A_m B_n^+)}{I_{s.s}(A_m B_n^+)} = \frac{F_{m,n}(\theta)\ S(\theta)}{F_{m,n}(s.s)\ S(s.s)} \qquad (118)$$

A comparison of the calculated values (eq.118) to experimental results is given in fig.53 assuming $S(\theta)\wedge S(s.s)$. The agreement is rather good. As a comparison, calculations of (118) have been performed assuming that atoms are distributed at random in the θ phase. It is seen that the agreement is less good.

Expression 118 allows one to explain the differences in the ion clusters intensities (fig.16) of the Al Cu alloy (2 at % of Cu) in solid solution and after the precipitation of copper into small grains of θ phase |54|.

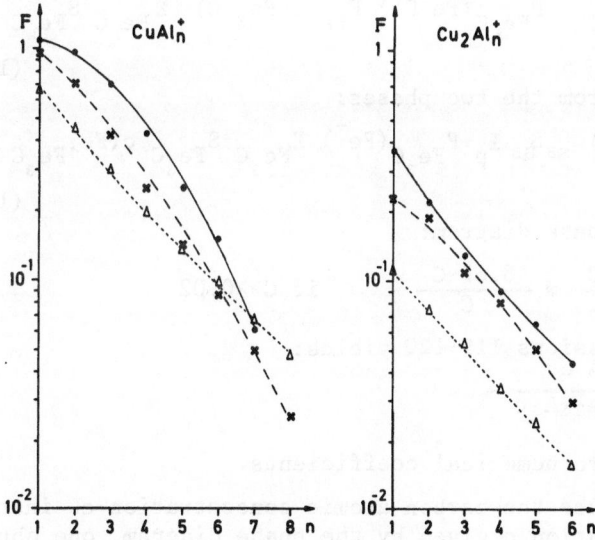

Fig.53 - Ratio of the structural factors of Cu Al$_n^+$ and Cu$_2$ Al$_n^+$ ions emitted by a pure θ phase (Al$_2$ Cu) and a Al Cu solid solution (2 at % of Cu) |from 54|.

● Experimental results.
X Calculations taking the structure of the θ phase into account
Δ Calculations assuming that atoms are distributed at random in the θ phase.

V-4 Ion emission from multi-phased systems:

V-4-1 application to a large grain phase: Fe-C alloys |90|.

The phase diagram of Fe-C alloys, schematically represented in fig.38, exhibits two phase limits corresponding respectively to $c_\phi = 0.02\%$ and $c_{\phi'} = 6.67\%$ in weight carbon concentrations. When the carbon concentration $0.02 < c < 6.67$ the alloy is composed of a solid solution of carbon in iron and large precipitates (several μ) of cementite $Fe_3 C$. In that case, the change in surface composition is completely negligible during an experiment (see sect. III.2.2.2).

The ion mass spectra of the two phases show that carbon is not detectable in the solid solution:

solid solution Fe^+ Fe_2^+

cementite C^+ Fe^+ FeC^+ FeC_2^+ Fe_2^+ C^- C_2^-

When a large area, containing both solid solution and cementite, is analysed, the intensities of the most important ions $Fe\ C^+$ and Fe^+ are given by:

- $Fe\ C^+$ ions come only from cementite, therefore:

$$I(Fe\ C^+) = f_{Fe,C} \cdot P_{Fe_3C}(Fe\ C^+)\ F_{Fe,C}(Fe_3\ C)\ \Sigma_{Fe_3C}\ S_{Fe_3C}\ I_p$$

(119)

- Fe^+ ions come from the two phases:

$$I(Fe^+) = P_{s.s}(Fe^+)\Sigma_{s.s}S_{s.s}\ I_p + P_{Fe_3C}(Fe^+)\ \Sigma_{Fe_3C}\ S_{Fe_3C}\ c(Fe)_{Fe_3C}\ I_p$$

(120)

According to the phase diagram:

$$\frac{\Sigma_{s.s}}{\Sigma_{Fe,C}} \simeq \frac{6.67-C}{C-0.02} \neq \frac{6.67-C}{C} \qquad \text{if } C \gg 0.02$$

The ratio of expressions 119-120 yields:

$$\frac{I(Fe\ C^+)}{I(Fe^+)} = \frac{A_1 C}{A_2 + A_3 C}$$

where A_1, A_2, A_3 are numerical coefficients.

If we introduce the carbon atomic concentration c_o instead of in weight concentration c given by the phase diagram, one obtains numerically:

$$\frac{I(Fe\ C^+)}{I(Fe^+)} = \frac{4\ c_o}{3 + 40\ c_o}$$

(121)

This expression is well supported by experiment |90|.

V-4-2 application to a small grain phase:

The sample consists of a copper single crystal containing small $Al_2\ O_3$ amorphous particles of 50 to 200 Å in diameter |123|. This is a case where the surface composition changes rapidly in time when operating under dynamic conditions (see section III.2.2.1). There is no Al in the copper matrix and no Cu in $Al_2 O_3$ precipitates.

Let c be the aluminium atomic concentration in the sample and $c_\phi = 0.4$ the aluminium atomic concentration in $Al_2 O_3$ precipitates. It is easy to show that exps. 50 yield:

$$\frac{D_m}{D_\phi} = \frac{C_\phi}{C} \gg 1$$

when c is small ($\lesssim 1$ at.%). Therefore simplified expressions 54, 55, 57 can be used.

Ion emission from the copper matrix:

As aluminium and oxygen concentrations are very small, the copper concentration is near unity. Therefore:

$$I_m(M^+) = P_m(M^+) S_m I_p \tag{122}$$

for any species emitted from the matrix.

Ion emission from $Al_2 O_3$ precipitates:

This emission is composed of ions characteristic of the presence of alumina (Al^+, Al_2^+; $Al O^+$, $Al_2 O^+$ ...). However we will just be interesting here in the emission of Al^+ ions. According to exps. 41, 42, 43 the Al^+ ion emission coming from precipitates is given by:

$$I_\phi(Al^+) = P_\phi(Al^+) c_\phi n_s \sigma S_\phi I_p \tag{123}$$

where ϕ stands for $Al_2 O_3$. The surface density of precipitates n_s varies exponentially in time according to eq.52. When the equilibrium surface density n_s^e is reached (eq.54), the intensity becomes:

$$I_\phi^e(Al^+) = P_\phi(Al^+) c S_m I_p \tag{124}$$

The dependence of exp.123 on the surface density has been verified experimentally by rotating the sample as indicated in section III.1.2. As $Al_2 O_3$ is amorphous S_ϕ remains constant under a rotation while S_m varies. It results in a change of the surface composition according to exp.52, with a characteristic time T given by 55. It has been verified that T was proportional to the diameter 2R of the particles (eq.56) (assuming the particles are spherical) and inversely proportional to the primary density N_p (fig.54).

Furthermore, these experiments have demonstrated that it exists a specific ion emission of the matrix-precipitate interface |123|. This emission is characterized by the presence of ions such as Cu Al^+, Cu AlO^+... which can't originate from the bulk of the Cu matrix or the bulk of $Al_2 O_3$ precipitates, since there is no copper inside the precipitates and no Al or O in the copper matrix. On the other hand, an enhancement of the Cu^+ ion emission is observed. This enhancement is explained by the influence of oxygen present in precipitates on the ion emission of the copper layer which surrounds each precipitate. It is a case where the superposition principle |63| is no longer valid.

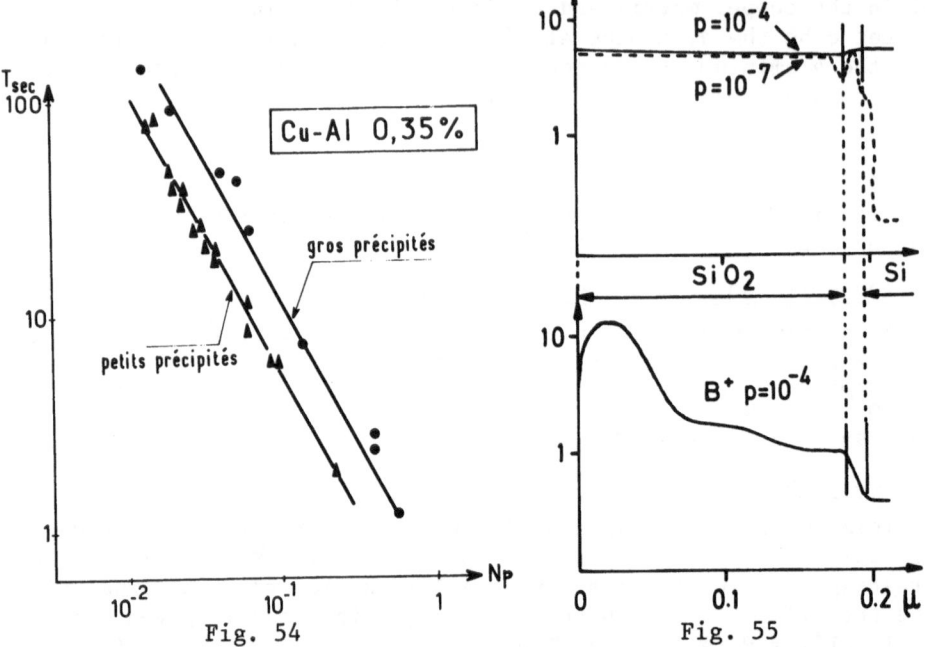

Fig. 54   Fig. 55

Fig.54 – Time T characteristic of the change of the surface composition in a copper matrix containing small precipitates of
$Al_2 O_3$, as a function of the primary beam density N_p (N_p is
in arbitrary units), for small precipitates ($\sim$100 Å) and big
precipitates ($\sim$200 Å) |from 123|.

Fig.55 – In-depth analysis of boron in Si O_2 – Si interface |from 92|.
Dashed curve Si^+ intensity under good vacuum conditions. Full
curves Si^+ and B^+ signals under 10^{-4} oxygen pressure.

V-5 In-depth analysis:

V-5-1 in the monolayer range:

        In-depth analysis in the monolayer range has been developped by
Benninghoven |11,19,36,69-76| on the basis of the ideal erosion model
described in sect.III.3. From eq.67, the intensity $I(M^+)$ of an element
M present at a concentration $c(M)$ in a surface structure which does
not exceed one monolayer is given by:

$$I(M^+) = P(M^+) \; c(M) \; S_T \; \exp- \frac{t}{\tau} \; I_p \qquad (125)$$

where S_T is the sputtering yield of this structure. Bombarding the
sample with a constant primary current I_p, the decrease of $I(M^+)$ must
be exponential in time (fig.39). This argument was used by Benninghoven to interpret the oxidation of metal surfaces. The author concluded
that, in general, metal oxidation proceeded in two steps: in the first
step a thick oxide layer , phase I, is built, then a second oxide
structure, phase II, composed of an oxygen monolayer fixed on the underlying oxide layer, is formed. Phase I should be characterized by

large intensities of negative ion species (MO_2^-, MO_3^- ...) and phase
II by positive ones (MO^+ ...). The sputtering yields deduced from measurments of τ should be $S_I \sim 2$ and $S_{II} \sim 20$ respectively.

In fact, the interpretation is open to criticism for many rea-
sons |65,66|. At first, the model is over-simplified with regard to
the diversity of metal-oxygen interactions. For example, an oxygen
layer at the surface of alkaline earth metals is not acceptable since oxygen adsorption always produces a decrease of the work function
of these metals, which clearly indicates the presence of metal atoms
at the surface |173|. On the other hand, the sputtering yield $S_{II} \sim 20$
of phase II seems to be much too high regarding the general behaviour
of oxides |116|. Furthermore, such a result should have a curious
consequence: it should be possible to increase the sputtering yield
of a metal by a factor ~ 10 just by maintaining it completely covered
with oxygen under bombardment. In fact, it is observed experimentally
that the oxidation of a metal always decreases the sputtering yield.
At last, experiments on oxides show that the relative importance of
positive molecular ions with respect to the negative ones of the same
species depends mainly on the nature of the metal and not on the composition of the oxide: for example Cr O^+ ions will always be more in-
tense than Cr O^- ones in all chromium oxides |63|; on the contrary,
in nickel oxides, Ni O^+ ions are less intense than Ni O^- |46|. There-
fore, the distinction between phases I and II by negative and positive
ions is in contradiction with the results obtained on oxides.

In fact, the behaviour of ion intensities during the removal
of first few atomic layers is very complicated as it appears in the
two examples presented in fig.56 |71,74|. The intensities of various
ions emitted by magnesium and chromium metals exposed to high oxygen
dose (saturation coverage) are registered as a function of the primary current exposure (I_p x time). Each ion species has its proper behaviour, therefore, it is very difficult to interpret the results correctly. In particular it should be erroneous to interpret the exponential decrease of $Mg_2 O^+$ ions as due to a monomolecular oxide layer.
After 12000 L exposure, the oxide layer largely exceeds the monolayer
range on this readily oxidizable metal.

In fact, the evolution of the intensity of each ion species
during the removal of the oxide layer (fig.56) appears to be very similar to the reverse evolution of intensities as a function of oxygen
exposure |65|(oxygen pressure x time) (fig.23 for example). This ob-
servation suggests that the surface under bombardement passes by the
same stages as those obtained during oxygen exposure. This can only
be explained by a surface diffusion process which continuously changes the structure of the surface while the oxide layer is removed.
This process is not included in Benninghoven's erosion model, therefore, it is advisable to be very careful in the interpretation of in-
depth analysis in the monolayer range.

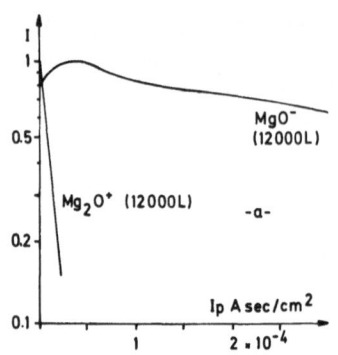 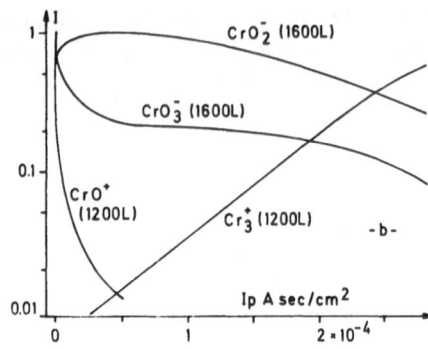

Fig.56 a - Removal of the oxide layer formed on a magnesium sample
exposure to oxygen (1200 L) |from 74|. Variation of MgO^-
and Mg_2O^+ ion intensities as a function of primary current
exposure (I_p x time).

b - Removal of the oxide layer formed on a chromium sample ex-
posed to oxygen (1600 and 1200 L) |from 71|. Variation of
CrO^+, Cr_3^+, Cr_3^+, CrO_2^- and $Cr\ O_3^-$ ion intensities as a function
of primary current exposure (I_p x time).

V-5-2 to a large extent:

       In-depth analysis to a large extent is one of the most popular
application of Ion Microanalysis. The early works dealt with the study of diffusion profiles but, more recently a number of investigators
have applied the method to thin films characterization and implant
profiles determination. For details the reader is referred to a recent
review by J.A. Mc Hugh |15| which provides a very comprehensive view
of the problem of in-depth analysis by ion emission and the various
applications.

       The principle of in-depth analysis by ion emission is simple.
Atomic layers located deeper and deeper in the sample are successively brought to the surface as the erosion proceeds. Assuming the erosion is parallel to the surface, the thickness dx to be removed during a time dt is :

$$\frac{dx}{dt} = a\ S_T(x)\ \frac{N_p}{N_0} = v(x) \qquad (126)$$

with $S_T(x)$ sputtering yield at depth x.

       N_0 atomic surface density.

       a   distance between atomic layers parallel to the surface.

       N_p density of the primary beam per unit time.

       $v(x)$ erosion rate (usually measured in Å/s).

The intensity of an ion species M^+ which concentration is $c(M,x)$ at
depth x is given by: $I(M^+,x)=P(M^+,x)\ c(M,x)\ S_T(x)\ I_p \qquad (127)$

If $S_T(x)$ and $P(M^+,x)$ are constant during the erosion, the ion intensity profile in directly proportional to the depth concentration

profile of the element M. Considering sect. III.1, III.2.1 and IV.4
it appears that the only instance where this situation can be achieved is for low concentration elements in amorphous-or single crystal-
one-phased matrices where the major constituents are homogeneously
distributed. Discussing the instrumental and physical factors which
limit the capability of the method in that case. Mc Hugh concluded
that the depth resolution and sensitivity inherent in Ion Microanalysis were < 50 Å and $< 10^{17}$ atom/cm^3.

   It is much more difficult to determine depth profiles on material composed of dissimilar layered phases since both parameters P
and S_T will vary from one phase to another. Procedures for quantita-
ting the ion signal have to be used |104|. An example is given in
fig.55 concerning the determination of the boron concentration pro-
file at SiO_2-Si interface |92|. In this experiment the boron concen-
tration is low so that the sputtering yield is imposed by the matrix,
that is SiO_2 on one side of the interface and Si on the other. The
sputtering yields of the two phases are different |116|: $S_T(SiO_2) \neq$
$S_T(Si)$. On the other hand, the presence of oxygen in SiO_2 enhances
the positive ion yields of elements present in this compound (see
sect. II.1.2), therefore $P_{SiO_2}(B^+)$ and $P_{SiO_2}(Si^+)$ are much greater than
$P_{Si}(B^+)$ and $P_{Si}(Si^+)$. When the experiment is carried out under good
vacuum conditions ($2 \cdot 10^{-7}$ toor) with A^+ bombarding ions, a sharp decrease of Si^+ and B^+ intensities mainly due to the variation of the
ionization probabilities, is observed at the interface (dashed curve
in fig.55 for Si^+). If the same experiment is now carried out under
an oxygen pressure of 10^{-4}torr which ensures an oxide layer on the
surface of silicon (see sect.II.2.3), the Si^+ signal is equal for the
two phases (full line in fig.55). This is not surprising since it was
concluded in sect.II.2.3 that the oxide layer formed on silicon exposed to oxygen was SiO_2 ; therefore:

$$S_T(Si) + O_2 = S_T(SiO_2)$$

and
$$P_{Si+O_2}(Si^+) = P_{SiO_2}(Si^+)$$

As the matrix has now the same composition on both sides of the interface, it is sure that the ionization probability of boron also is the
same. Therefore the variation of the B^+ signal indicates in fig.55
represents now the concentration profile of this element at the interface.

   This procedure which consists of flooding the surface of the solid with oxygen in order to make in-depth analysis easier was very
successful in a number of analytical applications on semiconductor
devices |see 15|. However it results from sect. II.2.3 on the nature
of the oxide layer formed by oxygen adsorption and from sect.IV.4 and
V.1 on the problem of quantitative analysis that is not a general procedure for quantitating the secondary ion intensities. Therefore, caution must always be exercised in the interpretation of depth concentration profiles.

The author is grateful to Pr. G. Slodzian, Pr. G. Morrison and Dr. P. Williams for their advice and comments on the manuscript.

REFERENCES

1 - G. Slodzian, THèse de Doctorat d'Etat-Orsay; Masson ed. Paris 1963.
2 - R. Castaing and G. Slodzian, J. Microscopie 1-395 (1962).
3 - G. Slodzian, Ann. Phys. Paris 9-591 (1964).
4 - R. Castaing and J.F. Hennequin, Advances in Mass Spectrometry, Vol.5-419; A. Quale ed. Institute of Petroleum, London (1971).
5 - R. Castaing, Electron Microscopy in Material Science, Academic Press Inc., New York (1971).
6 - G. Slodzian, Surf. Sc. 48-161 (1975).
7 - H. Liebl, J. Appl. Phys. 38-5277 (1967).
8 - H. Liebl, Anal. Chem. 46-22A (1974).
9 - H. Liebl, J. Phys. B 8,10-797 (1975).
10- C.A. Andersen and J.R. Hinthorne, Science 175-853 (1972).
11- A. Benninghoven, Surf. Sc. 28-541 (1971).
12- A. Benninghoven and E. Loebach, Rev. Sc. Instrum. 42-49 (1971).
13- Ya M. Fogel, Int. J. Mass Spectrom. Ion Phys. 9-109 (1972).
14- R.E. Honig, Advances in Mass Spectrometry, Vol. 6-337; A.R. West ed. Appl. Science Publ., Berking, Essex, England (1974).
15- J.A. Mc Hugh, Methods of Surface Analysis, S.P. Wolsky and A.W. Czanderna, ed. Elsevier, New York N.Y. (1975).
16- G.H. Morrison and G. Slodzian, Anal. Chem., Vol.47-11-932A (1975).
17- N.B.S. Proceedings of a Workshop on Secondary Ion Mass Spectrometry and Ion Microprobe Mass Analysis, Gaithersburg Md (1974) - ed. K.F.J. Heinrich and D.E. Newbyry.
18- R. Hernandez, P. Lanusse and G. Slodzian : Compte-Rendu Acad. Sc. Paris B 271-1033 (1971).
19- A. Benninghoven : Surf. Sc. 35-427 (1973).
20- D.K. Bakale, B.N. Colby, C.A. Evans; Anal. Chem. 47-9-1532 (1975).
21- J.F. Hennequin; J. Phys. 29-957 (1968).
22- H. Kerkow and M. Trapp; Int. J. Mass Spectrom. Ion Phys. 13-113 (1974).
23- G. Blaise and G. Slodzian; Rev. Phys. Appl. 8-105 (1973).
24- P. Joyes; J. Phys. 30-365 (1969).
25- J.F. Hennequin; J. Phys. 29-1053 (1968).
26- E. Dennis and R.J. Mac Donald; Radiation effects 13-243 (1972).
27- M. Bernheim, G. Slodzian and N. Soyris; to be published.
28- J.F. Hennequin; J. Phys. 29-655 (1968).
29- Z. Jurela and B. Perovic ; Canad. J. Phys. 46-773 (1968).
30- G. Staudenmaier; Radiation Effects 13-87 (1972).
31- K. Wittmaack; Nucl. Inst. Meth. 132-381 (1976).
32- Z. Jurela; Int. J. Mass Spectrom. Phys. 18-101 (1975).

33- R. Laurent, G. Blaise and G. Slodzian; Compte-Rendus Acad. Sc. 278B-11 (1974).
34- R. Laurent, G. Blaise and G. Slodzian; to be published.
35- J.F. Hennequin; Rev. Phys. Appl. 1-273 (1966).
36- A. Benninghoven; Surface Science 53-596 (1975).
37- Z. Jurela; Radiation Effects 13-167 (1972).
38- Z. Jurela; Atomic Collision Phenomena in Solids.Ed. D.W. Palmer, M.W. Thompson, P.D. Towsend Amsterdam (1970).
39- G. Blaise and G. Slodzian; Comptes-Rendus Acad. Sc. 271B-1216 (1970).
40- V.E. Krohn Jr: J. Appl. Physics 33-12-3523 (1962).
41- I.A. Abroyan et al.; Soviet Phys. Sol. State 7-2557 (1966).
42- B. Navinsek; J. Appl. Phys. 36-1678 (1965).
43- J.F. Hennequin; Compte-Rendus Acad. Sc. 264B-1127 (1967).
44- J.F. Hennequin; Compte-Rendus Acad. Sc. 263B-1246 (1966).
45- G. Blaise; unpublished.
46- J.C. Pivin, C.Roques-Carmes and P. Lacombe; Revue Française de Métallurgie (1975), to be published.
47- A. Havette; Thèse 3° cycle (1974).
48- G. Blaise and G. Slodzian; J. Physique 35-243 (1974).
49- G. Blaise; Radiation Effects 18-235 (1973).
50- G. Blaise and G. Slodzian; J. Physique 35-237 (1974).
51- J.Z. Schelten; Naturforsch 23a-109 (1968).
52- H.E. Beske; Z. Naturforsch 22a-459 (1967).
53- G. Blaise and M.C. Cadeville; J. Physique 36-545 (1975).
54- G. Blaise, R. Laurent and G. Slodzian; to be published.
55- G. Blaise and G. Slodzian; Compte-Rendus Acad. Sc. 266B-1525 (1968).
56- R.F.K. Herzog, W.P. Poschenrieder and F.G. Satkiewiez; Proc. Intern. Conf. Ion Sputtering and Related Phenomena-Garching (Munich) 173- (1972).
57- M. Leleyter and P. Joyes; J. Phys. B 7..4-516 (1974).
58- M. Leleyter and P. Joyes; Radiation Effects 18-105 (1973).
59- G. Hortig and M. Muller; Z. Physik 221-119 (1969).
60- M. Leleyter; Thèse de Doctorat d'Etat, Orsay (1975).
61- H. Rodriguez-Murcia and H.E. Beske; Berichte der Kernforschung-sanlage, Jülich n°1292 (1976).
62- J. Maul and K.W. Wittmaack; Surface Science 47-358 (1975).
63- H.W. Werner, H.A.M. de Grefte and J. van de Berg; Advances in Mass Spectrometry, Vol.6,Ed.A.D. West (Appl. Sc. Publ., Barking, Essex) (1974) p. 673.
64- R. Graf; Thèse, série A n°2789 - Paris (1955).
65- G. Blaise and M. Bernheim; Surf. Sc. 47-324 (1975).
66- G. Blaise; Bull. Soc. Chem. Belg. 84-6-617 (1975).
67- P.M. Holloway and J.B. Hudson; Surface Science 43-123 (1974).
68- W.H. Krueger and S.R. Pollack; Surface Sc. 30-263 (1972).
69- A. Benninghoven and A. Muller ; Thin Solid Films 12-439 (1972).
70- A. Benninghoven, E. Locbach, C. Plog and N. Treitz; Surf. Sc. 39-397 (1973).
71- A. Benninghoven and A. Muller; Surf. Sc. 39-416 (1973).

72- A. Muller and A. Benninghoven; Surf. Sc. 39-427 (1973).

73- A. Benninghoven, C. Plog and N. Treitz; Intern. J. Mass. Spectrom. Ion Phys. 13-415 (1974).

74- A. Benninghoven and L. Wiedmann; Surf. Sc. 41-483 (1974).

75- A. Muller and A. Benninghoven; Surf. Sc. 41-493 (1974).

76- E. Stumpe and A. Benninghoven; Phys. Status Solidi.(a) 2-479 (1974).

77- V.F. Rybalko, V. Ya. Kolot and Ya. M. Fogel; Soviet Phys. Solid State 11-1142 (1959).

78- Ya. M. Fogel , R.P. Slabospitskii and J.M. Kamankhov; Zh. Teor. i Eksperim. Fiz. 30-824 (1960); Soviet Phys. Tech. Phys. 5-777 (1961).

79- V.I. Shvachko, B.T. Nadykto, Ya. M. Fogel, B.M. Vasyutinskii and G.M. Kartmarzov; Soviet Phys. Solid State 7-1572 (1966).

80- Ya. M. Fogel; Soviet Phys. Usp. 10-17 (1967).

81- V. Ya. Kolot, V.I. Tatus, V.F. Rybalko and Ya. M. Fogel; Soviet Phys. Techn. Phys. 15-1934 (1971).

82- V. Ya. Kolot, V.I. Tatus, V.F. Rybalko, Ya. M. Fogel, V.V. Vodvlazhchenko and V.M. Evseev; Soviet Phys. Solid State 13-1275 (1971).

83- D. Guénot; D.E.S. Orsay (1966).

84- G. Slodzian and J.F. Hennequin; Compte-Rendus Acad. Sc. 263B-1246 (1966).

85- D. Brochard and G. Slodzian; J. Phys. (Paris) 32-185 (1971).

86- M. Bernheim and G. Slodzian; Intern. J. Mass Spectrom. Ion Phys. 12-93 (1973).

87- M. Bernheim and G. Slodzian; Surface Sc. 40-169 (1973).

88- J.F. Hennequin, G. Blaise and G. Slodzian; Compte-Rendus Acad. Sci. 268B-1507 (1969).

89- P. Joyes; Radiation Effects 19-235 (1973).

90- E. Darque; Thèse de 3° cycle, Orsay (1976).

91- G. Blaise and G. Slodzian; Surface Sci. 40-3-708 (1973).

92- B. Blanchard, N. Hilleret and J.B. Quoirin; J. Radioanalyt.Chem. 12-85 (1972).

93- C.A. Andersen; J. Mass Spectrom. Ion Phys. 2-61 (1969).

94- C.A. Andersen; J. Mass Spectrom. Ion Phys. 3-413 (1970).

95- M. Bernheim and G. Slodzian; Int. J. Mass Spectrom. Ion Phys. 20-295 (1976).

96- P. Sigmund; Phys. rev. 184-383 (1969) and 187-768 (1969).

97- P. Sigmund; Rev. Roum. Phys. 17-823; 17-969 and 17-1079 (1972).

98- H.H. Andersen and H.L. Bay; Radiation Effect 18-(1973).

99- J. Lindhard, V. Nielsen and M. Scharff; Mat. Fys. Medd. Dan. Vid. Selsk 36-10 (1968).

100-H. Oechsner; Appl. Phys. 8-185 (1975).

101-V.A. Molchanov, V.G. Tel'kovskii; Soviet Phys. Dokl. 6-137 (1961).

102-P.K. Rol, J.M. Fluit, J. Kistemaker Physica 26-1000 (1960).

103-H. Oechsner; Z. Physik 261-37 (1973).

104-M. Bernheim; Thèse Orsay (1973).

105-G.K. Wehner; Phys. Rev. 102-690 (1956).

106-R.H. Silsbee; J. Appl. Phys. 28-1246 (1957).

107-G. Leibfreid; J. Appl. Phys. 30-1388 (1959).

108-D.D. Odintsov; Sov. Phys. Solid State 5-4-813 (1963).

109-M. Bernheim; Radiation Effects 18-231 (1973).

110-A.L. Southern, W.R. Willis and M.T. Robinson; Journal Appl. Phys. 34-1-153 (1963).

111-Don E Harrison Jr.; W.L. Moore Jr.; H.T. Holcombe; Radiation Effects 17-167 (1973).

112-J. Nizam and N. Benazeth-Colombie; Rev. Phys. Appl. 10-4-183(1975).

113-G. Holmen, A. Buren and P. Hogberg; Radiation Effects 24-51 (1975). and 24-39 (1975).

114-G. Holmen, S. Peterstrom and A. Buren; Radiation Effects 24-45 (1975).

115-J.C. Bourgoin, J.F. Morhange and R. Beserman; Radiation Effects 22-205 (1974).

116-R. Kelly and Nghi Q. Lam; Radiation Effects 18-   (1973).

117-N. Laegreid and G.K. Wehner; J. Appl. Phys. 32-365 (1961).

118-Hazune Shimizu, Masatoshi Ono and Katsuya Nakayanna; Surface Sc. 36-817 (1973).

119-S.D. Dahlgren, E.D. Clanahan; J. Appl. Phys. 43-1514 (1972).

120-B. Navinsek; Thin Solid Films 13-367 (1972).

121-W.L. Paterson, G.A. Shirn; J. Vac. Sci. Techn. 4-343 (1967).

122-M. Ono, Y. Takasu, K. Nakayama, T.Y. Yamashina; Surf. Sci. 26-313 (1971).

123-C. Lyon, C. Roques-Carmes and G. Blaise; Surface Sci., to be published.

124-C. Roques-Carmes, G. Slodzian and P. Lacombe; Canad. Metall. Quarterly. Vol. 13-1-99 (1974).

125-M. Delesse; Annales des Mines 13-379 (1848).

126-A. Benninghoven; Z. Physik 230-403 (1970).

127-G.E. Rhead Surface Sci. 47-207 (1975).

128-G. Hortig, M. Muller; Z. Physik 221-119 (1969).

129-G.P. Konnen, A. Tip, A.E. De Vries; Radiation Effects 21-269(1974). G.P. Konnen; Thèse, Amsterdam (1974).

130-V.I. Veksler and M.B. Ben'Iaminovich; Sov.Phys.Tech.Transl.1(1956).

131-L.N. Dobretsov; Electron and Ion Emission-NASA-Tech. Transl.(1963).

132-C.A. Andersen and J.R. Hinthorne; Analytical Chem. 45-1421 (1973).

133-M.N. Saha; Phil. Mag. 40-472 (1920); Z. Phys. 6-40 (1921), and J. Eggert; Z. Phys. 20-570 (1919).

134-M. Bernheim, G. Blaise, G. Slodzian; Int. J. Mass. Spectrom. Ion Phys. 10-293 (1972).

135-E.S. Parilis and L.M. Kishinevski; Sov. Phys. Solid State 3-885 (1960).

136-K.H. Krebs; Fors. Physik 16-419 (1968).

137-J.N. Coles; Surface Sci. 55-721 (1976).

138-B. Djafari-Rouhani; Thèse de 3° cycle, Orsay (1974).

139-P. Joyes and G. Toulouse; Phys. Lett. 39A-267 (1972).

140-A. Blandin, A. Nourtier, D.W. Hone; J. Phys. 37-369 (1976).

141-A. Nourtier, to be published.

142-L.V. Keldish; Sov. Phys. JETP 20-1018 (1965).

143-P.W. Andersen; Phys. Rev. 124-41 (1961).

144-G. Blaise and G. Slodzian; J. Phys. 31-93 (1970).

145-R.J. Mac Donald; Surface Sci. 43-653 (1974).

146-Z. Sroubek; Surface Sci. 44-47 (1974).

147-J. Antal; Phys. Lett. 55A-8-493 (1976).

148-J.M. Schroeer, T.N. Rhodin and R.C. Bradley; Surf. Sci. 34-571
     (1973).

149-M. Cini; Surf. Sci. 54-71 (1976).

150-C.B. Kerkdijk and R. Kelly; Surf.Sci. 47-294 (1975).
     C.B. Kerkdijk; Thesis - Amsterdam (1975).

151-G.E. Thomas, E.E. de Kluizenaar and M. Beerlage; Chemical Phys.
     7-303 (1975).

152-G.E. Thomas, E.P. Kluizenaar; Inter. J. Mass Spectrom. Ion Phys.
     15-165 (1974).

153-J.P. Mériaux, R. Goutte, C. Guillaud; Appl. Phys. 7-313 (1975).

154-E.U. Condon and G.H. Shortley; The theory of atomic spectra (cam-
     bridge 1957).

155-G. Blaise and G. Slodzian; Rev. Phys. Appl. 8-247 (1973).

156-H.D. Hagstrum; Phys. Rev. 96-336 (1954).

157-J. Friedel; J. Phys. 23-692 (1962).

158-P. Joyes and M. Leleyter; J. Phys. B. 6-150 (1973).

159-M. Leleyter and P. Joyes; J. Phys. 35-L85 (1974).

160-P. Joyes; J. Phys. Chem. Solids 32-1269 (1971).

161-L. Salem; The Molecular Orbital Theory of Conjugate System
     (1966) Benjamin - New York.

162-B. Pullman, A. Pullman; Les Théories Electroniques de la Chimie
     Organique (1952) Masson & Cie p.176.

163-J.P. Perrier, B. Tissier and R. Tournier; Phys. Rev. Lett. 24-
     313 (1970).

164-V. Jaccarino and L.R. Walker; Phys. Rev. Lett. 15-258 (1965).

165-J.C. Pivin, C. Roques-Carmes a,d G. Blaise; Surf. Sci., to be
     published.

166-B. Window; J. Appl. Phys. 44-2853 (1973).

167-H. Rechenberg, L. Billard, A. Chamberod and N. Natta; J. Phys.
     Chem. Solids 34-1251 (1973).

168-J.P. Perrier, B. Tissier and R. Tournier; Phys. Rev. Lett. 24-
     313 (1970).

169-G. Blaise; Surf. Sci., to be published.

170-P.M. Holloway, J.B. Hudson; Surf. Sci. 43-123,141 (1974).

171-R.L. Park and H.E. Farnsworth; J. Appl. Phys. 35-220 (1964).

172-A.U. Mac Rae; Surf. Sci. 1-319 (1964).

173-F.C. Tompkins; The solid gas interface Vol.2, ed.E.A. Flood
     (1967).

ION INDUCED X-RAYS: GENERAL DESCRIPTION

Finn Folkmann

Gesellschaft für Schwerionenforschung

Postfach 541, D-6100 Darmstadt, West Germany

ABSTRACT

X ray production by ions with energies 0.5-10 MeV/amu is considered. The x rays are measured with semiconductor detectors or crystal spectrometers. The production for light ions is theoretically understood by BEA and PWBA calculations. For heavy ions also indirect production processes of selective nature contribute and a satellite structure is built on the lines,which may give chemical information. Background radiation from secondary electron bremsstrahlung and proton bremsstrahlung is dominant and leads to analytical sensitivities, which for concentration is 10^{-6}-10^{-7}, for absolute amounts 10^{-9} - 10^{-16}g and for lateral resolution 1 cm-1 μm. Enhancement is discussed and calculations are made for the ionization by secondary electrons. Analytical problems are illustrated by the newly proposed evidence for superheavy nuclei.

1. INTRODUCTION

The interest in ion induced x rays in the last couple of years has been stimulated by two factors. One is the perspective of employing the x ray emission as a novel and powerful tool for multi-element trace element analysis of small smaples. Another is the flow of new information which the study of x rays originating from ion collision has lead to by the observation and explanation of many interesting phenomena. Working with atomic physics experiments at the UNILAC heavy ion accelerator at GSI (1.4 - 10 MeV/amu Ar, Kr, Xe and U) I am biased for the latter subject, but as this meeting deals with analytical methods I will mainly discuss the first prob-

lems based on what our physical understanding of the involved processes implies for the analytical capability and which new ideas
might have bearings on future improvement on the analytical method.
The recent development was reflected on a conference last week in
Lund, Sweden, on Particle Induced X ray Emission (PIXE) and its Analytical Applications.[1]

The analytical method employs ion energies in the range
0.5-10 MeV/amu and makes mainly use of solid state Si(Li) x ray detectors, which implies that elements $Z \gtrsim 12$ are seen in the working
range 1-100 keV, as the K x ray characterizing an element with
atomic number Z has an energy $\sim Z^2 \times 0.010$ keV. With crystal spectrometers or windowless semiconductor detectors the working range can
be extended down to $Z > 4$. We are dealing with an atomic process
where inner shell electrons have a good chance of being ejected as
shown schematically in fig. 1 and later an x ray quantum can be
emitted by the filling of the vacancy. Its energy is characteristic
of the bombardment atom and a spectrum accumulated within a few minutes, as shown in the right part of fig. 1, then tells which elements
are present in the analyzed sample.

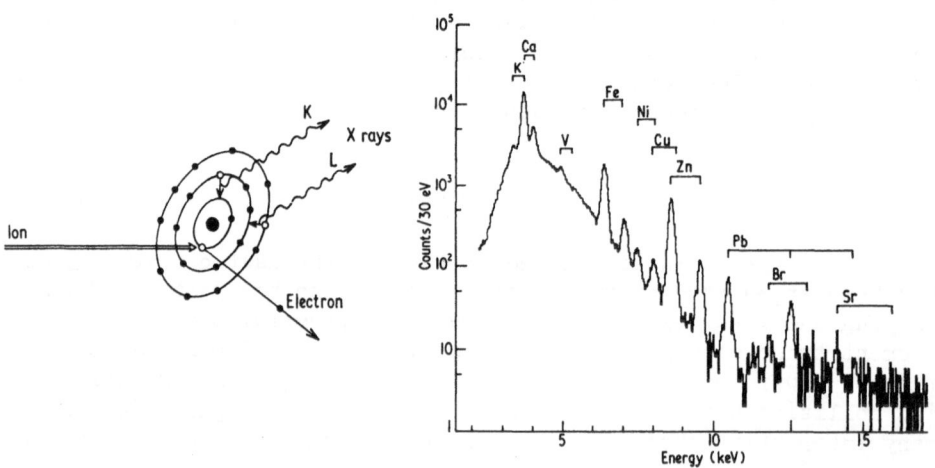

Fig. 1 X ray production schematically and an x ray spectrum
from a sample of dust from air bombarded with 3 MeV protons
indicating the lines from various elements[2].

2. X RAY DETECTION

The experimental setup used for this kind of analysis is very simple with an ion beam-commonly protons or He ions of 1-4 MeV/amu-bombarding a sample and an x ray detector situated in the vicinity registering the emitted radiation[2]. Most often the sample is placed in vacuum and the detector outside the scattering chamber with windows and an optional absorber between, but it is also possible to have the detector in vacuum or to let the beam pass a window so the analyzed specimen is kept in air or e.g. in a He atmosphere.[3,36]

The break-through of the analytical method was initiated by the development of semiconductor detectors with good energy resolution in the late 1960'ith. These Li drifted Si or Ge -Si(Li) and Ge(Li) - or intrinsic Ge diodes have high detection efficiency ($\sim$1) over the active surface of 10-200 mm^2, which means that total x ray detection probabilities of $\sim 10^{-2}$ are feasible. The energy resolution ΔE (Full Width at Half Maximum - FWHM) for such a detector including its amplification electronics is

$$\Delta E = (\Delta E_n^2 + 2.35^2 \, F\epsilon E)^{1/2} \tag{1}$$

where the last term comes from the statistics of the charge collection process in the active layer of the detector. As sketched in fig. 2 an x ray (of energy E) creates electron-hole pairs with a mean excitation energy ϵ (=3.81 eV for Si and =2.96 eV for Ge), which are dissipated into detectable quanta according to the Fano factor F (=.12 for Si and =.08 for Ge). The noise contribution ΔE_n arising in the detector and the preamplifier was lowered to a level near 100 eV FWHM mainly due to the improvements by Goulding[4] and his group in detector construction and amplifier design.[5] The lower part of fig. 2 shows a semiconductor detector in which the electron hole pairs created by an x ray are collected and further processed. They may for optimum resolution go to a preamplifier through a low noise Field Effect Transistor which is photosensitive, and repetively is drained by a pulse from a Light Emitting Diode (LED). After a main amplifier and often also a pileup rejector, which can reject pulses separated by more than $\sim$200 ns, signals proportional to the original x ray energy are produced and can be processed into ADC's of multichannel analyzers or computers for accumulation. All x rays absorbed in the active detector layer, which is typically $\sim$3 mm thick, can be recorded, i.e. in an energy region 1-30 keV for Si and 1-100 keV for Ge, and will be stored in one run, where the lower energy depends on the windows and absorber. A resulting x ray spectrum from one element is drawn at the bottom of fig. 2, and additionally to the two lines (K_α and K_β) a tail region extends to lower pulse hights with an intensity $\sim 10^3$ times lower resulting from incomplete charge collection in the edge regions of the detector (sketched in fig. 2). This

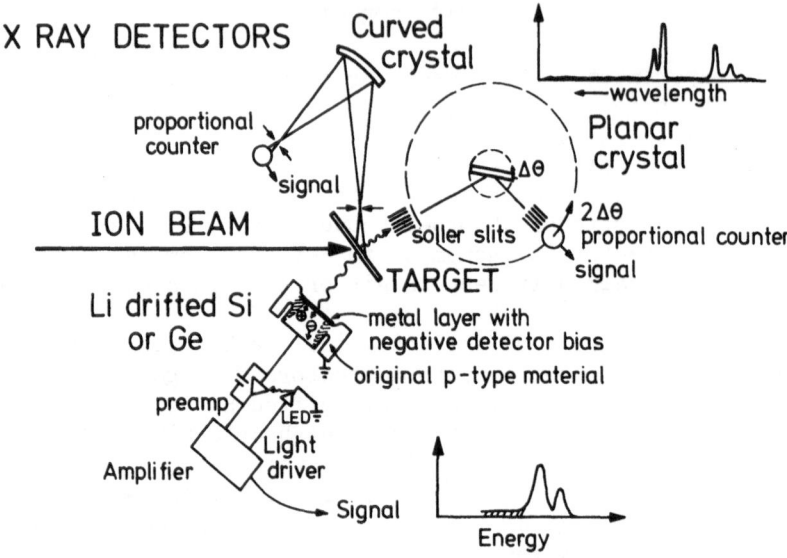

Fig. 2 X ray detectors used for ion induced x rays.
Looking on target is drawn a semiconductor diode
and a flat and a curved crystal. Diagrams of the
resulting spectra as function of increasing x ray
energy is represented. For crystal spectrometers
the energy variation is achieved by turning the
crystal and simultaneously the proportional counter.

effect can be suppressed by using a collimator in front of the detector or by more sophisticated guard ring or double guard ring
detectors[4,5]. Also it should be noted that instead of using pileup
rejection electronics one can deflect the ion beam away from the
target, when an x ray is detected, which can be done within ∿200ns
and has several advantages at high counting rates[6].

Crystal spectrometers offer another important tool for measuring x rays. The Bragg reflection in the crystal lattice with
parallel planes of distance d will let x rays of wavelength λ be
observed with an incident and outgoing angle θ to the plane
according to

$$n \cdot \lambda = 2d \cdot \sin\theta \tag{2}$$

where n = 1, 2, ... is the order of the reflection. As the x ray
energy is $E = 1.24 \text{ nm}/\lambda$ differentiation of (2) gives the energy
resolution

$$\Delta E = E \cdot \Delta\theta \cdot \cot\theta \tag{3}$$

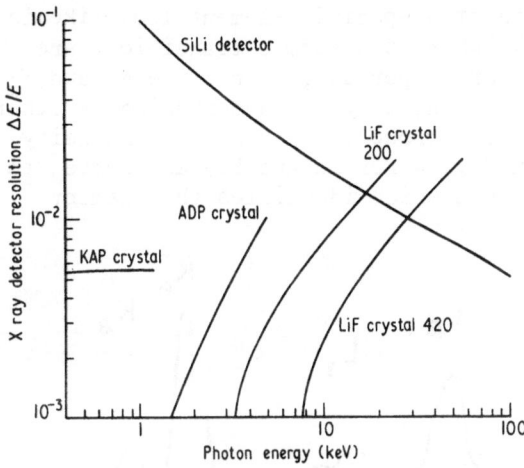

Fig. 3 Energy resolution of x ray detectors[2]. 2d for the crystals are LiF 200: .4028 nm, LiF 420: .1802 nm, Ammonium Dihydrogen Phosphate: 1.064 nm and Potassium Acid Phthlate: 2.663 nm.

$\Delta\theta$ is determined by crystal imperfections and by the angular defi-nition of the x rays onto the crystal ($\Delta\theta\sim 0.1°$). An experimental arrangement with a planar crystal is schematically shown in the right part of fig. 2. Soller slits define θ and $\Delta\theta$ for x rays hit-ting the crystal and a symmetrically arranged proportional counter with an optional set of soller slits in front of it register the reflected quanta[7]. The energy or λ of the x rays is for fixed d determined by θ and may be varied by turning the crystal $d\theta$ and the counter correspondingly $2d\theta$. The spectrometer has an energy resolution, eq. (3), which for low energies is significantly better than for semiconductor detectors, as shown in fig. 3 for various crystals. Only one energy can be measured at a fixed spectrometer position and the detection probability is very low (10^{-5}–10^{-6}) due to reflection loss and $\Delta\theta$. This last limitation can be circumven-ted using a curved crystal as reflector as shown in the upper part of fig. 2. A point source ($\sim$mm) is focussed onto a point in front of the proportional counter by the bent crystal. This gives an increased effective solid angle by 1–2 orders of magnitude rela-tive to the flat crystal. The analyzed x ray energy is geometri-cally fixed, but can be varied changing the position of the re-flecting crystal and the counter in a more complicated way.

A comparison of x ray detectors gives an energy resolution, which is in favour of crystal spectrometers at low x ray energies according to fig. 3. The efficiency is highest for semiconductor detectors and decreases by at least 2 orders of magnitude for curved crystal spectrometers and further for planar crystals. Semiconductor detectors have the feature of simultaneously recording a broad range of x ray energies whereas a wavelength dispersive crystal is me-chanically fixed to one energy being able to measure the intensity at only one point of the spectrum, i.e. sensitive to a single ele-ment and not effected by radiation at other energies. Several crystals, tuned for different elements, may be used for multielement analysis.

How x rays characteristic of a specific element look like is
schematically shown in fig. 4, where the atomic transitions are
indicated. Mainly a few lines are important; either the K_α and K_β
line or the L_α, L_β and L_γ lines but they are accompanied by other
less intensive lines. Within each group (K,L or M) the intensity
ratios are constant (given in fig. 4 for Zn and Pb) and serve to
establish the fingerprint pattern which identifies the element.

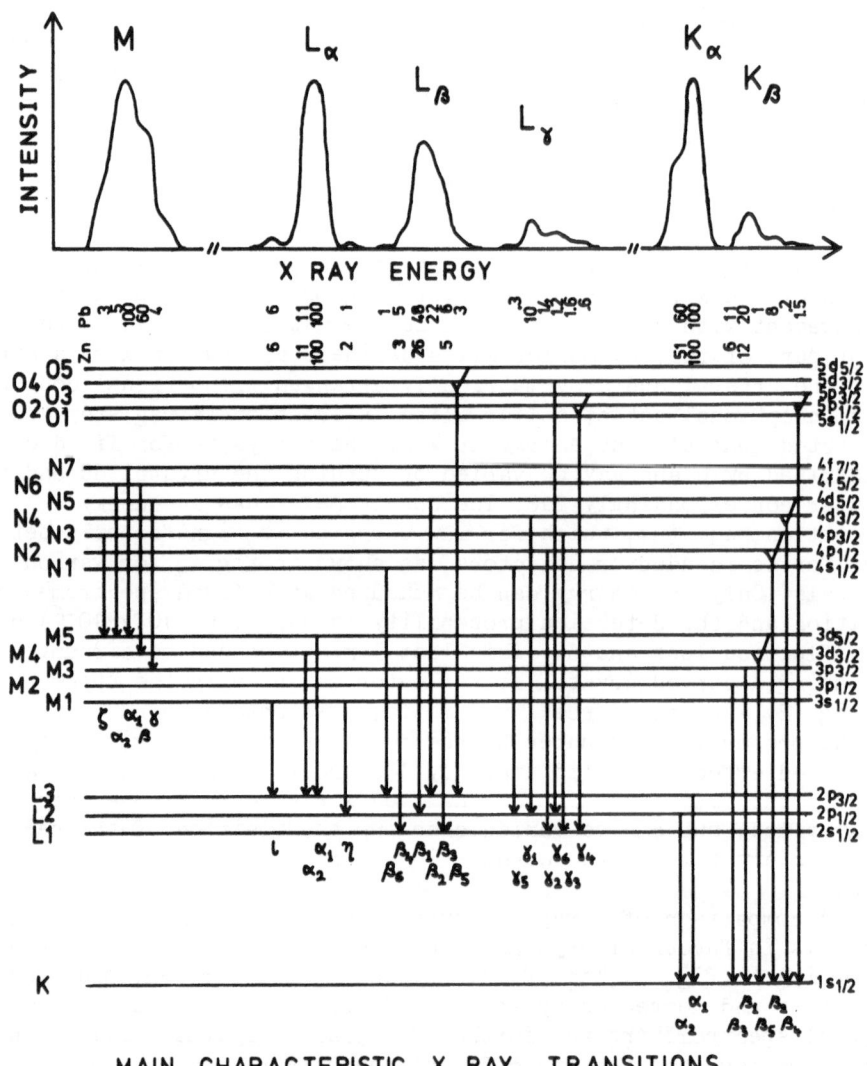

MAIN CHARACTERISTIC X RAY TRANSITIONS

Fig. 4 X ray transitions with indication of involved atomic
levels. Intensity ratios are given for Zn and Pb within each
group of transitions to a certain shell.

The main feature is the shift in x ray energy (with $\sim Z_2^2$) for various target elements. The major appearance of these characteristic x rays depends not on the exciting ion (or electron or photon) and can thus be used to identify elements in a sample. However, the relative and absolute intensities of the various groups (K, L, M) are sensitive to the exciting ion and its energy and will be discussed in the next chapter. Also details of line ratios and shifts are influenced by the projectile parameters and may be seen and checked according to the resolution of the x ray spectrometer. They will be mentioned in chapter 4.

3. X RAY PRODUCTION BY LIGHT IONS

Analytically protons and He ions are the most popular ions to excite x rays and the understanding of the physical processes generating the x rays for these projectiles is pretty good. The ejection of inner shell target electrons is governed by the Coulomb interaction between the nucleus of the ion (of atomic number Z_1, mass A_1M and energy E_1) and each of the atomic electron (of mass m) in the target. This vacancy production can be calculated classically in the 'Binary Encounter Approximation'[8] (BEA) which will hold in many cases also quantum mechanically and which gives a very simple result for the cross section of producing K vacancies

$$\sigma_K^{BEA} = \frac{Z_1^2}{U_K^2} \cdot f\left(\frac{m \cdot E_1}{A_1 \cdot M \cdot U_K}\right) \tag{4}$$

In this scaling law U_K is the binding energy of the target K electron containing the Z_2^2 dependence and the values of the general function f can be read from fig. 5. The argument of f in eq. (4) can be written if we note that $U_K = \frac{1}{2} mu^2$ and $E_1 = \frac{1}{2} A_1 Mv^2$ as $(v/u)^2$, i.e. the crucial parameter for the cross section is the ratio of the projectile velocity to the mean velocity of the atomic electrons. Fig. 5 shows how experimental results for proton bombardment give values close to the very useful eq. (4). Similar results apply to the higher L and M shells. When ions (Z_1, A_1) are used instead of protons, eq. (4) states that the cross section for the same target Z_2 and thereby same U_K

$$\sigma_{Z_1 A_1}(E_1, Z_2) = Z_1^2 \cdot \sigma_{proton}\left(\frac{E_1}{A_1}, Z_2\right) \tag{5}$$

is that for protons of the same velocity multiplied by Z_1^2.

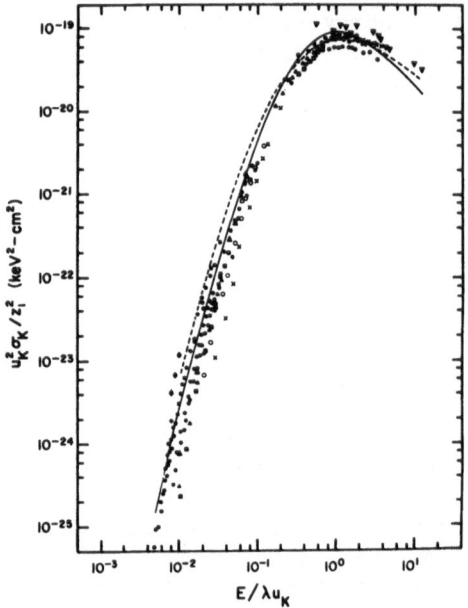

Fig. 5 K vacancy production
for proton bombardment (from
ref. 8) showing the general
relationship, eq.(4), for the
BEA (full drawn curve).
The dashed curve gives a PWBA
calculation, and the points
represent experimental data.

Perturbation theory offers another way of treating the same
process, and especially the 'Semi Classical Approximation'[9] (SCA)
and of the 'Plane Wave Born Approximation'[10] (PWBA) provide a more
general fundament for calculating the vacancy production, which give
results very similar to those of the BEA in eq. (4). The dimensionless parameters used for the SCA and PWBA are based on the hydrogenic
description of target electrons. In the following are stated some useful simplified quantitative relations for the K shell. The charge Z_K
by which the target K shell electrons are effected, may be the full
nuclear charge Z_2 or including the 'inner screening' from the other
electron reduced by the Slater value[9] to

$$Z_K = Z_2 - 0.3 \tag{6}$$

The corresponding energy is $E_K = Z_K^2 \cdot 0.0136$ keV, the mean velocity $u_K = (2E_K/m)^{1/2}$ and radius $a_K = a_o/Z_K$, where a_o is the Bohr radius. Measured by these quantities is the energy parameter

$$\eta = (v/u_K)^2 = \frac{m}{A_1 M} \cdot \frac{E_1}{E_K} = \frac{E_1}{A_1 \cdot Z_K^2 \cdot 25 \text{ keV}} \tag{7}$$

The actual binding energy of the electrons is taken into account
by the 'outer screening', described by the parameter

$$\theta_K = U_K/E_K = U_K/(Z_K^2 \cdot 0.0136 \text{ keV}) \tag{8}$$

and the K vacancy production in the PWBA can then be written

$$\sigma_K^{PWBA} = \frac{8\pi Z_1^2}{\theta_K Z_K^4} \; a_o^2 \; F \; (\frac{\eta}{\theta_K^2}), \qquad \text{with } a_o = 5.29 \cdot 10^{-11} m \quad (9)$$

where the function F is given in fig. 6. As the BEA function f in eq. (4) depends on $\eta/\theta_K = (v/u)^2$ the difference for the argument of the general PWBA scaling function F in eq. (9) is only a division by θ_K and for the factor in front of it a multiplication by θ_K. The PWBA (and SCA) results from eq. (9) are thus very similar to those of the BEA in eq. (4) and e.g. the projectile scaling in eq. (5) will apply for PWBA and SCA as well.

Calculations using PWBA[11] or SCA[9] constitute a good basis for taking into account perturbations which occur to the ionization process and which will effect the resulting cross sections. Physical factors which have a bearing on the quantitative results for the vacancy production are

1. Increased binding (formation of molecular orbitals)
2. Coulomb deflection (of the projectile)
3. Polarization (of the electron distributions in the collision)
4. Relativistic wave functions (for heavy targets)
5. Screening (in the target wave functions)

The first three factors represent additional physical processes for which an estimate can be given whereas the two last state the features of the wave function which may not have correctly treated in the simple result, eq. (9). Quantitative results for the first two effects are given below based on the paper by Basbas, Brandt and Laubert[12] as these perturbations are most important for vacancy production by the 1-8 MeV protons or He ions of highest analytical interest.

The binding energy of an electron will be increased when it is attracted not only by the target nuclear charge, but also by the charge of the projectile, which occurs preferentially when the projectile penetrates deeply into the target atom, i.e. at high projectile velocities. The increased effective binding energy will imply a lower cross section, which can be accounted for by replacing θ_K in eq. (9) by $\varepsilon\theta_K$, where the velocity dependent factor ε on the screening parameter can be approximated by

$$\varepsilon = 1 + \frac{2Z_1}{Z_K\theta_K} \cdot \frac{1+5\xi+7.14\xi^2+4.27\xi^3+.947\xi^4}{(1+\xi)^5} \qquad (10)$$

with $\xi = \frac{2v}{\theta_K U_K} = 2\eta^{1/2}/\theta_K$ as the velocity parameter.

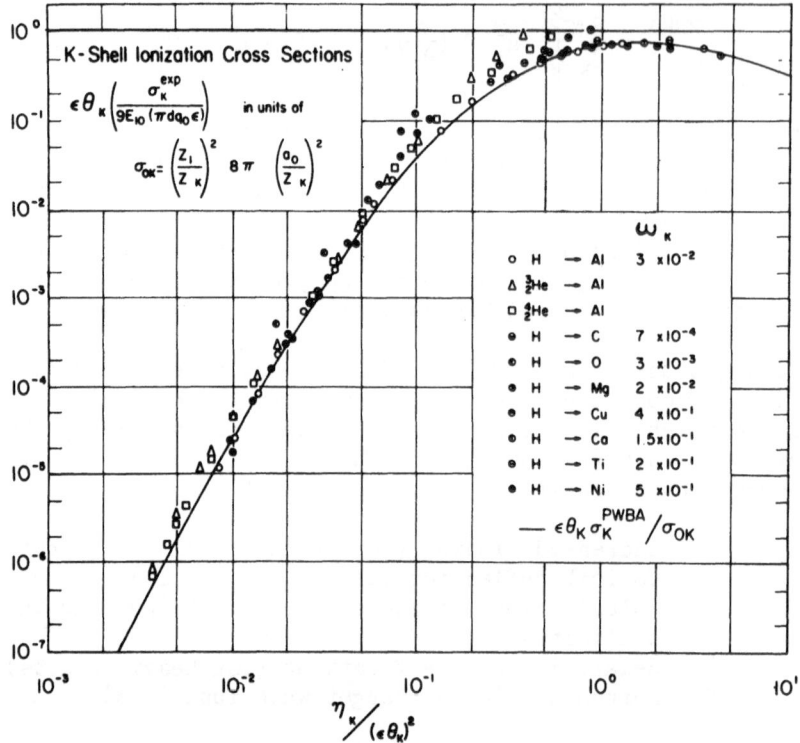

Fig. 6 K vacancy production from PWBA calculations including the perturbations from increased binding and Coulomb
deflection, eqs. (6)-(11). The diagram (from ref. 12) shows
the function F in eqs. (9) and (11). For the deflection
factor $9E_{10}$ depending on $x=\pi dqo\epsilon$ is in eq. (11) given a
simple approximation. The points indicate experimental results for H and He ions.

The deflection of the projectile in the electric field of
the target atom will also lead to lower cross sections due to the
repulsion of the two nuclei giving a larger effective distance
(impact parameter) for which a straight line trajectory will account
for the results. The effect will be most active at low projectile
velocities and at higher Z_2 and can be included by an extra factor
on the PWBA result, which then with binding energy and Coulomb deflection is

$$\sigma_K = \frac{9 \cdot e^{-x}}{9+x} \cdot \frac{8\pi Z_1^2}{\epsilon \theta_K Z_K^4} \; a_o^2 \; F \left(\frac{\eta}{(\epsilon\theta_K)^2}\right) \tag{11}$$

with $x = \dfrac{\pi m Z_1 \epsilon}{2A_1 M \theta_K^2} \cdot \left(\dfrac{\eta}{\theta_K^2}\right)^{-3/2}$

The factor $9 \exp(-x)/(9+x)$ is the approximate value[12] of the function $9E_{10}(x)$ mentioned in fig. 6.

The polarization represent a deformation of the electron bound states during the collision[13], which is most active for high projectile velocities and which will increase the cross section. Also the proper treatment of the target electron wave function will make the cross sections a little higher for the heavier elements including relativistic motion. The screening due to outer electrons, introduced by eq. (8), can be better incorporated using selfconsistent Hartree Fock wave functions.

How the above theoretical results compare with experimental values[14] is shown in fig. 7, where BEA, eq. (4), and PWBA, eq. (9), lie above the data points, whereas the Binding energy and Coulomb deflection corrected PWBA results from eq. (11) reasonably describe them. The last approach is in fig. 7 labled PWBA BC and is often referred to as a Perturbed Stationary State calculation (PSS)[15]

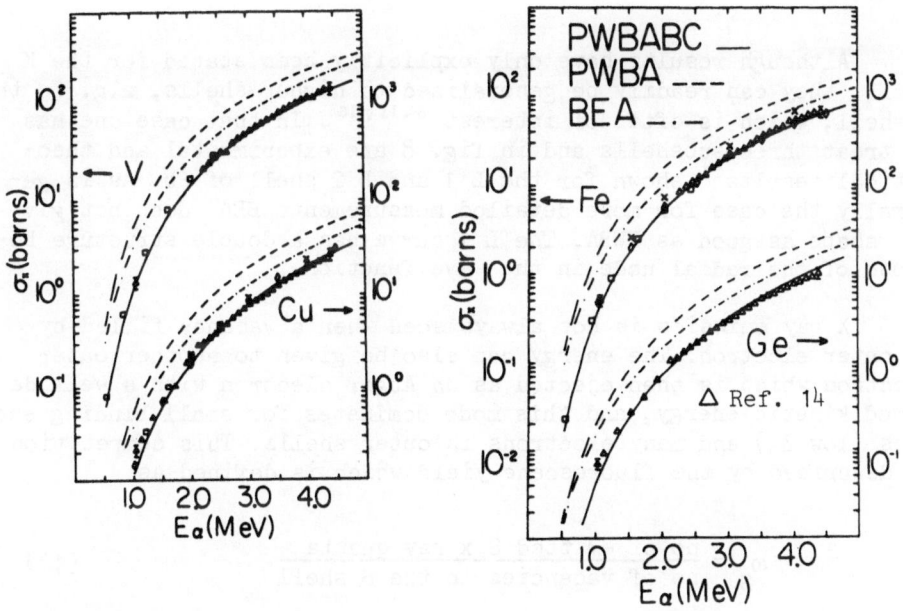

Fig. 7 K x ray production cross sections for 0.5-4.4 MeV He ions on $_{23}$V, $_{26}$Fe, $_{29}$Cu and $_{32}$Ge (from ref. 14). Theoretical values are calculated with PWBA eq.(9) upper curve, BEA eq. (4) middle curve, and PWBABC eq. (11), where the inclusion of increased binding is the main reason for the lower curve which fits well to the experimental data.

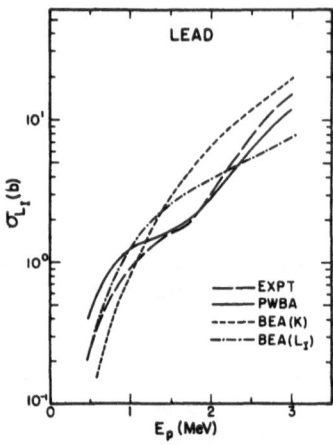 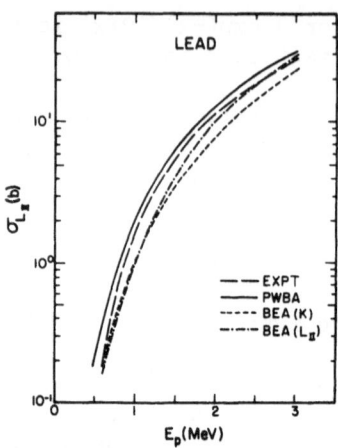

Fig. 8 L subshell ionization for protons on Pb (from ref.
17). Experimental and theoretical values are shown for
the L1 and L2 subshell. Note that PWBA fits the data
better than BEA and reproduce the observed kneepoint for
the L1 shell due to the radial node in the 2s wavefunction.

Although results have only explicitly been stated for the K
shell, they can readily be generalized to higher shells, e.g. to the
L shell, which is often of interest [8,11,16]. In that case one has
to treat three subshells and in fig. 8 are experimental and theoretical results[17] shown for the L 1 and L 2 shell of Pb. As is generally the case for more detailed measurements BEA does not give
the shape as good as PWBA. The L 1 curve has a double structure because of the radial node in the wave function.

X ray emission is not always seen when a vacancy filled by
an outer electron. The energy can also be given to another outer
electron which is then ejected as an Auger electron with a well defined kinetic energy, and this mode dominates for small binding energies (low Z_2) and many electrons in outer shells. This competition
is described by the fluorescene yield which is defined as

$$\omega_S = \frac{\text{no of emitted S x ray quanta}}{\text{no of vacancies in the S shell}} \qquad (12)$$

where S stands for the shell in question (S = K, L, M). The cross
section for emission of K x rays is thus that for producing vacancies, discussed in eqs. (4) to (11) multiplied by ω_K

$$\sigma_S^{\text{emission}} = \omega_S \cdot \sigma_S^{\text{vacancies}} \qquad (13)$$

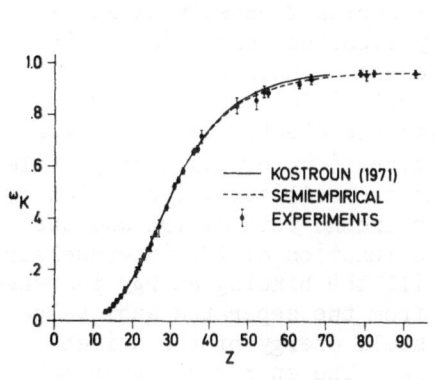

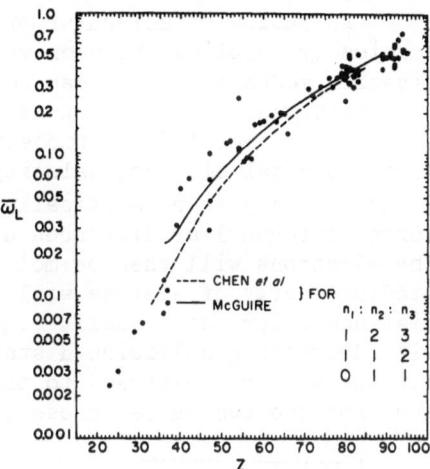

Fig. 9 K shell fluores-
cence yields, according
to eq. (12), (from ref. 18).

Fig. 10 L shell mean
fluorescence yields
(from ref. 18).

Values of ω_K and ω_L can be found in ref. 18 which are valid for
bombardment by light ions as H and He and are shown in figs. 9 and
10.

Although all essential physical processes seem to be
quantitatively understood I warn you to believe them to a too detailed level and will rather recommend for analytical calibra-
tions to measure x ray yields and branching ratios and only use
the theoretical results as a help in interpolation procedures and
physical argumentation.

4. X RAY PRODUCTION BY HEAVY IONS

For heavy bombarding ions the emission of x rays is governed
by the factors mentioned in the previous chapter but also new
effects occur which influence the intensity and the more detailed
appearance of the x rays. The increasing complexity of the x radiation is physically interesting and may have some perspectives
for applications but represent in most cases a complication in
the analysis of the material in a sample.

High vacancy production cross sections arise generally
due to the Z_1^2 scaling of the direct Coulomb excitation from eq.
(5). That applies not only to the shell, to which x ray emitting
transitions are studied but also to outer shells, effecting
details in the x rays.

Formation of molecular orbitals (MO) and transfer of vacancies by couplings between various orbitals constitute an excitation mechanism which can be very important and which exhibit resonances and a pronounced threshold behaviour at low
energies. Especially for projectile velocities small compared
to the orbital electron velocity v<<u the electron orbitals will
be able to adjust adiabatically to situations of fixed projectile-
target internuclear distances during the collision. The motion of
the electrons will then be molecular around both centra and the
binding energy of a state will be a function of the internuclear
distance[8]. For the K shell, e.g., will the binding energy increase
with decreasing collision distance from the separated atom value
(of the higher Z partner) to the K shell energy of the united
atom for the two nuclei close together. The energy of other mole-

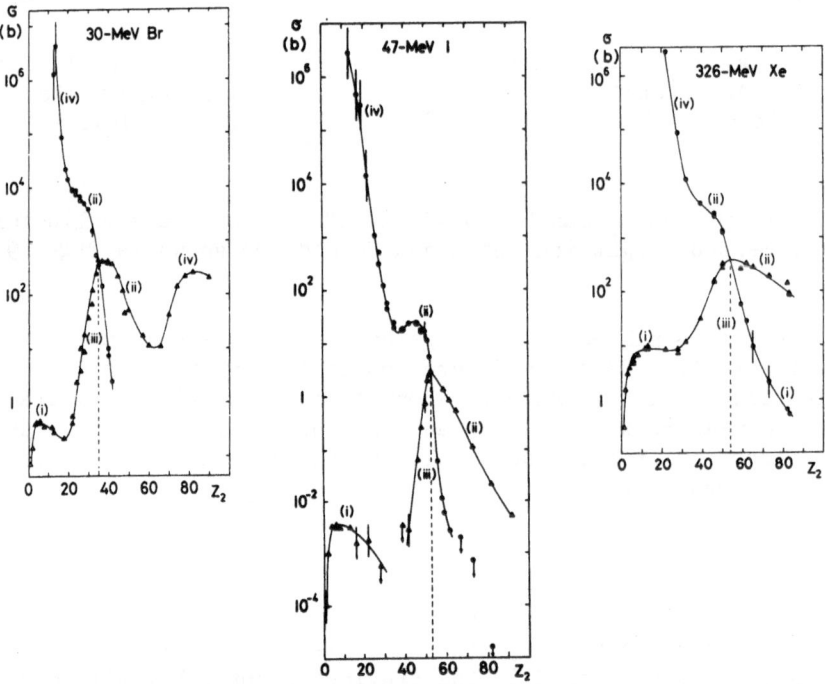

Fig. 11 Heavy ion K vacancy production (from ref. 19).
Target x rays (circles) and projectile x rays (triangles)
were measured for 0.37 MeV/amu Br and I and for 2.5 MeV/amu
Xe. Note the increase of the target vacancy production
close to symmetry $Z_2 \sim Z_1$. The curves are to guide the eye,
and the roman numbers indicate various features: (I) simple
excitation from the deepest 1sσ molecular orbital (MO),
(II) excitation via 2pπ MO and 2pσ MO coupling to the 1sσ
MO, (III) vacancy sharing increasing the higher Z cross
section close to symmetry, (IV) level matching between K
shell of lower Z and L shell of higher Z collision partner.

cular orbitals follow different curves which may approach or cross at certain distances, and which lead in the separated atom limit – before or after the collision – to e.g. also L or M shells of target or projectile. In such cases vacancies from higher shells, which are produced with high probabilities, may be transfered to lower shells. Important parameters for such coupling strengths are the difference in binding energy of the involved levels and the collision velocity, and the effect is often clearly seen for $Z_2 \sim Z_1$. For near symmetric collisions, e.g. L vacancies can be converted to K vacancies[19] and a peaklike structure in the production of target K x rays is seen for targets with atomic numbers close to that of the projectile as indicated in fig. 11 superimposed on the direct Coulomb cross section decreasing with increasing Z_2 for fixed Z_1. This feature may have implications for analytical use of heavy ion induced x rays. For lower projectile energies such resonances are more clearly seen and have thresholds for the first ionization process depending strongly on the projectile target combination, which makes good analytical separation of various elements possible[20], but suffers from low cross sections and missing quantitative data and has not been followed much up in applied work.

Capture of electrons into vacancies of the projectile will for heavy ions contribute to ionization of the target. This drainage of electrons to the projectile may be as important as the ejection to the continuum (described by BEA or PWBA). Calculations generally overestimate the effect, e.g. by a factor 10 for F, S and Cl ions[21]. It depends sensitively on the charge state of the ion and is higher the more vacancies the ions bring with them. For solid targets beam excitation takes place in the first layers, and when the later target excitation depends on the projectile charge state, as for electron capture, the resulting vacancy production cross section can depend on target thickness. Also the main material of the sample may influence the projectile excitation f. ex. by MO excitation mechanisms. Recoiling atoms in the sample may obtain high velocities by the big momentum transfer from incident heavy ions and can give another indirect contribution to the excitation cross sections via high MO excitation for $Z_2 \sim Z_1$, systems especially at lower energies.

The fluorescence yield, eq. (12), which determines the intensity of the observed x rays, can for heavy ion collisions have a value which is higher than what is seen from figs. 9 and 10. This is a consequence of the high degree of ionization. When the outer shells of the target are highly depleted for electrons, there is a smaller probability that the filling of a vacancy will take place under ejection of another Auger-electron compared to x ray emission. With only one outer electron left, an inner shell vacancy can only be filled unter x ray emission,

Fig. 12   Ne K x ray (lower) and
Auger electron (upper) cross
sections for 50 MeV Cl bombardment of a Ne gas (from ref. 22).
The total vacancy production
$\sigma_{tot} = \sigma_x + \sigma_{Auger}$ and $\omega_K = \sigma_x / \sigma_{tot}$,
see eq. (12), increase with
the Cl charge state.

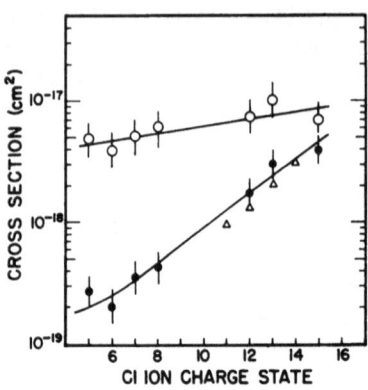

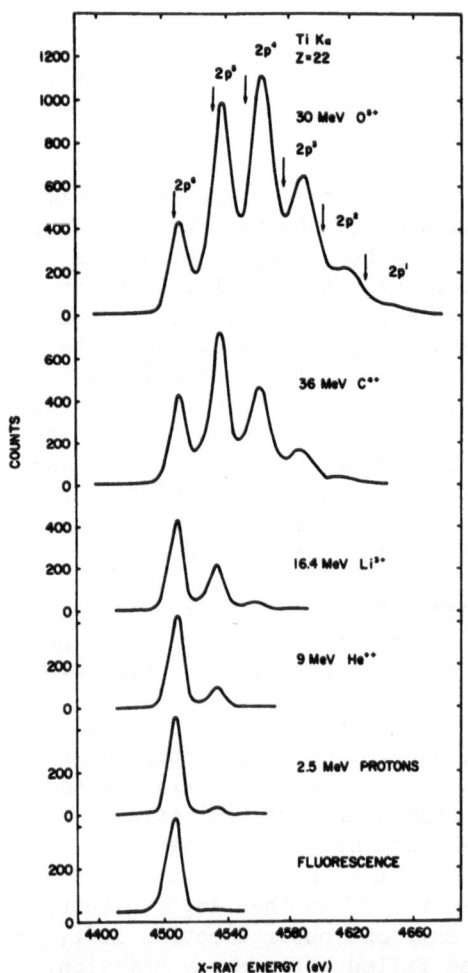

Fig. 13   Ti K α satellite spectra from bom-
bardment of a thick Ti
target with photons and
H, He, Li, C and O ions
(from ref. 24).With
higher Z_1 is observed
an increasing no of sa-
tellite lines correspon-
ding to additional L va-
cancies. The arrows in-
dicate calculated ener-
gies with various no of
2p L vacancies but no M
or N vacancies.

i.e. $\omega = 1$. Generally one has to use mean fluorescence yields,
which are higher than the values of figs. 9 and 10, and which
vary with projectile, its energy and charge state. In fig. 12 the
K x ray emission from Ne (lower points) has a ratio to the Auger
emission (upper points) which increase dramatically with the
charge state of the bombarding Cl ion[22], as does then also
$\omega_K = \sigma_x / (\sigma_x + \sigma_{Auger})$.

The fine structure is another important feature besides the
intensity of the ion induced x rays treated so far. Due to the
high degree of ionization in ion collisions multiple vacancies
exist, which result in a shift in x ray energy, so that a number
of satellite lines occur additionally to the normal lines[23]. A
crystal spectrometer will normally have such a good resolution,
that the single lines can be seen, and in fig. 13 are shown
spectra taken with a planar crystal[24] of the Ti K_α line bombard-

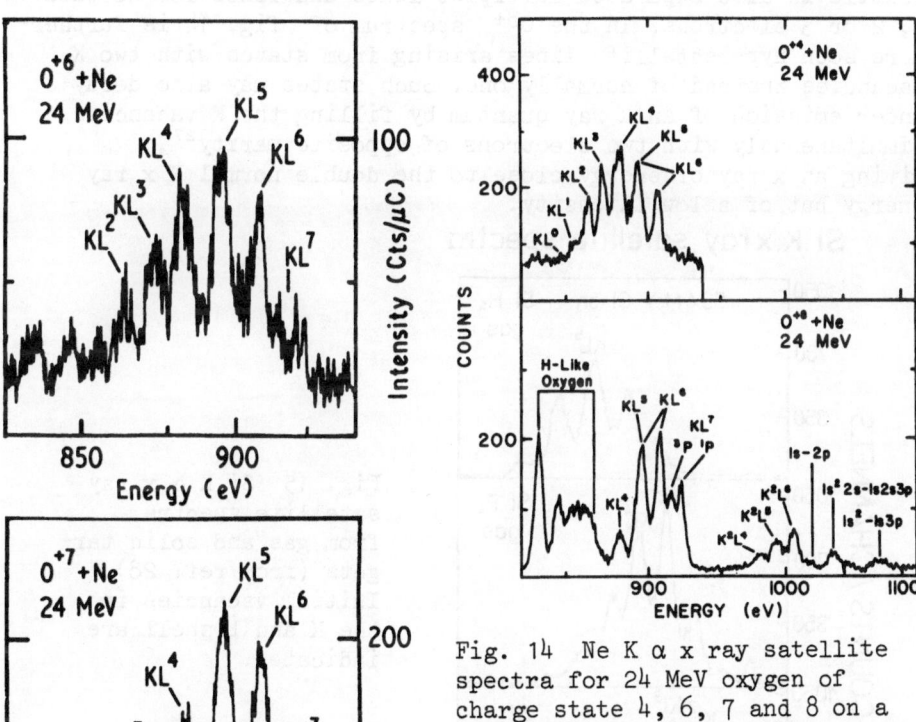

Fig. 14 Ne K α x ray satellite
spectra for 24 MeV oxygen of
charge state 4, 6, 7 and 8 on a
Ne gas target (from refs. 25 and
26). Electron defect configurations in the K and L shell are
indicated. For O^{8+} are seen
hyper-satellite lines with two
initial K vacancies and also
transitions within 1, 2 and 3
electron states.

ing a Ti target with photons and ions of increasing Z_1 from H
to O. There is a systematic increase with Z_1 of the intensities and
the complexity of the satellite lines, which are attributed to
states with additional vacancies in the L shell. For the highest
Z_1 there is also a mean number M shell vacancies,which shift the
lines towards higher energy. If the same spectra were measured
with a semiconductor detector, the resolution would only allow
to note a shift in centroid energy and an increasing width of
the lines.

The charge state of the projectile is important for the
satellite structure of the target x rays as seen in fig. 14.
Here the Ne K x ray spectrum is shown for bombardment with 24
MeV O of charge 4, 6, 7 and 8+, i.e. with 4, 2, 1 and 0 electrons[25,26] Satellite lines arise from states with increasing no
of L vacancies for higher charge states, and for the bare projectile is also separated multiplet lines and lines for Ne with
1, 2 or 3 electrons. In the O^{8+} spectrum of fig. 14 is furthermore seen hypersatellite lines arising from states with two K
vacancies instead of normally one. Such states may also decay
under emission of an x ray quantum by filling the K vacancies
simultaneously with two electrons of opposite parity[27],
giving an x ray of energy close to the double normal K x ray
energy but of a low intensity.

Si K x ray satellite spectra

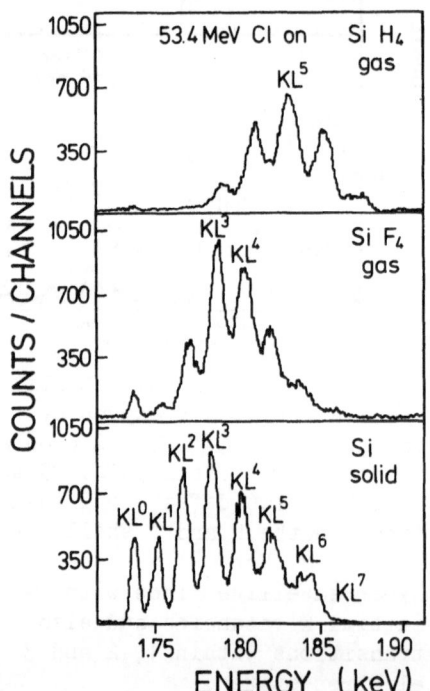

Fig. 15 Si K α x ray
satellite spectra
from gas and solid tar-
gets (from ref. 28).
Initial vacancies in
the K and L shell are
indicated.

The satellite spectra from solid targets are determined
not only by the initial ionization, but also by relaxation of
the outer shells after the collision, but prior to the inner
shell decay. Fig. 15 displays Si K x ray spectra for two gas
targets and for one solid sample [28]. For the SiH_4 gas target,
which best approximates an undisturbed collision with Si, the K x
rays originate from states which in mean are missing 5 L electrons.
With a SiF_4 gas target they are missing 3-4 L electrons which means
that rearrangement, e.g. capture of some of the nearby available F
valence electrons into some of the vacancies. This is even more pronounced for the solid target, where a broad range of lines appear
and are shifted to even lower no of L vacancies, but with relative
intensities which are not quite easy to understand. From such a
distribution one can only simply extract the width and the centroid
value, f.ex. measured in units of the apparent average L vacancy
fraction p_L[29] (see figs. 16 and 17).

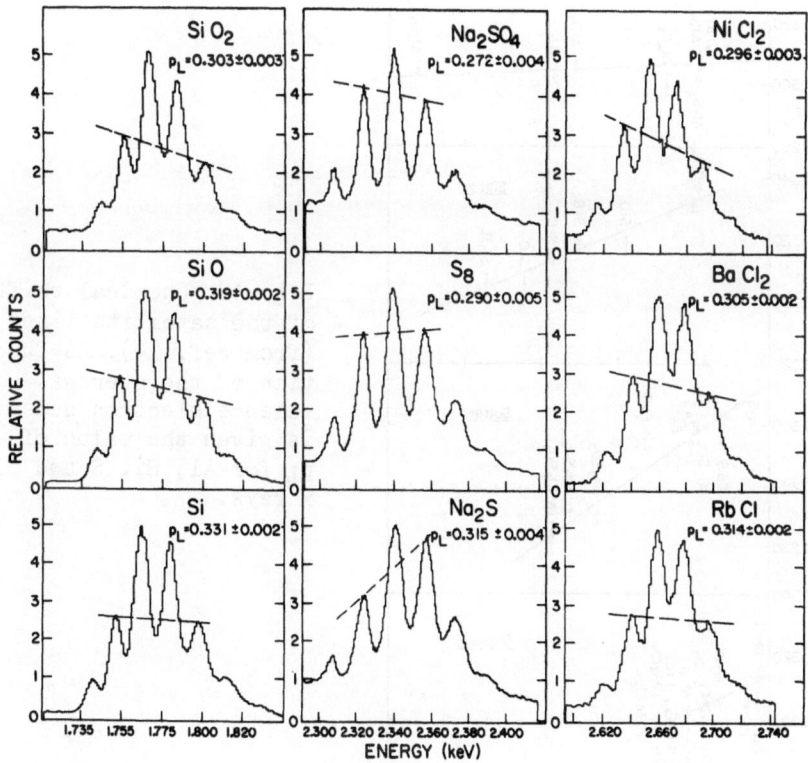

Fig. 16   K α satellite spectra for Si, S and Cl from
various chemical compounds bombarded with 32.4 MeV O
ions (from ref.29). Satellite patterns, quantified by
p_L (see fig. 17),depend on the chemical environment of
the element.

Fig. 17 Energy shift from Si K α satellite lines for 0 ions of various
energies on Si and SiO_2 targets (from
ref. 29). Open circles are with thin
targets and filled symbols with thick
targets. p_L is defined by
$p_L = (\Sigma n \cdot f_n)/N_L$, where f_n is the fraction of the total K α x ray yield in
the n'th satellite peak and N_L is the
no of L shell electrons in the ground
state atom.

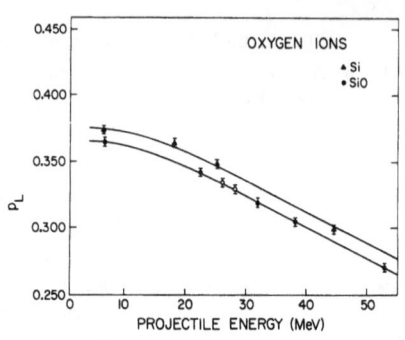

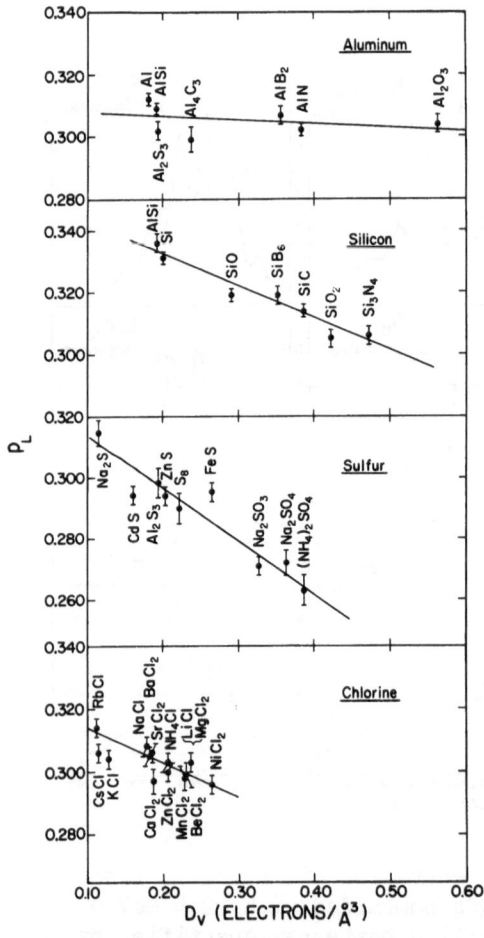

Fig. 18 Chemical shift
of the satellite lines
(from ref. 29). As func-
tion of the average
valence electron density
is given the value of
p_L for Al, Si, S and Cl
x rays.

Analysis of satellite spectra can as shown by Watson et al[29],[30] give essential information on the atoms of a sample, based on the observed value of p_L With increasing beam energy the satellite lines move towards lower no of L vacancies, i.e. lower p_L as shown in fig. 17 for the K x rays of Si under bombardment with O ions. This means that one for a thick target can tell the mean energy of the beam or the mean depth from which the x rays originate. More important is the observation that various chemical environments give different shifts as shown in fig. 17 for Si and SiO. This is clearly demonstrated in fig. 16 where Si, S and Cl K x rays are compared for each 3 different compounds and the extracted p_L values are stated. A simple explanation of these spectra has been given by Watson et al[29] in terms of the density of valence electrons. Their compilation of results, shown in fig. 18, exhibits a systematic decrease of p_L with increasing valence electron density. These electrons may be captured into the initially produced target vacancies, analog to the situation in fig. 15 explaining the observed trend. These observations are very interesting, because they open up the possibility from simple x ray measurements not only to get elemental but also chemical information.

Although the use of heavy ions offer some perspectives over light ions as generally high cross sections, selective MO excitation and information by fine structure studies, complications in the interpretation of the results are also imposed, e.g. by difficulties in quantifying the many excitation contributions and by the broadening effect of the many satellite lines.

5. BACKGROUND RADIATION

The importance of ion induced x rays is not mainly due to the production cross sections for the characteristic lines which have been discussed in the first chapters, but rather to the low level of the background radiation[31],[2]. Traditionally x rays have been excited by incident photons, e.g. from x ray tubes, or electrons, and as seen in fig. 19 the characteristic K cross sections are generally of the same order of magnitude for frequently used energies. Similar projectile fluxes are often employed in the three cases and one reason for this is the counting rate limitation determined by the detector. The use of ion excitation has the advantage of a favourable background, which e.g. is much smaller than the bremsstrahlung of the electrons. However, as already seen from fig. 1 the background, over which the peaks are observed, is not negligible.

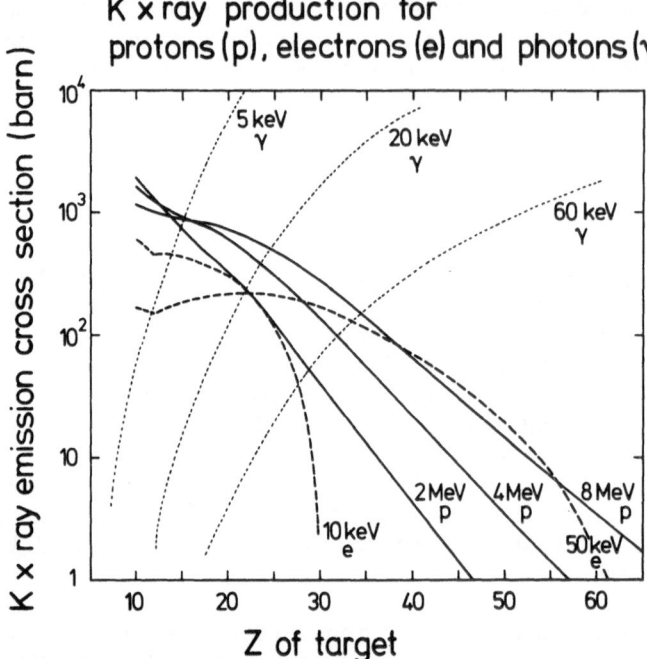

Fig. 19 K x ray emission by protons, electrons and
photons. For photons and electrons the beam energy
represent an upper limit for the K shell binding
energy. For lower Z_2 values the photon cross section
decreases whereas the electron cross section increases - and in a manner which is similar to the
proton excitation. Calculations are made with the
PWBA approach for protons and the electron production from ref. 31 and with photo absorption tables.

The main sources of background radiation induced by ions
are listed in table 1 with an indication of the x ray and ion
regime in which they are most important. The first three in
table 1 mentioned mechanisms are sketched in fig. 20 together
with the kind of spectra, they produce in a semiconductor detector covering the x ray region.

The secondary electron bremsstrahlung (SEB) is generally
the dominant continuum radiation at low x ray energies and extends up to an energy, T_m, where it decreases rapidly[31] with a
power of the x ray energy $\sim$10. It is produced by a two step
process, see fig. 20, where the first is the ejection of secondary electrons of energy higher than the radiation energy E_r and
the second is the bremsstrahlung of such electrons, which
happens in later collisions in competition with their slowing
down. The important feature is the probability of ejecting elec-

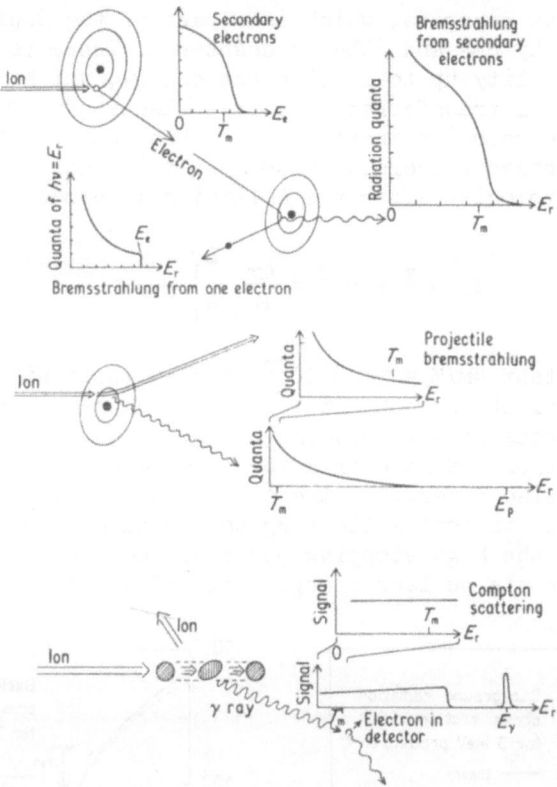

Fig. 20 Schematic representation of the three major
background processes: secondary electron bremsstrahlung (SEB), projectile bremsstrahlung (PB) and Compton scattering of nuclear γ rays[2].

Continuum radiation	x ray energy	ions	ion ve- locity
Secondary Electron Bremsstrahlung	low	all	all
Projectile Bremsstrahlung	high	all	low
Compton scattering of γ rays	high	all	high
Quasi molecular radiation	medium	heavy	low
Radiative Electron Capture	low	heavy	high
Inverse and inner bremsstrahlung	low	heavy	high

Table 1.   Ion induced background radiation in the x ray region

trons of various energies, which is a part of the Coulomb ionization treated by BEA and PWBA in chapter 3. There is a high
ejection probability up to an electron energy, which is the
maximal energy T_m transferred to a free electron[31]. T_m is determined by the velocity increase in a 'head on collision',
which in the projectile-electron CM system is nearly v, so the
lab. electron velocity after the collision is close to 2v and
thus

$$T_m = \frac{m}{2}(2v)^2 = \frac{4m}{M}\frac{E_1}{A_1} \tag{14}$$

where the constant $4m/M = 2.18 \cdot 10^{-3}$. For energies higher than
T_m only bound electrons can contribute and always with a low
intensity, increasing with binding energy[31]. The production of
secondary electrons is made from the bulk material of the
sample, which also determines the succeeding bremsstrahlung.
This takes place in most solid targets and also in many gas
targets due to the high stopping power of the electrons[32]. By
acceleration in the nuclear charge field of atoms, which the

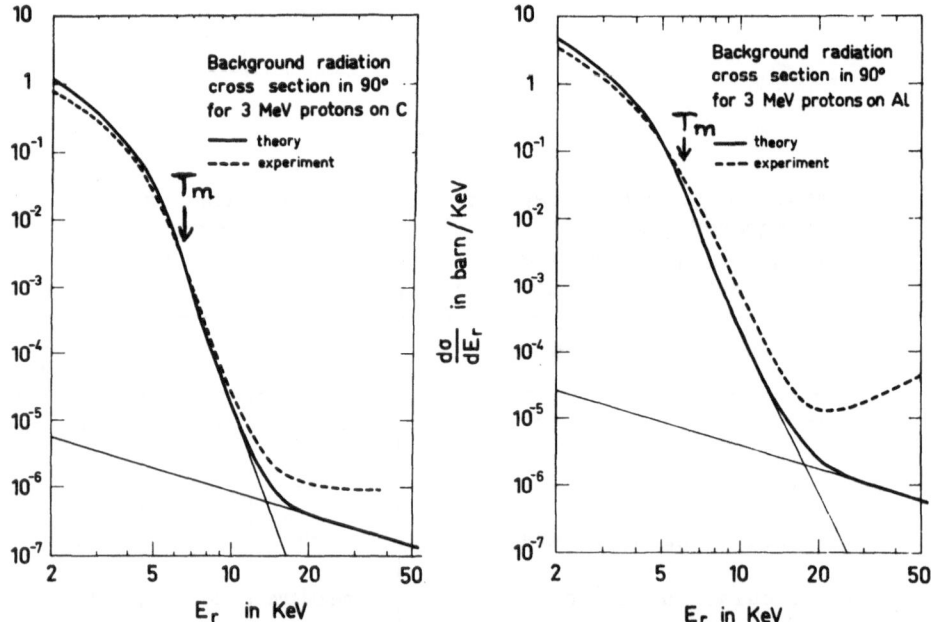

Fig. 21 Background radiation cross sections for 3 MeV protons on C and Al as function of the radiation energy E_r.
Theoretical curves are given for SEB using BEA and for PB.
The experimental cross sections are corrected by the detection efficiency of a Si(Li) detector, which for Al gives an
artifical increase to higher E_r due to a pretty constant
Compton scattering contribution[31].

ejected electrons encounter on their way, bremsstrahlung quanta
may be emitted ranging up to their full kinetic energy. The
resulting radiation probability can be calculated, e.g. with
the first ejection describen in BEA[31] , and the theoretical and
experimental results for 3 MeV protons on C and Al are shown in
fig. 21. Due to the fact that the same process is active as for
the vacancy production in chapter 3 exactly the same projectile
(Z_1,A_1 and E_1) scaling applies to the SEB continuum from a given
matrix as to the characteristic x ray emission from every given
element according to eq. (5).

Projectile bremsstrahlung (PB) quanta may be emitted by acceleration of the projectile nucleus in the nuclear electric field of
a target atom as indicated in fig. 20 and is also called nucleus-
nucleus bremsstrahlung (NNB). This process is most important for
low projecile energies and the theoretically estimated contribution
is shown in fig. 21. The dipole bremsstrahlung, which normally
dominates, depends on the difference in charge to mass ratio for
projectile and target[31] and may thus tend to vanish for near symmetric collision systems. In these cases quadrupole bremsstrahlung
contribute and will also interfere with small dipole terms. This
has been calculated semiclassically[33] and demonstrated experimentally[34] and the correspondence is nicely shown by fig. 22 for two
heavy ion collision systems.

Nuclear γ rays frequently arise in energetic heavy ion collisions. They may themselves lie in the x ray region or via Compton
scattering in the detector material produce electrons of kinetic
energies which form a continuum in the interesting energy region.
This process is sketched in fig. 20 and is often the major source
of background at high radiation energies and higher projectile velocities. Its magnitude is difficult to calculate,but for important sample matrices it can be measured. In semiconductor detectors
the direct radiation is seen, but also secondary scattering may occur
and can f.ex. influence the proportional counter of a crystal
spectrometer in a complicated - but essential- way.

Quasi molecular radiation can occur in heavy ion collisions
when molecular orbitals (MO) are formed (see chapter 4) and the
lifetime of produced vacancies is shorter than or comparable to the
collision time. In such cases radiation will be emitted going to
a level with an energy depending on internuclear distance. Averaging over this distance during the collision, a resulting band of
radiation will extend up to the united atom energy - and also
above owing to the uncertainty principle. In fig. 22 quasimolecular bands stand up over the PB (NNB) in the central part of the
spectra.

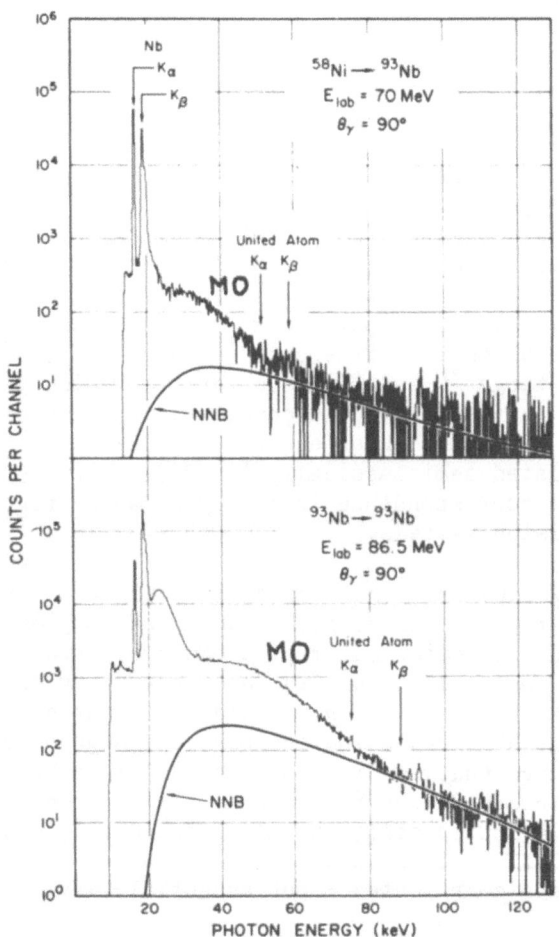

Fig. 22 X ray spectra from heavy ion collisions showing the high
energy continua from quasimolecular radiation (MO) and projectile
bremsstrahlung (nucleus-nucleus bremsstrahlung, NNB) for 1.21 MeV/
amu Ni and 0.93 MeV/amu Nb on Nb. The solid line shows a theoretical calculation[33] folded with absorption and detection efficiency.
(from ref. 34).

Radiative electron capture (REC) appears for heavy ions of
high velocity where electrons from the target can jump into projectile inner shell vacancies and thereby can win both the
binding energy and their translational energy $1/2\ mv^2$ relative
to the ion. The radiation will thus be centered $T_m/4$ higher
than the projectile binding energy, i.e. a little higher than
the observed characteristic projectile x rays and it will have a
width which is determined by the velocity distribution of the
outer target electrons[35]. The REC width is proportional to the
mean velocity of the target electrons and to the projectile velocity. For the lighter ions $Z \sim 10\text{-}20$ will REC form a nice peak and
for heavier ions or targets a broader structure which gradually
merges into quasi molecular continua.

Inverse bremsstrahlung can as REC only been seen for high
velocity heavy ions and is due to nearly free atoms scattered in
the moving projectile field. It will only extend up to the maximum energy $T_m/4$. By inner bremsstrahlung an ejected electron is
emitting a quantum in the field of the centers of the ver collision system from which it was produced. It is, so to say, a first
step in SEB and is known from ß decay studies where there is
only a single simple field. Both these bremsstrahlung processes
are restricted to heavy ion bombardment and are even there of
meager analytical importance.

The classification of the ion induced background radiation,
table 1, has led to a quantitative understanding of the two first
major sources SEB and PB. The third general source, important at
high energies, can be determined experimentally and so can the
specific heavy ion MO and REC continua. Only the direct ion
induced background has been treated and not other experimental
problems as e.g. interfering lines from beam or impurities.
Also the essential comparison with competing techniques, mentioned
in the beginning of this chapter will be postponed until conclusions about the ion induced x rays are more clearly drawn, based
on the involved physical processes.

6. ANALYTICAL SENSITIVITIES

The basic question to be asked is how good one can measure
the presence of a given element. We will answer that in various
ways, but in all cases we consider a trace element problem, where
the element we want to detect is embedded in a matrix. The
matrix is responsible for the background over which we look for
the x ray signal. To ensure it is clearly seen we demand that the
no of characteristic x ray counts N_T from a trace element exceeds
the statistical uncertainty on the background counts N_B within the

width of the peak by

$$N_T \geq 3\sqrt{N_B} \tag{15}$$

Employing this criterion and using theoretical and experimental information on atomic processes and beam properties we come to the
sensitivity figures given in table 2.

Concentration	$10^{-6} - 10^{-7}$		
Absolute amounts	for matrix	conc.	beam area
10^{-9} g	1 mg/cm² C	10^{-6}	defocused 1 cm²
10^{-12} g	100 µg/cm² C	10^{-6}	focused 1 mm²
10^{-16} g	100 µg/cm² C	10^{-5}	micro (10µm)²
Range (effective)	for matrix	with protons	of full range
2 mg/cm²	carbon	2 MeV	8 mg/cm²
4 mg/cm²	carbon	3 MeV	16 mg/cm²
Lateral resolution	for beam	in the range	
10 mm	defocused	20 - 5 mm	
0.5 mm	focused	2 - 0.1 mm	
4 µm	micro	10 - 1 µm	

Table 2. Practical sensitivities in ion induced x ray analysis[32].

The concentration of a trace element, which can be measured
is determined by eq. (15). To evaluate the detection limit in this
way one has to specify the matrix and know both the background radiation it produces and the characteristic x ray production for
various trace elements. This has been discussed in chapters 5 and
3, but to get actual yields one also must specify the sample
thickness, the no of incident ions and the detection efficiency
of the x ray detector. A calculation for the various physical
process can now be made, and in fig. 23 is shown the characteristic
x ray yield from 1 ng/cm² of elements 15<Z<95 in a detector covering a solid angle of 0.038 sr of internal efficiency 1 (and no
absorbers), when the target is bombarded with 100 µC protons of
energy 4 MeV (f.ex. 150 nA in 11 min) Fig. 23 also exhibits the
theoretical yield[31] of the background radiation due to SEB and PB

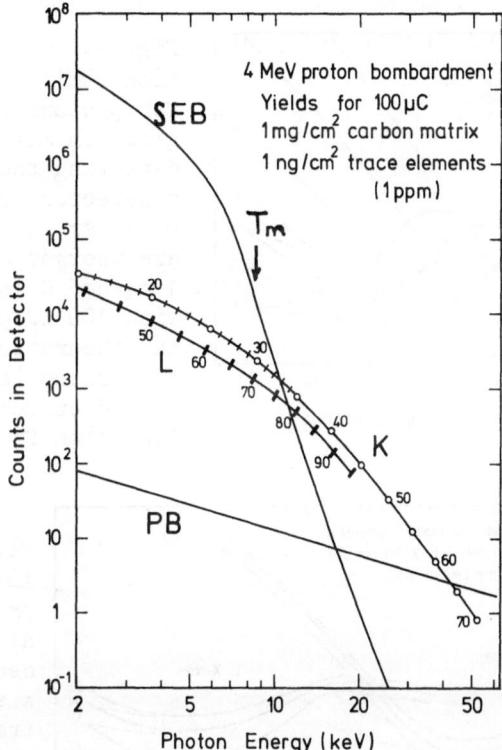

Fig. 23 Calculation of x ray yeilds in a detector from 4 MeV proton bombardment. K and L x ray intensities are shown for Kα and Lβ energies with Z_2 indicated along the curves for 1 ng/cm². Bremsstrahlung background yields are calculated per 200 eV for 1 mg/cm² C. The detector solid angle is 0.038 sr and the proton dose 100 μC.

per energy interval of 200eV for a 1 mg/cm² C matrix. Based on these assumptions on no of ions and detector efficiency,eq. (15) together with the theoretical K x ray emission and the background from SEB and PB lead to concentration sensitivities in a 1 mg/cm² C matrix, which are shown in fig.24 for protons of different energies. The concentration detection limit is as normally in applied work given in mass of the trace element per mass unit of the total matrix. From fig. 24 is noted, that the optimum sensitivity is in the region of 10^{-6} to 10^{-7} within a Z range which

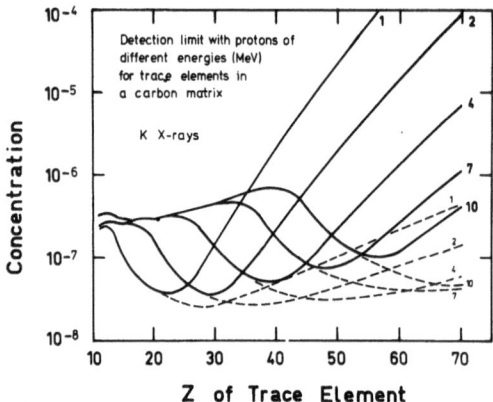

Fig. 24 Concentration detection limit, from eq. (15), for protons of various energies (in MeV along curves) detecting the K x rays with a detector of solid angle 0.038 sr. The trace elements are thought embedded in a 1 mg/cm^2 C matrix bombarded with 100 μC protons. Only the theoretical SEB and PB background is included. The dashed curves give the continuation from SEB alone[32].

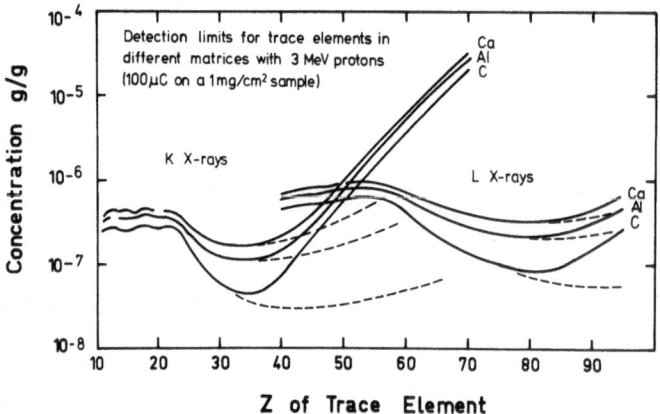

Fig. 25 Detection limit for 3 MeV protons with C, Al and Ca matrices using both K and L x rays from the trace elements[32]. Same basis as for fig. 24.

can be shifted with beam energy. The high Z part of the curves is determined by the PB background. For low E_1 it may be raised due to poor counting statistics, if an additional constriction like $N_T \geq 10$ or $N_T \geq 100$ is added to eq. (15), and for high E_1 the Compton scattering for γ rays will cause a higher level. An optimum all round analytical energy will thus be between 2 and 4 MeV per amu and the detection limit for a fixed proton energy of 3 MeV, but for both K and L x rays and for three different matrices C, Al and Ca is shown in fig. 25. How the detection limit changes with experimental parameters is given by eq. (5) in ref. 32. For heavy ions one shall use the same velocity, or E_1/A_1 as for protons in figs. 24 and 25. This comes from the fact that the low Z_2 part of the curve is determined by SEB, which scales with projectile just as the x ray production, eq. (5). As stated in table 2 of ref. 32 there is as a rule nothing to win in analytical sensitivity by using heavy ions instead of light projectiles, and from a more complete background discussion the same conclusion will also often be drawn.

The absolute amount of trace elements detectable is easily deduced from the concentration sensitivity, which has just been analyzed. One just has to multiply by the matrix thickness (in mass per area) and by the beam area. It is here assumed that either the trace element is homogeniously distributed within the lombarded area or the beam is homogenious. The undestructive nature of the analysis assures that one by reducing the beam area obtain the same reduction of the amount of material which gives a signal of a given height for a certain charge. For a typical concentration sensitivity of 10^{-6}, target thicknesses of 1 to 0.1 mg and beam areas of 1 cm^2 to 1 mm^2 10^{-9} to 10^{-12}g are measurable. With a 10 μm diameter beam so high currents can be maintained,[36] that the concentration 10^{-6} is achievable and even 10^{-16}g may be seen.

The effective range is a parameter which describes the depth sensitivity for thick samples. It is connected with the range of the ion beam, which for protons is shown in fig. 26, but is also influenced by the variation of the x ray production with projectile energy along the ion trajectory in the sample. A mean value of the depth, from which x rays are seen, is often around 1/4 times the total range[37], depending somewhat on how fast the vacancy production varies with E_1, i.e. depending on v/u. A more detailed depth resolution is generally poor. To probe the location in depth of trace elements in a thick sample, the following possibilities exist: 1) Variation of incident ion energy. 2) Variation of probed depth via the angle between the target surface and the beam direction – with the detector viewing angle to target kept constant. 3) Variation of the absorption through the angle between target surface and detector. 4) Observation of the absorption effect on the K_α/K_β ratio[38] or other well known line ratios. 5) Observation from the satellite structure[29] of the mean ion energy which has produced the radiation. Tilting the target represents a combination of 2) and 3) where the detector position is fixed[39]. However, the mathematical treatment of the results is complicated by error propagation[39], e.g. due to a steep decrease of σ_x with E_1 and due to uncertainty in attenuation calculations, so often one can only extract one parameter with confidence from such measurements. F.ex. one can determine the thickness of a surface layer or judge whether a distribution depends linearly or exponentially on the distance from surface. With some effort it might be possible to extract two parameters to describe a distribution, but due to averaging of the slowly varying atomic cross sections no good profiling like by Rutherford backscattering can be expected. A special penetration phenomenon is seen in a channeling direction of an aligned crystal, where anomalous x ray line intensities and line ratios are observed, mainly due to changes in stopping power and absorption, but also effected by changes in vacancy production for the outermost electrons.

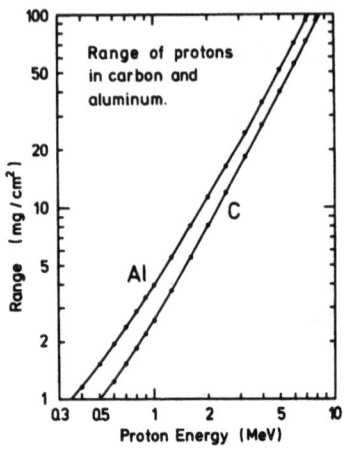

Fig. 26 Ranges of protons
in C and Al, which also
apply for He ions of same
velocity (or E_1/A_1).
The mean distance from which
x rays originate is re-
duced - often by a fac-
tor 4 - relative to the
full range[37].

The lateral resolution is determined by the beam size. For conventional accelerators is the beam diameter readily varied from 20
to 0.1 mm. With properly designed slit systems or apertures, beams
can be prepared down to μm dimensions[40], but with low intensity.
The ion intensity is improved by a final focusing before the target
with a carefully designed magnet. Such systems are operating in several laboratories, e.g. in Harwell[36], Heidelberg[40,41], Zürich and
Uppsala. The beam diameter is in Zürich 10 μm and down to 2 μm in
Harwell and Heidelberg[41] - with an intensity ∿ 0.2 nA (50 pA/μm^2).
The intensity problem is twofold as the target may often only stand
a certain energy dissipation, and one has to sweep the beam in one
or two dimensions.

7. ENHANCEMENT FROM X RAYS AND SECONDARY ELECTRONS

For quantitative analysis a good experimental calibration of
the total analysis system is essential, and although the measured
yields are physical constants, it is recommendable to compare results with known international standards and check the system calibration by interlaboratory comparisons[42]. However, secondary excitation of characteristic x rays from trace elements may take place due to the interaction of the beam with the matrix and has to be
paid attention to. For heavy ions is already mentioned that recoiling atoms and beam excitation in the matrix may lead to an additional x ray production. Also for the important proton and alpha beams
a general enhancement can be seen. Two major sources, seen in
fig. 19, are 1) photons, being created in the matrix, which have
a high probability of ejecting electrons, especially when the electron binding energy is only a little lower than the photon energy,
and 2) electrons (of kinetic energy $E_e = \frac{1}{2} m\, v_e^2$) arising from
collisions, which will also be able to make inner shell vacancies

when the binding energy U is smaller than E_e, but with a maximal
probability for $u = v_e/2$, i. e. $U = E_e/4$. In the case of heavy
ion bombardment these two effects may directly be observed by an
increase in the first - unperturbed - satellite peak with a crystal
spectrometer. The most important photon source is the flux of characteristic x rays of higher Z main components of the matrix, and
secondary electrons ejected during the ion impact on matrix atoms
represent the main flux of electrons in the sample, for which quantitative values of the enhancement will be given.

Enhancement from characteristic x rays of major trace elements
in the matrix can be significant. This secondary excitation is well
known from x ray fluorescence (initial photon excitation) and has
been calculated by Ahlberg[43] for proton excitation of a Fe matrix,
As the photo-excitation may have a long effective range compared
to available sample thicknesses, the enhancement is target thickness dependent and has a maximal value for thick targets. Values of
the % increase in the x ray emission of trace elements are given
in table 3 for a thick 100% Fe matrix. The effect is largest when
the electron binding energy (absorption edge) is just below the energy of the exciting radiation. If a sample contains heavy elements
like Fe in concentrations lower than 100%, their enhancement effect
will be correspondingly reduced. For matrices with less than 5% of
heavy elements the effect will thus not exceed a few per cent and
will for most elements be below 1%.

photon source	Fe K_α	Fe K_β
trace element	2.5 MeV p on 100 % Fe	
17 Cl	.83	.09
19 K	2.7	.30
20 Ca	4.9	.54
21 Sc	8.6	.95
22 Ti	14	1.6
23 V	24	2.8
24 Cr	41	4.7
25 Mn	–	7.8
26 – 92	–	–

Table 3 Enhancement
in percent due to x
rays in a 100% Fe
matrix for the K x
rays of trace ele-
ments. The numbers
from ref. 43 are
calculated for
2.5 MeV protons in-
cident on a thick
Fe sample.

The enhancement due to secondary electrons was emphasized by
a presentation on the PIXE conference[1] by Vis[44] of calculations,
which indicated high contributions. I will here give the results
of some simple estimates which yield per cent values, and make some
comments on the general behaviour. A calculation is made as for SEB[31]
replacing the bremsstrahlung cross section by the electron vacancy
production cross section (eq. (15) in ref. 31) in the integration
over the secondary electron production and the later slowing down
(eqs. (4) to (8) in ref. 31). This vacancy production via excitation by secondary electrons is given in table 4 in % relative to
the normal direct   production for the same trace element (by the
PWBA procedure of ref. 31). From table 4 is noted: 1) that the
effect is in the per cent region, 2) that it for fixed projectile
energy is nearly constant for various matrices, 3) that it is only
important for such small Z_2 values, where the corresponding binding
energy $U \leq T_m/2$, and 4) that it increases with projectile energy.
The at first sight surprising independence of the matrix atomic no
Z_M for the enhancement in the important low Z_2 range $U \lesssim T_m/2$ is due
to the fact that both the stopping power and the production of secondary electrons in the classical region $U < T_m$ are proportional to an
effective no of outer electrons ($\propto Z_M$). This does not hold for the Z_2

Matrix	C	Al	Ca	Fe	Zr	Aluminium matrix				
Z_2	Proton energy: 2.5 MeV					2	3	4	6	8
11	4.0	3.7	3.4	3.3	3.0	2.2	5.4	9.6	20	32
13	2.4	2.2	2.1	2.0	1.9	1.2	3.6	6.8	15	26
15	1.3	1.2	1.1	1.1	1.1	.54	2.1	4.6	12	20
18	.31	.35	.36	.38	.40	.12	.79	2.2	7.1	14
21	.04	.07	.09	.11	.14	.03	.20	.86	3.9	8.9
25		.01	.02	.03	.04	.01	.03	.15	1.4	4.3
29			.01	.01	.02	–	.01	.03	.32	1.6
35					.01	–		.01	.03	.18

Table 4   Enhancement in percent due to secondary electrons.
Calculations are made for 2.5 MeV protons on 5 different
matrices and for proton energies 2–8 MeV incident on an Al
matrix with the procedure and formulae of ref. 31 modified
to electron x ray production. The numbers are relative to
the theoretical PWBA x ray production cross section for the
trace element Z_2 and apply to thin targets. Z_2 values are
chosen so the underlined values in the right part correspond to $Z=T_m/2$, two lines higher to $U=T_m/4$ and two lines
lower to $U=T_m$ in each column.

region $U \gtrsim T_m$, where the high energy electron production increases faster with Z_M, and hence give more enhancement, but at a level which is small and therefore insignificant. From the E_1 variation in the right part of table 4 scaled with the characteristic energy T_m is seen that the enhancement has a constant - but small - value independent of E_1 for $U = T_m$ (which increases with Z_M, see left part). The effect grows for lower Z_2 to be important around $U = T_m/2$ and increases further also for $U \lesssim T_m/4$. The level for same U/T_m ratio increases with projectile energy. At high E_1 the ionization owing to secondary electrons can not be ignored relative to the direct ionization, which is important for cross section measurements. Although the enhancement has a high value, especially for high E_1, an experimental determination of x ray production from solid targets will assure that the effect has been included in the efficiency determination, and the Z_M independence will then imply that no big error is introduced going to another matrix. That means that with a calibration procedure outlined in the beginning of this chapter, the secondary electron enhancement will have an effect which is small for applied work, and the size is measured not directly by the numbers in table 4 but by the smaller matrix variation indicated.

8. ANALYTICAL CHARACTERIZATION

From the previous description of the sensitivities and other features of PIXE we can now summarize its main analytical characteristics in a few words:

> Elemental identification $Z_2 \gtrsim 12$
> Small amounts and concentration can be analyzed
> Many elements recorded in one run
> The sensitivity is smoothly varying with atomic no
> Microbeams can be used
> Nondestructive analysis
> Quantitative and absolute calibrated method
> Rapid analysis - and fast computer generated results
> Low cost - competitive, but accelerator necessary
> Suited for routine, survey or special analysis
> Can be combined with other techniques

Numbers, e.g. in table 2, have illustrated most of these statements. Experimental conditions as beam intensity can normally be chosen so the analysis is nondestructive, but this has to be checked for special volatile elements, for which also the chemical state has an influence. Results are normally reliable to the 5% level. For effective and fast operation a suited computer program is essential, as the amount of data is high. The cost for large scale analysis has been set as low as 5.5 $ per sample[45] but may be higher for non standard analysis. Techniques which are often applied simultaneously with PIXE are Rutherford scattering and analysis with γ rays.

Based on this characterization working fields should be
selected, where one makes the most of the special properties
of the technique. On the other hand, first sample types have
been chosen, the experimental conditions should be optimized
to meet the demands of that concrete task. As a good description of such a procedure I recommend the considerations of
Cahill[45] for the system in Davis, which is operating in the
very important field of air pollution control, where PIXE is
now one of the major analytical tools. A general problem is the
reduction of the background, and as the characteristic x rays
are emitted isotropically and the SEB background is peaked between 90° and 50° relative to the beam direction [32,46] it is advantageous to place a detector in backward angles.

Comparison with competing techniques helps to clarify strong
and weak points of an analytical method. As ion beam based techniques are carefully treated at this meeting I will confine myself to some other important established techniques. The ion (or
particle) induced x ray emission (PIXE) has analytical capabilities which can be described as intermediate between those of 1)
x ray fluorescence (XRF), 2) the electron microprobe and 3) neutron activation. The two first employ x rays for elemental identification and are thus similar to PIXE. XRF is normally used for
high precision determination of major elements in bigger samples,
but has also capacity of trace analysis down to the ppm level. The
electron microprobe uses a well focused electron beam probing regions of μm dimensions for major elements. By neutron activation
several elements can be measured with a superior sensitivity. The
novel features of PIXE is that it offers a fast and survey-like
method compared to neutron activation, has the potential of nearly same spatial resolution as the electron microprobe, but better
concentration sensitivity [31], and compared to XRF adds the beam
characteristics combined with good concentration sensitivity and
an x ray excitation of another nature. The most important comparison is that with XRF and I will here refer to earlier papers [2,32],
to a discussion by Goulding [47] (later than ref.4) based on optimum
performance of both systems and to an empirical comparison [48], where it was noted that although PIXE has a concentration sensitivity
significantly better that XRF, this advantage is reduced for heavy
matrices. Also two other analytical methods have sensitivities in
the ppm region but are best used for only one or a few fixed elements, namely optical emission spectroscopy and atomic absorption.
They may be used with advantage, if a customer is not interested
in the multielement feature of PIXE.

9. EVIDENCE FOR SUPER-HEAVY NUCLEI?

Instead of summarizing analytical problems and surveying fields of application I will discuss a single experiment, which is right now in the highlight, namely the report be Gentry, Cahill et al.[49] on evidence for superheavy nuclei. It represents a nice example of using the special analytical capabilities of PIXE, and to judge what confidence one can have in the result, a discussion of analytical problems is appropriate especially in the light of the far reaching consequences of the proposed findings.

The samples analyzed in this experiment[49,50] were small inclusions of monazite in mica from Madagascar. They had been found by Gentry to be surrounded with halos, ascribed to radiation damage by α particles from radioactive nuclei. But a few halos were so big that he could not naturally explain them, and the problem was whether unknown elements could occur in these inclusions with giant halos. The size of the grains ranged from 40 to 150 μm in diameter, which for a specific density of $5g/cm^3$ gives thicknesses 20-75 mg/cm^2. The elemental composition of the monazite (Ce, La, Th) PO_4, is 40% rare earths (Ce,La,...) 8% actinides (Th,U), 3% Y, 15% P, 30% O and traces of other elements (Fe, Pb,...?).

To look for heavy elements these samples were bombarded with proton beams of energy 4.7 and 5.7 MeV focused down to a diameter around 30 μm and the emitted x rays were observed with a 80 mm^2 x 3 mm Si(Li) detector. By this procedure was profited from the survey feature of PIXE for a broad range of unknown elements, its good sensitivity and the pinpointing by the ion beam (by means of a quadrupole magnet and a simple collimator). The proton energy was chosen so high to get large x ray cross sections and penetration through the grains and low enough to avoid too many γ rays from nuclear reactions.

X ray spectra from a normal Th/U inclusion and from one with a giant halo are shown in fig. 27. The spectra are dominated by the K x rays from the rare earths and the L x rays from Th,U. From these lines is seen a similar composition of the two samples. At higher energies is observed a background, which is highest for the smaller "giant" halo, and between the two line-groups is a valley where a closer examination shows some minor differences as indicated in fig. 28. The small lines which are only present in the "giant" spectrum coincide with the calculated position of L x rays

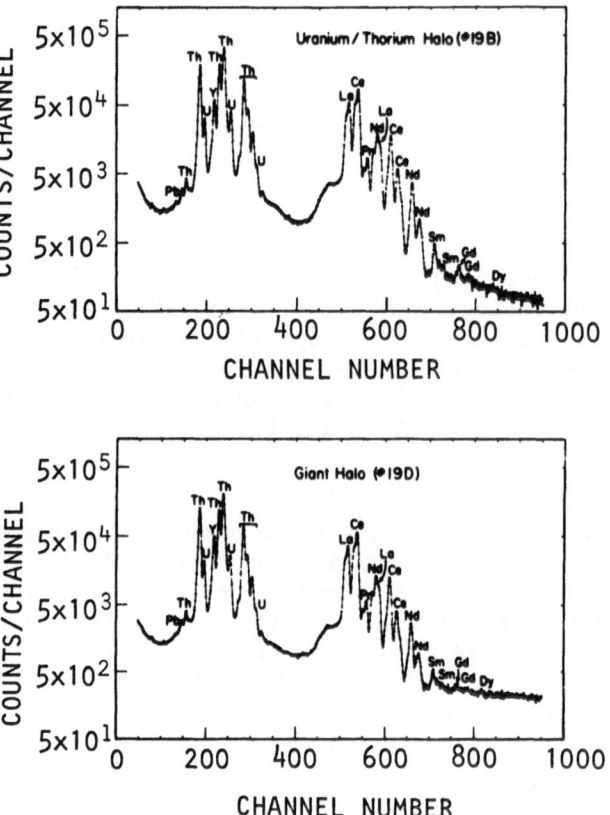

Fig. 27 X ray spectra from monazite inclusions (from ref. 49) bombarded with 5.7 MeV protons. Sample 19 B (2 µg) had a normal Th/U halo and sample 19 D (400 ng) had a giant halo. K x rays from Y, La to Dy and L x rays from Pb, Th, U are indicated with the main composition of the two samples being the same. Differences are seen in the window region 22-30 keV displayed in figs 28 and 29.

from superheavy atoms with Z = 126, 124 and 116. The identification
of these small lines is the key problem, and in fig. 29 is shown
another "giant" spectrum in the same energy region exhibiting a
peak at 27.23 keV corresponding to the $L_{\alpha1}$ line of Z=126. However,
there may also be other and more natural explanations for such a
peak, which have to be carefully studied.

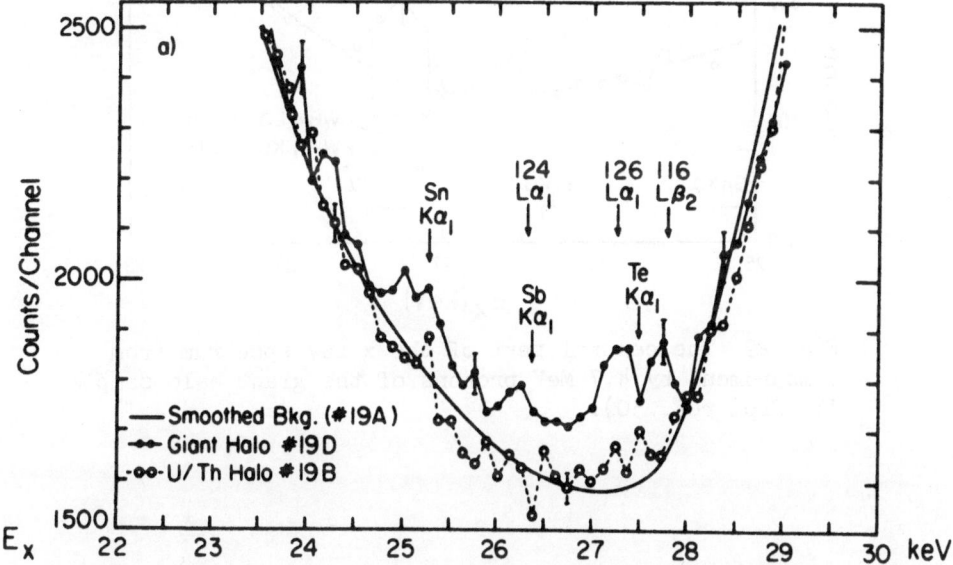

Fig. 28   The central part of the x ray spectra in fig. 29
normalized to same La Kα yield. Additional peaks are seen in the
'giant' sample 19 D and the possible origin from x ray lines is
indicated (from ref. 49). Sample 19 A was small (300 ng).

Interferences often occur when a line from one element
coincides with a line from another element. In such cases also
other lines from the involved elements are expected (see fig.4),
and if they are seen, their intensity imply a certain contribution to the unresolved peak. The 27.23 keV line in fig. 29 is lying close to the K lines of Te, but the shift in centroid energy
is significant and the observed peak has only the width of an isolated line whereas the Te K_{α} line should be broader due to the
two unresolved α_1 and α_2 components. The peak in fig. 29 could
also be attributed to In $K_{\beta1}$, the $L_{\beta2}$ line of Z=115, the $L_{\alpha1}$ line
of Z=126, a γ line from the $^{140}Ce(p,n)^{140}Pr$ reaction and various
x ray line effects. All these possibilities were quantitatively
examined[50] for correspondances with other lines seen – or not
seen – in the spectrum as illustrated in table 5. Certain of the
associated lines lie in steeply decreasing or increasing parts of
the spectrum, which makes it difficult to see whether they are

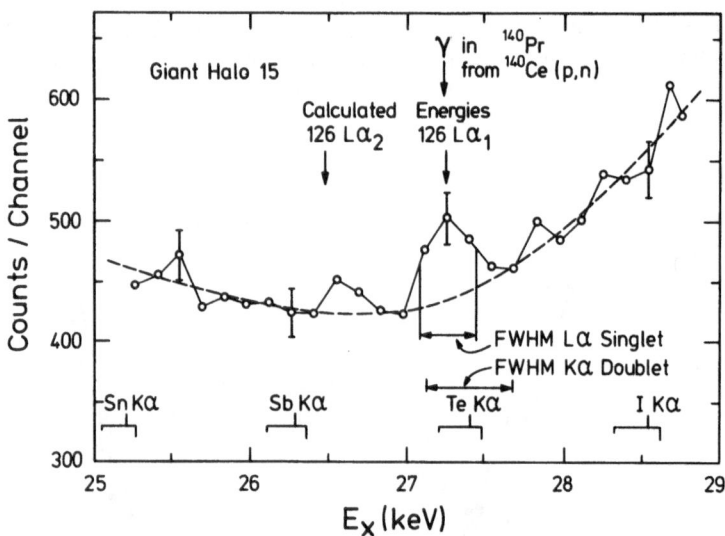

Fig. 29 The central part of the x ray spectrum from
bombardment by 4.7 MeV protons of the giant halo sample
15 (from ref. 50).

present and may also cause negative peak areas. This kind of interference unravelling is a standard procedure in analytical work
together with statistical analysis of the found peaks. In the chosen example the most probable interpretation seems to be the $L\alpha$
line of Z = 126 (with a concentration 200 ppm). However, it should
be noted, that the γ line from ^{140}Pr is also a candidate[51], which
in the present example will contribute, but with a low value[50] due
to the proton energy of 4.7 MeV. In most of the spectra from ref.49
5.7 MeV protons were used, and in that case the contribution is
higher[51] and can account for a larger fraction. This is especially
important if the Ce was concentrated on the surface of the monazite[51], because the nuclear excitation of the γ ray decreases faster
with decreasing proton energy in the sample than the x ray production from Ce. It thus is essential to investigate the contribution
from the γ line further, as the "Z = 126 line" is the most pronounced line and the other more dubious lines may be explained by K
lines from Sn and Sb.

    Other techniques will be employed to analyze these and similar
samples - which still exist after the nondestructive PIXE measurement - but until now I have heard of no supplementary positive
outcomes.

E = 27227 ± 30 eV, FWHM = 360 ± 30 eV, N = 168 ± 44 counts				
possible interpretation	energy eV	width eV	associated peak, counts	contribution to N
In Kβ_1	27276	340	In Kα − 74	−15 ± 16
Te Kα	27400	505	Te Kβ_1 −91	−500±700
Z=115 Lβ_2	27249	340	115 Lα_1 18	2 ± 5
Z=126 Lα_1	27253	340	126 Lα_2 36	300±350
^{140}Pr γ	27230	340	Ce Kα 61200	7 ± 2

Table 5 Interpretation scheme for the peak at 27.23 keV in fig. 29 from sample 15 (from ref. 50). Pile up from U L α, escape peaks and radiative Auger emission contribute less than 5 counts to N.

Improvements with PIXE are certainly possible. The intense K lines seen from the rare earths can be relatively suppressed by using a thin Si(Li) detector with a strongly decreasing peak efficiency in this region and an absorber foil with an absorption edge properly chosen. Also the huge peak tails to lower energies may be reduced by collimation in front of the detector and the first analysis tend to favour a lower proton energy. The microbeam handling is to be improved so it can run stable for longer time than the one hour runs made in the first experiment[49]. Alone with PIXE the sensitivity may be extended and more definite results obtained on these special but very interesting indications of novel trace elements.

References

1. Proceedings of the 'International Conference on Particle Induced X ray Emission and its Analytical Applications', Lund,
   Sweden, 23-26 Aug. 1976, which will be published as a special
   volume of Nucl. Instr. Meth. (early 1977) - later referred to
   as Nucl. Instr. Meth. (PIXE)
2. F. Folkmann, J. Phys. E: Sci. Instr. $\underline{8}$, 429 (1975)
3. G. Deconninck, Nucl. Instr. Meth. (PIXE)
4. F.S. Goulding, Nucl. Instr. Meth. (PIXE)
5. G. Bertolini and G. Restelli, Atomic Inner Shell Processes, ed.
   B. Crasemann (Academic Press, New York, 1975) Vol. 2,p. 123
6. W. Koenig, F.W. Richter, U. Steiner, R. Stock, R. Thielmann
   and U. Wätjen, Nucl. Instr. Meth. (PIXE)
7. Y. Cauchois and C. Bonnelle, Atomic Inner Shell Processes, ed.
   B. Crasemann (Academic Press, New York, 1975) Vol. 2 p. 83
8. J.D. Garcia, R.J. Fortner and T.M. Kavanagh, Rev. Mod. Phys.
   $\underline{45}$, 111 (1973)
9. J.M.Hansteen, O.M. Johnson and L. Kocbach, Atomic Data Nucl.
   Data Tables $\underline{15}$, 305 (1975)
10. D.H. Madison and E. Merzbacher, Atomic Inner Shell Processes, ed.
    B. Crasemann (Academic Press, New York, 1975) Vol. 1 p. 1
11. G.S. Khandelwal, B.H. Choi and E. Merzbacher, Atomic Data $\underline{1}$,
    103 (1969); Atomic Data $\underline{5}$, 291 (1973)
12. G. Basbas, W. Brandt and R. Laubert,Phys. Rev. $\underline{A\ 7}$, 983 (1973)
13. W. Brandt, Proc. Int. Conf. on Inner Shell Ionization Phenome-
    na and Future Applications, Atlanta, USA, eds. R.W. Fink et al.
    (USAEC, Oak Ridge, 1973) p. 948
14. C.G. Soares, R.D. Lear, J.T. Sanders and H.A. Van Rinsvelt,
    Phys. Rev. $\underline{A\ 13}$, 953 (1976)
15. G. Basbas, W. Brandt and R.H. Ritchie, Phys. Rev. $\underline{A\ 7}$, 1971 (1973)
16. W. Brandt and G. Lapicki, Phys. Rev. $\underline{A\ 10}$, 474 (1974)
17. D.H. Madison, A.B. Baskin, C.E. Busch and S.M. Shafroth, Phys.
    Rev. $\underline{A\ 9}$, 675 (1974)
18. W. Bambynek, B. Crasemann, R.W. Fink, H.U. Freund, H. Mark,
    C.D. Swift, R.E. Price and P.V. Rao, Rev. Mod. Phys. $\underline{44}$, 716(1972)
19. W.E. Meyerhof, R. Anholt, T.K. Saylor, S.M. Lazarus, A. Little
    and L.F. Chase, Phys. Rev. $\underline{A\ 14}$, 1653 (1976)
20. J.A. Cairns and L.C. Feldman , New Uses of Ion Accelerators,
    ed. J.F. Ziegler (Plenum Press, New York, 1975) p. 431
21. J.R. Macdonald, M.D. Brown,S.J. Czuchlewski, L.M. Winters,
    R. Laubert, I.A. Sellin and J.R. Mowat, Phys. Rev. $\underline{A\ 14}$(dec 1976)
22. D. Burch, N. Stolterfoht, D. Schneider, H.Wieman and J.S.Risley
    Phys. Rev.Lett. $\underline{32}$, 1151 (1974)
23. P. Richard, Atomic Inner Shell Processes, ed. B. Crasemann
    (Academic Press, New York, 1975) Vol. 1 p. 73
24. K.W. Hill, B.L. Doyle, S.M. Shafroth, D.H. Madison and R.D.Des-
    lattes, Phys. Rev. $\underline{A\ 13}$, 1334 (1976)

25. R.L. Kauffman, C.W. Woods, K.A. Jamison and P. Richard, J. Phys.
    B: Atom. Molec. Phys. 7, L 335 (1974)
26. R.L. Kauffman, C.W. Woods, K.A. Jamison and P. Richard, Phys.
    Rev. A 11, 872 (1975)
27. T. Åberg, K.A. Jamison and P. Richard,Phys. Rev. Lett. 37,63(1976)
28. F. Hopkins, A. Little, N.Cue and V. Dutkiewicz, to be published
29. R.L. Watson, A.K. Leeper, B.I. Sonobe, T. Chiao and F.E. Jenson,
    Phys. Rev. A to be published
30. R.L. Watson, A.K. Leeper and B.I. Sonobe, Nucl.Instr.Meth. (PIXE)
31. F. Folkmann, C. Gaarde, T.Huus and K. Kemp, Nucl. Instr. Meth.116,
    487 (1974)
32. F. Folkmann, Ion Beam Surface Layer Analysis, eds. O. Meyer,
    G. Linker and F. Käppeler (Plenum Press, New York, 1976) p. 695
33. J. Reinhardt, G. Soff and W. Greiner, Z. Phys. A 276, 285 (1976)
34. H.P. Trautvetter, J.S. Greenberg and P. Vincent, Phys. Rev. Lett.
    37, 202 (1976)
35. H.-D. Betz, M. Kleber, E. Spindler, F. Bell, H. Panke and
    W. Stehling, The Physics of Electronic and Atomic Collisions,
    eds. J.S. Risley and R. Geballe (University of Washington Press,
    Seattle, 1976) p. 520
36. J.A. Cookson and F.D. Pilling, Phys. Med. Biol. 20, no 4,to be
    published (1976)
37. F. Folkmann, Ion Beam Surface Layer Analysis, eds. O. Meyer,
    G. Linker and F. Käppeler (Plenum Press, New York, 1976) p. 747
38. M. Ahlberg, Nucl. Instr. Meth. 131, 381 (1975)
39. W. Pabst, Nucl. Instr. Meth. 124, 143 (1975)
40. R. Nobiling, Y. Civelekoglu, B. Povh, D. Schwalm and K. Traxel,
    Nucl. Instr. Meth. 130, 325 (1975)
41. R. Nobiling, F. Bosch, Y. Civelekoglu, B. Martin, B. Povh,
    D. Schwalm and K. Traxel, Nucl. Instr. Meth. (PIXE)
42. D.C. Camp, A.L. Van Lehn, J.R. Rhodes and A.H. Pradzynski, X-
    Ray Spectrometry 4, 123 (1975)
43. M. Ahlberg, Nucl. Instr. Meth. (PIXE)
44. P.M.A. van der Kam, R.D. Vis and H. Verheul, Nucl. Instr. Meth.
    (PIXE)
45. T.A. Cahill, New Uses of Ion Accelerators, ed. J.F.Ziegler (Ple-
    num Press, New York, 1975) p. 1
46. K. Ishii, S. Morita and H. Tawara, Phys. Rev. A 13, 131 (1976)
47. F.S. Goulding, Nucl. Instr. Meth. (PIXE)
48. J. Scheer, L. Voet, U. Wätjen, W. Koenig, F.W. Richter and
    U. Steiner, Nucl. Instr. Meth. (PIXE)
49. R.V. Gentry, T.A. Cahill, N.R. Fletcher, H.C. Kaufmann,
    L.R. Medsker, J.W. Nelson and R.G. Flocchini, Phys. Rev. Lett.
    37, 11 (1976)
50. T.A. Cahill, R.G. Flocchini, J.W. Nelson, N.R. Fletcher, H.C.
    Kaufmann and L.R. Medsker, Nucl. Instr. Meth. (PIXE)
51. J.D. Fox, W.J. Courtney, K.W. Kemper, A.H. Lumpkin, N.R. Flet-
    cher and L.R. Medsker, Phys. Rev. Lett. 37, 629 (1976)

THE EVOLVING USE OF ELECTRONS, PROTONS AND HEAVY IONS IN THE

CHARACTERISATION OF MATERIALS

J. A. Cairns

Metallurgy Division, AERE Harwell, Didcot, Oxfordshire,

OX11 ORA, England

ABSTRACT

Bombardment of a solid with electrons, protons or heavy ions produces one common effect: the generation of X-rays. However, these three probes offer quite different prospects for the examination of materials. This article considers how some of their relative merits have gradually become apparent. It ranges in scope from the original methods of generating X-rays, through advanced ion beam probes, to the modern sophistication of Controlled Atmosphere Electron Microscopy.

INTRODUCTION

One of the best ways to become familiar with a subject is to consider how it progressed gradually to its present state. It is instructive to attempt to understand why changes in direction occurred at certain stages (for example due to a sudden advance in theoretical understanding, or to significant improvements in instrumentation, or simply to a change in the prevailing scientific climate). This then leads us to appreciate more clearly the current state of a subject, and to be in a better position to anticipate future trends.

In the present case we shall be concerned mainly with the use of electrons, protons and heavy ions to generate X-rays from solid targets (although this will lead us eventualty to a specialised type of electron microscopy) and so it is appropriate to begin with the original discovery of X-rays and trace the evolution of the subject from that point.

ELECTRON-INDUCED X-RAYS

It is obvious to us now, with the benefit of hindsight, that
Rontgen's discovery of X-rays in 1895 was due simply to the
bombardment of a solid target by energetic electrons. Rontgen's
genius was apparent not simply in elucidating the fundamental
properties of X-rays, but also in applying some of these
properties soon after, and thereby initiating the subject of
X-radiography, with its subsequent spectacular success in medicine.
Incidentally, few discoveries have generated such an immediate
response among the general public (fig. 1). This was followed
later by Bragg's use of the wavelength properties of X-radiation
to reveal the atomic architecture of crystals. Hence we see
that at first the generation of X-rays from a solid by electron
bombardment was not envisaged as a means of identifying the
constituents of the solid, although it is interesting to record
that Bragg (1) in 1913 had noted that the spectrum from an X-ray
tube consisted of a continuous band, on which were superimposed a
number of peaks which were <u>characteristic</u> of the target material.

THE NEW ROENTGEN PHOTÓGRAPHY.
" LOOK PLEASANT, PLEASE."

Fig. 1 Cartoon, from <u>Life</u>, 27 February 1896

The same year saw one of the first ion induced X-ray experiments, when Chadwick (2) bombarded various materials with energetic alpha particles from radioactive sources and identified the resultant characteristic X-rays. At this time naturally the main motivation was to study the structure of matter.

These early beginnings take us a long way towards more recent studies, because they led to the development of particle accelerators, which constitute the main tool in current studies involving the use of protons and heavy ions for X-ray generation.

PROTON INDUCED X-RAYS

There are several ways of approaching this topic, but one means of demonstrating its logical progression is to start with the period when Khan (3) and his collaborators at Livermore started measuring X-ray cross-sections from a number of pure elements on proton bombardment. One of the advantages which they appreciated was the potential for improved analytical sensitivity, because of the high peak-to-background signal, as compared to X-rays generated by electron bombardment, which were accompanied by a more troublesome continuous background bremsstrahlung radiation. This led Poole and Shaw (4) to propose the construction of a proton microprobe, which, they argued, could offer some distinct advantages in elemental sensitivity over the electron microprobe. The resultant instrument has been applied to a variety of problems by Cookson et al (5).

The same general conclusions as to the potential analytical sensitivity of proton induced X-rays were of course appreciated by others. For example, Johansson et al (6) demonstrated the sensitivity for atmospheric trace element analysis by bombarding samples, collected on carbon foils, with 2.5 MeV protons, and detecting the X-rays with a Si(Li) detector to obtain an instantaneous multi-element spectrum (fig. 2). (In this work there was no fine scale resolution, i.e. they did not employ a microbeam).

These concepts were carried to their logical conclusion by the work of Cahill et al (7) in analysing trace elements present in atmospheric particulates, with very high sample throughput. He took advantage of a very significant feature of the proton probe: protons in the few MeV energy range lose only a small fraction of their energy in passing through these particulates, which are typically only a few microns in diameter. Hence the X-rays are generated by protons of a fairly well defined energy, and so the elemental composition can be quantified. It should be explained that by this time the theoretical models for calculating X-ray cross-sections, which were based on the concept of a direct Coulomb interaction between the incident projectile and the bound target electron, were now well established (8).

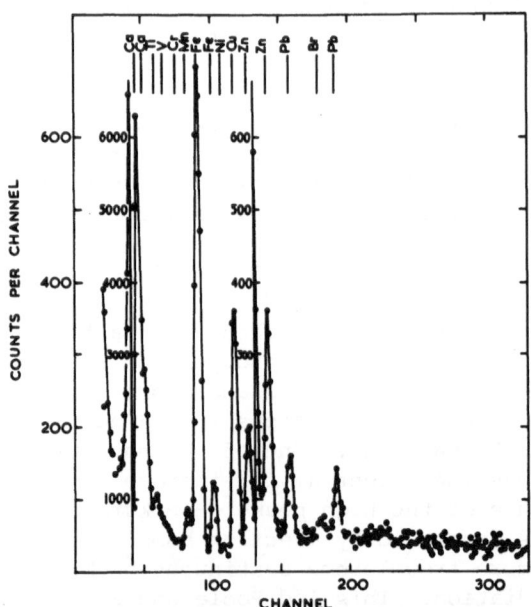

Fig. 2 A pulseheight spectrum from analysis of airborne particles with a diameter of 5 - 9 μm. The background has not been subtracted. Mn, Zn, Br, and Pb are present in the particles. The other elements are impurities in the carbon foil. From Johansson et al (6).

A considerable number of experimentally measured cross-sections had been accumulated and demonstrated the general validity of the theory, as is apparent from fig. 3.

The concepts used by Cahill have been employed by Cairns et al (9) to measure the elemental composition of supported metal catalysts. Advantage was taken of the fact that the support material (typically alumina, silica or carbon) could be ground down to a particle size of a few microns, then dispersed in a liquid medium, and applied as a small drop to a thin carbonaceous foil, and dried. Subsequent bombardment with 2-3 MeV protons yielded an X-ray spectrum which revealed not only the main catalytic elements, but also trace elements present, even within the support (see fig. 4).

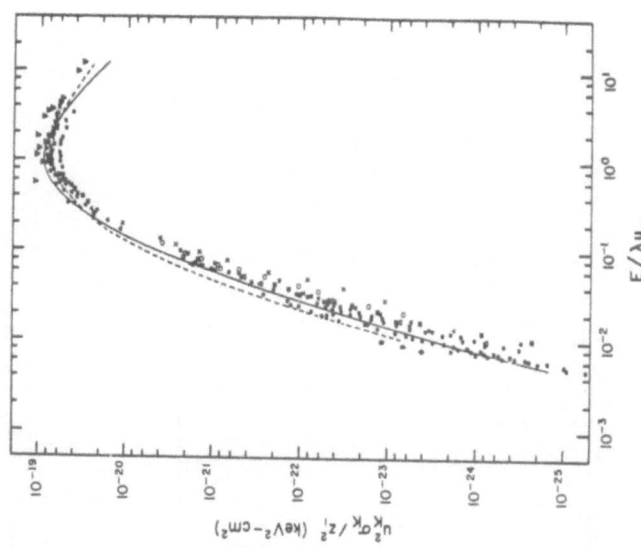

Fig. 3 (30) General comparison of Born approximation (dashed curve) and binary encounter approximation (solid curve) with experimental K-shell X-ray cross-sections by proton impact.

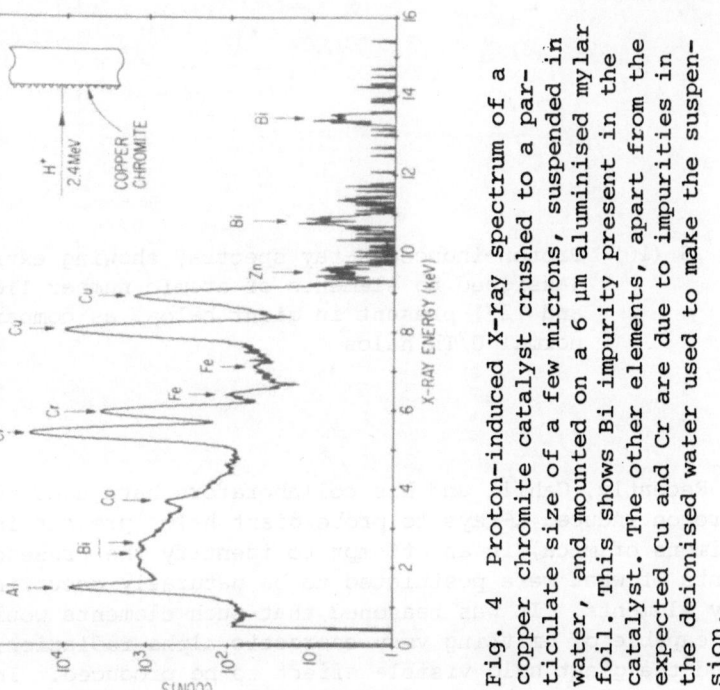

Fig. 4 Proton-induced X-ray spectrum of a copper chromite catalyst crushed to a particulate size of a few microns, suspended in water, and mounted on a 6 μm aluminised mylar foil. This shows Bi impurity present in the catalyst. The other elements, apart from the expected Cu and Cr are due to impurities in the deionised water used to make the suspension.

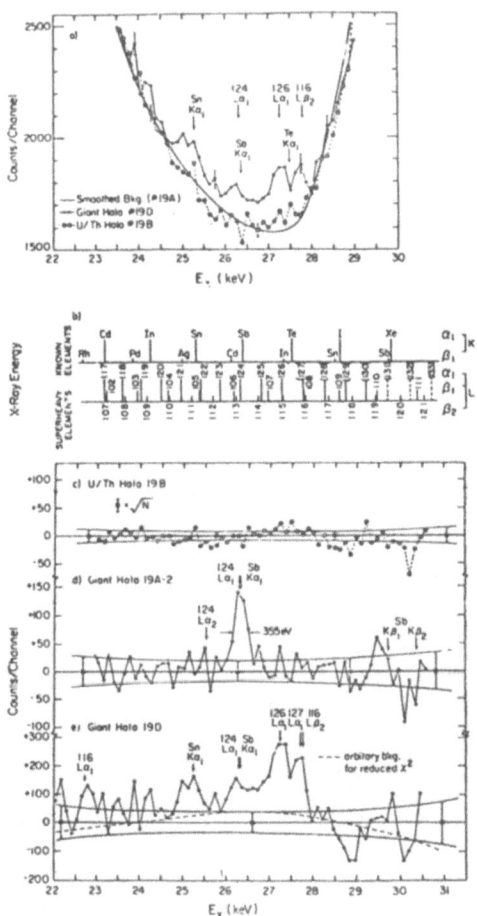

Fig. 5 (10)   Proton-induced X-ray spectra, showing extra peaks
              (assigned to elements of atomic number 116, 124, 126
              and 127) present in giant halos, as compared to
              normal U/Th halos

Recently, Cahill and his collaborators have used the technique
of proton induced X-rays to probe giant halos present in some
specimens of mica, in an attempt to identify the presence of trace
amounts of what were postulated to be naturally occurring superheavy elements. It was reasoned that such elements would have
been capable of emitting very energetic alpha radiation, which
caused the giant halo visible effect to be produced. In the first
announcement of their findings (10) the X-ray spectra (fig. 5)
from giant halos are indeed found to contain some extra peaks which

have been assigned to elements having atomic numbers 116, 124, 126
and 127. This work rather ingeniously took advantage of an X-ray
'window' which exists between the last L transition of uranium
at 21.54 keV and the first K transitions of the rare earths at
33.03 keV, and within which the L X-rays from elements between
Z = 105 and Z = 134 would be expected to occur. It certainly
represents one of the most exciting current areas of investigation.

HEAVY ION INDUCED X-RAYS

Although the generation of X-rays by heavy ion bombardment
uses essentially the same equipment as does proton irradiation, this
topic has taken a completely different direction. This has been
dictated really by the mechanism by which X-rays are produced from
solids on heavy ion bombardment. In the first place, the Direct
Coulomb excitation mechanism, which we saw (fig. 3) was quite

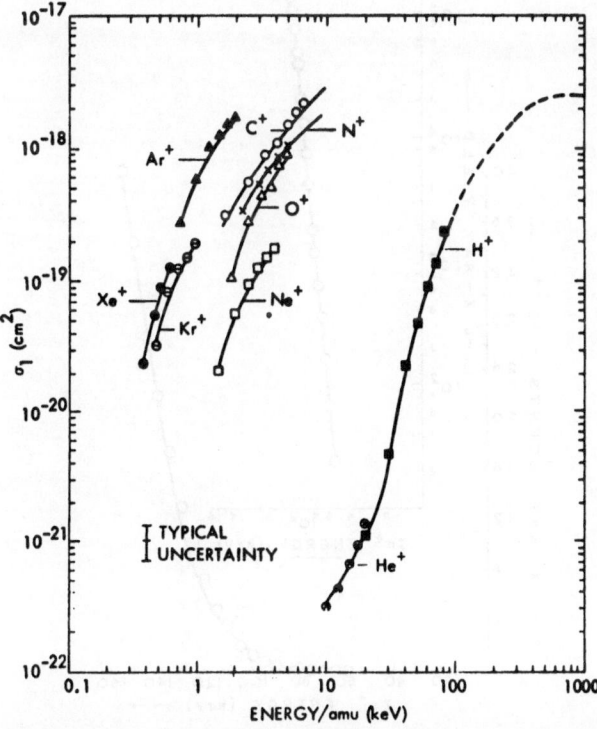

Fig. 6 (11)   Carbon K-shell excitation cross-section as a function
of incident ion energy per amu

adequate to explain proton X-ray induced cross-sections, was
inappropriate.   This emerged when it was found that the X-ray
cross-sections on bombardment of a series of solid targets by
heavy ions of modest energy were orders of magnitude higher than for
protons of the same velocity, as may be verified by reference to
fig. 6 (11).   There were other differences: one could measure an
effective threshold energy, below which certain characteristic
X-rays were generated with negligable probability;   then as this
projectile energy increased above the threshold limit, the X-ray
yield often showed a spectacular increase (12) (fig. 7),
particularly when the target and projectile were capable of
emitting X-rays of similar energy.

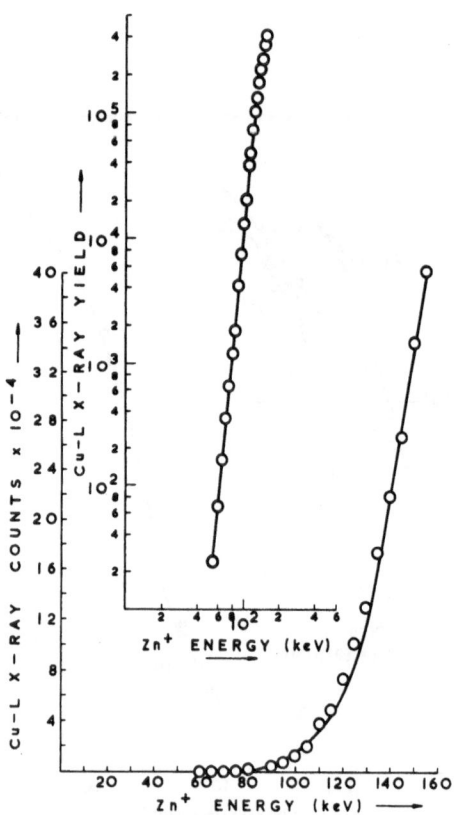

Fig. 7   Cu L X-ray yield, generated by bombardment of a solid
         copper target with Zn$^+$ ions

All of these apparent difficulties in interpreting the X-ray
mechanism largely disappeared when it was postulated (13) that
when a projectile, moving at a relatively slow velocity (with
respect to the orbital electron velocity) collides with a solid
target, it may form a pseudomolecule. During the lifetime of
such an entity, inner shell electrons can be promoted to
unoccupied levels of lower binding energy by means of 'level
crossings', so that when the target and projectile separate after
the collision there may be inner shell vacancies in one or other
or both, with the consequent emission of characteristic X-rays.
This model accounts not only for the high cross-sections, since
electron promotions can be very probable under favourable conditions,
but also explains the existence of the 'cut-off' energy: this

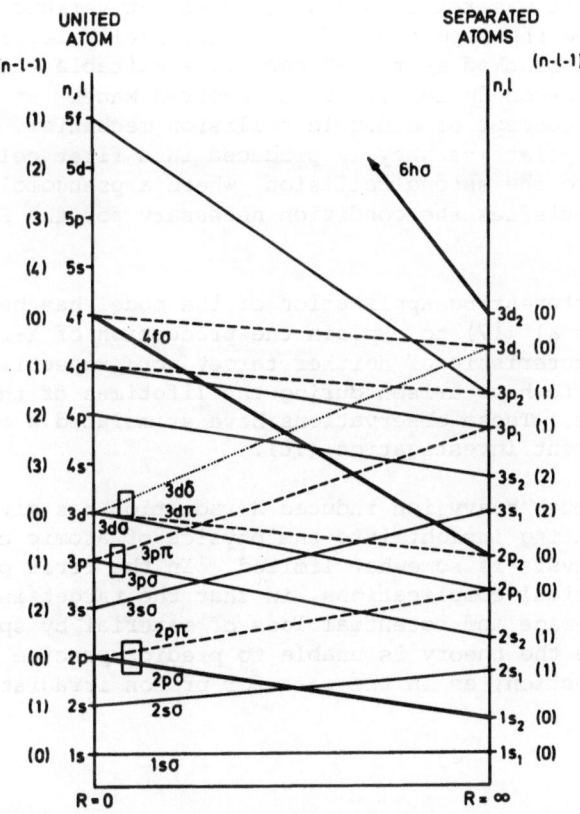

Fig. 8 (27)   MO correlation diagram constructed using the Barat-
      Lichten rule. This diagram is for the case of Z_1 slightly
      larger than Z_2. The solid lines correspond to σ states, the
      dashed lines to π states, and the dotted lines to δ states.

simply represents the minimum distance of closest approach
between the target and projectile which is necessary to form the
pseudomolecule with reasonable probability.

The existence of an effective projectile cut-off has been
used (14) to generate X-rays from a specific element within a
target by choosing the most appropriate projectile.

In more general terms, the 'pseudomolecule' concept has proved
to be very useful. There are simple rules (15) which allow a
molecular orbital diagram to be constructed from the particular
combination of target/projectile (fig. 8) (27) and this allows an
estimate to be made of the likelihood of certain X-rays being
generated.

However, when silicon K X-rays were found to arise during
bombardment of a silicon target by argon ions of a few tens of
keV in energy, it was realised that this was at variance with
the model, since the promotion of the inner shell electron from
the silicon was blocked by the absence of a suitable vacancy in
the projectile argon 2p level. This inspired Macek et al (16)
to suggest the concept of a double collision mechanism, by
which the appropriate vacancy is produced in a first collision,
and carried into the second collision, where a pseudomolecule is
formed which satisfies the condition necessary for the Si K vacancy
production.

Another interesting application of the model has been its
use by Saris et al (17) to explain the production of X-radiations
which are characteristic of neither target nor projectile, and
are postulated to have arisen during the lifetimes of the appropriate
pseudomolecules. These observations have stimulated a great
deal of subsequent investigation (18).

Thus although heavy ion induced X-radiation has given rise
to much fascinating insight into the physics of atomic collisions,
its use in analysis is somewhat limited. In the first place,
there are practical complications, in that the target is subject
to radiation damage and potential loss of material by sputtering;
and in addition the theory is unable to predict precise cross-sections
for X-ray production, as in the case for proton irradiation.

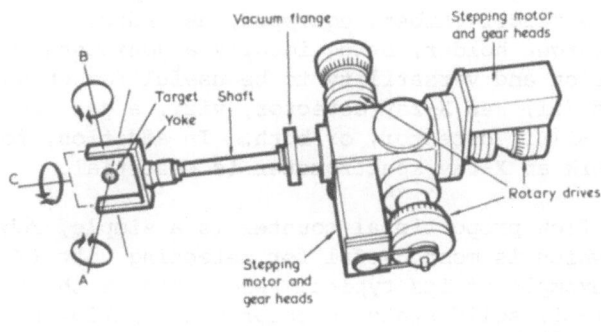

Fig. 9 (19)   The basic elements of a 3-axis goniometer

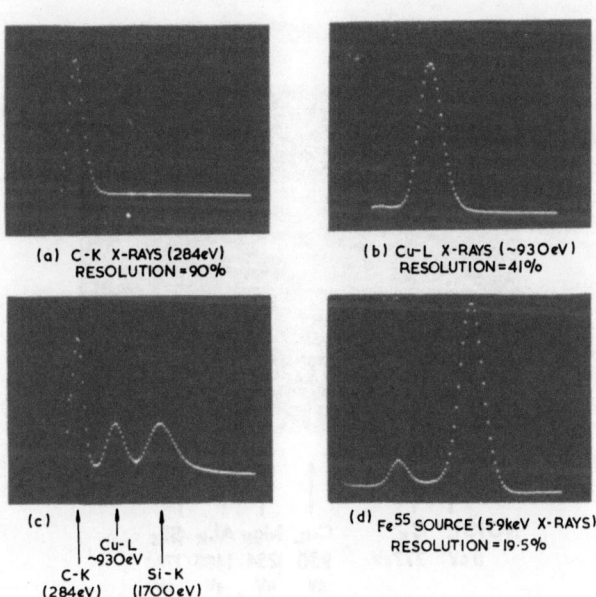

Fig. 10 (21)   Typical X-ray resolution obtainable with a
              gas flow proportional counter

INSTRUMENTATION

Anyone working in this field will use, apart from an
accelerator, a target chmber, equipped, as indicated in fig. 9 (19)
with (i) a target holder, being ideally a goniometer with
enough precision and versatility to be useful for channelling
work (20) and (ii) an X-ray detector, viz., a gas flow proportional
counter or a Si(Li) detector, or both. In addition, for high
resolution work an X-ray spectrometer is essential.

The gas flow proportional counter is a simple, robust
instrument, which is most useful for detecting soft (< 1 keV)
X-rays. An example of its typical resolution is shown in fig. 10
(21). The Si(Li) solid state detector has superior resolution,
but is equipped usually with a beryllium window, which restricts
its detection to X-rays above ∿1 keV. However, with suitable
design this window can be replaced by a thin plastic one, and then
the detector may be used for even soft X-ray applications (22),
as shown in fig. 11 (28). The great attraction of both of these
detectors is that they give an instantaneous X-ray spectrum of
the generated X-rays.

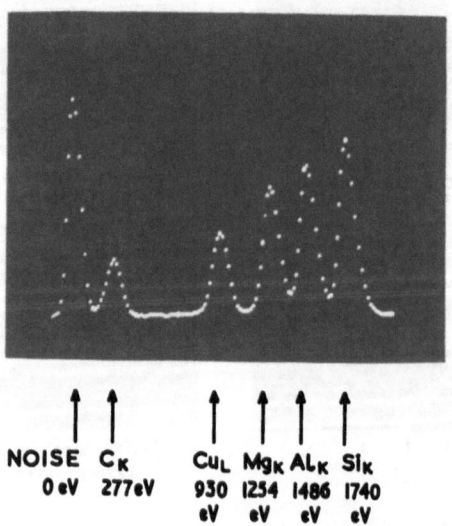

NOISE C$_K$ Cu$_L$ Mg$_K$ Al$_K$ Si$_K$
0 eV   277eV   930 1254 1486 1740
                   eV   eV   eV   eV

Fig. 11 (28)   Soft X-ray resolution exhibited by a Si(Li) detector
               fitted with a 6 μm Al/mylar window

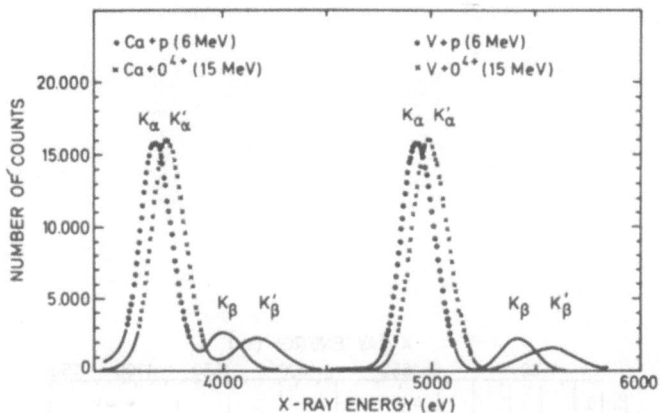

Fig. 12 (23)   X-ray spectra obtained by proton and oxygen bombardment
of Ca (left) and V (right). The heavy ion bombardment
is seen to have caused a shift to higher energies

On the other hand, the X-ray spectrometer, which uses a
diffracting element, such as a ruled grating or an appropriate
crystal, must be scanned through the region of interest, but
it then exhibits vastly improved resolution. As an example of
how such resolution may become essential, fig. 12 (23) shows a
comparison between proton bombardment and oxygen ion bombardment
of two targets, namely calcium and vanadium, measured with a
Si(Li) detector. For each target the heavy ion bombardment has
caused a shift in the characteristic X-ray positions to higher
energy values. This effect was postulated to be due to the
production of multiple vacancies in the inner shells of the target
atoms. These ideas received elegant confirmation when Knudson
et al (24), using a crystal spectrometer, compared the X-ray
spectra obtained by bombarding an aluminium target with 5 MeV protons

and nitrogen ions. They showed (fig. 13) that the heavy ion
spectrum contained a series of satellite lines which could be
attributed to the simultaneous production of up to five aluminium
L-shell vacancies by the nitrogen projectile.

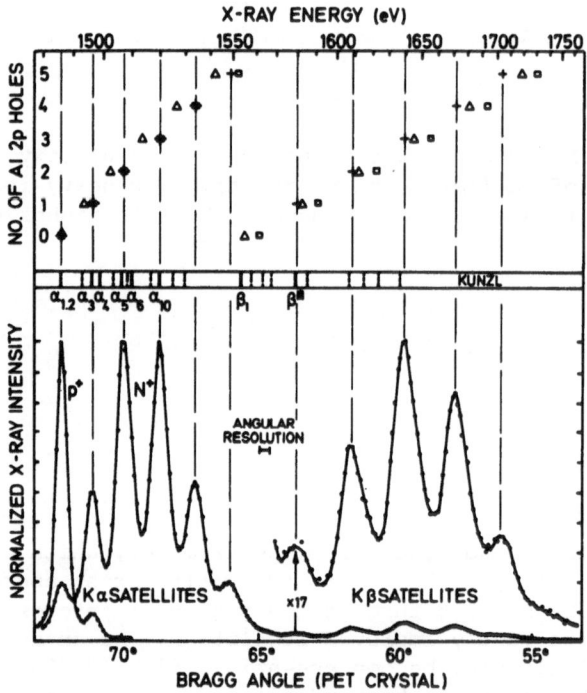

Fig. 13 (24) Al K X-ray spectra obtained by bombardment of an
aluminium target with protons and nitrogen ions.
The satellites produced by the heavy ion bombardment
are shown to coincide in energy with the calculated
values (top) of X-ray shift which arise from the
production of from one to five 2p vacancies in the
aluminium atom

FUTURE TRENDS: ASSOCIATION WITH
COMPLEMENTARY TECHNIQUES

Inevitably, as the subject has developed, and investigators
have sought to extract increasingly sophisticated information, it
is becoming apparent that new insights can best be obtained by
incorporating other allied techniques.   Naturally Auger Electron
Spectroscopy and X-ray Photoelectron Spectroscopy (ESCA) come
readily to mind, but here another approach is suggested, namely
the use of Controlled Atmosphere Electron Microscopy.   This
technique uses a transmission electron microscope, within which
the specimen is mounted, and surrounded by a gaseous environment,
as indicated in fig. 14 (25).   Hence the behaviour of the specimen
can be observed in the presence of various gases and as a function
of temperature.   This information can be recorded on video tape.

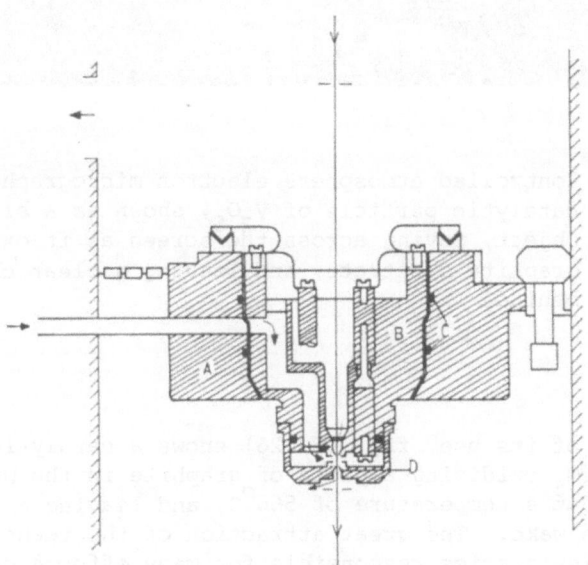

Fig. 14 (25)   A schematic representation of the JEM electron
               microscope gas reaction cell and stage.   The stage A
               is permanently attached to the specimen translate
               mechanism of the microsocpe, whilst the cell B, which
               can be inserted in a similar manner to a conventional
               specimen holder, seats on two 'O' rings C, and
               thereby creates an annular channel to which gas gains
               access via a duct in the stage.   The specimen on its
               heater is mounted at point D.

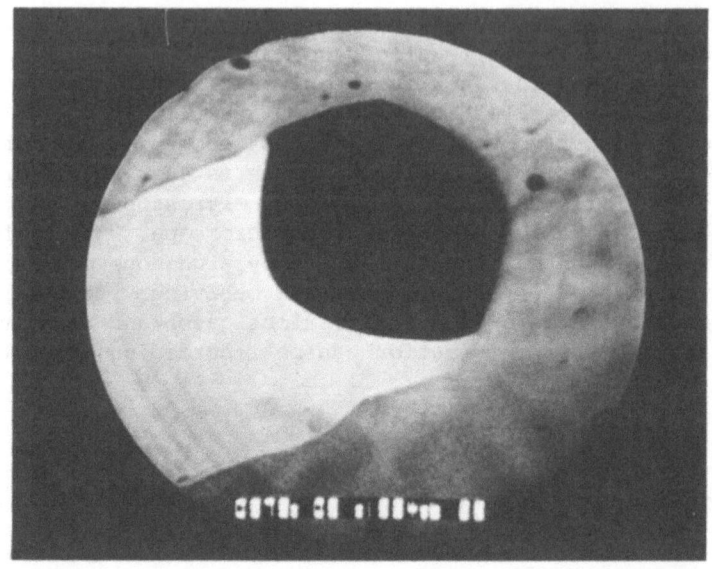

Fig. 15 (26)   Controlled atmosphere electron micrograph of a
               catalytic particle of V_2O_5, shown as a black
               object, moving across the screen as it oxidises the
               graphite substrate, and leaving a clear channel
               behind

An an example of its use, fig. 15 (26) shows a catalytic
particle of V_2O_5 oxidising a piece of graphite in the presence
of oxygen gas at a temperature of $560^{\circ}C$, and leaving a transparent
channel in its wake. The great attraction of the technique is
that the precise species responsible for many effects can be
observed in action. For example, it has been mentioned
earlier that the technique of proton-induced X-rays can be
used (9) to identify the elements present in a supported metal
catalyst. However, this is only part of the information which
is necessary for the catalyst chemist to make progress. Ideally,
he would like to identify the precise species which is
responsible for a given catalytic behaviour under certain conditions
of temperature and pressure. The technique of Controlled
Atmosphere Electron Microscopy allows this to be achieved.

CONCLUSION

This assessment began with Rontgen's observations, which arose from the use of a basic electron bombardment system, and ended with a sophisticated, high resolution form of electron bombardment. In between, it has examined the relative merits of two ion beam techniques which have now become generally established for providing vital information in the continuing quest to reveal the composition, structure and potential behaviour of solid materials.

FOOTNOTE

One manifestation of the establishment of the subject of inner shell vacancy production by ion bombardment has been the gradual appearance of review articles (27, 28, 29, 30) over the past few years. These provide a general compilation of the progress made to date.

REFERENCES

1.   W. H. Bragg, Proc. Roy. Soc. <u>89</u>, 246, (1913)
      W. L. Bragg, Proc. Roy. Soc. <u>89</u>, 248, (1913).

2.   J. Chadwick, Phil. Mag. <u>25</u>, 193, (1913).

3.   J. M. Khan and D. L. Potter, Phys. Rev. <u>133</u>, A890, (1964).
      J. M. Khan, D. L. Potter and R. D. Worley, Phys. Rev. <u>139</u>, A1735, (1965).

4.   D. M. Poole and J. L. Shaw, in <u>5th Intern. Congr. on X-ray Optics and Microanalysis</u>, Springer, New York, p319, (1968).

5.   J. A. Cookson, A.T.G. Ferguson and F. D. Pilling, J. Radioanal. Chem. <u>12</u>, 39, (1972).

6.   T. B. Johansson, R. Akselsson and S.A.E. Johansson, Advances in X-Ray Analysis, <u>15</u>, 373, (1972).

7.   T. A. Cahill, in 'New Uses of Ion Accelerators', J. F. Ziegler (Ed.) Plenum Press p1, (1975).

8.   E. Merzbacher and H. W. Lewis, in <u>Encyclopaedia of Physics,</u> S. Flugge (Ed.) <u>34</u>, Springer, Berlin, p166, (1958).

      J. D. Garcia, Phys. Rev. A<u>1</u>, 280, (1970).

      J. D. Garcia, Phys. Rev. A<u>1</u>, 1402, (1970).

9.   J. A. Cairns, A. Lurio, J. F. Ziegler, D. F. Holloway and
     J. Cookson, J. Catalysis, to be published (1976).

10. R. V. Gentry, T. A. Cahill, N. R. Fletcher, H. C. Kaufmann,
     L. R. Medsker, J. W. Nelson and R. G. Flocchini, Phys.
     Rev. Lett. 37, 11, (1976).

11. R. C. Der, T. M. Kavanagh, J. M. Khan, B. P. Curry and
     R. J. Fortner, Phys. Rev. Lett., 21, 1731, (1968).

12. J. A. Cairns, D. F. Holloway and R. S. Nelson, in 'Atomic
     Collision Phenomena in Solids', D. W. Palmer, M. W. Thompson
     and P. D. Townsend (Eds.), North Holland Publ. Co.,
     Amsterdam, p541, (1970).

13. U. Fano and W. Lichten, Phys. Rev. Lett., 14, 627, (1965).

14. J. A. Cairns, D. F. Holloway and R. S. Nelson, Radiation
     Effects, 7, 167, (1971).

15. M. Barat and W. Lichten, Phys. Rev. A, 6, 211, (1972).

16. J. Macek, J. A. Cairns and J. S. Briggs, Phys. Rev. Lett.,
     28, 1298, (1972).

17. F. W. Saris, W. F. van der Weg, H. Tawara and R. Laubert,
     Phys. Rev. Lett., 28, 717, (1972).

18. J. R. MacDonald and M. D. Brown, Phys. Rev. Lett., 29, 4,
     (1972).

     J. R. MacDonald, M. D. Brown and T. Chiao, Phys. Rev. Lett.,
     30, 471, (1973).

     G. Bissinger and L. C. Feldman, Phys. Rev. A, 8, 1624, (1973).

     J. A. Cairns, A. D. Marwick, J. Macek and J. S. Briggs,
     Phys. Rev. Lett., 32, 509, (1974).

     G. Bissinger and L. C. Feldman, Phys. Rev. Lett., 33, 1,
     (1974).

     J. S. Briggs, J. Phys. B, 7, 47, (1974).

     J. Macek and J. S. Briggs, J. Phys. B, 7, 1312, (1974).

     A. L. Lurio, J. A. Cairns, J. F. Ziegler and J. Macek,
     Phys. Rev. A, 12, 498, (1975).

19. J. A. Cairns, B. Benneworth and R. C. Yeates, Nucl. Inst. and Meth., $\underline{96}$, 89, (1971).

20. J. A. Cairns and R. S. Nelson, Phys. Lett., $\underline{27A}$, 14, (1968).

21. J. A. Cairns, C. L. Desborough and D. F. Holloway, Nucl. Inst. and Meth., $\underline{88}$, 239, (1970).

22. J. A. Cairns, A. D. Marwick and I. V. Mitchell, Thin Solid Films, $\underline{19}$, 91, (1973).

23. D. Burch and P. Richard, Phys. Rev. Lett., $\underline{25}$, 983, (1970).

24. A. Knudson, D. Nagel, P. Burkhalter and K. L. Dunning, Phys. Rev. Lett., $\underline{26}$, 1149, (1971).

25. R.T.K. Baker and P. S. Harris, J. Phys. E, $\underline{5}$, 793, (1972).

26. R.T.K. Baker, R. B. Thomas and M. Wells, Carbon, $\underline{13}$, 141, (1975).

27. Q. C. Kessel and B. Fastrup, Case Studies in Atomic Physics, M.R.C. McDowell and E.W. McDaniel (eds.), North Holland Publ. Co., Amsterdam, $\underline{3}$, 137, (1973).

28. J. A. Cairns and L. C. Feldman, 'New Uses of Ion Accelerators', J. F. Ziegler (Ed.), Plenum Press, p431, (1975).

29. P. Richard, 'Atomic Inner-Shell Processes', B. Crasemann (Ed.), Academic Press, New York, p73, (1975).

30. J. D. Garcia, R. J. Fortner and T. M. Kavanagh, Rev. Mod. Phys., $\underline{45}$, 111, (1973).

BACKSCATTERING OF IONS WITH INTERMEDIATE ENERGIES

H.Verbeek

Max-Planck-Institut für Plasmaphysik, EURATOM

Association,   8046 Garching b.München, Germany

INTRODUCTION

In this paper material characterization by backscattering of ions in the energy range between a few keV and several hundred keV will be discussed. This energy range lies between those energies generally used for low energy ion surface scattering (ISS) and the high energies of light ions normally used for Rutherford backscattering (RBS) analysis. It will be shown that the descriptions used for ISS as well as RBS are applicable also at energies larger or lower respectively than those normally used for these techniques.

In a simple picture (Fig.1) the principles of backscattering in the different energy regions are demonstrated: In ISS and in RBS backscattering takes place predominantly in one single elastic collision with a target atom. At low energies (ISS) this occurs at the very surface, while in RBS the particle is backscattered from an atomic nucleus at a depth up to several μm from the target surface. In addition to the elastic energy loss from the nuclear collision the particle looses energy quasi continuously to the electrons along its nearly straight trajectories inside the target.

There is an intermediate energy region (denoted MEIS for medium energy ion scattering in Fig.1),where neither single elastic collisions nor straight trajectories can be assumed. In this region governed by multiple collisions the elastic and electronic energy losses are of the same order of magnitude.

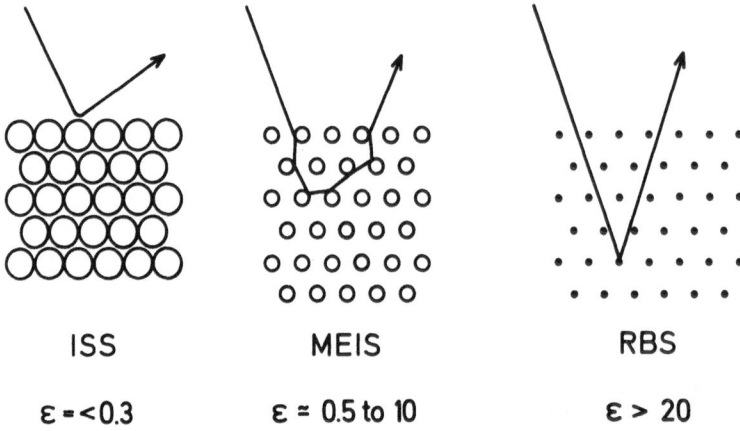

$$\varepsilon = <0.3 \qquad \varepsilon = 0.5 \text{ to } 10 \qquad \varepsilon > 20$$

FIG.1.: Principles of backscattering in different regions of the reduced energy ε.

Additionally a large part of the incident ions comes to rest in the near surface region to be analyzed. Also collision cascades are initiated in this region which cause radiation damage and sputtering. This is in contrast to high energies were the range of the incident particles is much larger than the analyzed surface layer. Thus backscattering of ions with intermediate energies is more complex and difficult to apply for material characterisation.

As will be shown, Ar ions with 80 keV show backscattering spectra typical for ISS, while 50 keV protons are backscattered with the Rutherford cross section. Thus the particle energy alone is not appropriate to characterize the different scattering regimes. More suitable is the Thomas-Fermi scaling with Lindhard's reduced energy ε /1/. It is

$$\varepsilon \;=\; \frac{M_2\, a}{(M_1+M_2)Z_1 Z_2 e^2}\, E$$

where $\qquad a \;=\; 0.468\,(Z_1^{2/3}+Z_2^{2/3})^{-1/2} \;\overset{o}{A}$

is the screening length in the TF-potential. $\epsilon = 1$ characterises
the energy at which in a head-on collision the distance of closest
approach is equal to the screening length a. In Fig.1 a rough
classification in terms of ϵ is given. $\epsilon = 0.3$ corresponds to Cu
bombarded with 2 keV He, which is usually the upper limit for
ISS work but also to 30 keV Ar ions, while $\epsilon = 20$ corresponds
to 60 keV protons which can be used for RBS. The boundaries given
in Fig.1 are not well defined; they depend also on the scattering
angle and on what the demands of a specific experiment are. In
the following these limits will be justified i.e. it will be shown
what the high energy limits for ISS and the low energy limits for
RBS are. The possibilities of these methods and their advantages
and disadvantages in the present energy range will be demonstrated.

For a discussion of the basic principles the reader is referred to the papers of W.K.Chu, H.H.Brongersma, and J.W.Mayer
in these proceedings. Additional information may be found on ISS
in the recent reviews by Buck /2/ and Taglauer and Heiland /3/,
on intermediate energies by Molchanov /4/ and on RBS in a paper
of Buck and Wheatley /5/ and in the Yorktown Heights /6/ and
Karlsruhe /7/ conference proceedings.

EXPERIMENTAL EQUIPMENT

Accelerators for energies up to several hundred keV are much
less expensive than MeV van de Graaff machines. In many laboratories are these acceleration already available. For example, mass
separators and most ion implantation machines are suitable to do
backscattering work, either to analyse surfaces by ISS or for depth
analysis near the surface by RBS. Then one has a means to investigate the implanted layers in the same machine and in situ.

There is only little additional equipment necessary to do
backscattering work with an existing ion accelerator as an ion
implantation machine. Of course, there is some space necessary in
the target chamber.

Assuming that the primary beam from the accelerator is already
mass analysed one usually needs additional apertures to confine
the incoming particles to a small spot on the target and to have
a small angular divergence. A good collimation of the primary beam
is necessary to define the scattering geometry. The target holder
should be insulated from ground to be able to measure the current
onto the target. For measurements on single crystals the target
has to be mounted on a goniometer head. The beam intensity can be
measured also with a beam monitor such as a rotating vane wheel /8/

from which a fraction of the beam is scattered into a reference
detector. Both the target current and the monitor have to be
calibrated by a Faraday cup which can be moved into the beam.

Different techniques are used to energy analyse the backscattered particles. A mass analysis is usually not necessary
since backscattered particles can easily be discriminated from
sputtered particles, which have very low energies. In the case of
light ions surface barrier detectors which are used at MeV energies can also be used in this energy range. The advantages of surface barrier detectors are their simplicity, their capability of
simultaneous measurements of all energies and their sensitivity
to all charge states of the detected particles. For heavy ions
they are less suitable, because of large energy losses in the
detector window, poor resolution and rapid radiation damage of
the detector. For light ions their ultimate limit at low energies
is at about 6 keV for protons, and 12 keV for He ions, when the
signal amplitudes reach the noise level. With cooled detectors
and preamplifiers an energy resolution of 1.5 keV can be achieved
/9/. But the energy resolution is independent from the energy and
therefore the relative energy resolution $\Delta E/E$ becomes large at
low energies.

Another method to determine energy distributions is the measurement of times of flight. For this method the primary beam has
to be pulsed. The velocity of the backscattered particles
is determined from the time which elapses from the start signal
given by the beam pulse and the detection in a multiplier in a
certain distance. Here, as with solid state detectors, the system
is always open for the detection of particles with any energy.
But to avoid preferential counting of the fastest particles the
primary beam has to be accomodated such that not more than one
particle per beam pulse is counted. The repition rate of the
pulses must be adjusted to the longest times of flight, i.e.
the slowest particles to be measured. Thus this method requires
rather long measuring times to achieve sufficient counting statistics. Buck et al. /10/ measured backscattered Ar and He particles
(ions and neutrals) between 4 and 32 keV with this method.

The best energy resolution can be obtained using electromagnetic energy analyzers. These can be magnetic sector fields
/11,12/ or electrostatic analyzers as spherical or cylindrical
condensers /5,13-16/. In principle, the energy resolution can be
made as high as necessary since it depends only on the geometric
dimensions. Magnetic analyzers are favourable for high energies.
With a position sensitive detector a large part of the spectrum
can be measured simultaneously. They are, however, expensive
and heavy. Electrostatic analyzers (ESA) can easily be built with

practical dimensions for energies up to 300 keV. For design
principles of magnets and ESA's the reader is referred to Ref./17/.
Electromagnetic analyzers are, naturally, only sensitive to charged
particles. At low energies - below 40 keV for hydrogen -, however,
the majority of the backscattered particles is neutral. As will be
discussed later the charge states depend on different parameters
as the exit energy and on the surface cleanliness. This has to be
taken into account for quantitative measurements and is a principal disadvantage of the electromagnetic analyzers.

For the measurement of an energy distribution with an ESA
its voltage has to be scanned over the energy range of interest.
The ESA is only sensitive to particles within the specific energy
window which corresponds to the ESA voltage and its resolution.
Thus the measuring time is much longer as compared with the solid
state detector. For all types of ESA the energy resolution ΔE
is proportional to the energy E, or the width of the energy
window through which particles are transmitted depends on the
energy. This distorts the spectra and has to be taken into account,
but it gives good energy resolution also at low energies. The
particles transmitted through an ESA can be detected with any open
electron multiplier such as a channeltron connected to standard
electronics.

For many problems it is desirable that the analyzer for the
backscattered particles (surface barrier detector or ESA) can be
swivelled around the target, especially when single crystals are
investigated. Then the scattering angle can be chosen as is
required for a specific experiment.

In many surface investigations it is not possible to
achieve unique answers without having another method for surface
characterization. Thus it is often desirable to have an additional
LEED and/or AUGER system in the scattering chamber.

In Fig.2 systems for backscattering measurements are shown
schemetically; in Fig. 2a) using a solid state detector and
in 2 b) with the help of an ESA.

As one decreases the primary energy surface effects become
increasingly important and the vacuum in the scattering chamber
needs more consideration. In an oil pumped system surface layers
of carbon are easily created. These may severely impaire the
depth resolution of backscattering methods near the surface.
Turbomolecular pumps in connection with Ti sublimation onto
cooled surfaces deliver clean vacuum and high pumping speeds.
Ion getter pumps have the disadvantage of creating metastable
atoms especially when noble gas ions are used. Metastables are

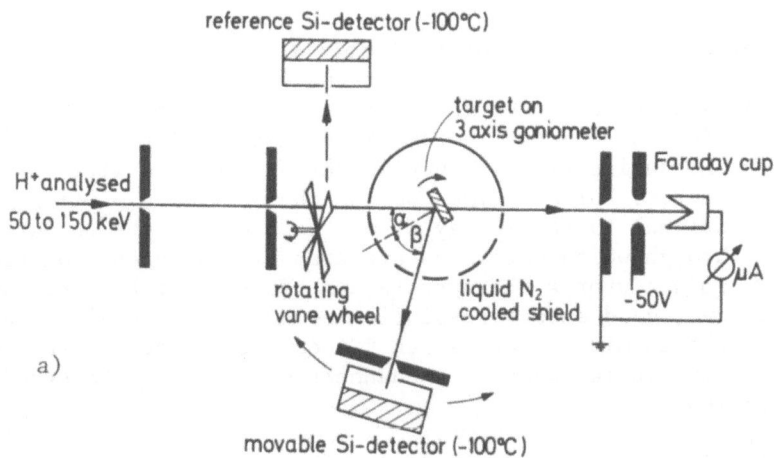

a)

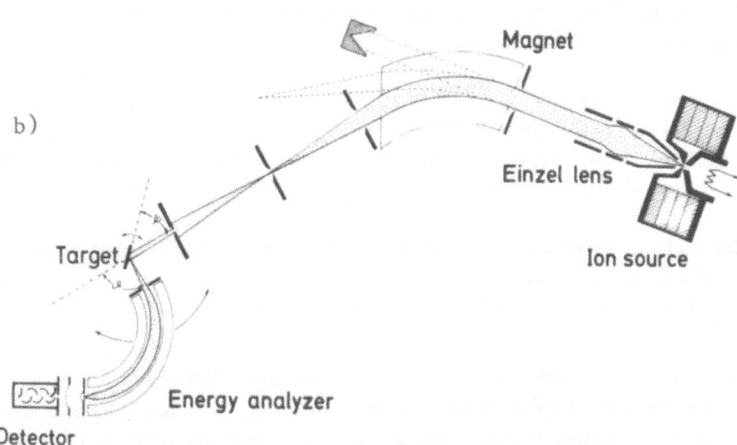

b)

FIG.2.: Backscattering experiments using a) surface barrier
         detectors and b) an ESA.

often the reason for a large background in the detectors. An oil-
free pumping system, however, brings little advantage when all
other sources of contamination like sliding O-rings etc. cannot
be avoided. On the other hand only moderate vacuum conditions
(10^{-7}Torr) are sufficient in RBS studies if not the ultimate
depth resolution is necessary /5/.

MEDIUM ENERGY ION SCATTERING AS A LIMIT FOR ISS

ISS in fact was at first observed at much higher energies
than those used today for this method. One of the classic papers
on noble gas ion scattering from solids is that of Datz and Snoek
/11/. They bombarded Cu, Ag, and Au targets with Ar^+ ions of
40 - 80 keV and analyzed the backscattered particles with a 60°
sector magnet. In Fig. 3 their basic results are represented: There
are sharp peaks due to backscattering in binary elastic collisions
with surface atoms of the target, which can be attributed to the
different charge states of backscattered Ar ions. The energy E_1
of the backscattered Ar^+ ions can be calculated according to the
kinematic formula $E_1 = KE_o$, with

$$K = \left[\frac{M_1\cos\vartheta}{M_1+M_2} + \left\{ (\frac{M_1\cos\vartheta}{M_1+M_2})^2 + \frac{M_2-M_1}{M_1+M_2} \right\}^{1/2} \right]^2 \qquad (1)$$

Also target recoil ions from binary collisions with different
charge states can be seen. The energy E_2 of the recoils is

$$E_2 = \frac{4\,M_1 M_2}{(M_1+M_2)^2} \cos^2\vartheta\, E_o \ ,$$

which is also a purely kinematic formula. Also at energies as
high as 80 keV the particles are backscattered according to the
same rules as in the usual ISS energy range. The shoulders at
the peaks correspond to multiple collisions, which are also observed at low energies. The relative contribution of multiple
collisions to the spectrum decreases with ionic charge. Thus
also with these high energies, in principle, it is possible
to determine the surface composition from the energy of the ions
and the sacttering angle. As in ISS for best mass resolution the
ions should be in the same mass region as the species which one
wants to detect on the target surface /3//18/.

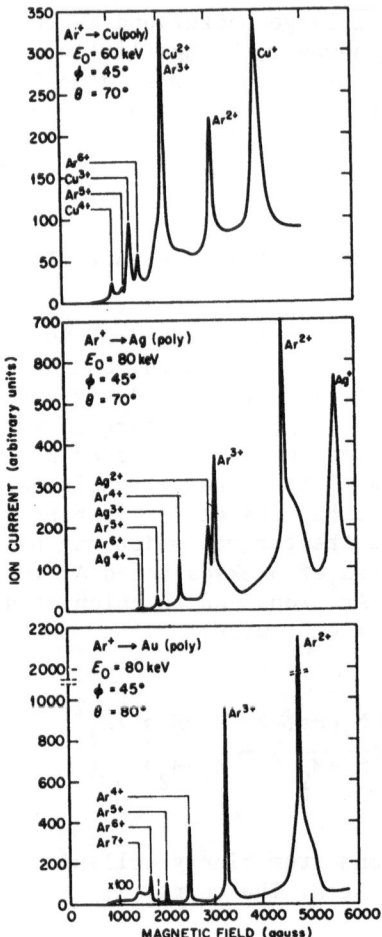

FIG.3.: Backscattering spectra of Ar⁺ from Cu,
Ag, and Au (from Ref.11).

Considering the ε-scaling, however, 80 keV Ar onto Cu corresponds to ε = 0.75. This is in the region denoted "MEIS" in Fig.1.
In the experiment only those backscattered particles are seen,
which are scattered from the surface since these retain their
charge. The majority of the particles, however, penetrate to a
mean depth of ~500 Å into the target. They initiate collision
cascades in the surface region which cause radiation damage and
sputtering.

Sputtering is the ejection of surface atoms by the collision cascadescreated in the solid by the ion bombardment. According to Sigmund's theory /19/ the sputtering yield is proportional to the energy deposited into nuclear motion in the surface region and inversely proportional to the surface binding energy. Experimental sputtering yields for various ions on stainless steel are shown in Fig. 4 from Ref./20/. It is seen, that the sputtering yields for Argon are larger than 1 in the interesting energy region. That means that with energies above 200 eV much more target atoms are sputtered away than are scattered. This shows drastically that this method is not at all a nondestructive method.

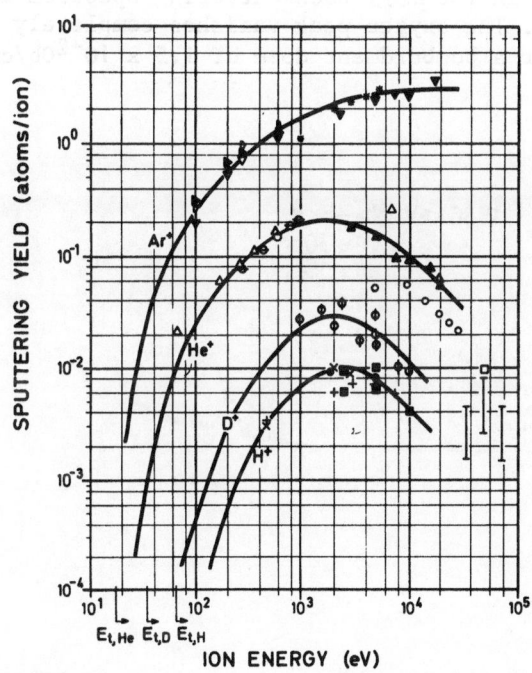

FIG.4.: Sputtering yields versus ion energy for H^+, D^+, He^+, and Ar^+ ions on stainless steel. (from Ref.20).

On the other hand, one can think of a depth profiling
method: Subsequent layers of the material under investigation are
removed by sputtering while the surface composition is continuously monitored by energy analyzing the backscattered part of the
bombarding ions. One difficulty can be ion mixing /21/,/22/
and recoil implantation /23/. It is seen from Fig.4 that the
sputtering yields for the light ions are much lower. For protons
a maximum value of 10^{-2} is reached at 2-3 keV. At the higher
energies discussed in this paper the yield is much smaller, and
sputtering generally is negligible for protons and He ions. But
for experiments with large ion doses sputtering should be recognized.

To demonstrate the influence of sputtering as an example
the scattering of 15 keV Ne ions from a Vanadium sample (ε =0.43)
is shown in Fig.5. The target was contaminated by oxygen which
is clearly seen in the Neon backscattering spectrum in the upper
trace of Fig. 5. The oxygen peak vanishes completely due to
sputtering after a bombardment dose of 1.5 x 10^{-2}Cb/cm^2, while

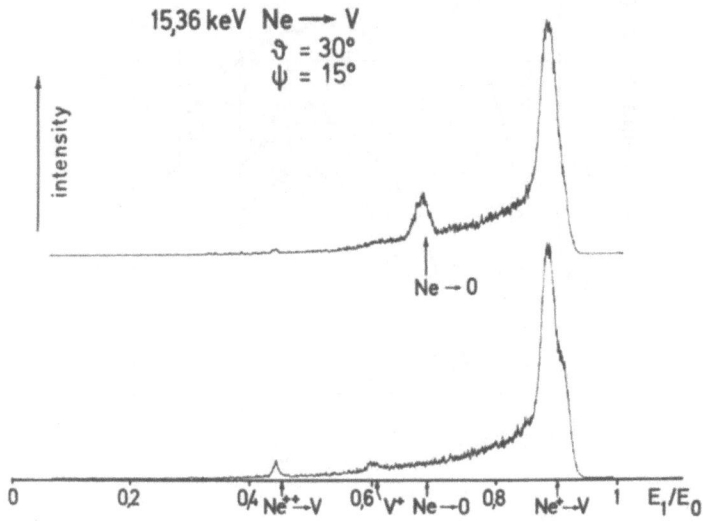

FIG.5.: Energy distribution of 15.36 keV Ne backscattered from
        V contaminated by oxygen before (upper trace) and after
        (lower trace) bombardment.

$2 \times 10^{-3} Cb/cm^2$ was used to record the spectrum. At the same time
a peak of V-recoil ions appeared. This effect is frequently used
to clean surfaces. With the neon scattering one has a tool to
see the result of the cleaning procedure. Another feature is the
long tail of the main peak. This indicates penetration deeper into
the material and insufficient neutralization of the ions backscattered from deeper layers.

The latter effect is more characteristically demonstrated
by Fig.6 from Ball et al./14/. Spectra of He-ions with primary
energies 2 to 25 keV (ε = 0.09 to 1.12) backscattered from Au
are shown. For analytical problems in ISS normally He ions are
used because all masses $M_2 > 4$ can be detected at all scattering
angles. At the highest energy (25 keV) a surface peak is no
longer visible and a broad energy distribution extends to zero
energy. Thus a scattering peak from a potential light mass contamination on the surface is obscured by this large background.

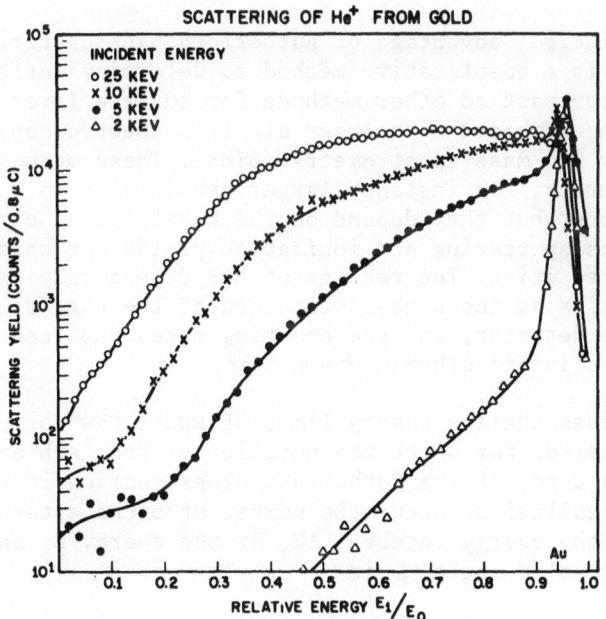

FIG.6.: Backscattering spectra of He from Au for incident
energies E_o = 2,5, 10, and 25 keV (from Ref.14).

On the other hand surface contaminations can be seen with high
sensitivity if their mass number is larger than that of the substrate. Buck et al./5/,/14/ were able to detect 10^{12} Au atoms per
cm^2 on Si by 100 keV He scattering. With single crystal targets
in channeling and/or blocking conditions also light impurities
can be detected /5/,/24/. For more details on channeling experiments see the contribution of J.A.Davies in these proceedings.

In a number of laboratories especially in the USSR the backscattering of ions with several 10 keV energy from single crystals
has been investigated /4/. The spatial distributions of ions
backscattered from single crystals can be correlated to the
crystal structure. One feature is the "ion focusing effect"
which occurs when ions are reflected in a sequence of collisions
from a so called semi-channel. There are attemps being made to
develop on this basis a method to investigate the crystalline
structure of the surface /25/. This aim, however, is still not
reached because quantitative results are still missing.

RUTHERFORD BACKSCATTERING OF LIGHT IONS

The principal advantage of Rutherford backscattering (RBS)
is, that it is a quantitative method to determine surface layers.
This is in contrast to other methods for surface layer analysis
as sputtering combined with Auger electron spectroscopy (AES)
or secondary ion mass spectrometry (SIMS). These methods have
their own merits, for instance larger sensitivity to low mass
contaminations, but they depend on the knowledge of certain
quantities as sputtering and ionization yields for which generally
only estimates exist. The results of RBS depend only on measure-
able quantities as the bombardment dose of the target, the solid
angle of the detector, and the counting rates. RBS can therefore
be used to calibrate other methods /26/.

To discuss the low energy limit of RBS a few formulae
must be repeated, for which the notation of Behrisch and Scherzer
/26/,/27/ is used. If the Rutherford cross section is valid and
only single collisions occur the number of backscattered particles Δn in the energy interval ΔE_1 at the energy E_1 and in
the detector solid angle $\Delta \Omega$ is

$$\frac{\Delta n}{n_o} = N \frac{z_1^2 \, z_2^2 e^4}{16 \, dE/dx} (K + \cos \alpha / \cos \beta)(E_1 + E_o \cos \alpha / \cos \beta)^{-2} f(\vartheta) \, \Delta E_1 \Delta \Omega \tag{2}$$

$$f(\vartheta) = 4 \left[\cos\vartheta + \left\{ 1 - (\frac{M_1}{M_2}\sin\vartheta)^2 \right\}^{1/2} \right]^2 \sin^{-4}\vartheta \left\{ 1 - (\frac{M_1}{M_2}\sin\vartheta)^2 \right\}^{-1/2}$$

K is the kinematic factor according to (1), $f(\vartheta)$ the angular dependence of the Rutherford cross section, N the atomic density of the target, Z_1, M_1 and Z_2, M_2 nuclear charge and mass number of projectiles and target atoms, α and ß the angles of incidence and emergence with respect to the target normal, n_o the number of ions with primary energy E_o hitting the target, and $\overline{dE/dx}$ the mean differential energy loss. For the derivation of (2) it was assumed that the differential energy loss dE/dx does not depend on energy.

If the backscattering is from a thin film on thickness x, the backscattered particles have energies E_1 between $E_1 = K.E_o$ and E_1^*. E_1^* is related to the thickness x and the differential energy loss dE/dx:

$$E_1^* = (E_o - \frac{x}{\cos\alpha} \frac{dE}{dx}) K - \frac{x}{\cos ß} \frac{dE}{dx} \tag{3}$$

Integration of (1) between $E_1 = KE_o$ and E_1^* gives the number of all particles n backscattered from the film into $\Delta\Omega$.

$$\frac{n}{n_o} = N x \frac{Z_1^2 Z_1^2 e^4}{16 E_o E_1} \frac{f(\vartheta)}{\cos\alpha} \tag{4}$$

$\overline{E}_1$ is a mean energy closely related to E_o:

$$\overline{E}_1 = (E_o \cos\alpha/\cos ß + E_1^*) (K + \cos\alpha/\cos ß)^{-1}$$

where E_1^* has to be taken from the experiment.

Formula (4) is independent from dE/dx. As the projectile (H^+ or He^+) and the scattering geometry ($\alpha, \vartheta, \Delta\Omega$) can be properly chosen, and the numbers n and n_o can be measured, one can absolutely determine Nx i.e. the number of atoms per surface area.

One advantage of going to lower energies in RBS is immediately seen from (4): The number of backscattered particles n increases rapidly with decreasing primary energy E_o. Thus the sensitivity of RBS for surface layers is increasing, when the energy is decreased. Naturally (4) depends on the validity of

the Rutherford cross section. It is clear, however, that for low
energies and depending on the scattering angle the distances
of closest approach in the single scattering events between ions
and target atoms become so large that the interaction is influenced by the atomic electrons. The Coulomb field of the nucleus
causing the Rutherford interaction is screened by the atomic
electrons. This means that Thomas-Fermi type interaction potentials like the Bohr or Molière potentials, which contain a
screening length have to be applied. These have been largely
investigated /1//28/. However, no simple formulae can be given.
Thus, for practical reasons, it is desirable to find experimentally the energy ranges where the Rutherford law is valid.

This has been investigated by a number of authors. Van
Wijngaarden et al./29/ studied the total backscattering from,
thin Au films of various known thicknesses. This is a very
simple experiment. Besides the surface barrier detector with
its amplifier only an integral discriminatior is necessary
whose level is so adjusted that all counts from the film are
registered. (No multichannel analyzer is needed). In Fig.7
the backscattering yields from films of various thicknesses
versus energy are shown. The solid lines go through the experimental points. It was found that the backscattering yield was
not proportional to the film thickness, which was believed to be
accurate within $\pm$ 5 Å. This discrepancy was attributed to small
angle multiple scattering. With an estimate for this effect
the dashed (corrected) curves were found. As one expects the
correction for multiple scattering effects increases with increasing film thickness and ion mass. From the corrected curves
the energy dependence of the scattering cross section could be
calculated. For protons onto Au the authors found that the
Rutherford scattering law was valid down to 50 keV. This corresponds to ε = 4.7.

To avoid the complication due to multiple scattering,Gold-
berg et al./30/ scattered protons and He ions from thin surface
layers with thicknesses of the order of less than a monolayer.
They compared their results with a Thomas-Fermi interaction
according to Lindhard et al. /28/ and found good agreement.
The Thomas-Fermi cross section coincides with the Rutherford
cross section for large energies and/or large deflection angles.
The results in Fig.8 were plotted versus a scaling parameter
$t^{1/2} = \varepsilon \sin \vartheta/2$, which takes the influence of the scattering
angle into account. Agreement with Rutherford was found for He
when $t^{1/2} > 8$, for protons when $t^{1/2} > 6$. With an Au target and
ϑ = 150° this corresponds to E_o = 190 keV for He and E_o = 60 keV
for protons, which is somewhat higher than the findings of
van Winjngaarden /29/.

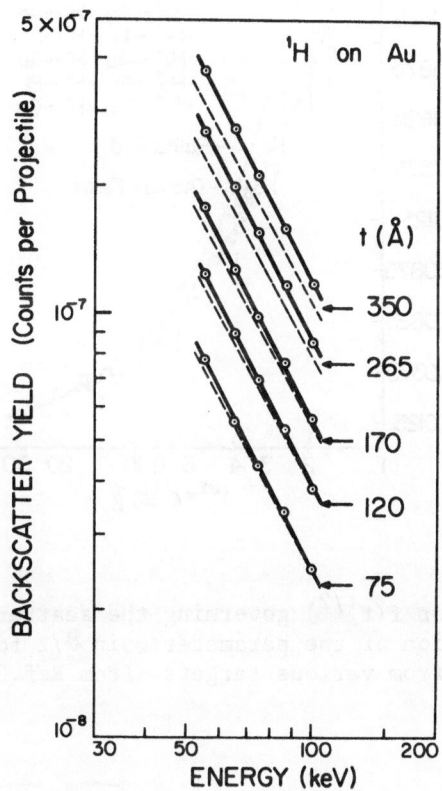

FIG.7.: Backscattering yields of protons from Au films
of various thicknesses as a function of energy
(from Ref.29).

The increase of the scattering cross section with decreasing
energy has two effects on the backscattering yields: 1) With
decreasing energy and increasing penetration into the solid the
probability for scattering by any angle increases. This causes an
attennuation of the bean on its way in and out of the target.
This effect was considered by Buck et al./10/. 2) The assumption
of backscattering in one single large angle deflection breaks down
at low energies.

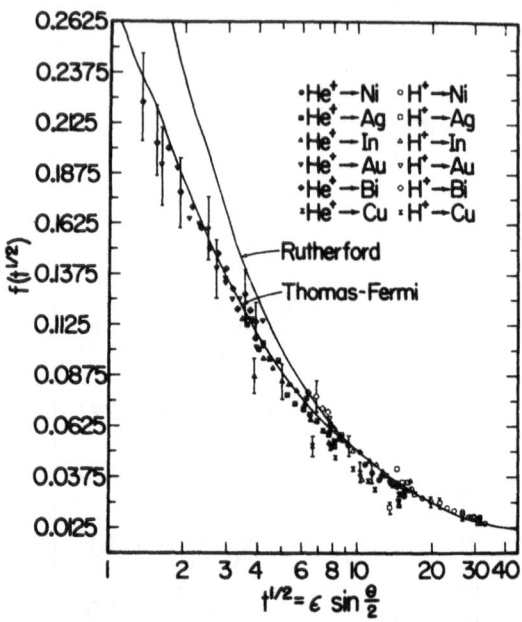

FIG.8.: The function $f(t^{1/2})$ governing the scattering cross section as a function of the parameter $\varepsilon \sin \vartheta/2$ for H^+ and He^+ ions scattered from various targets (from Ref.30).

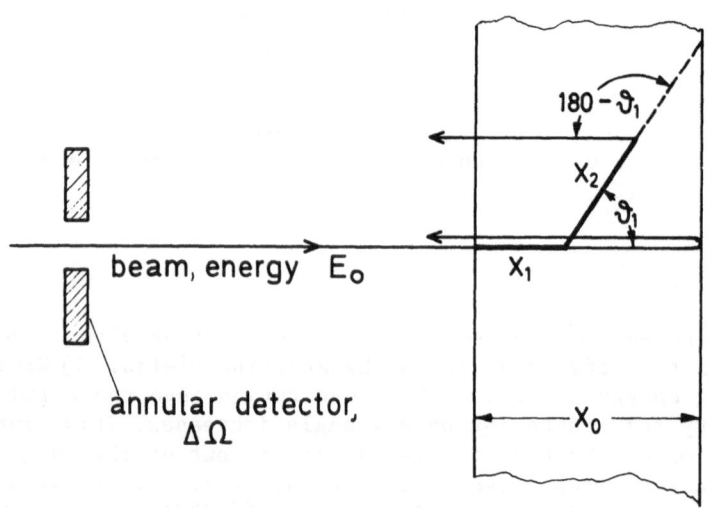

FIG.9: 180° scattering geometry

I have tried to find an estimate at which energy more than one large angle collision contribute significantly to the back-scattering yield. Consider a foil of thickness x from which normal incident particles are backscattered by $\vartheta \stackrel{\circ}{=} 180^{\circ}$. In this arrangement the largest depth x_o can be investigated. Experimentally, this can be accomplished with an annular detector, which may subtend the solid angle $\Delta\Omega$. For simplicity the validity of the Rutherford cross section and the fact that the cross section and $\frac{dE}{dx}$ do not depend largely on the energy in the foil is assumed. According to formula (4) the probability of backscattering from the foil is:

$$P_1 = x_o \ N \ \frac{z_1^2 z_2^2 e^4}{16 \ E_o^2} \ \Delta\Omega \tag{4}$$

since $f(\vartheta) = 1$ for $\vartheta = 180^{\circ}$. Now let us assume that the particle is scattered with $\vartheta_1 > \vartheta_{min}$ in the angle interval $2\pi \sin\vartheta_1 \ d\vartheta_1$, at a depth x_1 in the foil. If a second deflection with $\vartheta_2 = 180^{\circ} - \vartheta_1$ occurs in a distance x_2 from the first deflection which falls into $\Delta\Omega$, this particle is detected. The probability for this process is

$$P_2 = \int\limits_{\vartheta_{min}}^{\pi} \int\limits_{o}^{x_o} dx_1 \ N \ 2\pi \sin\vartheta_1 \ d\vartheta_1 \ \frac{z_1^2 z_2^2 e^4}{16 \ E_o^2} \ f(\vartheta_1)$$

$$\times \ N \ x_2 \ \frac{z_1^2 z_2^2 e^4}{16 \ E_o^2} \ f(180 - \vartheta_1) \ \Delta\Omega \tag{5}$$

All small angle deflections with $\vartheta_1 < \vartheta_{min}$ are not considered. ϑ_{min} may be related to the $\Delta\Omega$ of the detector. ($\vartheta_{min} = 5^{\circ}$ corresponds to an $\Delta\Omega$ subtended by an annular detector of 1.7 cm diameter in a distance of 10 cm). The distance of x_2 is somewhat arbitrarily restricted to $x_2 \leq x_o$. Then the integration over x_1 yields x_o^2. For $M_1 \ll M_2$ is $f(\vartheta)^o = \sin^{-4}\vartheta/2$. With this assumption the integration over ϑ_1 yields a function

$$F(\vartheta_{min}) = \int\limits_{\vartheta_{min}}^{\pi} \frac{d\vartheta}{\sin^3\vartheta} \ .$$

With this the integration of (5) yields:

$$P_2 = N^2 2\pi \left(\frac{Z_1^2 Z_2^2 e^4}{16 E_o^2} \right)^2 F(\vartheta_{min}) \, x_o^2 \, \Delta\Omega \tag{6}$$

and $$\frac{P_2}{P_1} = 2\pi N \frac{Z_1^2 Z_2^2 e^4}{16} F(\vartheta_{min}) \frac{x_o}{E_o^2} \tag{7}$$

For $\vartheta_{min} = 5^o$ and for protons on Ni and Au the relative probability of double collisions as a function of foil thickness is shown in Fig.10 for several primary energies. This can serve as rough estimate.

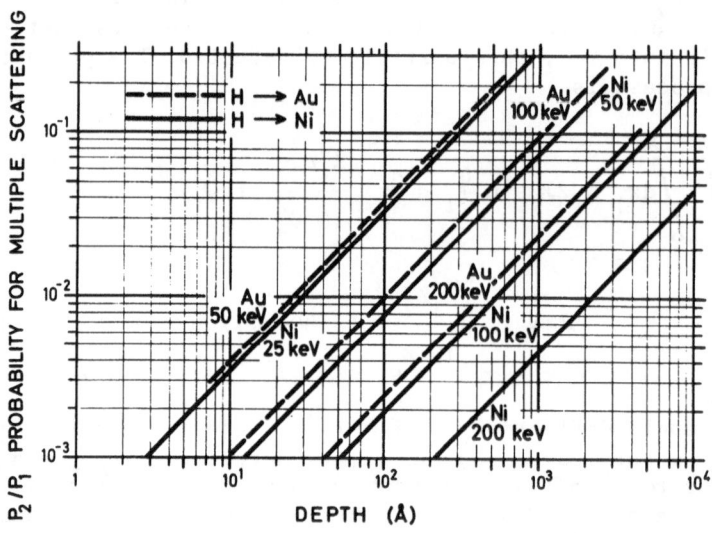

FIG.10. Relative multiple scattering probability versus depth for protons of various energies in Ni and Au.

As shown earlier, Goldberg et al./30/ demonstrated the validity of the Rutherford scattering law for protons on Au down to 60 keV. But, as seen from Fig.10, already at a penetration depth of 200 Å about 5 % double scattering contributes to the

backscattering yield at this energy. To availuate the low energy
limit for RBS this must be born in mind. As a rule of thumb one
can say:

*Multiple collisions become important at roughly the same ener-
gies, where the Rutherford scattering law breaks down.*

At energies below 20 keV for protons the whole backscattering is dominated by multiple large angle deflections. The energy
loss in the elastic collision according to (1) is larger for
a single collision with scattering angle ϑ than that of more
collisions with angles $\vartheta_1 \ldots \vartheta_n$ with $\Sigma \vartheta_n = \vartheta$. On the other
hand, the electronic energy loss is still proportional to the
length of the particle's trajectory. Fig.11 from Ref.32 shows
a computer simulation of 15 keV protons onto a Ni single crystal
in a two dimensional model. On some of the trajectories the
energies are notated when the particles leave the surface. It is
seen that particles with very different pathlengths have nearly
the same energy. This effect distorts the depth resolution drastically even in the surface region.

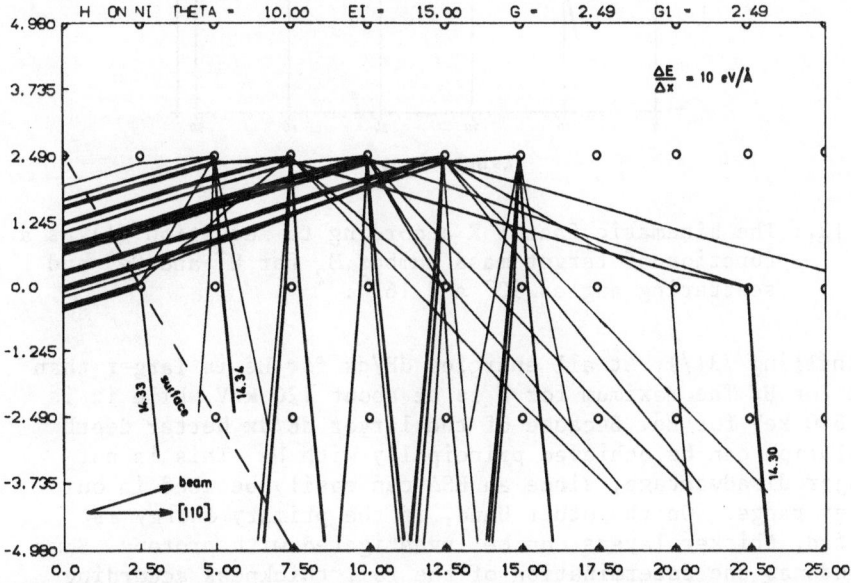

FIG.11.: Computer simulation of protons penetrating into a Ni
single crystal (from Ref.32).

    In a short intermission I want to discuss the relative merits
of protons and He ions for backscattering work, if one has an
accelerator which goes up to not more than a few hundred keV.
In Fig.12 the kinematic factor K as a function of the target mass
number M_2 for H and He and 2 different scattering angles is shown.
K governs the mass resolution, which can be achieved. It is seen
that in this respect He is much superior. The differential
energy loss for H and He on Ni as a function of energy is shown
in Fig.13. The curves are taken from the tables of Northcliffe

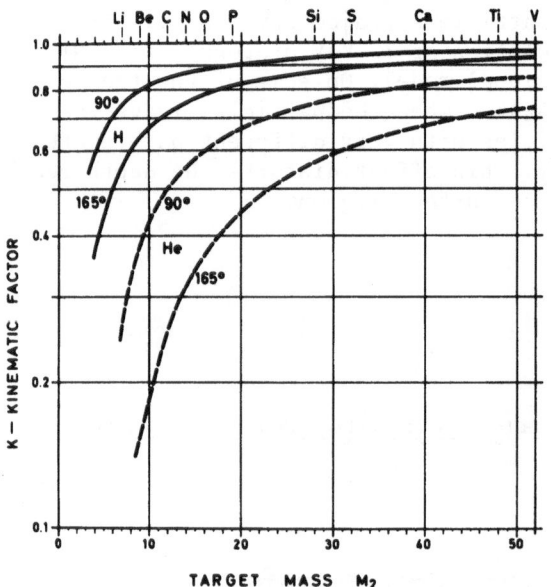

FIG.12.: The kinematic factor K according to equation (1) as a
         function of target mass number M_2 for H^+ and He^+ and
         scattering angles 90° and 165°.

& Schilling /31/). At all energies dE/dx for He is larger than
that for H. The maximum for H is at about 120 keV while it is
at 800 keV for He. Because of the larger dE/dx better depth
resolution can be achieved principally with He. This is not
a major disadvantage, since an ESA can easily be used in our
energy range. On the other hand, if the primary energy is
limited, thicker layers can be investigated with protons. Some
methods as the determination of the foil thickness according
to formula (4) or the shape of the backscattered energy distribution (formula (2)) depend on the fact that dE/dx is independent
from the energy. The assumption dE/dx = const. is fairly well

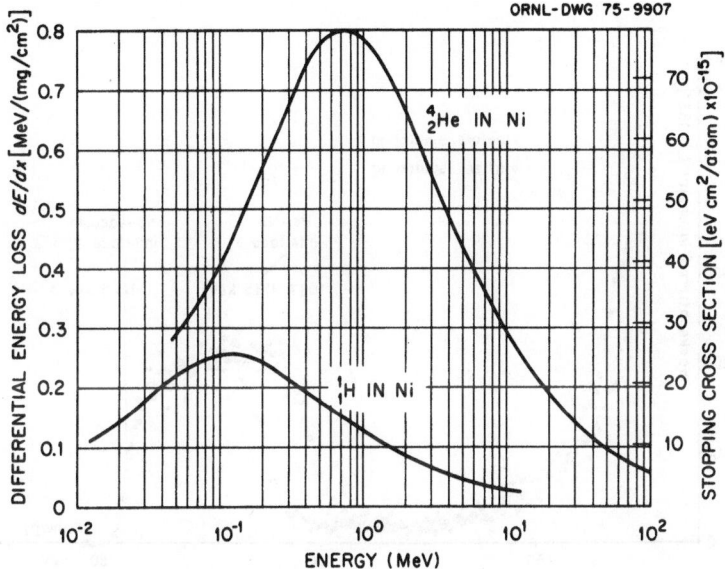

FIG.13.: Differential energy loss dE/dx for H$^+$ and He$^+$ in Ni as
a function of energy (data from Ref. 31).

fulfilled in our energy range for protons but not for He (see
Fig.13). Another important advantage of protons compared to He
ions is that the Rutherford scattering law and the single collision model are valid down to much lower energies than for He ions.

That RBS can successfully be used for material characterization also with energies in the hundred keV energy range will
now be shown in a few examples.

The determination of film thicknesses with backscattering
of 150 keV protons is shown in Fig.14. Here the sputtering of
Nb-films on Be-substrates by D-ions with energies of 3 to 8 keV
was investigated /33/. The films were bombarded in a different
apparatus and the reduction in film thickness due to sputtering
was measured by backscattering. From the spectra in Fig.14 the
film thicknesses can be evaluated in two manners. When the primary ion dose and the detector solid angle are known the film
thickness can be calculated from the total backscattering yield.
If, on the other hand, the absolute values are not measurable,
one can evaluate the thickness from the energy difference
between front and leading edges of the Nb peak.

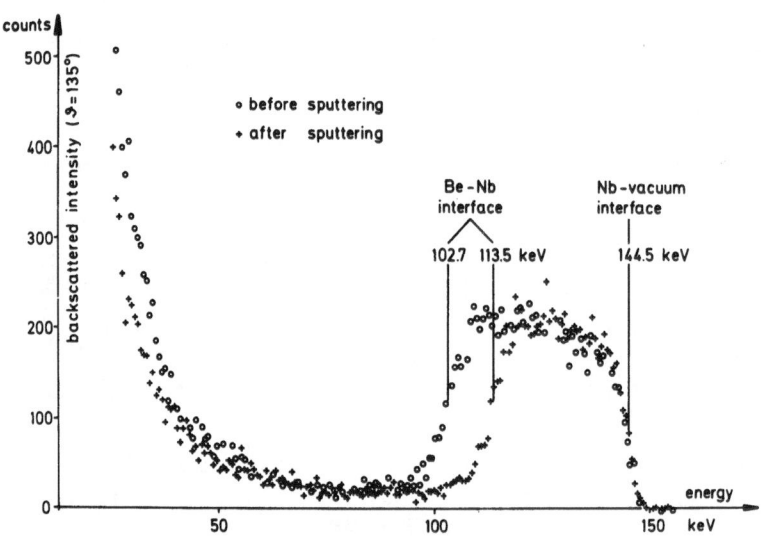

FIG.14.: Backscattering spectra of 150 keV protons from a Nb
film on a Be substrate before and after sputtering with
D$^+$ ions (from Ref.33).

This requires the knowledge of dE/dx of the material in
question. In the present case it turned out that we had to
assume a rather large oxygen content in the film to achieve
agreement of the results of the two methods. (The oxygen enhances
the stopping cross section considerably according to Bragg's
rule).

In another example from Ref.27 an Al-Be-Nb sandwich was
investigated by two methods: i) sputter etching with Ar$^+$ions
and subsequent determination of the surface composition by
Auger electron spectroscopy (AES) and ii) by RBS of protons.
Fig.15 shows the backscattering spectra. It is remarkable that
with all energies used, 60, 100, and 150 keV, the evaluated
number densities per surface area of Nb agree within 5 %. This
indicates that the Rutherford cross section is valid and single
collisions govern the backscattering even at energies as low as
60 keV. The Nb peak, however, is better separated from the substrate edge at higher energies. The intensity in the valley is
due to oxygen contamination of the film.

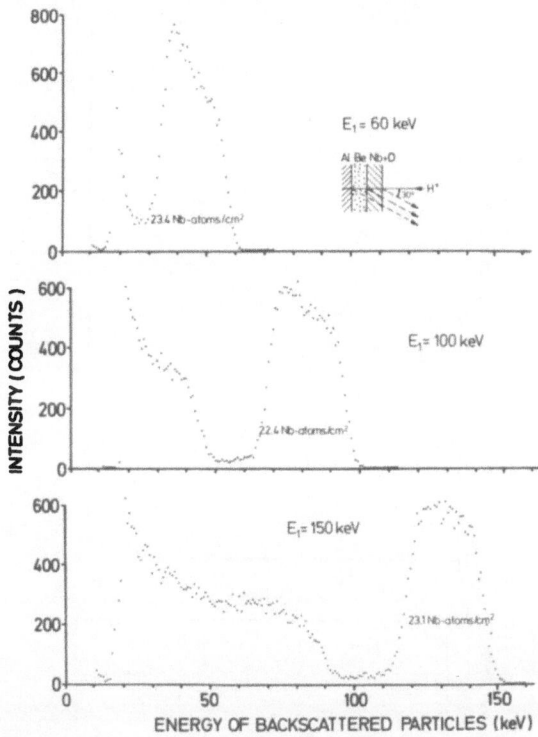

FIG.15.: Backscattering spectra of 60, 100, and 150 keV protons
from a Al-Be-Nb sandwich target (from Ref.27).

    In Fig.16 (upper part) a spectrum of 150 keV protons from
a thick Nb sample is shown / 34/. The sample had been implanted
in the surface region by 4 keV He ions. The step in the spectrum
is due to the enhancement of the stopping cross section by the
implanted gas. Using formula (2) dE/dx could be determined at
each energy E_1 (lower part of Fig.16). This was related to
the depth and with Bragg's rule the helium concentration versus
depth could be determined. For this example the relatively low
energy was especially advantageous because of the independence
of dE/dx from the energy for protons in this energy range.

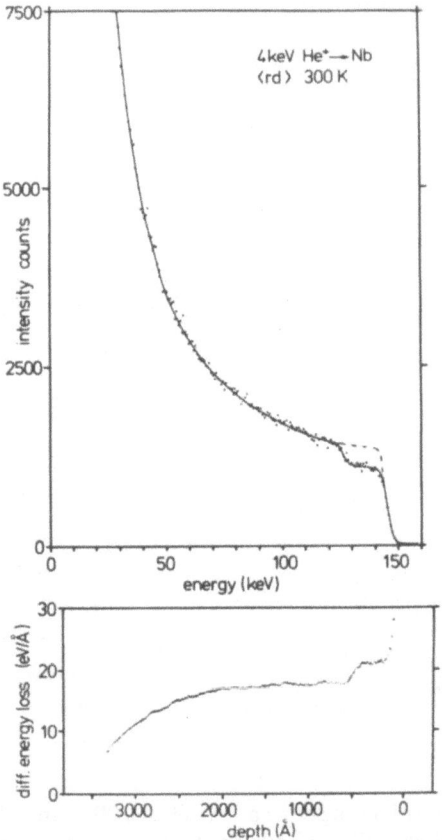

FIG.16.:   Backscattering spectrum of 150 keV protons from a thick
           Nb target, which had been implanted with 4 keV He ions.
           dE/dx calculated from the spectrum is shown in the lower
           curve (from Ref.34).

       As a last example Fig.17 shows the backscattering spectrum
of 250 keV He from a very thin (35 Å) Au layer measured by
Feuerstein et al./16/ with an ESA. This spectrum gives an idea
what the achievable depth resolution is. There is a number of
effects which contribute to the energy spread of the backscattered
particles :

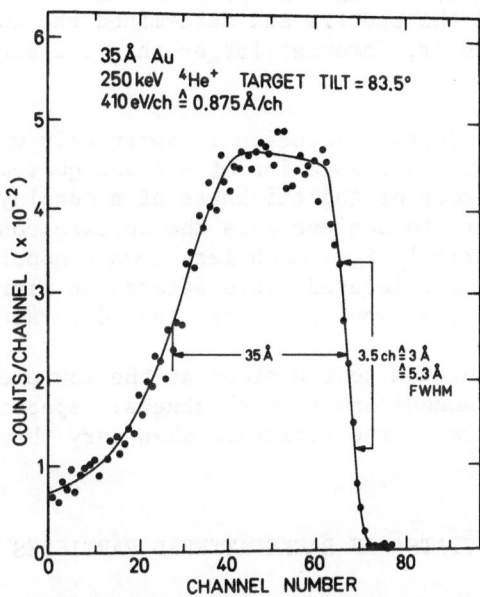

FIG.17: Backscattering spectrum of He$^+$ ions from a 35 Å thick
Au film (from Ref.16).

a) the finite analyzer resolution
b) the angular spread of the beam and the finite
   solid angle of the detector
c) impurity layers on the target surfaces
d) energy straggling of the ion, which increases
   with depth.

Basically only the last effect limits the achievable depth
resolution, while the first three can be controlled in the experiment. When an ESA is used the analyzer resolution and the solid
angle in principle can be made so small that their effect is
negligible as compared to energy straggling. The effect of
surface impurities is often overlooked. A thin surface layer of
some light material i.e. carbon shows no backscattering but
causes a substantial decrease of the slope of the spectra at
the high energy edge /35/.

In Fig.17 the authors attributed the slope at the high
energy edge (the surface side) of the peak alone to the finite

analyzer resolution and that of the trailing edge to the analyzer
resolution plus energy straggling. By assuming Gaussian distributions they unfolded the spectra and determined the energy straggling. This was, however, somewhat larger than the authors expected from theory.

The achievable depth resolution is apparently much better
near the surface. Here a resolution of 5 Å was quoted. A depth
resolution in the order of the thickness of a monolayer should
be possible. However, to achieve this the surface conditions have
to be controlled strictly i.e. much less than a monolayer of
contaminations can be tolerated. This affords an ultra high
vacuum system where the samples can be cleaned in situ.

Another contribution to the slope at the low energy side
is properly due to nonuniform film thickness. Especially Au
tends to form islands on the substrate when very thin films are
evaporated.

CHARGE STATES OF BACKSCATTERED PARTICLES

In low energy ion surface scattering (ISS) generally only
the charged particles are analyzed and detected. It is assumed
that particles penetrating deeper into the solid are preferential
neutralized. Buck et al./10/ measured also neutral argon atoms
backscattered from Au (Fig.18). They found indeed that the
fraction of positive ions is highly peaked for those ions backscattered from the surface. There are, however, also some Ar^+ ions
backscattered from deeper layers. The ion fractions for these ions
are depending on the primary energy and are different for equal
scattered energies E_1. Thus the ion fractions depend on the depth
from which the particles are backscattered.

For protons it was observed that the charged fraction
depends only on the velocity of the emerging particles and not
on the depth /36//37/. Here no dependency on the primary energy
was observed. Fig. 19 from Ref. 38 shows that no large
differences for different metals could be found. There is,
however, at energies below 20 keV a significant influence of
surface contaminations like oxygen, which is demonstrated in
Fig.20 showing the H^+ fraction backscattered from Au before and
after cleaning. The Au target was as built in contaminated by
oxygen and carbon. After cleaning by sputtering no more O and
C could be detected on the surface by AES.

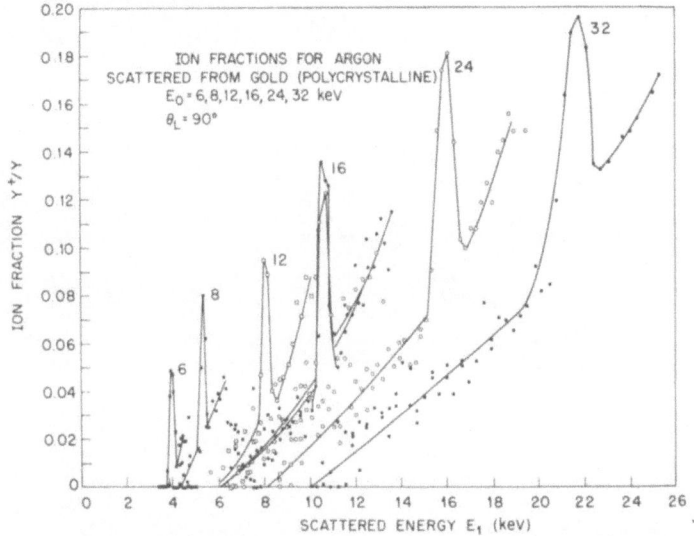

FIG.18: Charged fraction of Ar ions with various primary energies scattered from Au (from Ref. 10).

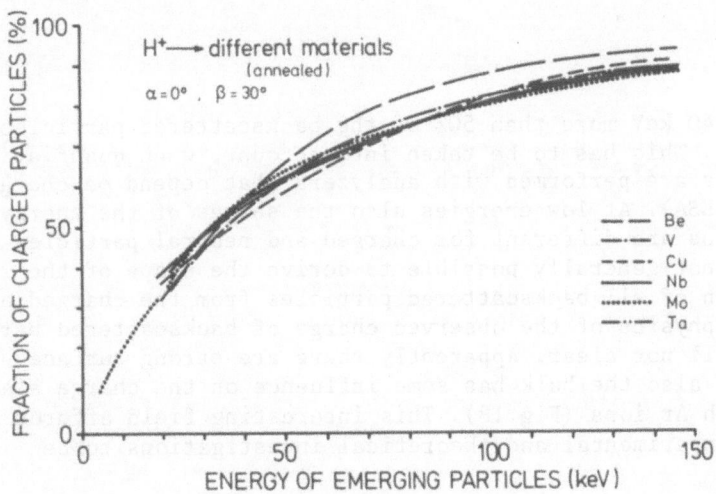

FIG.19: Charged fraction of hydrogen scattered from various
targets (from Ref. 38).

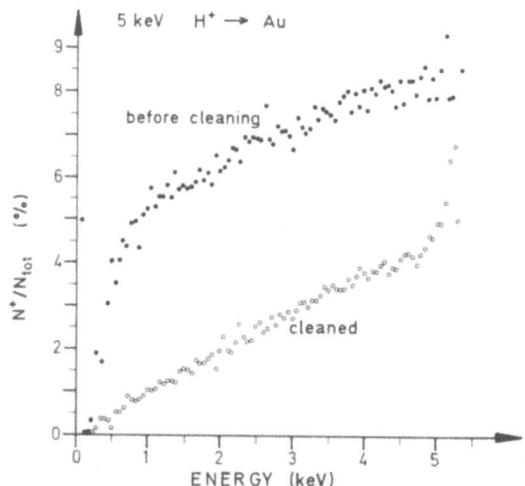

FIG.20.: Charged fraction of hydrogen scattered from Au which
was atomically clean (lower curve) and contaminated
by O and C (upper curve).

Below 40 keV more than 50% of the backscattered particles
are neutral. This has to be taken into account, when quantiative
measurements are performed with analyzers that depend on charged
particles (ESA). At low energies also the shapes of the energy
distributions are different for charged and neutral particles /37/.
Thus it is not generally possible to derive the shape of the
distribution of all backscattered particles from the charged ones
alone. The physics of the observed charge of backscattered particles is still not clear. Apparently there are strong surface
effects but also the bulk has some influence on the charge states
as seen with Ar ions (Fig.18). This interesting field affords
much more experimental and theoretical investigations to be
clarified.

ACKNOWLEDGEMENTS

For his suggestions and his constructive criticism I am
indebted to R.Behrisch, who also had the idea for Fig.1. Thanks
are due to W.Eckstein, B.M.U.Scherzer and H.Vernickel for numerous discussions.

REFERENCES

1) J.Lindhard, M.Scharff, H.E.Schiøtt, Kgl.Dan.Vid.Selsk.Mat. Fys.Medd.33, No 14 (1963).

2) T.M.Buck in "Methods of Surface Analysis", ed.A.W.Czanderna Elsevier, Amsterdam 1975, page 75.

3) E.Taglauer, W.Heiland,Appl.Phys. 9, 261 (1976).

4) E.S.Mashkova, V.A.Molchanov, Rad.Eff.16, 143 (1972), and V.A.Molchanov, Rad.Eff.23, 197 (1974).

5) T.M.Buck, G.H.Wheatley, Surf.Sci.33 35, (1972).

6) J.W.Mayer, J.F.Ziegler (eds), Ion Beam Surface Layer Analysis, Elsevier Sequoia, Lausanne 1974.

7) O.Meyer, G.Linker, F.Käppler, Ion Beam Surface Layer Analysis, Plenum,New York 1976.

8) R.Behrisch, B.M.U.Scherzer, H.Schulze, Rad.Eff.13, 33 (1972).

9) H.Schmidl, Report IPP 9/3, Max-Planck-Institut für Plasmaphysik (1971).

10) T.M.Buck, Y.S.Chen, G.H.Wheatley, W.F. van der Weg, Surf.Sci.47, 244 (1975).

11) S.Datz, C.Snoek, Phys.Rev.134 A 347 (1964)

12) E.Bøgh, Rad.Eff.12,

13) W.Eckstein, H.Verbeek, Vacuum 23, 159 (1973).

14) D.J.Ball, T.M.Buck, D.McNair,G.H.Wheatley, Surf.Sci.30, 69 (1972).

15) A.van Wijngaarden, B.Miremadi,W.E.Baylis, Canad.J. Phys.49, 2440 (1971).

16) A.Feuerstein, H.Grahmann, S.Kalbitzer, H.Oetzmann, in Ion Beam Surface Layer Analysis, Vol.1, ed.O.Meyer, G.Linker, F.Käppler, Plenum 1976,p.471.

17) A.Septier (ed.) Focusing of Charged Particles Acad.Press. N.Y.1967.

18) W.Eckstein, H.G.Schäffler, H.Verbeek, Report IPP 9/16, Max-Planck-Institut für Plasmaphysik, Garching,Jan.1974.

19) P.Sigmund, Phys.Rev.184, 383 (1969).

20) H.v.Seefeld, H.Schmidl, R.Behrisch, B.M.U.Scherzer, J.Nucl.Mat. in press.

21)   P.Staib, Rad.Eff.18, 217 (1973).

22)   T.Ishitani, R.Shimizu, Appl.Phys.6, 241 (1975).

23)   R.A.Moline, G.W.Reutlinger, J.C.North in Atomic Coll.
        in Solids ed. S.Datz, B.R.Appleton, C.D.Moak, Plenum N.Y.
        1974 and
        O.Christensen, H.Bay, Appl.Phys.Lett.28, 491 (1976).

24)   W.Turkenburg, W.Soszka, F.W.Saris, H.H.Kersten,
        B.G.Colenbrander, Nucl.Instr.Meth.132, 587 (1976),
        see also W.C.Turkenburg, Thesis, Univers. Amsterdam (1976).

25)   S.H.A.Begemann, A.L.Boers, Surf.Sci.30, 134 (1972).

26)   R.Behrisch, B.M.U.Scherzer, Thin Solid Films 19, 247(1973).

27)   R.Behrisch, B.M.U.Scherzer, P.Staib, Thin Solid Films 19,
        57 (1973).

28)   J.Lindhard, V.Nielson, M.Scharff, Kgl.Dan.Vid.Selsk.Mat.
        Fys.Medd.36, No 10 (1968).

29)   A.van Wijngaarden, E.J.Brimmer, W.E.Baylis, Can.J.Phys.48,
        1835, (1970).

30)   H.J.Goldberg, H.E.Jack, E.B.Dale, Phys.Rev.A 12, 908 (1975).

31)   L.C.Northcliffe, R.F.Schilling, Nucl.Data Tables A7, 233
        (1970).

32)   W.Eckstein, H.G.Schäffler, H.Verbeek Rad.Eff.18, 263 (1973).

33)   W. Eckstein, B.M.U.Scherzer, H.Verbeek, Rad.Eff.18, 135
        (1973).

34)   J.Roth, R.Behrisch, B.M.U.Scherzer, Appl.Phys.Lett.25, 643
        (1974).

35)   W.Eckstein, H.Verbeek, J.Vac.Sci.Techn.9 (1972)612

36)   T.M.Buck, G.H.Wheatley, L.C.Feldman, Surf.Sci.35, 345,
        (1973).

37)   P.Meischner, H.Verbeek, J.Nucl.Mat. 53, (1974),276.

38)   R.Behrisch, W.Eckstein, P.Meischner, B.M.U.Scherzer,
        H.Verbeek, Atomic Collisions in Solids, ed.S.Datz, C.D.Moak,
        B.R.Appleton, Plenum Press.N.Y.1974, p.315.

BACKSCATTERING ANALYSIS WITH MeV ^{4}He IONS

J.W. Mayer and M-A. Nicolet

California Institute of Technology
Pasadena, California 91125, USA

W.K. Chu
IBM, East Fishkill Facility
Route 52, Hopewell Junction, New York 12533, USA

INTRODUCTION

In this review we apply the formulae developed in reviews of backscattering spectrometry (Chu, et al, 1973; Ziegler, 1975; Mayer, et al, 1977) to examples which illustrate the analytical approach and the magnitude of such quantities as spectrum heights, typical energy losses and measurable amounts of impurities in the samples. This review was taken from Chapter 5 of Backscattering Spectrometry (Chu, et al, 1977). The examples were chosen to illustrate analytical methods rather than to describe applications; therefore, many of them are academic rather than practical. Often two or more different approximations are applied to the same example. Comparison of the results will help the reader decide whether to make a zero-th order approximation to get a quick answer, or make a detailed second-order analysis.

For most of the examples we used a 2 MeV ^{4}He ion beam with normal incidence on the target and with scattering geometry at 170°. The scattered particles are analyzed with a solid-state detector located about 10 cm from the target and with a solid angle, Ω, between 3 and 4 msr. The detector resolution is between 15 and

20 keV and the multi channel analyzer is set up with a channel
width E between 3 and 5 keV. The spectra were obtained under
routine experimental conditions; no special effort was made to
optimize those conditions.

In these examples, we use stopping cross section values
tabulated by Ziegler and Chu (1974). Although these values may
be found to be in error, as more refined measurements are made,
they serve as a basis for demonstrating different approaches to
numerical calculations.

SURFACE IMPURITY ON AN ELEMENTAL BULK TARGET

System Calibration

Backscattering can be used to detect surface impurities
on a light-element substrate. For example, a carbon substrate
is often used as a control sample to check the quality of a
vacuum-deposited layer. Any surface impurity with an atomic mass
greater than that of carbon will be visible in a backscattering
spectrum. These samples can also be used to determine the channel
width, E, of the multichannel analyzer.

Figure 1 shows a spectrum for 2 MeV ^{4}He ions backscattered

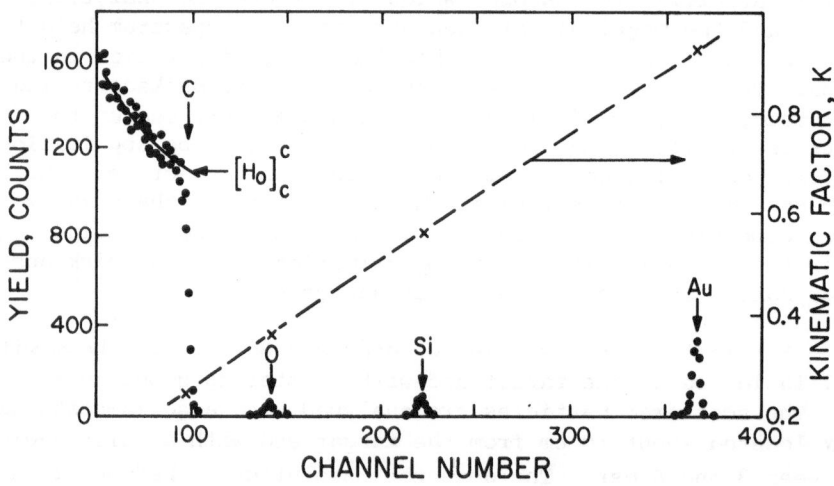

Fig. 1 Backscattering spectrum for 2.0 MeV ^{4}He ions incident on a
carbon substrate containing a surface layer of O, Si and Au.
Scattering geometry for normal incidence with $\theta=170°$,
$\Omega = 4.11$ msr and beam current of 15 nA for $Q = 10$ μC.

at an angle θ of 170° from a carbon target on which oxygen, silicon, and gold are present as surface impurities. There is a thick
target signal from the carbon substrate with its leading edge at
channel number 98 (located at half-height) and three peaks for the
impurities at channel numbers 141, 222 and 366 (located at the
midpoints of the full width at half the peak heights). The scale
at the right side gives kinematic values with K = 0.2526, 0.3625,
0.5655 and 0.9225 for scattering from targets with masses of 12,
16, 28 and 197, the masses of C, O, Si and Au, respectively. The
slope of the dashed line gives a value of 5 keV for the channel
width and the intercept is at 18 keV. Hence the detected energy
E_1 is obtained from 5 keV times the channel number plus 18 keV.

It is necessary to ensure that the contaminants are on
the surface rather than buried under it. This can be done by
tilting the target; signals from species on the surface will not
be affected, but signals from species below the surface will shift
to lower energies.

Number of Impurities/cm^2

The number, Nt, of impurities/cm^2 can be calculated directly from the area of the signals, A, given by the total number
of counts integrated over the region of interest. For a given
impurity denoted by the subscript i, the area A_i can be expressed by

$$A_i = \sigma_i \Omega Q (Nt)_i \tag{1}$$

where σ_i is the average differential scattering cross section.
With

$$\sigma = \frac{1}{\Omega} \int (\frac{d\sigma}{d\Omega}) \cdot d\Omega \tag{2}$$

and

$$\frac{d\sigma}{d\Omega} = (\frac{Z_1 Z_2 e^2}{4E})^2 \frac{4}{\sin^4 \theta} \frac{\{\sqrt{1 - (\frac{M_1}{M_2}\sin\theta)^2} + \cos\theta\}^2}{\sqrt{1 - (\frac{M_1}{M_2}\sin\theta)^2}} \tag{3}$$

with the subscript 1 denoting the incident particle, the subscript
2, the target atom and θ the scattering angle. The value of σ_i
can be calculated from Eq. 3 for an impurity with atomic number
Z_2. The scattering cross sections given in Table 1 were obtained
using the conversion factors $e^2 = 1.44 \times 10^{13}$ MeV cm and csc^4
$(\theta/2) = 1.0154$ for $\theta = 170^\circ$. The solid angle Ω is determined
from the experimental setup, and the amount of charge Q collected

during the measurement is read from a current integrator. For
the spectrum in Fig. 1, a solid-state detector was used with an
active area of 49 mm^2, at a distance of 109.2 mm from the target;
the solid angle is 4.11 msr. The total charge collected was 10
μC, Q = 6.25 x 10^{13} ions. The total number of impurity atoms per
unit area $(Nt)_i$, for a given species on the surface, can be calculated by use of Eq. 1; values are given in Table 1.

Table 1 Analysis of Fig. 1 for experimental conditions of
E_0 = 2.0 MeV, θ = 170°, Ω = 4.11 msr, E = 5.0 keV, and
$[\mathcal{E}]_C$ = 42.4 x 10^{-15} eV-cm^2

$$K_i = K_{M_2} = \{\frac{\sqrt{M_2^2 - M_1^2 \sin^2\theta} + M_1 \cos\theta}{M_2 + M_1}\}^2$$

Element	Mass (amu)	K_i	$K_i E_0$ (keV)	σ_i (x 10^{-24} cm^2)	A_i (counts)	$(Nt)_i - (10^{15}$ at/cm$^2)$	
						Eq. 1	Eq. 4
C	12	0.2526	505	0.037	1100*		
O	16	0.3625	725	0.074	390	20.5	20.9
Si	28	0.5655	1131	0.248	460	7.2	7.4
Au	197	0.9225	1845	8.150	1400	0.67	0.68

*H_C, the surface height of the carbon signal, is also written as
$[H_0]_C^C$.

The value of Ω was determined from the solid angle subtended by the area of the detector under the assumption that there
was no "dead" spots on the detector surface. Under prolonged exposure to energetic particles, the detector can degrade and de-
velope dead regions. The charge collection Q, is based on suppression of secondary electrons. Total suppression is sometimes
difficult to achieve. One can eliminate ΩQ from Eq. 1, and express the amount of impurity from the height H_C of the carbon
spectrum and the stopping cross section factor $[\varepsilon]_C$ (see the following section)

$$(Nt)_i = \frac{A_i}{H_C} \frac{\sigma_C}{\sigma_i} \frac{E}{[\varepsilon]_C} \qquad (4)$$

where the subsript, C, refers to the carbon substrate. Values of
the number of impurities/cm^2 are given in Table 1. The energy
loss factor $[\varepsilon]_C$ for 2.0 MeV He ions in a carbon substrate has a
value 42.4 x 10^{-15} eV-cm^2 using the surface energy approximation
in which the incident energy E_0 and the energy $K_C E_0$ are used for
the energies along inward and outward tracks. Values for the
stopping cross sections are given by Ziegler and Chu (1974). The

height, $[H_0]_C^C = H_C = 1100$ counts, of the carbon signal was found by drawing a line through the scatter of points for the signal from the carbon substrate and extrapolating the line to channel 98 $(E_1 = K_C E_0)$.

Figure 1 and Table 1 give us some estimate of the sensitivity of backscattering in the determiation of impurities (a monolayer is of the order of 10^{15} atoms/cm^2). In principle, in a low-noise system, that is one in which there is no background counts, one can detect infinitesimal amounts of impurities (Nt), on the surface simply by increasing Q without limit. In reality, however, the larger the Q the larger the background noise. Any value quoted for sensitivity will have to depend on the experimental conditions and the criteria used to define the sensitivity. In a routine operation such as the present one, the sensitivity of backscattering for 2 MeV ^{4}He ions can be estimated on a purely empirical basis by

$$(Nt)_i = [\frac{Z(substrate)}{Z(impurity)}]^2 \times 10^{14} \text{ impurity atoms/cm}^2 \qquad (5)$$

This equation gives an estimate of the minimum amount of surface impurities on a lighter substrate, $Z(impurity) > Z(substrate)$, that can be detected by 2 MeV ^{4}He backscattering. For gold as an impurity on carbon, Eq. 5 gives the detectable minimum as 10^{12} atoms/cm^2 or $\approx 10^{-3}$ monolayers.

A trace amount of impurity on a substrate that has a mass number larger than that of the impurity cannot be detected because the signal from the impurity is buried under that from the thick target yield. By using channeling to depress the thick target yield, the ratio of the impurity signal to the background can be improved.

ELEMENTAL SAMPLES CONTAINING UNIFORM CONCENTRATIONS OF IMPURITIES

Low Impurity Concentrations

The amount of surface impurity was expressed as $(Nt)_i$, the number of impurity atoms per unit area. In this section, dealing with uniform impurity in a bulk sample, we will use N_i to denote the number of impurity atoms per unit volume. Further, we will use the surface energy approximation since we assume uniform concentrations.

For silicon uniformly doped with arsenic (Fig. 2), the height of the arsenic signal is proportional to the density of the arsenic in the silicon, $(H_{As} \propto N_{As})$. The signal height for

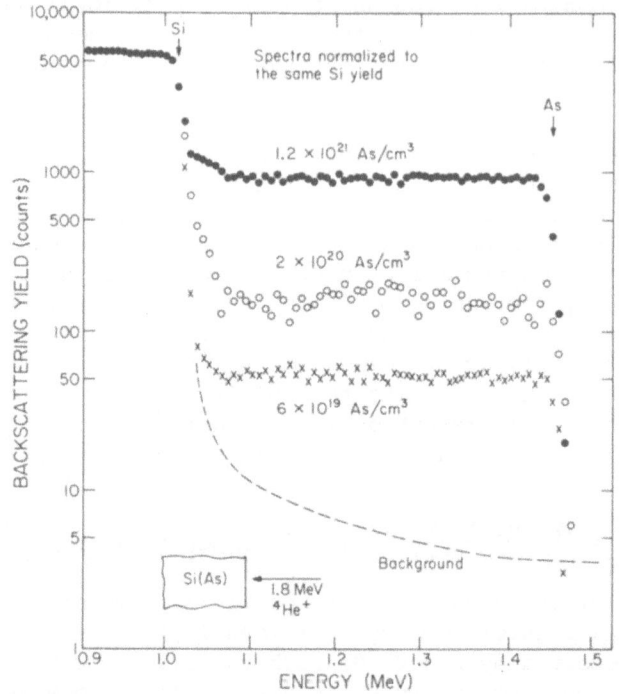

Fig. 2   Composite backscattering spectra for 1.8 MeV [4]He ions
         incident on Si samples containing different concentrations
         of As atoms. The spectra were normalized to the Si signal
         height.

the arsenic atoms, H_{As}, can be written (Chu, et al, 1973)

$$H_{As} = \sigma_{As} \Omega Q \frac{N_{As}}{N_{Si}} \cdot \frac{E}{[\varepsilon]_{As}^{Si}} \qquad (6)$$

where $[\varepsilon]_{As}^{Si}$ is defined as the stopping cross section factor for
arsenic in a silicon matrix; in the surface energy approximation:

$$[\varepsilon]_{As}^{Si} = K_{As} \varepsilon(E_0) + \frac{1}{|\cos\theta|} \varepsilon(K_{As}E_0) \qquad (7)$$

where ε is the stopping cross section in the stopping medium, here
silicon. The small amount of arsenic does not influence ε, but
does enter $[\varepsilon]$ through the collision kinematics. The stopping
cross sections are taken at energies E_0 and $K_{As}E_0$. The

cross section factor is labeled with a subscript to denote the
scattering atom, and a superscript to denote the stopping medium,
for example, $[\varepsilon]_{As}^{Si}$. When the scattering atom and the stopping
medium are the same, we use only the subscript. Thus for an analysis of a thick silicon target the notation would be $[\varepsilon]_{Si}$, im-
plying that both the scattering atom and the stopping medium are
silicon.

        As in the case of the surface impurity, we can eliminate
the values of Ω and Q by taking the height ratio:

$$\frac{N_{As}}{N_{Si}} = \frac{H_{As}}{H_{Si}} \frac{\sigma_{Si}}{\sigma_{As}} \frac{[\varepsilon]_{As}^{Si}}{[\varepsilon]_{Si}^{Si}} \tag{8}$$

The $[\varepsilon]$ ratio can be calculated from values given in Zielger and
Chu (1974) for 2 MeV ^{4}He ions at $\theta = 170°$ to give:

$$\frac{[\varepsilon]_{As}^{Si}}{[\varepsilon]_{Si}^{Si}} = \frac{95.27 \times 10^{-15} \text{ eV-cm}^2}{92.59 \times 10^{-15} \text{ eV-cm}^2}$$

The $[\varepsilon]$ ratio is within 3% of unity in this case. In general, for
MeV He ions the ratios are within 10% of unity for a wide variety
of impurities and substrates.

        The ratio H_{As}/H_{Si} can be directly measured from Fig. 2,
and therefore the concentration ratio can be obtained. With
$N_{Si} = 4.98 \times 10^{22}$ atoms/cm^3, $\sigma_{As} = 1.425 \times 10^{-24}$ cm^2 and $\sigma_{Si} =$
0.248×10^{-24} cm^2, the value of the concentration given by crosses
in Fig. 2 was calculated to be 6×10^{19}/cm^3.

        The sensitivity of backscattering for measuring bulk sam-
ples depends on the problem and the experimental conditions. For
typical conditions we have found that the detectable height of an
impurity signal is one-thousandth of the height of the substrate
target signal; that is, $H_{impurity}/H_{substrate} = 10^{-3}$. If we assume
that the $[\varepsilon]$ ratio is unity, and that the σ ratio is approximately
equal to the Z^2 ratio, Eq. 8 then gives a sensitivity limit,

$$\frac{N_{impurity}}{N_{substrate}} \gtrsim \frac{Z^2_{substrate}}{Z^2_{impurity}} \times 10^{-3} \tag{9}$$

As an example, the amount of impurity that can be detected in
silicon is $\approx 10^{19}$ atoms/cm^3 if the impurity is arsenic, or 1.5 x
10^{18} atoms/cm^3 if it is gold. If the substrate is CdTe (Z = 50

as an average), the amount of gold impurity that can be detected
is 2 x 10^{19} atoms/cm^3.

High Impurity Concentration

If the amount of impurity is too high, the sample can no
longer be treated as a pure element and the value of the stopping
cross section changes. As a rough estimate, impurity concentrations above one atomic percent will make a detectable change in
the stopping cross section.

For an Al sample alloyed with Cu, we denote the mixture
$Al_{1-x}Cu_x$ where the atomic ratio of Cu to Al is given by x/1-x.
For simplicity, we abbreviate the nomenclature to AlCu. The stopping cross section is given by

$$\varepsilon^{AlCu} = (1 - x)\varepsilon^{Al} + x\varepsilon^{Cu} \tag{10}$$

where we assume Bragg's rule of linear additivity. The ratio of
Cu to Al signals (as shown in Fig. 3) is given by:

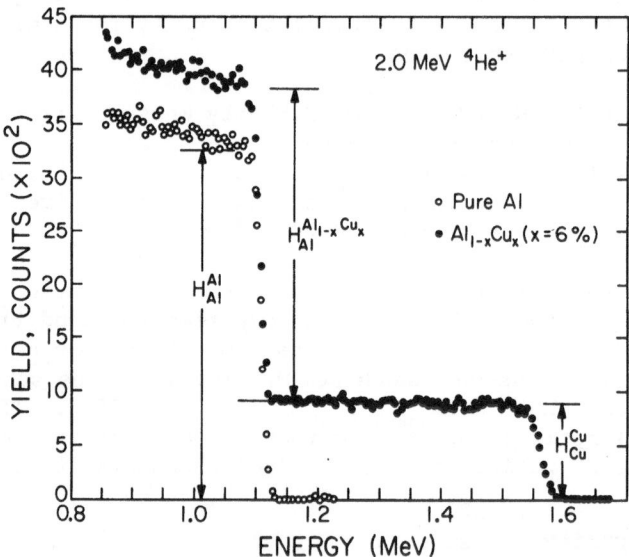

Fig. 3   Backscattering spectra for 2.0 MeV ^{4}He ions incident on an
         Al sample and an Al-6% Cu sample. Scattering geometry for
         normal incidence with θ = 170°, Ω = 4.11 msr, δE = 5 KeV
         and Q = 10 μC.

$$\frac{H_{Cu}^{AlCu}}{H_{Al}^{AlCu}} = \frac{x}{1-x} \frac{\sigma_{Cu}}{\sigma_{Al}} \cdot \frac{[\varepsilon]_{Al}^{AlCu}}{[\varepsilon]_{Cu}^{AlCu}} \qquad (11)$$

As in Eq. 7, the superscripts denote the stopping medium, AlCu, and the subscripts denote the collision partner Al or Cu, and hence the choice of the kinematic factor used in calculating $[\varepsilon]$. Similar to the treatment in the previous section, the $[\varepsilon]$ ratio for particles scattered from different elements in the same medium is close to unity and is not changed appreciably by changes in the atomic concentration ratios. Consequently, one determines zero-order values of x and 1-x by first setting the $[\varepsilon]$ ratio equal to unity. An improvement in the calculation can be made by using the zero-order values of x and 1-x to calculate $[\varepsilon]$ values which can then be used in Eq. 11 to give new values of x and 1-x. Generally, this first iteration does not change the values of x and 1-x by more than a few percent.

One can not treat the $[\varepsilon]$ ratio as unity when calculating ratios of signal heights for the same element in different matrices. For example, the ratio of the Al signal heights in AlCu to Al is given by

$$\frac{H_{Al}^{AlCu}}{H_{Al}^{Al}} = (1-x) \frac{[\varepsilon]_{Al}^{Al}}{[\varepsilon]_{Al}^{AlCu}} \qquad (12)$$

For the sample given by the spectra in Fig. 3, the composition corresponds to values of x = 0.06 (Howard, et al. 1976). However, the measured height ratio of H^{AlCu}/H_{Al} equals 0.90 which would imply a value of x = 0.10 if the $[\varepsilon]$ ratio were unity. That is, one could make an error of nearly a factor of two in assigning composition values if the change in stopping cross section factors was neglected. This procedure was also found to be necessary when evaluating the composition of GaAlAs when comparing the ratio H^{GaAlAs}/H^{GaAs} (Mayer, et al, 1973).

COMPOSITION OF HOMOGENEOUS SAMPLES
CONTAINING MORE THAN ONE ELEMENT

Two Elements

Backscattering can be used to analyze a bulk compound or a mixture. The method is straightforward, except that for some compounds containing both heavy and light elements the accuracy of the analysis is reduced because the signals from the light elements

are always superimposed on the signals from the heavy elements.
As in impurity analysis, the concentration is determined from
the signal height.

We will treat samples of known composition, Si and SiO$_2$, to
illustrate the method of calculating spectrum heights in the surface energy approximation. The silicon target has been damaged by
ion implantation to prevent channeling from affecting the yield.
The SiO$_2$ target is made of fused quartz, with a very thin metal
layer on the surface for charge integration. The silicon target
is not necessary to determine the composition of SiO$_2$; however, we
include the backscattering analysis of silicon in this example for
the sake of comparison. These samples are analyzed with an incident
beam of ^{4}He$^+$ ions at 2.0 MeV, with a beam current of 15 nA and a

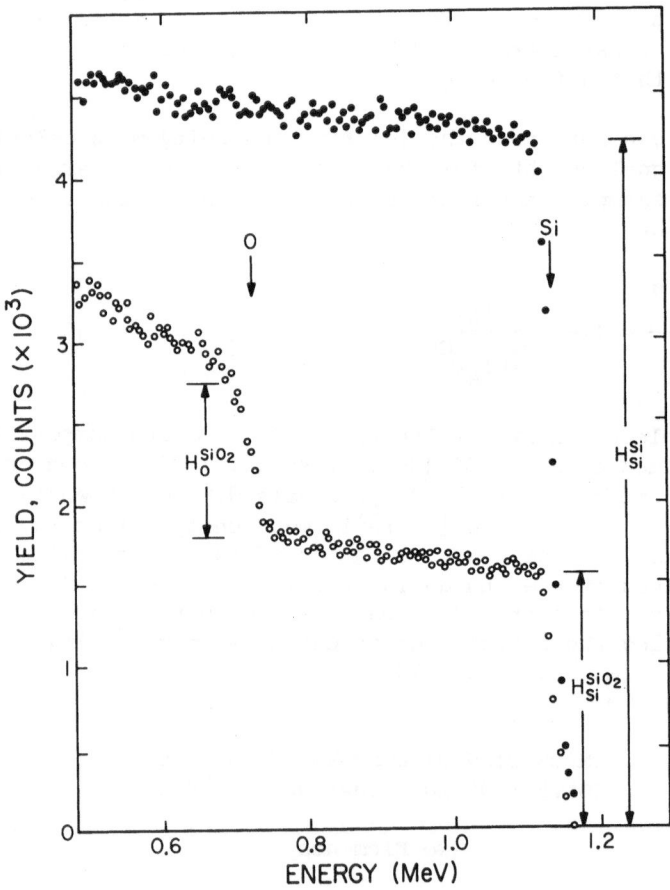

Fig. 4 Backscattering spectra for 2.0 MeV ^{4}He ions incident on a
        Si sample with an amorphous top layer and a sample of
        SiO$_2$. Normal incidence with $\theta = 170°$, $\Omega = 4.11$ msr,
        $Q = 10$ μC and $E = 5.4$ keV.

total dose of 10 μC, as measured by a Faraday cup with a current integrator.

The backscattering spectra thus obtained are plotted together in Fig. 4. Their heights are designated as

$$H_{Si}^{SiO_2}, \qquad H_O^{SiO_2}, \qquad \text{and} \qquad H_{Si}^{Si}$$

with values determined from the spectra to be 1500, 980 and 4200 counts respectively.

There are two equivalent methods of treating the stopping cross section in the medium using Bragg's rule of linear additivity: on a molecular basis

$$\epsilon^{SiO_2} = \epsilon^{Si} + 2\epsilon^O \tag{13}$$

where one considers the stopping cross section per molecule of SiO$_2$ or on an atomic basis

$$\epsilon^{Si_xO_{1-x}} = 0.33 \, \epsilon^{Si} + 0.66 \, \epsilon^O \tag{14}$$

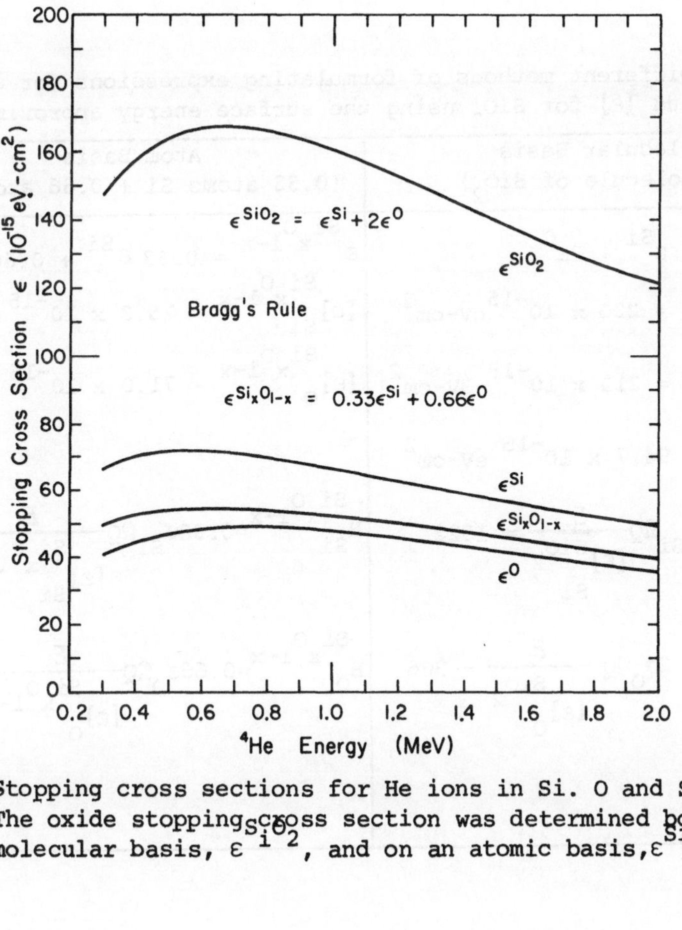

Fig. 5 Stopping cross sections for He ions in Si. O and SiO$_2$. The oxide stopping cross section was determined both on a molecular basis, ϵ^{SiO_2}, and on an atomic basis, $\epsilon^{Si_xO_{1-x}}$.

where x = 0.33 and one considers the effective cross section per atom within the molecule. Figure 5 shows the stopping cross section for the two methods. The spectrum heights for the oxygen component, for example, are given for the compound by

$$H_O^{SiO_2} = \frac{N_O^{SiO_2}}{N^{SiO_2}} \sigma_O \Omega Q \frac{E}{[\epsilon]_O^{SiO_2}} \tag{15}$$

where $(N_O^{SiO_2}/N^{SiO_2}) = 2$ since there are two oxygen atoms per SiO_2 molecule; and for the mixture by

$$H_O^{Si_xO_{1-x}} = \frac{N_O^{Si_xO_{1-x}}}{N^{Si_xO_{1-x}}} \sigma_O \Omega Q \frac{E}{[\epsilon]_O^{Si_xO_{1-x}}} \tag{16}$$

where $(N_O^{Si_xO_{1-x}}/N^{Si_xO_{1-x}}) = 0.66$.

Table 2   Different methods of formulating expressions for ϵ, H and [ϵ] for SiO_2 using the surface energy approximation.

Molecular Basis (one molecule of SiO_2)	Atom Basis (0.33 atoms Si + 0.66 atoms O)
1. $\epsilon^{SiO_2} = \epsilon^{Si} + 2\epsilon^C$	$\epsilon^{Si_xO_{1-x}} = 0.33\,\epsilon^{Si} + 0.66\,\epsilon^O$
2. $[\epsilon]_{Si}^{SiO_2} = 226 \times 10^{-15}\,eV\text{-}cm^2$	$[\epsilon]_{Si}^{Si_xO_{1-x}} = 75.3 \times 10^{-15}\,eV\text{-}cm^2$
$[\epsilon]_O^{SiO_2} = 213 \times 10^{-15}\,eV\text{-}cm^2$	$[\epsilon]_O^{Si_xO_{1-x}} = 71.0 \times 10^{-15}\,eV\text{-}cm^2$
$[\epsilon]_{Si}^{Si} = 94.7 \times 10^{-15}\,eV\text{-}cm^2$	
3. $H_{Si}^{SiO_2} = \sigma_{Si}\Omega Q \dfrac{E}{[\epsilon]_{Si}^{SiO_2}} = 1522$	$H_{Si}^{Si_xO_{1-x}} = 0.33\sigma_{Si}\Omega Q \dfrac{E}{[\epsilon]_{Si}^{Si_xO_{1-x}}} = 1522$
$H_O^{SiO_2} = 2\sigma_O\Omega Q \dfrac{E}{[\epsilon]_O^{SiO_2}} = 966$	$H_O^{Si_xO_{1-x}} = 0.66\sigma_O\Omega Q \dfrac{E}{[\epsilon]_O^{Si_xO_{1-x}}} = 966$

The different formulations are shown in Table 2. The
values in the table were calculated for 2.0 MeV ^{4}He ions, surface
energy approximation $\theta = 170^\circ$, $Q = 6.25 \times 10^{13}$ particles (10 μC)
$\Omega = 4.11$ msr, and $\bar{E} = 5.4$ keV. The cross section values are
$\sigma_{Si} = 0.248 \times 10^{-24}$ cm^2 and $\sigma_O = 0.742 \times 10^{-25}$ cm^2. The tabulated
values show that consistent spectrum heights can be obtained if
consistent values for $[\varepsilon]$ and N^{AB} are chosen. It also indicates
that the $[\varepsilon]$ ratio is close to unity for scattering from different elements in the same matrix.

Multi-Elemental Samples

We demonstrate the backscattering analysis of a bulk
sample from measurements of spectrum height, shown in Fig. 6 for
2.4 MeV ^{4}He ions incident in a nonchanneled direction from an

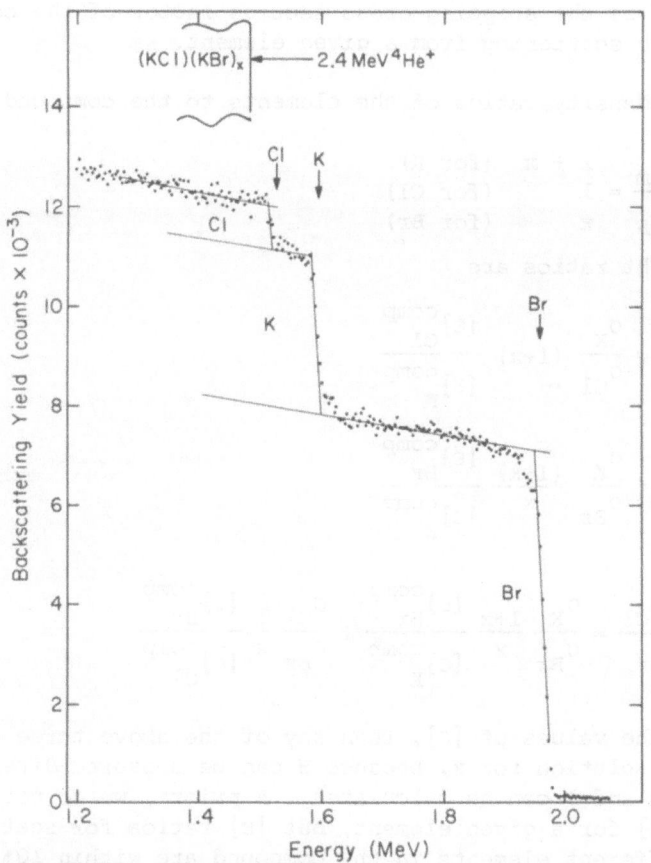

Fig. 6 Backscattering spectrum for 2.4 MeV ^{4}He ions incident on
an alkali-halide crystal of $(KCl)_1(KBr)_x$: normal incidence with $\theta = 170^\circ$.

alkali halide crystal made of $(KCl)_1(KBr)_x$. We solve for the unknown x by measuring the surface heights of the backscattering
signals due to chlorine, potassium, and bromine and assume that
there is a K atom associated with each Cl or Br atom. In Fig. 6
lines are drawn over the points on the spectrum to form a ladder.
The vertical positions of the ladder, that is, the half height at
the leading edges, are at 1.530, 1.596 and 1.967 MeV, corresponding to the energies of particles scattered from Cl, K and Br in
the surface layer of the compound.

The height of the signal due to scattering from a given
element is

$$H_{elem} = \sigma_{elem} \, \Omega Q \, \frac{N_{elem}}{N^{comp}} \cdot \frac{E}{[\varepsilon]^{comp}_{elem}} \tag{17}$$

Here $[\varepsilon]^{comp}_{elem}$ is the stopping cross section factor of the compound
material with scattering from a given element.

The density ratios of the elements to the compound are

$$\frac{N_{elem}}{N_{comp}} = 1 \quad \begin{array}{ll} 1+x & \text{(for K)} \\ 1 & \text{(for Cl)} \\ x & \text{(for Br)} \end{array} \tag{18}$$

and the height ratios are

$$\frac{H_K}{H_{Cl}} = \frac{\sigma_K}{\sigma_{Cl}} \, (1+x) \, \frac{[\varepsilon]^{comp}_{Cl}}{[\varepsilon]^{comp}_K} \tag{19}$$

$$\frac{H_K}{H_{Br}} = \frac{\sigma_K}{\sigma_{Br}} \, \frac{(1+x)}{x} \, \frac{[\varepsilon]^{comp}_{Br}}{[\varepsilon]^{comp}_K} \tag{20}$$

$$\frac{H_K + H_{Cl}}{H_{Br}} = \frac{\sigma_K}{\sigma_{Br}} \, \frac{1+x}{x} \, \frac{[\varepsilon]^{comp}_{Br}}{[\varepsilon]^{comp}_K} + \frac{\sigma_{Cl}}{\sigma_{Br}} \, \frac{1}{x} \, \frac{[\varepsilon]^{comp}_{Br}}{[\varepsilon]^{comp}_{Cl}} \tag{21}$$

If we know the values of $[\varepsilon]$, then any of the above three equations
will give a solution for x, because H can be measured directly from
the spectrum and σ can be calculated. A priori, we do not know the
values of $[\varepsilon]$ for a given element, but $[\varepsilon]$ ratios for scattering
from the different elements in the compound are within 10% of unity,
regardless of composition.

Approximating the $[\varepsilon]$ ratios by unity, and using calculated

values for σ and measured values of the spectrum height, we solve
for x and obtain three different values of x: x = 1.66, 1.54 and
1.57 with an average value of 1.59. These three values are zero-th
order approximations, because the $[\varepsilon]$ ratio terms in Eqs. 19 to 21
have been ignored. Now, for a zero-th order value of x, we can
calculated values for $[\varepsilon]$ and make a first-order calculation for x,
using the ratio of these $[\varepsilon]$ values. In calculating ε, we assume
Bragg's rule:

$$\varepsilon^{comp} = \varepsilon^{KCl_1(KBr)_{1.59}} = 2.59\varepsilon^K + \varepsilon^{Cl} + 1.59\varepsilon^{Br} \qquad (22)$$

The values of $[\varepsilon]^{comp}$ then follow by using Eq. 22 with the elemental ε values and the $[\varepsilon]$ ratios become

$$\frac{[\varepsilon]^{comp}_{Cl}}{[\varepsilon]^{comp}_{K}} = 0.999, \qquad \frac{[\varepsilon]^{comp}_{Br}}{[\varepsilon]^{comp}_{K}} = 1.010, \qquad \frac{[\varepsilon]^{comp}_{Br}}{[\varepsilon]^{comp}_{Cl}} = 1.011 \qquad (23)$$

We note that $[\varepsilon]$ ratio is not a sensitive function of x. For
example, when x changes from 0.1 to 10.0, the terms $[\varepsilon]^{comp}_{Cl}/[\varepsilon]^{comp}_{K}$
and $[\varepsilon]^{comp}_{Br}/[\varepsilon]^{comp}_{Br}$ change by 1% or less.

Substituting values of $[\varepsilon]$ and σ ratios into Eqs. 19 to 21
and solving for x, we obtain x = 1.66, 1.58 and 1.60. The average
values, $\bar{x}$ = 1.61, differs from the zero-th order value of 1.59 by
1.3%. This indicates that the zero-th order analysis is adequate
based on the assumption that the $[\varepsilon]$ ratio is unity.

A last example for bulk analysis by backscattering is
shown in Fig. 7. Here the sample is magnetic bubble material
grown on gadolinium gallium garnet (GGG). The bubble material is
a film 10 μm thick, which is thicker than the range of the 2.0
MeV ^{4}He beam, and thus in effect acts like a bulk material. Because the bubble material is an insulator, a thin film of Al was
deposited on it before the analysis to provide a return path for
beam current measurements.

We treat this example on a molecular basis, and, as in Eq.
15, we have

$$H^{comp}_A = \frac{N_A}{N} \sigma_A Q\Omega \frac{E}{[\varepsilon]^{comp}_A} \qquad (24)$$

Here the subscript A indicates one of the elements in the garnet,
N_A/N is the number of A atoms in a molecular unit and $[\varepsilon]^{comp}_A$ is
the stopping cross section on a molecular basis.

On the assumption that the $[\varepsilon]$ ratios are unity, the

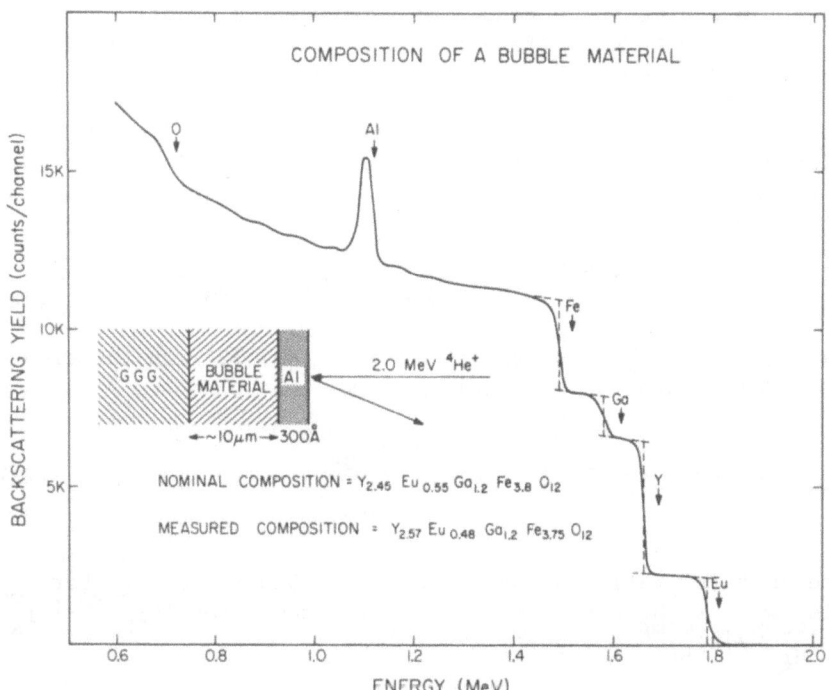

Fig. 7 Backscattering spectrum of 2.0 MeV ^{4}He ions incident on
a thick target consisting of a magnetic bubble material
with a thin surface layer of Al. The bubble material
was known to have the garnet composition X_8O_{12} with the
nominal ratios of the various materials as given in the
figure.

spectrum height ratios are:

$$\frac{H_A}{H_{A'}} = \frac{N_A}{N_{A'}} \frac{\sigma_A}{\sigma_{A'}}$$
(25)

where A and A' are any two of the four elements, iron, gadolinium
yttrium and europium contained in the garnet. For elements with
high atomic number, Z, one can simplify Eq. 25 to

$$\frac{H_A}{H_{A'}} = \frac{N_A}{N_{A'}} \frac{Z_A^2}{Z_{A'}^2}$$
(26)

The values of $N_A/N_{A'}$ can be obtained from the signal heights in
Fig. 7 which are 2820, 1280, 4340 and 2120 counts for Ge, Gd, Y

and Eu, respectively.

The garnet molecule is known to be X_8O_{12}. This gives an additional condition:

$$N_{Fe}/N + N_{Gd}/N + N_Y/N + N_{Eu}/N = 8 \qquad (27)$$

Substituting the values of N_A/N_{Fe} into Eq. 27, we have:

$$N_{Fe}/N(1 + 0.320 + 0.685 + 0.128) = 8 \qquad (28)$$

which gives a value $N_{Fe}/N = 3.75$. This value gives $N_{Gd}/N = 1.2$, $N_Y/N = 2.57$ and $N_{Eu}/N = 0.48$ in good agreement with the nominal compositions quoted by the garnet supplier.

IMPURITIES DISTRIBUTED IN DEPTH FOR AN ELEMENTAL SAMPLE

Ion Implantation Samples

The first major application of backscattering to semiconductor problems was in the investigation of ion implantation processes. Ion implantation has advanced rapidly over the past years, and implantation methods are firmly established in semiconductor technology. Backscattering and channeling offer independent methods of measuring the implantation dose, the range profile, and the lattice location of the impurities, and of studying damage; therefore, backscattering has become a major method of characterizing the implantation process. We will illustrate the method with a very simple example.

Figure 8 shows an energy spectrum (Sigmon, et al, 1975) of 2.0 MeV ^{4}He ions backscattered from a silicon target implanted with ^{75}As at 250 keV to a dose of 1.2×10^{15} As/cm^2. The silicon signal gives a step with leading edge at 1.13 MeV and the arsenic signal (plotted on an amplified scale) has a Gaussian distribution with the peak at 1.55 MeV and FWHM of 60 keV. The data from Fig. 8 is given in Table 3.

From Fig. 8, we can start a zero-th order analysis, that is, a surface energy approximation, and calculate the dose range and range distribution of arsenic in silicon. We assume that implantation to be so shallow that E_0 can be used to evaluate the stopping cross section and the differential scattering cross section. The implantation dose can be calculated from Eq. 4 with arsenic implants treated as a surface impurity. The dose of arsenic is

$$(Nt)_{As} = \frac{A_{As}}{H_{Si}} \frac{\sigma_{Si}}{\sigma_{As}} \frac{E}{[\varepsilon]_{Si}} = 1.2 \times 10^{15} \text{ As/cm}^2 \qquad (29)$$

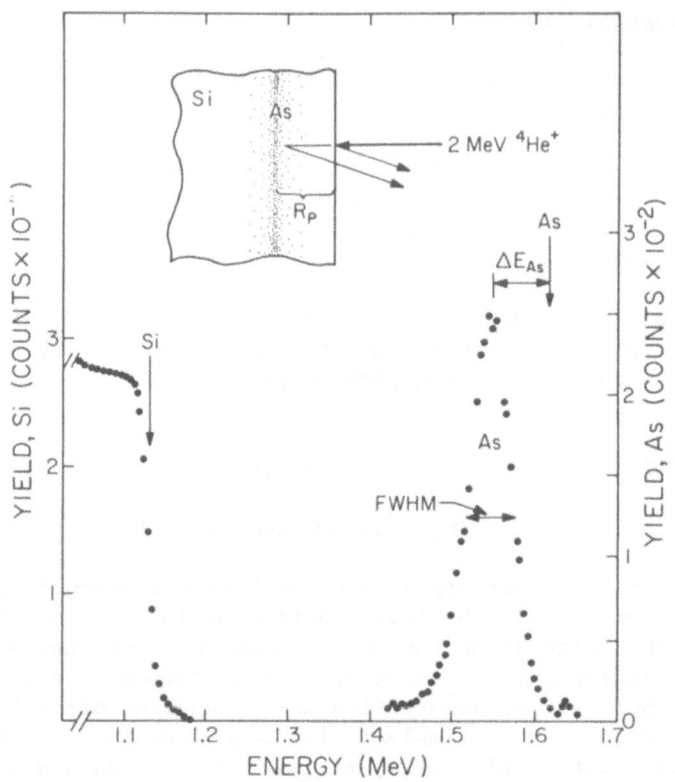

Fig. 8   Energy spectrum of 2 MeV ^{4}He ions backscattered from a
          silicon crystal implanted with 1.2 x 10^{15} As ions/cm^2 at
          200 keV. The vertical arrows indicate the energies of
          particles scattered from surface atoms of ^{28}Si and ^{75}As.

Table 3.   Data extracted from Fig. 8 with backscattering para-
            meters based on the surface energy approximation with
            E_0 = 2.0 MeV, θ = 170° and E = 5.0 keV.

Data	Parameters
H_{Si} = 27,000 counts	$[\varepsilon]_{Si}^{Si}$ = 92.6 x 10^{-15} eV-cm^2
H_{As}^{Si} = 250 counts (at peak)	$[\varepsilon]_{As}^{Si}$ = 95.3 x 10^{-15} eV-cm^2
A_{As} = 3350 counts	σ_{As}, σ_{Si} = 1.425, 0.2475 x 10^{-24} cm^2
ΔE_{As} = 68 keV, FWHM$_{As}$ =60 keV	K_{As}, K_{Si} = 0.809, 0.566

in agreement with the nominal value of the implanted dose.

The maximum concentration of arsenic in silicon can be
estimated from the peak height of the arsenic signal. Using the
formula derived for the bulk impurities, we have

$$\frac{N_{As}}{N_{Si}} = \frac{H_{As}}{H_{Si}} \frac{\sigma_{Si}}{\sigma_{As}} \frac{[\varepsilon]_{As}^{Si}}{[\varepsilon]_{Si}^{Si}} = 0.166 \text{ at } \% \tag{30}$$

or $N_{As} = 8.3 \times 10^{19}$ atoms/cm^3 using $N_{Si} = 4.98 \times 10^{22}$ atoms/cm^3.

To get a concentration profile, we need the stopping
cross section factor $[\varepsilon]_{As}^{Si}$, which gives an energy-to-depth conversion for scattering from arsenic in a silicon matrix. The
peak position of the arsenic is shifted by ΔE_{As} = 68 keV below
the surface position, and

$$N_{Si} R_p = \Delta E/[\varepsilon]_{As}^{Si} = 7.14 \times 10^{17} \text{ atoms/cm}^2 \tag{31}$$

where R_p is the projected range of the implanted arsenic. If the
arsenic profile is not symmetric, then R_p does not coincide with
the maximum concentration. A position that divides the arsenic
area into two equal areas will give a projected range. The value
of R = 1434Å assuming a bulk density for the implanted layer.
The depth scale in Å is more convenient than that in atoms/cm^2,
but the latter is an intrinsic unit in depth for backscattering.

When the implant distribution is Gaussian, the depth
profile can be described by a projected range and a range straggling ΔR_p, which is the standard deviation of the Gaussian distribution in depth. The standard deviation is related to the
FWHM of a Gaussian distribution by

FWHM = $2\sqrt{2\ln2}$ x (standard deviation)

= 2.355 x (standard deviation) (32)

The FWHM is the energy spectrum for arsenic is measured to be 60
keV. This FWHM contains not only the depth distribution of the
arsenic, but also the detector resolution of the backscattering
system and the energy straggling of the ^{4}He ions.

The detector resolution of the backscattering system can
be measured from the slope of the silicon step as in Fig. 8. If
we differentiate the step near the silicon surface, we get a
negative Gaussian (negative because the yield decreases when the
energy increases) the FWHM of this negative Gaussian is the system resolution of the backscattering system. It too can be

obtained easily, without differentiating the spectrum, by simply measuring the energy spread of the step from 12% to 88% of the step height. In Fig. 8, the energy spread of the silicon step from 12% to 88% of the height is 22 keV.

The energy straggling of ^{4}He ions in silicon has not been measured, but can be estimated from Bohr's theory to be about 3.2 keV in the implanted region. The FWHM of this energy straggling is then 2.355 x 3.2 = 7.5 keV.

From the measured FWHM of arsenic, it is necessary to de-convolute the measured FWHM detector resolution (22 keV) and the energy straggling (7.5 keV). Since all three distributions are assumed to be Gaussian, the deconvolution process is simply a subtraction in quadrature:

$$\text{FWHM (corrected)} = \sqrt{(60 \text{ keV})^2 - (22 \text{ keV})^2 - (7.5 \text{ keV})^2}$$

$$= 55.3 \text{ keV} \tag{33}$$

This value represents the real spread of arsenic in the energy scale. It can be readily converted into a depth scale by using Eq. 31 and 33:

$$\Delta R_p = \text{FWHM(corrected)}/2.355 \, N[\varepsilon]_{As}^{Si} = 500\text{Å} \tag{34}$$

Up to this point our analysis has been based on a surface energy approximation. This is, the stopping cross section $[\varepsilon]_{As}^{Si}$ is evaluated at a surface energy E_0 = 2.0 MeV using ε values shown in Fig. 9. If we use the mean energy approximation, then

$$[\varepsilon]_{As}^{Si} = k_{As}\varepsilon(\overline{E}_{in}) + \frac{1}{|\cos\theta|}\varepsilon(\overline{E}_{out}) \tag{35}$$

with
$$\overline{E}_{in} = \frac{1}{2}(E + E_0) \text{ and } \overline{E}_{out} = \frac{1}{2}(E_1 + KE) \tag{36}$$

where E is the energy before scattering. To obtain the average energies, $\overline{E}_{in}$ and $\overline{E}_{out}$, we use the symmetric mean energy approx-imation:

$$\overline{E}_{in} = E_0 - \frac{\Delta E}{4} = E_0 - \frac{1}{2} NR_p\varepsilon(E_0) = 1982 \text{ keV} \tag{37}$$

$$\overline{E}_{out} = E_1 + \frac{\Delta E}{4} = E_1 + \frac{1}{2} NR_p\varepsilon(E_1) = 1569 \text{ keV} \tag{38}$$

By substituting these energies into Eq. 35, we have $[\varepsilon]_{As}^{Si}$ = 97.1 x 10^{-15} eV/cm^2. The new stopping cross section factor is only 2% larger than that calculated from the surface approxi-mation. Therefore, the new range and range straggling will be 2% lower than the values obtained.

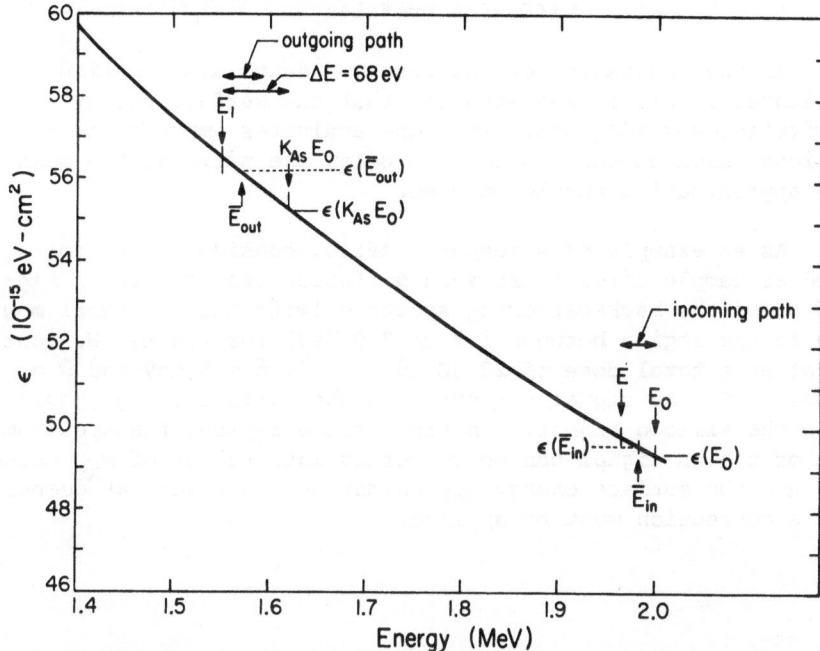

Fig. 9   Stopping cross section of [4]He ions in Si used to
evaluate the As depth profile given in Fig. 8.

The results of the two analyses, one performed by surface
energy approximation and the other by mean energy approximation,
are summarized in Table 4.   The differences are small.   We conclude
that for shallow depth analysis, where ΔE is small compared to E
and the change of ε over ΔE is not significant, the surface energy
approximation is adequate for the analysis of depth distributions.

Table 4.   Surface energy approximation used in the analyses of
Fig. 8 using ε values shown in Fig. 9.

Method	Incoming Energy (keV)	Outgoing Energy (keV)	$[\varepsilon]_{As}^{Si}$ $(10^{-15}$ eV/cm$^2)$	R_p (Å)	ΔR_p (Å)
Surface energy approximation	E_0 = 2000	$K_{As}E_0$ = 1617	95.3	1434	500
Mean energy approximation	$\overline{E}_{in}$ = 1982	$\overline{E}_{out}$ = 1569	97.10	1406	490

Diffusion Profiles

    In the discussion of the dose and depth distribution of
As implanted in Si, it was apparent that the surface energy
approximation was adequate. When one evaluates impurity distributions which extend 0.5 to 1 μm below the surface, the mean
energy approximation should be used.

    As an example of a deeper profile, consider an As im-
planted Si sample after a drive-in diffusion process step. Figure 10 shows the backscattering spectrum (with the As signal mag-
nified in the region between 1.5 to 2.0 MeV) for 2.4 MeV He ions
incident at a total dose of 20 μC ($\theta = 170^\circ$, $E = 5$ keV and $\Omega =$
4.11 msr). The As signal extends from the surface energy position
down to the silicon signal. In the surface region, the spectrum
height of the As signal can be converted into values of N_{As} using
Eq. 30 and the surface energy approximation. However, at deeper
depths a correction must be applied.

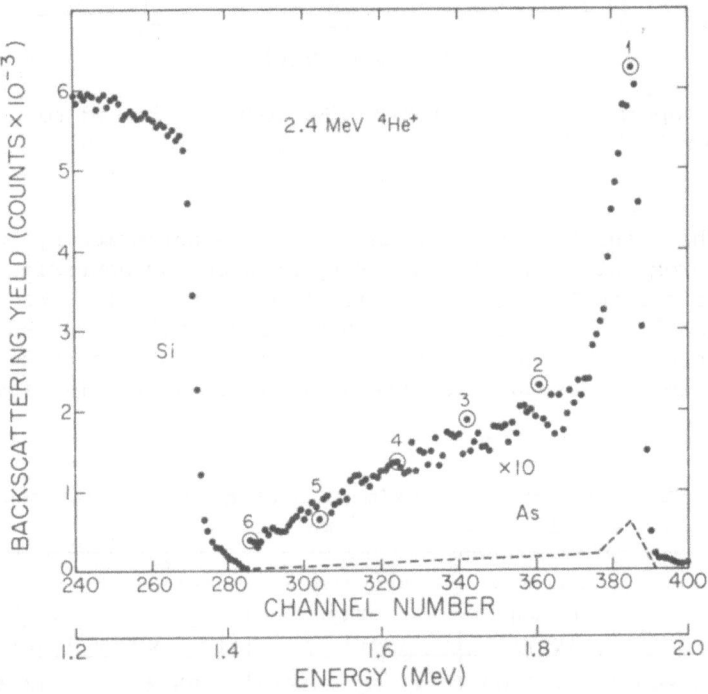

Fig. 10 Backscattering spectrum of 2.4 MeV ^{4}He ions incident
         on a silicon sample implanted with 3.4 x 10^{16} As/cm^2
         and the heat treated to produce a deep diffusion profile
         of As. Scattering geometry at normal incidence, $\theta =$
         170°, $\Omega = 4.11$ msr and $Q = 20$ μC.

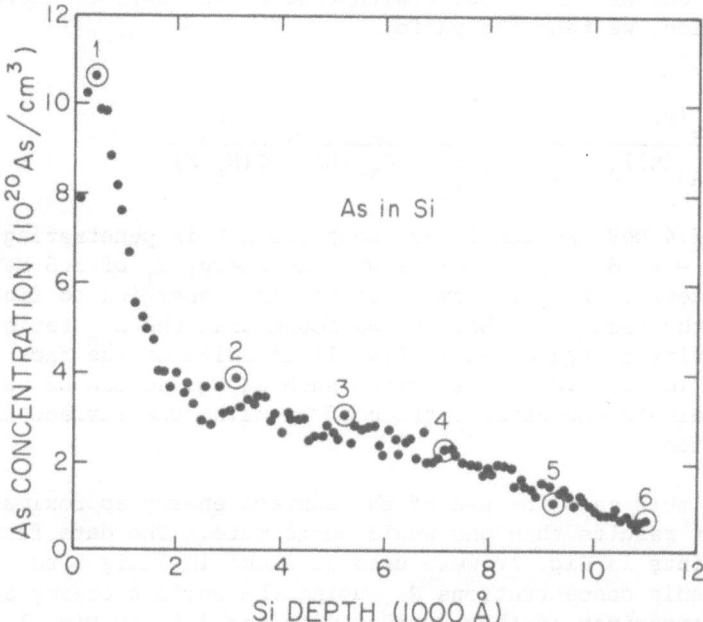

Fig. 11 Concentration of As vs depth obtained from the data
of Fig. 10 using the mean energy approximation to
convert from energy to depth. The encircled points
are concentrations based on the mean energy calculation. The calculation based on surface energy approximation for these encircled points of Fig. 10, are
given in Table 5.

Figure 11 shows $N_{As}(x)$ calculated using the mean energy
approximation (data points). In the latter case data conversion
from Figs. 10 and 11 was done numerically by a simple computer
program which calculates the energy loss and scattering cross
section at each energy E before scattering and the energy loss
after scattering. To evaluate the height $H_{As}(E_1)$ of the As signal
at a detected energy E_1, one should use Ziegler, et al (1975),
Mayer, et al (1977);

$$H_{As}(E_1) = \sigma_{As}(E)\Omega Q \, \frac{N_{As}(x)}{N_{Si}} \cdot \frac{E}{[\varepsilon(E)]_{As}^{Si}} \, \frac{\varepsilon(K_{As}E)}{\varepsilon(E_1)} \qquad (39)$$

where $[\varepsilon(E)]_{As}^{Si}$ is the stopping cross section factor evaluated at
the energy E before scattering and the ratio $\varepsilon(K_{As}E)/\varepsilon(E_1)$ corrects
for the change with depth of the thickness of a slab corresponding
to one channel.

To compare this result with that of the surface energy approximation, we take the ratio:

$$\frac{N_{As}(x)}{(N_{As}(x))_0} = \frac{[\varepsilon(E)]_{As}^{Si}}{[\varepsilon(E_0)]_{As}^{Si}} \cdot \frac{\sigma_{As}(E_0)}{\sigma_{As}(E)} \cdot \frac{\varepsilon(E_1)}{\varepsilon(K_{As}E)} \tag{40}$$

A 2.4 MeV ^{4}He ion loses about 220 keV in penetrating 1 μm into Si (E = 2.18 MeV) and exits with an energy E_1 of 1.5 MeV after scattering from As atoms. Using these energies to find the values of the factors in Eq. 40, we found that the N_{As} ratio is close to unity as indicated in Fig. 11 in spite of the fact that the σ_{As} ratio was 0.825. Relations such as Eq. 40 can be used to quickly evaluate the error introduced by using the surface energy approximation.

In this case the use of the surface energy approximation gave better results than one would anticipate. The data from the labeled points in Fig. 10 were used as shown in Table 5 to cal- culate arsenic concentrations N_{As} using the surface energy approxi- mation. Comparison of these values with the data in Fig. 11 in- dicates that the error in total depth incurred by using the surface energy approximation is about 4% at a depth of 1 μm. The overall correction in N_{As} is within 5% because the change in $\sigma(E)$ was compensated by the ε ratio. This result is not typical of all analyses and indicates the usefulness of Eq. 40 in evaluating the magnitude of corrections.

Table 5. Values of N_{As} calculated from data in Fig. 10 using the surface energy approximation.

Labelled Points	Channel No.	Difference[1] in Ch. No. ΔCh	ΔE[2] keV	Depth[3] (Å)	Height (counts)	N_{As}[4] (x 10^{20}/cm^3)
1	385	3	15	345	628	10.7
2	361	27	135	3103	232	3.97
3	342	46	230	5287	188	3.22
4	324	64	320	7356	136	2.33
5	304	84	420	9655	65	1.11
6	286	102	510	11724	40	0.68

1. Taking the position of As at the surface, $K_{As}E_0$, to be channel 388.
2. ΔCh times 5 keV, where the energy width E of on channel=5 keV.
3. Depth = ΔE/$[\varepsilon_0]$ where $[\varepsilon_0]$ = 43.5 eV/Å.
4. Using Eq. 6 with N_{As} = (H_{As}/5.846) x 10^{19}/cm^3.

THICKNESS OF THIN FILMS

Elemental Films

A very thin film is usually deposited on a thick substrate. If elements in the substrate are of lighter mass than
those in the film, the backscattering signal from the substrate
does not interfer with the signal from the film. When a film is
very thin, say from a fraction of a monolayer to a few hundred
angstroms, it can be treated as a surface contamination of the
substrate.

In this section, we treat thicker films (t > 100Å) and
consider different methods that can be used to find the number
of atoms/cm^2, Nt, from backscattering spectra. Figure 12 shows
seven backscattering spectra obtained by 2.0 MeV ^{4}He ion backscattering from seven targets of platinum evaporated onto a
silicon substrate. The thickness of the film, as determined by
backscattering, ranges from 125 to 4000Å. For simplicity, the
contribution of the silicon substrate to the spectra at low
energies is not plotted in this figure. The energy difference ΔE
between particles scattered from the surface of the platinum and
those scattered from the platinum-silicon interface is related to

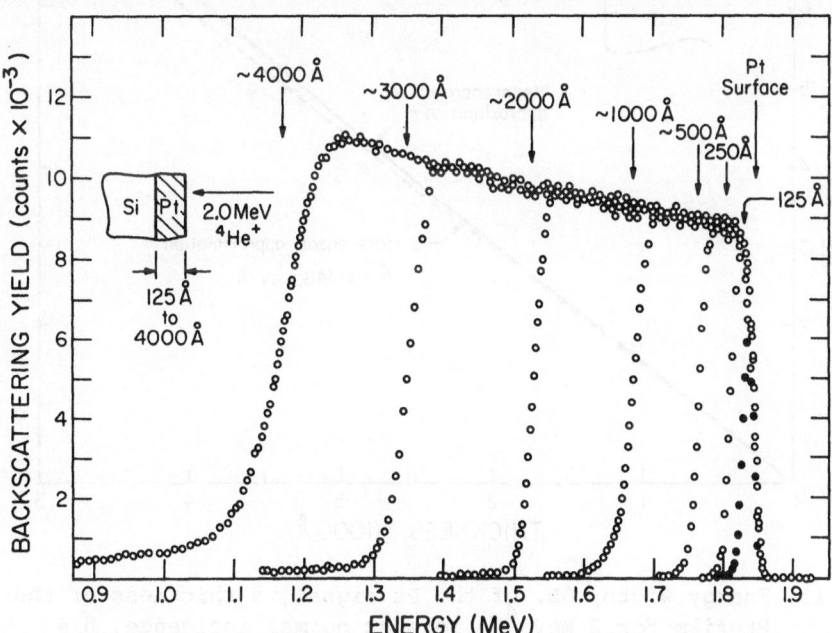

Fig. 12   Composite backscattering spectra for 2.0 MeV ^{4}He ions
          incident on seven targets of Pt deposited on Si sub-
          strates. The Si signal is now shown in the composite.

the thickness of the film by the energy loss factor; that is,

$$\Delta E = [\epsilon] Nt \simeq [\epsilon_0] Nt = (224.4 \times 10^{-15}) Nt \text{ eV} \tag{41}$$

where the stopping cross section factor $[\epsilon_0]$ was calculated using the surface energy approximation with $\theta = 170°$, $E_0 = 2.0$ MeV and $K_{Pt} = 0.9218$. If we assume that the density of the thin film is the same as that of bulk platinum, $N = 6.62 \times 10^{22}$ atoms/cm^3, the surface energy approximation provides a linear conversion between ΔE and t as shown by the straight line (dashed) with a slope of 148.5 eV/Å in Fig. 13. The solid curve represents the nonlinear relation between ΔE and t, as obtained by the mean energy approximation

$$[\bar{\epsilon}] = K_{Pt} \epsilon(\bar{E}_{in}) + \frac{1}{|\cos\theta|} \epsilon(\bar{E}_{out}) \tag{42}$$

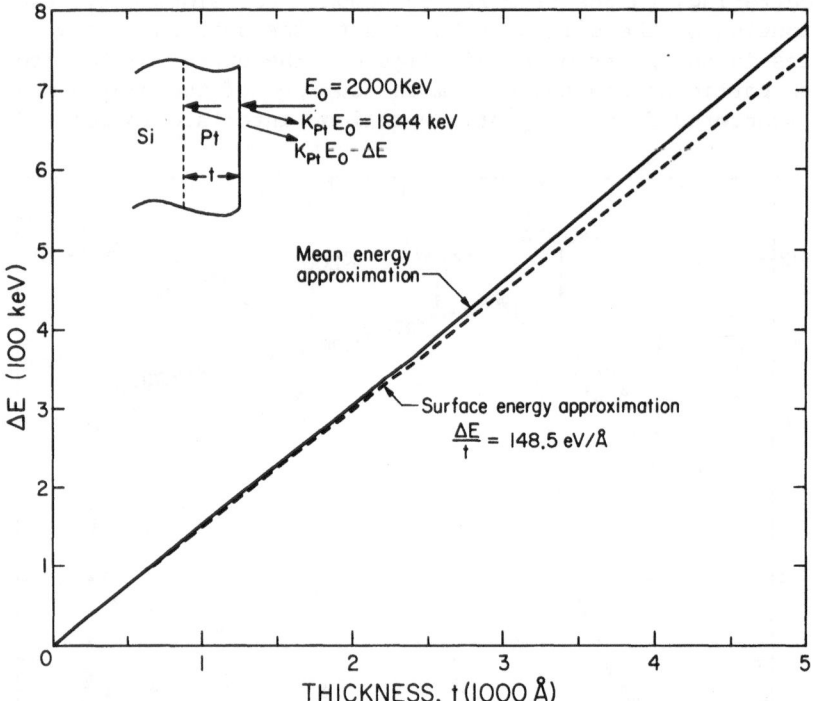

Fig. 13 Energy width, ΔE, of the Pt signal vs thickness of the
        Pt film for 2 MeV ^{4}He ions at normal incidence, $\theta = 170°$. The solid line is based on the mean energy
        approximation and the dashed line on the surface energy
        approximation assuming $N_{Pt} = 6.62 \times 10^{22}$ atoms/cm^3.

Here $\overline{E}_{in}$ and $\overline{E}_{out}$ depend on the thickness of the platinum layer. The value of $[\overline{\varepsilon}]$ depends on t. We should emphasize that $[\overline{\varepsilon}]$ is used only in evaluating thickness, never in evaluating spectrum height. For the spectrum height, the energy E at a given energy of the projectile immediately before scattering is needed and can be approximated from the symmetric mean energy approximation Eqs. 37 and 38.

Table 6 provides a comparison of thicknesses obtained from the energy width ΔE for the two different analytical approximations. Here E_1 is obtained by measuring the position of the trailing edge of each spectrum in Fig. 12. The trailing edge has a finite slope, which is due to the energy straggling of the helium ions, the detector resolution, and the nonuniformities in the film. The position of the trailing edge is defined as the half height of the step. At the lower end of the spectrum there is a finite background, which needs to be taken into account in determining the position of the half height. From the data shown in Table 6, we can conclude that the difference between the two calculations is about 1% for each 1000Å of film thickness.

Table 6. Comparison of surface energy approximation and mean energy for 2.0 MeV He ions scattered from Pt films.

E_1	(keV)	1170	1350	1526	1670	1765
ΔE	(keV)	674	494	318	174	79
Surface energy approximation	t (Å)	4540	3330	2140	1170	530
Mean energy approximation	t (Å)	4320	3200	2100	1150	530

The depth accessible with 2.0 MeV ^{4}ions depends on the energy loss of ^{4}He ions in the target. For example, ΔE = 500 keV will give Δt = 1/3 μm for platinum, but about 1 μm for silicon or aluminum. For a platinum film, the maximum thickness that can be analyzed with 2.0 MeV ^{4}He ions is about 0.5 μm.

Since protons lose much less energy in material than ^{4}He ions do, thicker films can be measured with proton backscattering. For the spectra in Fig. 14, ^{4}He ions and protons were backscattered on gold films deposited on a carbon substrate. The top part of Fig. 14 gives the backscattering spectrum for 1.4 MeV ^{4}He ions, and indicates that 0.6 μm is too thick for analysis by He ions. The lower part of Fig. 14, indicates that 1.4 MeV protons can easily measure films 3 μm thick. In this portion of the figure,

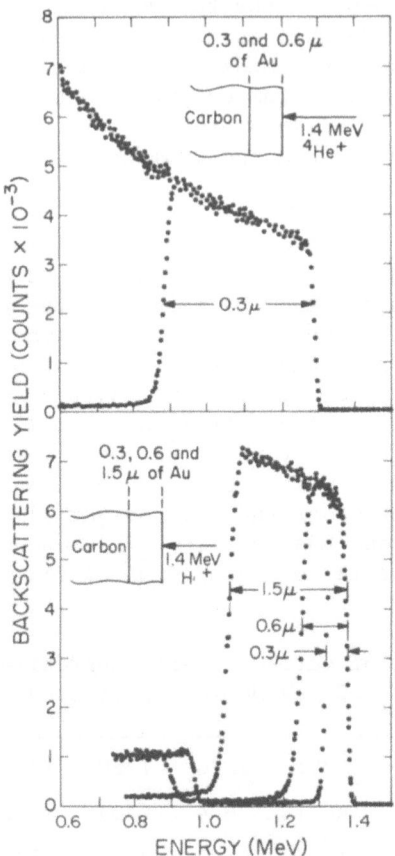

Fig. 14   Composite backscattering spectra for 1.4 MeV [4]He
ions (upper portion) and 1.4 MeV protons (lower
portion) incident normal to Au films deposited
on carbon substrates.

signals from the carbon substrate can also be seen at lower
energies. Backscattering with MeV [4]He ions, then, is useful for
analyzing layers less than 1 μm thick, whereas a beam of protons
is useful for analyzing layers from about 1 to 10 μm thick.

An independent method of obtaining thickness of a thin
film by backscattering without using the energy loss value is to
calculate the film thickness from the area, A, under the backscattering spectra as given in Eq. 1. If one takes into account
the energy loss of the incident beam as it traverses the target,
the scattering cross section will increase as the projectile
loses energy. One method of including the cross section correction is shown below. For a Pt film about a 1000Å thick, the

correction is about 3.8% for 2.0 MeV He ions:

$$(Nt)_0 = \frac{A}{\sigma(E_0)\Omega Q} = 6.62 \times 10^7 \text{ atoms/cm}^2 \tag{43}$$

and

$$Nt = (Nt)_0 [1 - \frac{\varepsilon(E_0)(Nt)_0}{E_0}] \tag{43}$$

where $\varepsilon(E_0) = 115 \times 10^{-15}$ eV-cm^2 for Pt. For fixed incident energy, the amount of correction is directly proportional to Nt and ε. For elements of low atomic number, ε is smaller and so is the amount of correction calculated. For example, for a 1000Å Si film, the amount of correction for 2 MeV helium backscattering is 1.2% rather than 3.8%.

Multi-elemental and Layered Films

For a multi-elemental thin film structure, we can obtain two pieces of information by backscattering: composition and film thickness. Composition analysis has been discussed earlier for bulk samples. Here we discuss thickness measurement. Figure 15 shows an energy spectrum of 2 MeV ^{4}He ions backscattered from SiO$_2$ film thermally grown on a silicon substrate. When the film is thick enough, the energy shift is well defined and the thickness of the film can be readily calculated by using the stopping cross section factor:

$$Nt = \Delta E_{Si}/[\varepsilon]_{Si}^{SiO_2} = \Delta E_{Si}/226 \times 10^{-15} \text{ eV-cm}^2$$

$$\tag{44}$$

$$Nt = \Delta E_{0}/[\varepsilon]_{0}^{SiO_2} = \Delta E_{0}/213 \times 10^{-15} \text{ eV-cm}^2$$

using a molecular basis and the surface energy approximation (Table 2). From Fig. 15 we have $\Delta E_{Si} = 262$ keV and $\Delta E_0 = 238$ keV. By using both $[\varepsilon]$ values in Eq. 44, we calculate Nt to be 1.16×10^{18} molecules/cm^2 for the first value, and 1.12×10^{18} molecules/cm^2 for the second. The mean value, 1.14×10^{18} SiO$_2$/cm^2 is equivalent to 5000Å SiO$_2$ film when a bulk density of 2.28×10^{22} SiO$_2$ molecules/cm^3 is assumed for the oxide film.

If a mean energy approximation (Eq. 42) is used $\bar{E}_{in}$ and $\bar{E}_{out}$ depend on the thickness of the film, and in this particular case we have $\bar{E}_{in} = 1930$ keV, $\bar{E}_{out,Si} = 956$ keV and $\bar{E}_{out,0} = 577$ keV. These energies give values of

$$[\varepsilon]_{Si}^{SiO_2} \qquad \text{and} \qquad [\varepsilon]_{0}^{SiO_2}$$

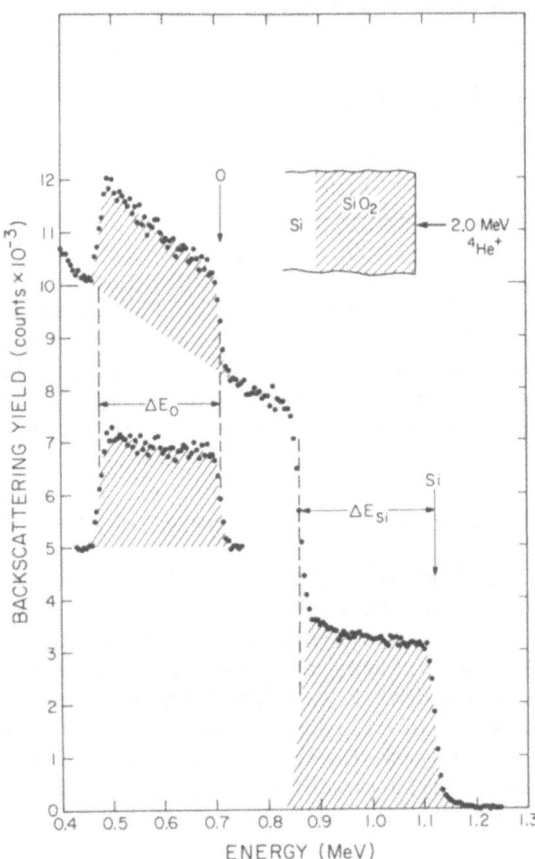

Fig. 15 Backscattering spectrum for 2.0 MeV ⁴He ions incident
         on a 5000Å thick layer of SiO₂ thermally grown on a
         Si substrate.

of 234 and 213 x 10^{-15} eV-cm², respectively, and an oxide thickness of 4910Å. The two approximations differ by 0.4% for every
1000Å of thickness.

        Figure 16 gives backscattering energy widths for SiO₂,
Si₃N₄, Al₂O₃, AlN and Ta₂O₅ films obtained with 2 MeV ⁴He ions at
normal incidence, θ = 170°. The density of the films are assumed
to be identical with that of the bulk compounds. The value of
[ε̄] is calculated by the mean energy approximation for scattering from the heavier two elements in the films.

        The energy loss ratio method of Lever (1976) provides an
estimate of the energy E before scattering. The method is based
on the fact that the ratio $\alpha = \Delta E_{out}/\Delta E_{in}$ does not change

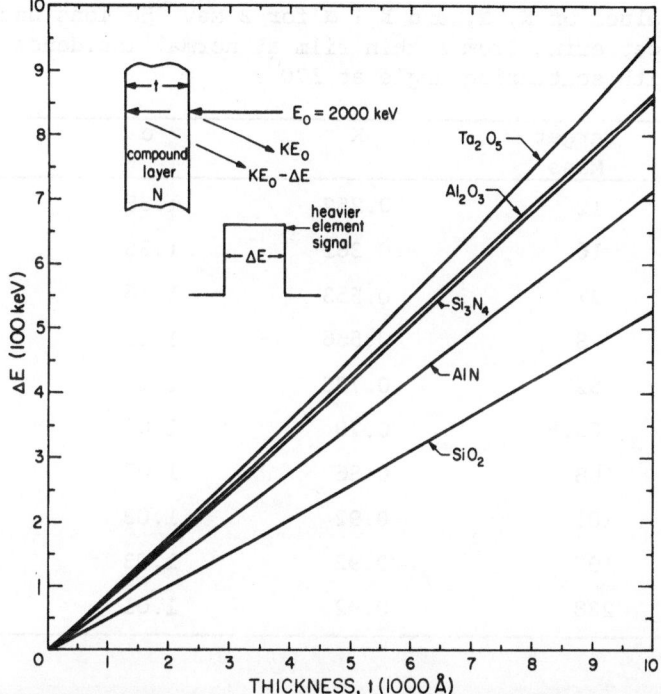

Fig. 16 Energy widths, ΔE, of the heavier element signal vs
         film thickness for different dielectric layers ob-
         tained for 2.0 MeV ^{4}He ions at normal incidence,
         $\theta = 170^\circ$, and assuming bulk density of the films.

appreciably with depth. Values of α are listed in Table 7. For
the layer of SiO_2 shown in Fig. 15, the value of E at the Si/SiO_2
is found to be:

$$E = \frac{E_1 + \alpha E_0}{K + \alpha} = 1.86 \text{ MeV} \tag{45}$$

where $\alpha = 1.29$ for Si and $E_1 = 869$ keV for particles scattered
from Si atoms at the Si/SiO_2 interface. As shown by the values
in Table 8, the nearly same energy E is obtained when the com-
putation is made on the basis of scattering from O atoms at the
interface.

The energy loss method can also be used with multilayer
films to determine the energies before scattering at the various
interfaces. Since the particle traverses exactly the same range
of compositions on the outward path as on the inward path, the
value of α will not change to a great extent even though the

Table 7. Values of K, α, and K + α for 2 MeV ^{4}He ions back-
          scattering from a thin film at normal incidence
          with scattering angle at 170°.

Target Element	Target Mass	K	α	K + α
C	12	0.252	1.35	1.60
O	16	0.362	1.35	1.71
Al	27	0.553	1.16	1.71
Si	28	0.566	1.29	1.86
Cr	52	0.736	1.11	1.85
Cu	63.5	0.78	1.06	1.84
Ag	108	0.86	1.07	1.93
Ta	181	0.92	1.03	1.95
Au	197	0.92	1.03	1.95
U	238	0.92	1.01	1.95

Table 8. Calculation of the energy E before scattering for the
          spectrum shown in Fig. 15 using the energy ratio
          method with α_{Si} = 1.29 and α_0 = 1.35.

E_1	E_1 (keV)	$E_1 + \alpha E_0$ (keV)	E (keV)
$K_{Si}E_0$	1131	3710	2000
$K_{Si}E_0 - \Delta E_{Si}$	869	3499	1858
$K_0 E_0$	725	3425	2000
$K_0 E_0 - \Delta E_0$	487	3187	1860

composition changes in the different layers. Figure 17 shows the
spectrum for a sample with Ni$_2$Si formed between Ni and the Si substrate (Tu, et al, 1975). The presence of the silicide layer can
be deduced from the step in the Ni and Si shoulders. The energies
E before scattering at the various interfaces are given in Table
9 for α_{Ni} = 1.11. Rather good agreement is found for the energies
at the Si/Ni$_2$Si interface as computed from scattering from Ni and
Si atoms.

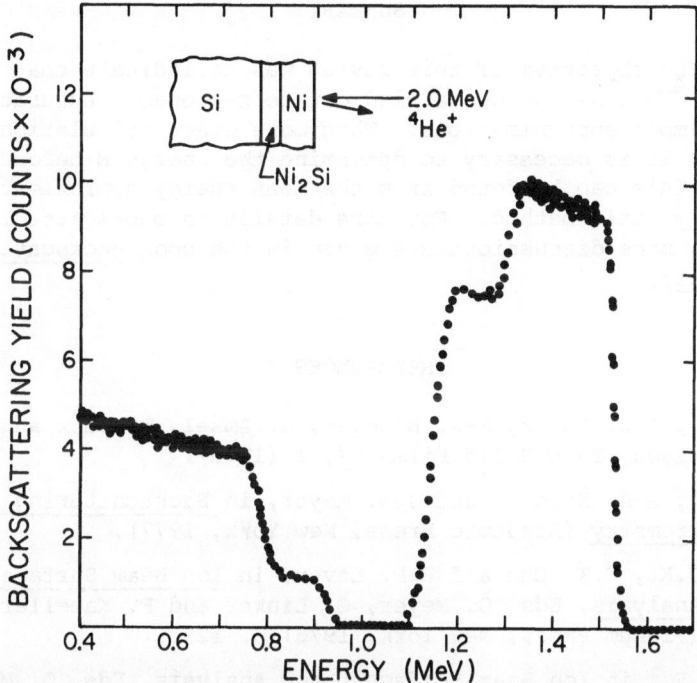

Fig. 17 Backscattering spectrum for 2.0 MeV ^{4}He ions incident
on a multilayer sample with Ni_2Si formed between Ni
and the Si substrate.

Table 9. Calculation of the energy E before scattering for
the spectrum shown in Fig. 17 using the energy
ratio method with $\alpha_{Ni} = 1.11$ and $\alpha_{Si} = 1.29$.

Position	Scattering element	E_1 (keV)	$E_1 + \alpha E_0$ (keV)	E (keV)
Ni surface	Ni	1525	3745	2000
Ni/Ni_2Si	Ni	1310	3530	1885
Ni_2Si/Si	Ni	1140	3360	1794
Ni/Ni_2Si	Si	930	3510	1896
Ni_2Si/Si	Si	780	3360	1810

SUMMARY

The objective of this review was to indicate that analysis with MeV ^{4}He ions can be carried out to reasonable accuracy using rather simple approximations. When more exact calculations are required, it is necessary to determine the energy E before scattering. This can be found from the mean energy approximation or the energy ratio method. For more details on backscattering analysis, more discussions are given in the book Backscattering Spectrometry.

REFERENCES

Chu, W.K., J.W. Mayer, M-A. Nicolet, G. Amsel, T. Buck and F. H. Eisen, Thin Solid Films 17, 1 (1973).

Chu, W.K., M-A. Nicolet and J.W. Mayer, in Backscattering Spectrometry (Academic Press, New York, 1977).

Howard, J.K., W.K. Chu and R.F. Lever, in Ion Beam Surface Layer Analysis, Eds. O. Meyer, G. Linker and F. Kapeller (Plenum Press, New York, 1976) p. 125.

Lever, R.F., in Ion Beam Surface Layer Analysis, Eds. O. Meyer, G. Linker and F. Kappeler (Plenum Press, New York, 1976) p. 111.

Mayer, J.W., J.F. Ziegler, L.L. Chang, R. Tsu and L. Esaki, J. Appl. Phys. Phys. 44, 2322 (1973).

Mayer, J.W., G. Foti and E. Rimini, in Handbook of Materials Analysis, Eds. J.W. Mayer and E. Rimini (Academic Press, New York, 1977) Chapter 2.

Sigmon, T.W., W.K. Chu, H. Muller and J.W. Mayer, Appl. Phys. 5, 347 (1975).

Tu, K.N., W.K. Chu and J.W. Mayer, Thin Solid Films 25, 403 (1975).

Ziegler, J.F., J.W. Mayer, B.M. Ullrich and W.K. Chu, in New Uses of Low Energy Accelerators, Ed. J.F. Ziegler (Plenum Press, New York, 1975) Chapter 2.

Ziegler, J.F. and W.K. Chu, Atomic Data and Nuclear Data Tables 13, 463 (1974).

MICROANALYSIS BY DIRECT OBSERVATION OF NUCLEAR REACTIONS

A. Cachard and J.P. Thomas

Département de Physique des Matériaux
Institut de Physique Nucléaire
Université LYON I, 69621 Villeurbanne, FRANCE

E. Ligeon

Département de Recherche Fondamentale
C.E.N.G. 85 X, 38041 Grenoble Cedex, FRANCE

INTRODUCTION

Among the different interactions taking place when charged particles impinge a target, those assigned as nuclear represent an important part of the means of characterization. The nuclear reaction products resulting from these interactions are essentially light charged particles, recoil nuclei and γ-rays of desexcitation. According to the life time of these radiations, one can distinguish the emission observed when the beam is hitting the target and after the bombardment. This last case requires a production of radio-isotopes of sufficiently long half-life to be counted after some delay (allowing the target to be removed and eventually chemically treated) and is classically called activation analysis. Being by itself a detailed subject of discussions, described on the other hand, in numerous related books, this case will be not treated here.

For the first case, the term "prompt radiation analysis" (P.R.A.) is often used but can also be applied to all the techniques described in this book."Direct observation of nuclear reaction products" (D.O.N.R.P.), sometimes reported, has a somewhat greater specificity. Standard equipments (target chamber, vacuum, detectors and electronic set-up) similar to those required for backscattering and ion-induced X-rays and described through this book, are most often used, as illustred in the figure 10 a of the last chapter.

Following a brief discussion of the characteristics of a nuclear reaction, we will outline the general features of this type of analysis and especially the selectivity (related to kinematics) and the sensitivity (related to the cross section). We will then indicate the kind of analytical problem which can typically be solved by the technique and what performances can be excepted, the analysis being global or in depth.

Together with a necessary compilation of the nuclear reactions of interest, typical areas of investigation will be reported, related to problems of particular current interest where the method has a definite advantage. More details will be given at last, upon some new trends of nuclear reaction analysis : the ability to identify, locate and profile the lightest elements including hydrogen. The increasing interest for such a non destructive analysis justifies a special development in this paper.

I – GENERAL CHARACTERISTICS OF NUCLEAR REACTION ANALYSIS

Chemical analysis from the detection of nuclear reactions products can be considered as originating near 1960. As for backscattering the very first attempts (1) were done using magnetic spectrometers. Nevertheless together with the development of semiconductor detectors nuclear reactions are extensively used since nearly fifteen years, with special mention to oxygen with ^{16}O and ^{18}O isotopic tracer analysis (2).

For broader information, the reader may refer to general paper and books presented in the chapter III. In what follows we shall present a general description of the subject.

The bombardement of nuclei of mass M_2 by MeV energy range ions of mass M_1 may induce a nuclear reaction which can be written as:

$$M_1 + M_2 \rightarrow M_3 + M_4 + Q$$

where M_3 is the emitted "particle" (which can be either a charged particle or a gamma ray or both), M_4 is the residual nucleus and Q is the defect mass energy of the reaction called the Q value of the reaction. When the reaction releases energy (exoenergetic) the Q value is positive and for endoenergetic reactions (energy absorbed) Q is negative. For any type of reaction the Q value is a vell defined quantity.

Then for a reaction induced by an incident particle M_1 of energy E_1 (fig. 1), the energy E_3 of the emitted particle M_3 in the direction θ (relative to the incident direction in the laboratory) is determined by the conservation of total energy and momentum, in non-relativistic case and exprimed by the formula:

$$E_3^{1/2} = A \pm (A^2 + B)^{1/2} \qquad\qquad (1)$$

where $\quad A = \dfrac{(M_1 M_3 E)^{1/2}}{M_3 + M_4} \cos\theta \quad$ and $\quad B = \dfrac{M_4 Q + E_1(M_4 - M_1)}{M_3 + M_4}$

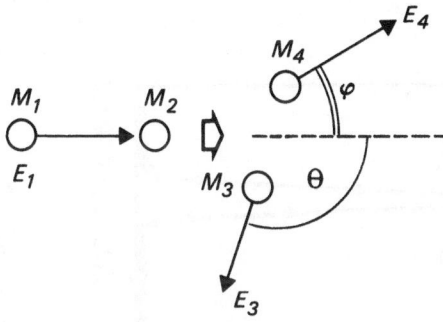

<u>Fig. 1</u> Schematic diagram of a nuclear reaction

Equations (1) show that E_3 is characteristic of the reaction for a given E_1 and θ. In fact the residual nucleus can be left in ground state or in excited states, each state corresponding to a different Q value for the same reaction, and hence to a different value of E_3. The energy spectrum of the emitted particles will exhibit a series of peaks, and is highly specific of the investigated reaction. Then it may be used to identify the presence of the given nucleus M_2.

Such an operation is obtained usually with solid state detectors; the energy of a peak allows the identification of the reaction (and hence of the nucleus M_2) and from the surface of the peak the amount of the specie M_2 can be determined.

Equations (1) can be approximated within a wide energy range by

$$E_3 = \alpha E_1 + \beta \qquad\qquad\qquad (2)$$

where α and β (as A and B in (1)) are specific of the reaction under study and depend on the detection angle θ. Thus the selectivity of the analysis is not easily discussed due to the specificity of the nuclear reactions. Usually the selectivity is very good; the only limitations are possible interferences between two peaks in the energy spectrum arising when several light nuclei are simultaneously present in the sample. The solution is to determine in each case the optimal nuclear reaction at the optimal energy. On an other hand this specificity is an advantage in the sense that two isotopes of the same element behave in a completely different way. In many cases only one of the isotopes has a positive Q value, the other being ignored regardless of its concentration. As an example for oxygen using a proton beam ^{18}O only is detectable through the $^{18}O(p,\alpha)^{15}N$ reaction; the $^{16}O(p,\alpha)^{13}N$ reaction having a negative Q value is not detectable (2). Isotope tracer experiments are thus feasible by this technique, in contrast to backscattering in which the kinematic factor is only slightly changed.

A typical energy spectrum (3) for the deuteron bombardment (E_d = 1700 keV) of a 1700 Å thick aluminium nitride film on nickel backing is given in fig. 2. The peaks of $^{27}Al(\alpha,p_{o+})$, ^{28}Al and

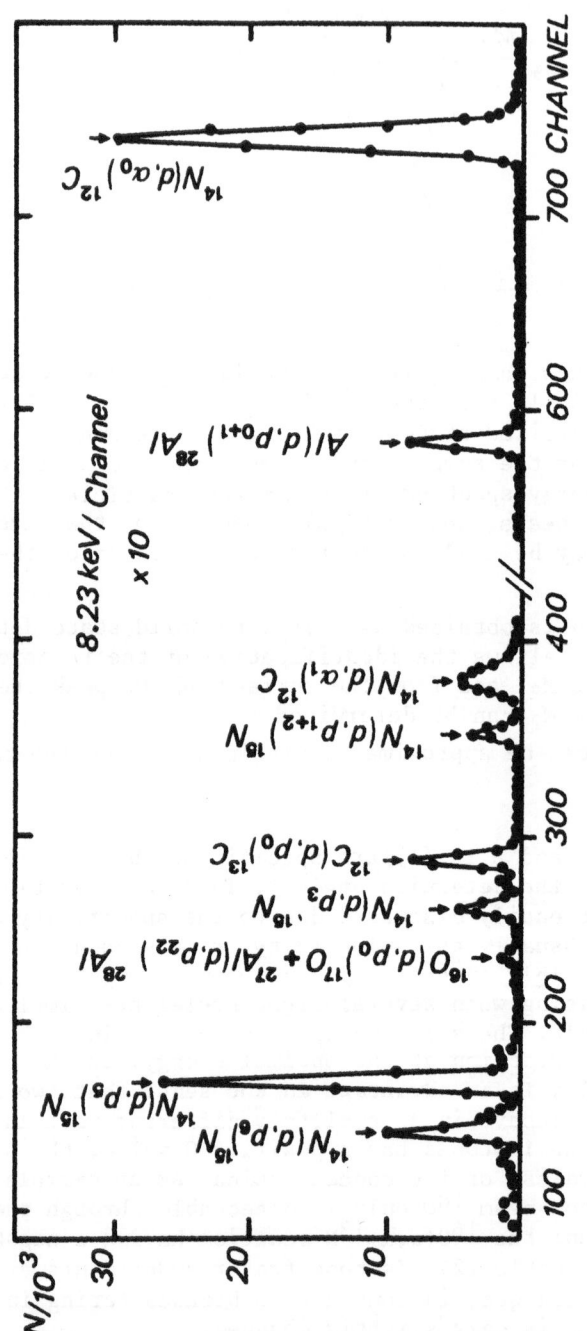

Fig. 2. A typical energy spectrum from the deuteron bombardment of a 1700 Å thick aluminium nitride film of nickel backing; E_d= 1700 keV; θ = 160° mylar absorber; 24 µm

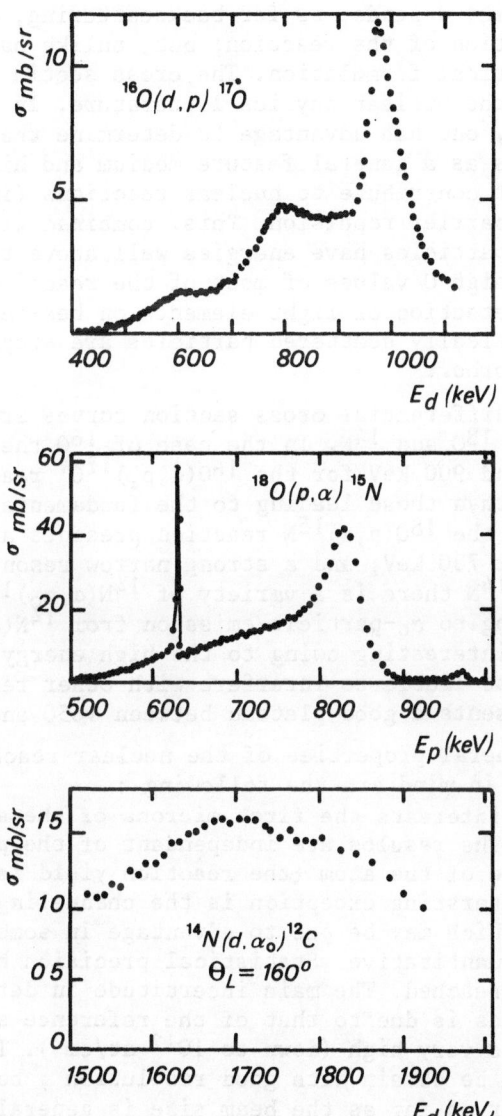

<u>Fig. 3</u>. Differential cross-sections of the reactions used to analyse ^{16}O and ^{18}O (at $\theta = 165°$) and ^{14}N (at $\theta = 160°$). ^{16}O and ^{18}O curves are taken from ref.4 and ^{14}N from ref. 3.

$^{14}N(d,\alpha_0)^{12}C$ reactions are located in a background free part of the spectrum. The oxygen and carbon contaminants are also visible.

The sensitivity depends, as for backscattering, on the differential cross-section of the reaction; but, unlike backscattering, there is no analytical formulation. The cross section curves can be obtained from the nuclear physics litterature. In fact for precise measurements, one has advantage to determine them experimentally. Nevertheless as a general feature medium and high Z nuclei practically do not contribute to nuclear reactions (in MeV range) owing to Coulomb barrier repulsion. This, combined with the fact that the emitted particles have energies well above that of the beam (due to the high Q values of most of the reactions), allows background free detection of light elements on heavier substrate provided the elastically scattered particles are stopped in a suitable thin absorber.

Examples of differential cross section curves are shown in figure 3 for ^{16}O, ^{18}O and ^{14}N. In the case of ^{16}O there is a plateau between 800 keV and 900 keV for the $^{16}O(d,p_1)^{17}O^*$ reaction, more efficient to use than those leading to the fundamental level of ^{17}O (2). For ^{18}O, the $^{18}O(p,\alpha)^{15}N$ reaction presents a stationnary cross section near 730 keV, and a strong narrow resonance at 629 keV (2). For ^{14}N there is a variety of $^{14}N(d,\alpha_0)^{12}C$ reaction, but the one leading to α_0-particle emission from $^{14}N(d,\alpha_0)^{12}C$ reaction is the most interesting owing to the high energy of the α_0 particle, thus less liable to interfere with other reactions. Its cross section presents a good plateau between 1650 and 1730 keV (3).

The other general properties of the nuclear reaction analysis that must be kept in mind are the following :
The analysis interests the first microns of the sample and is non destructive. The results are independant of the physical or the chemical state of the atom (the reaction yield is only nuclear dependent). An intersting exception is the channeling effect in single crystals which may be put to advantage in some special cases. The results are quantitative. Statistical precision better than 1% can be easily reached. The main incertitude in determining absolute quantities is due to that of the reference standards. The sensitivity is very high (down to 10^{12} at/cm^2). The depth distributions can be obtain with good resolution ; but the lateral resolution is poor as far as the beam size is generally larger than 0.1 mm.
In the MeV range the bombarding particles are mainly protons and deuterons. Heavier nuclei can be used but with higher energies.

II - QUANTITATIVE ANALYSIS

In this chapter we will developp the general principles for quantitative analysis using nuclear reactions. Classical example will illustrate the method.

II.1 - Total number of atoms $(at.cm^{-2})$

The area of the peak of characteristic reaction is proportional to the total number N (cm^{-2}) of atoms in the surface region of a solid

$$A_{(counts)} = N \ q \ \Omega \ \frac{d\sigma}{d\Omega}$$

where q is the total number of incident particles, Ω the detector solid angle and $d\sigma/d\Omega$ the differential cross-section. This formula supposes that $d\sigma/d\Omega$ is constant through the thickness of interest of the sample. This condition imposes to choose an energy for which there is a plateau in the excitation curve of the reaction. Then the total number N is given by :

$$N \ (at.cm^{-2}) = \frac{A_{(counts)}}{q\Omega \ \frac{d\sigma}{d\Omega}}$$

The cross-section and the solid angle have to be precisely determined, in order to achieve a good absolute accuracy. In fact it is better to use a standard of known composition N_S and to compare the peak areas A_S and A under the same analysis conditions.

$$N \ (at.cm^{-2}) = N_S \ (at.cm^{-2}) \cdot \frac{A}{A_S}$$

This possibility is unique to nuclear microanalysis, because the reaction yield depends only on nuclear cross-section, regardless of the physical or chemical state of the atom under investigation.

As an example, by using $^{16}O(d,p_1)^{17}O^*$ reaction, less than a monolayer of oxygen $(10^{14}$ atoms/cm$^2)$ can be detected within minutes with a precision of a few percent (4).

Another example is the analysis of boron in silicon using the $^{11}B(p,\alpha)^{\alpha\alpha}$ reaction (with $E_p = 660$ keV) (5). The cross-section varies of less than 5% for 1.5 µm of analysed depth. The detection limit is of the order of 10^{12} $at.cm^{-2}$

The composition of compound can be obtained in the same way (especially for oxides, nitrides, fluorides of hight elements boron aluminium, magnesium, silicon). In figure 2 the ratio of the number of nitrogen to the number of aluminium can be calculated.

$$\frac{N_N}{N_{Al}} = \frac{A_{N(\alpha_0)}}{A_{Al(p_{0+1})}} \cdot (\frac{d\sigma}{d\Omega})_{Al(p_{0+1})} \cdot (\frac{d\sigma}{d\Omega})^{-1}_{N(\alpha_0)}$$

using the reactions : $^{14}N(d,\alpha_0)^{12}C$ and $^{27}Al(d,p_{0+1})^{27}Al$. The ratio of the cross-section is determined from a standard of known nitrogen/aluminium ratio (3) (in this case aluminium oxinate). The composition can be measured within a few percent precision even for films less than 100 Å thick. Such a method has been used also for the anlaysis of thin films (6) of Al_2O_3, SiO_x, Si_3N_4, SiC and LiF.

II.2 – Concentration profile determination

II.2.1 – Energy spectrum analysis

When the analysed specie extends over an important depth, the characteristic peak is no longer narrow nor symmetric as for backscattering.

For the energy E_0 of incident particles, the energy of the detected species, resulting from a reaction at the surface, is

$$E'_0 = \alpha E_0 + \beta$$

(see fig. 4).

At a depth x, the incomming particles have an energy :

$$E_1 = E_0 - \left| \int_0^x \left(\frac{dE}{dx}\right)_{in} dx \right| \simeq E_0 - S\,x$$

where $S = \left(\frac{dE}{dx}\right)_{E_0}$

At this depth the emitted particles have an energy

$$E'_1 = \alpha E_1 + \beta$$

and they escape from the target with an energy :

$$E'_x = E'_1 - \left| \int_0^{x/\cos\theta} \left(\frac{dE}{dx}\right)_{out} dx \right| \simeq E'_1 - S'\left|\frac{x}{\cos\theta}\right|$$

where $S' = \left(\frac{dE}{dx}\right)_{E'_x}$

Then the energy difference between the detected particles originating from the surface or from the depth x is :

$$\Delta E = E'_x - E'_0 \approx x\left[\alpha S + \frac{S'}{|\cos\theta|}\right]$$

This formula gives a correspondance between the depth scale and the energy scale. If the cross section is known the concentration profile can be deduced in principle from the shape of the experimental spectrum, S and S' being functions of the composition. By using Bragg's additivity law, we can calculate the shape of the spectrum for different profiles and by comparison to that of the experimental spectrum we can deduce the concentration profiles within errors set by statistics and energy resolution.

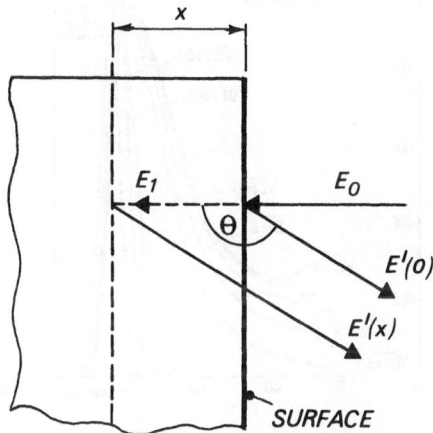

Fig. 4. Particles trajectories in a thick target

Such a procedure can be illustrated by the study of the diffusion of oxygen in zirconium (7). Calculation are made to obtain
theoretical spectra for different concentration profiles corresponding to different values of the diffusion coefficient, assuming that
they are given by complementary error function. The comparison with
the experimental spectrum gives the determination of the diffusion
characteristics of the element. The Fig. 5 shows typical spectra
calculated for different $(Dt)^{1/2}$ values (in μm) in the case of an
error function diffusion of oxygen in zirconium : the reaction is
$^{16}O(d,p_1)^{17}O^*$ for a deuteron energy E_d = 823 keV.

The depth resolution is rather poor in this case due to the
energy straggling of the detected protons in the absorber used to
stop the backscattered deuterons. The energy resolution is of the
order of 40 keV, which combined to the low value of stopping power
for deuterons or protons corresponds to 0.25 μm for zirconium.

Nevertheless in some cases, as for thin oxyde film at the surface
of a sample, it is possible to extract the contribution of this film
from the contribution of the continuum (8). The fig. 6 is an example
of such a procedure. The spectrum due to protons emitted from

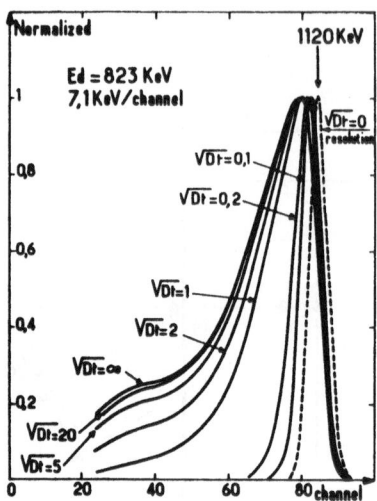

Fig. 5. Spectra calculated theoretically for different $(Dt)^{1/2}$
values (in μm)for oxygen concentration curves in
zirconium following an error function law; $^{16}O(d,p_1)$
$^{17}O^*$ reaction for E_d = 823 keV. (Ref. 7).

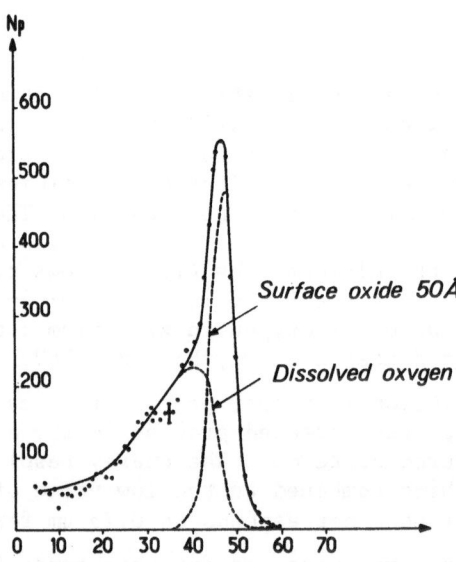

Fig. 6. Proton spectrum of $^{16}O(d,p_1)^{17}O^*$ reaction for chemically polished titanium sample. The calculated curve
(—) is the sum of the two partial curves (---).(Ref. 8).

$^{16}O(d,p_1)^{17}O*$ reaction on chemically polished titanium sample shows clearly the two contributions of the 50 Å thick surface oxyde and of the bulk dissolved oxygen.

In an other hand it is possible to improve the depth resolution by use of the telescope technique (9). The inactive absorber (mylar) is replaced by a thin detector which is sufficient to stop the deuterons scattered by the substrate but induces only a small energy loss to the energetic protons. These particles are stopped in a second large thickness detector. The detection in coincidence and the summation of the two peaks from the protons eliminate the straggling. The energy resolution can then be reduced to 20 keV which represents 0.12 μm in zirconium. Pile-up effect in the thin detector is the main limitation.

Another specific possibility is the use of the $^{10}B(n,\alpha)^7Li$ reaction (10,11) for the analysis of boron in silicon. There is no background from the substrate and no needs for absorber. The depth resolution is then very good (200 Å). But one has to use a rather high flux neutron source (of the order of 10^8 neutrons cm^{-2} s^{-1})

II.2.2 - Concentration profiles from resonances

When the cross section curve $(d\sigma/d\Omega)(E_0)$ exhibits a strong narrow resonance at an energy E_R, concentration profiles can be extended from the measurement of the yield curve $N(E_0)$ of the emitted particles as a function of the energy E_0 in the vicinity of the resonance energy (2, 4).

If $E_0 = E_R$ the detected particles are mainly due to reactions at the surface of the sample. If $E_0 > E_R$, due to the slowing down of the incomming particles, the energy E_R is reached at a depth roughly given by $x = (E_0-E_R)/S$ and the yield is then characteristic of the concentration $C(x)$. The yield curve $N(E_0)$ is thus an image of the concentration profile $C(x)$, and allows a rapid rough estimate of $C(x)$ as illustrated on the fig. 7. For a precise determination of $C(x)$ we must take into account the energy straggling effects, and the correct actual relation between $N(E_0)$ and $C(x)$ is :

$$N(E_0) = \{ \int_0^\infty \int_0^\infty C(x) \frac{d\sigma}{d\Omega} (E) \, P \, (E_0,E,x) \, dx \, dE\} \times c^{te}$$

where $P(E_0,E,x)$ is the probability that a particle of incident energy E_0 has an energy between E and $E + dE$ at a depth x. Methods to solve such an equation imply the calculation of $P(E_0,E,x)$ and of the stopping power of the incident particles; they can be found in ref. 12 and 13. Such an analysis technique has been extensively used by Amsel and coworkers especially for ^{18}O tracing to study oxygen transport during anodic oxidation of metals (14). They used the 629 keV resonance of the $^{18}O(p,\alpha)^{15}N$ reaction which allows a 200 Å depth resolution in Ta_2O_5.

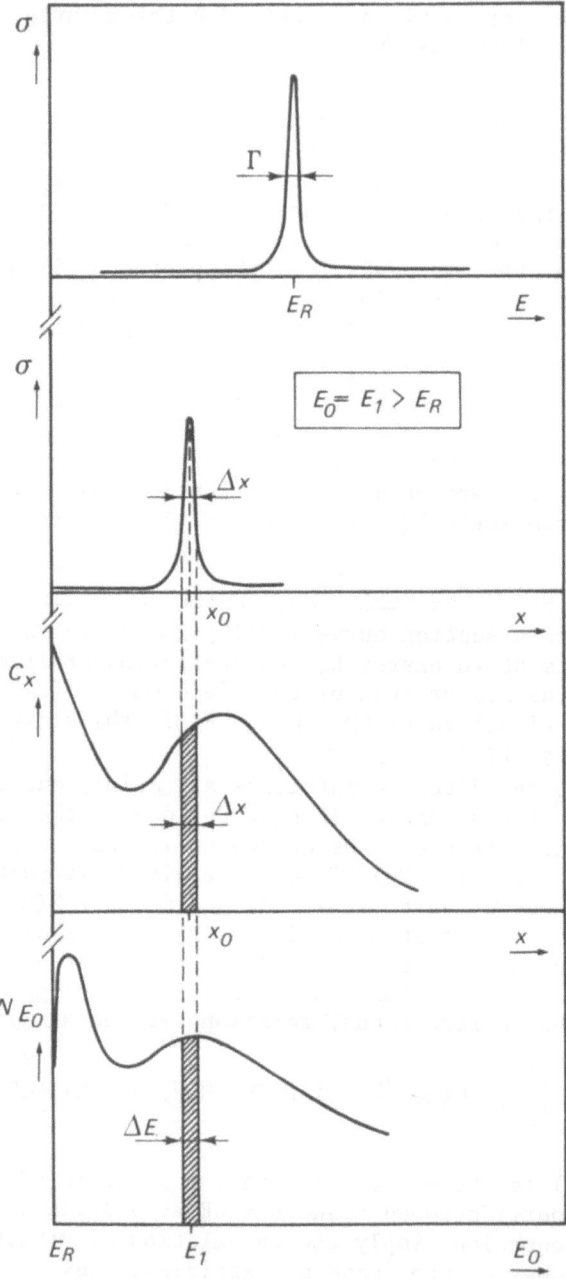

<u>Fig. 7.</u> Principle of concentration profile measurements using resonant reaction

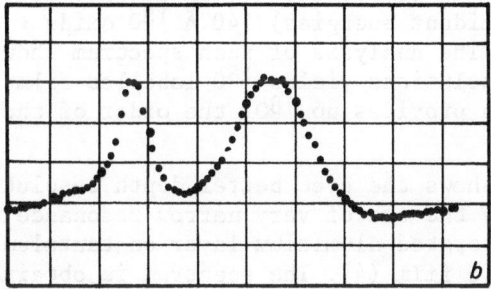

Fig. 8. Typical yield curve using the 629 keV resonance of the $^{18}O(p,\alpha)^{15}N$ reaction for an $^{18}O-^{16}O-^{18}O$ Ta_2O_5 "sandwich" target; 390 eV/channel and 5.10^3counts/division. (Ref. 4).

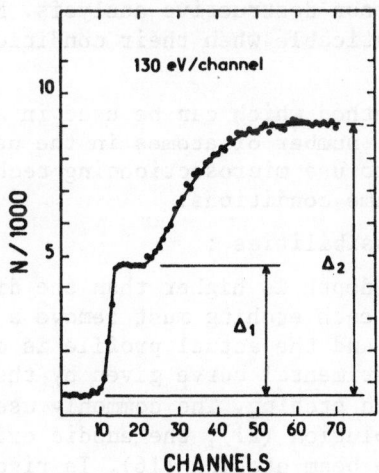

Fig. 9. Typical excitation curve of the $^{27}Al(p,\gamma)^{28}Si$ reaction near the 992 keV resonance for a 475 Å anodic oxide film grown on a 0.8 μm aluminium layer on tantalum; 40 μc beam dose per channel. (Ref. 4).

The figure 8 shows a typical spectrum obtained with an anodic
tantalum oxide film of composition (from surface towards the metal
i.e. increasing incident energies) 140 Å ^{18}O oxide + 840 Å ^{16}O oxide
+ 520 Å ^{18}O oxide. The analysis of such spectrum shows that oxida-
tions in labelled solutions yields ^{18}O labelled films, whereas that
in natural solution provides no ^{18}O, the order of the oxygen atoms
being conserved.

The figure 9 shows the even better depth resolution achieved
near the surface by the use of very narrow resonance. The sample is
a 0.8 µm thick evaporated aluminium layer on tantalum backing with
a 475 Å thick anodic film (4). The spectrum is obtained using the
resonant ^{27}Al$(p,\alpha)^{28}$Si reaction (E_R = 992 keV). The sharp rise and
the good plateau correspond to a depth resolution of 60 Å in Al$_2$O$_3$.
The ratio of the plateau heights Δ_1 in Al$_2$O$_3$ and Δ_2 in aluminium
indicates an actual Al$_2$O$_3$ composition for the oxide film. The same
reaction on aluminium has been used by Dunning (15) to determine
the concentration profile of aluminium implanted into silicon carbide.

II.2.3 - Concentration profiles determination using
destructive techniques

The extraction of a concentration profile from an energy
spectrum or from the excitation curve of a resonant reaction has
the advantage to be a non destructive analysis. Nevertheless in many
cases they are not praticable when their condition of application
cannot be fullfilled.

A more general method which can be used in almost all cases is
to determine the total number of atomes in the near surface region
of the material, and to use microsectionning techniques with successive counting in the same conditions.

There are two possibilities :

First the probed depth is higher than the distribution depth
of the element. Then each etching must remove a thickness lower
than the probed depth and the actual profile is obtained by the
derivation of the experimental curve given by the successive
measurements after each etching. The commonly used etching techniques
are the chemical dissolution (21), the anodic oxidation (129) or the
in situ low energy ion beam etching (16). In rigourously controled
conditions a depth resolution of 300 Å can be obtained with a good
reproducibility.

The second case is encountered in large depth diffusion
profiles. Here the probed depth is lower than the profile and the
concentration of the element can be considered as a constant in the
region seen by the beam. The etched thickness must be larger than
the probed depth. Then the successive countings give directly the

image of the concentration profile (8). Such a procedure has been applied to the study of the oxygen diffusion in titanium and zirconium at high temperature (17).

III - SCOPE OF NUCLEAR REACTION ANALYSIS

Due to the limited space available for such a task the aim of the authors is essentially to give enough background materials and references. As useful review articles, one must refer to WOLICKI (18) and AMSEL (4)(19). Comparison of this technique with others described through this book, have also been made at the first "Ion Beam Surface Layer Analysis" conference   (20)(21).

III.1 - Data of interest

If one excepts the compilations regularly made by nuclear physicists (22)(23)(24)(25)(26)(27), the most recent attempt to select the data of specific use in nuclear microanalysis is due to PICRAUX et al (28) in the "Handbook of materials analysis". One can also refer to the compilation of LORENZEN and BRUNE (29). To present as much as possible references we have adopted here a table form for the most used reactions. As the most significant information, typical incident energy and cross section (for large detection angle unless indicated and according to the nature of the reaction, resonant (R) or not) will appear. The first reference of each reaction is extracted from the nuclear physics litterature while the others list the most significant applications which have eventually lead to more accurate excitation functions determination. Being presented in the last chapter, lighter elements, from hydrogen to helium, are not included.

Although proton and deuterons are the most commonly used particles to induce nuclear reactions, interesting possibilities have been shown when other ions are available. ^{3}He ions can also allow carbon and oxygen analysis (98) including the ^{17}O isotope (99)(100). Mainly used in activation analysis, tritons are also an interesting alternative for oxygen analysis (101). At high energy ($\geqslant$ 4 MeV) α particles can also induce nuclear reactions, mainly (α,p) : boron, sodium and aluminium are then determined (102) while fluorine requires (α,$\alpha'\gamma$) (α,pγ) and (α,nγ) near 8 MeV (achievable energy for doubly-charged particles on a 4 MeV accelerator) (103).

The (p,p'γ) reactions are not only useful, as indicated in the table, when inelastic scattering occurs (fluorine and sodium analysis is also performed in such a way (104)) but heavier elements are also detectable, provided the incident energy is sufficient to induce Coulomb excitation (E > 2 MeV). For this last process, the cross section is smoothy increasing like for Cr, Mn, Se, Rh, Ag, Pt and Au (105) the sensitivity being generally weak ($\geqslant$ 1000 ppm) as well as for Zn (92) and Ge (97). For this last element and also for As (97), (p,nγ) reactions are even more efficient. If one account space results on Cr (106) and Ni (107) determination using (α,p) reactions

Table I

Elements	Most useful reaction	Q value (MeV)	recommended. Ei (KeV)	dσ/dΩ (mb /st)		References
^{6}Li	(p,α)	4.02	1900	~ 20	(30)	(31) (32)
	(d,α$_0$)	22.36	1200	~ 4		(33) (34)
^{7}Li	(p,α)	17.347	4000	~ 1.5	(35)	(31) (33) (36)
	(p,γ)		441 (R)	~ 6	(37)	(36)
^{9}Be	(d,α$_0$)	7.153	2300	~ 20 (90°)	(38)	
	(p,γ)	6.587	992 (R)	~ 0.5	(37)	(36)
^{10}B	(d,p) (0 to 10)	11.049↓ 2.832↓	1500 to 4000	2 to 4	(39)	(40) (41)
^{11}B	(p,α)	5.681	660 (R)	~ 90	(42)	(5) (43)
	(p,γ)		{163 (R) 469 (R)		(37)	(44)
^{12}C	(d,p)	2.712	1000 (R)	~ 30	(45)	(41) (46) (47) (48)
	(p,γ)	1.944	459 (R)	~ 0.13	(37)	(44) (49)
^{13}C	(p,γ)	7.55	1748 (R)	340	(37)	(44) (49)
^{14}N	(d,p(0 to 5))	8.615↓ 1.305	1700	~ 7 (d,p5) ~ 1,6	(50)	(3) (36) (51) (52) (46) (53)
^{15}N	(d,α$_0$)	6.701	1700	~300 (total 90°)	(3)	(51) (52) (54) (55) (56)
	(p,αγ)	0.535	429 (R)		(57)	(44)

Elements	Most useful reaction	Q value (MeV)	recommended Ei (KeV)	$d\sigma/d\Omega$ (mb/st)	References	
^{16}O	(d,P$_o$ and 1)	1.919 1.048	1490 850	~5 ~6	(58)	(2)(5)(7)(8)(14)(17)(47) (59)(60)(61)(62)
^{18}O	(d,α$_o$)	3.116	930 1250	~8 ~6	(58)	(2)(63)(64)
	(p,α)	3.973	629 (R) 1776 (R)	~60 ~100	(64)	(2)(4)(14)(65)(66) (67)(68)(69)(70)
^{19}F	(p,α$_o$)	8.119	1250 1355 (R)	~0.6 ~3	(71)	(33)(72)(73)(74)
	(p,αγ)	2.064	872 (R) 1275 (R) 2000	~18 ct ~13 (90°) ~30 to 10	(71)	(33)(44)(73)(74)(75)(76) (77)(78)(79)(80)
^{23}Na	(d,α)	10.033	1500	0.15	(81)(33)(36)	
	(p,γ)	11.694)	1417 (R) 1778 (R)		(37)	(15)(82)
^{25}Mg	(d,P$_o$ to 2)	8.890	1800		(83)	
^{26}Mg	(p,γ)	8.272	1548 (R)		(84)	(85)
^{27}Al	(p,γ') (p,p'γ)	11.583 {-0.843 {-1.368	992 (R) ~2000		(86) (88)	(4)(15)(82)(85)(87) (82)
	(d,P$_{o+1}$)	5.499	~1700	0.26	(89)	(3)(56)(90)
^{28}Si	(d,P$_o$ to 10)	6.250 1.320	> 1300	~5(d,p10)	(91)	(61)(62)

Table I continued

Elements	Most useful reaction	Q value (MeV)	recommended E_i (KeV)	$(d\sigma/d\Omega)$ (mb/st)	References
^{28}Si	$(p,p'\gamma)$	-1.779	3100 (R) 3335 (R)		(92)
^{31}P	$(p,p'\gamma)$	-1.266	>2150		(88) (87)
	(p,α)	1.917	1892 (R)	~15	(93) (5)(16)
^{32}S	$(d,p_o \text{ to } 7)$	6.417 3.197 ↓	2000	~0.1 to 0.6	(94) (95)
	$(p,p'\gamma)$	-2.237	3379 (R) 3716 (R)		(96) (92)(97)

at 3.5 MeV, it appear that nuclear reaction could be adequate
for a large range of elements, even, as already indicated, the
method is more suited for the lighter. At last one cannot conclude
omitting nuclear reactions dealing with neutrons, either produced
((d,n) reactions) requiring time of flight techniques for detecting
carbon, oxygen, nitrogen (108) as well as fluorine (109) or inducing
charged particles (ex : $^{11}B(n,\alpha)$ (11) or γ-rays (110)).

Some promising developments are expected from the availability
of heavier ions at higher energy. As a matter of fact, beams of
^{16}O (15-20 MeV) for oxygen and magnesium analysis or ^{35}Cl for (40 MeV)
for potassium and copper analysis (111) are not yet routinely used.
More accessible are ^{6}Li ions (5-10 MeV), promising in the analysis
of carbon, oxygen, fluorine and neon (112).

III.2 - Typical areas of investigations

From the references quoted in the table and the examples previously given, one can realise that nuclear reactions are now applied
to many fields where the analysis of light elements in the near-
surface is required. The technique has often a definite advantage
and is sometimes routinely used :

- In solid state electrochemistry

The growth through chemical on electrochemical reactions, of
native films on substrates exposed to liquid or gaseous external
media has been one of the first spectacular application of the use
of nuclear reactions. Pioneering in the field is the PARIS group of
AMSEL, most references being found in (18) and (145). Oxyde formation studies offer a typical example, the process being thermal or
anodic. Representative of the first case is the silicon for which
kinetics determination is of obvious technological interest (116)
(117). The second case involves many materials (Al, Ta, Nb, Zr, Si)
(14)(118)(119)(120)(121)(65) among them aluminium is a good example
fo the important informations given by the technique. Not only growth
laws, but also stoichiometry and anionic movement with or without
growth have been determined either for porous or barriers layers
(14)(118)(146)(147). The powerfull application of the use of isotope
^{18}O is desmonstrated through these investigations. Of course the
technique is applicable to natural oxide layers even very thin, the
sensitivity being very high as already reported (fraction of a monolayer). As it will be seen for the metallurgy applications, this
great sensitivity is not only encountered for oxygen but various
contaminants (C, N, F, etc...) are also determined in trace amounts
in correlation with the various surface treatments of the materials
(preparation, exposures to various media, etching, etc...) (73)(74)
(122)(124)(59)(85).

   – In metallurgy :

    The development of polishing procedures benefits of the control of the natural oxide film formed almost instantaneously at the surface during various treatments as well as of the determination of light impurity incorporations (C, F, Al, Si, etc...) (65)(74)(85) (123). A typical illustration is the following (148) : Fig. 10 b shows the $0^{16}(d,p)0^{17*}$ proton spectrum for a nickel sample similar to that of Fig. 10 a, but electrochemically polished for 10 minutes in a H_2SO_4 solution at 400 mA/cm². The more than 100 fold decrease in the oxygen content of the surface region and the narrow peak obtained illustrates the effectiveness of the polishing procedure chosen. The remaining oxygen, 4.10^{15} at/cm², is most probably located at the surface and is equivalent to about 3 atomic layers. More generally, corrosion studies have an extended use of nuclear microanalysis as shown in the review article of DEARNALEY (113).

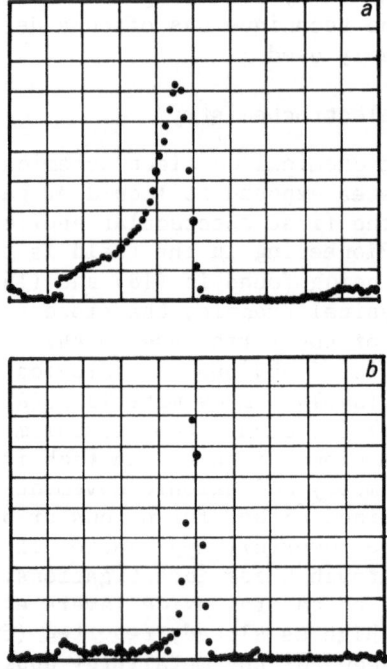

Fig. 10 : Proton energy spectra for the reaction $0^{16}(d,p)0^{17*}$ observed on nickel sample (a) unpolished (b) polished 17 keV/channel. Vertical scale : (a) 2000 counts/div; (b) 20 counts/div. (Ref. 148).

- In thin film elaboration

For deposit and growth of thin films of various nature
(insulators, semiconductors, supraconductors, magnetic, amorphous
etc...) nuclear reactions offer unique features : stoichiometry
and thickness determination are achieved with high level of confidence (ex : SiO_2, SiO, Al_2O_3, Si_3N_4 , AlN etc...) (61)(62)(125)
(54)(56). Selectivity problems are very limited in the case of films
of less than 5000 Å, the various peak corresponding to different
possible groups being most often well separated (see Fig. 2).
Fundamental studies of deposit mechanisms have been made for reactive and non reactive sputtering, thermal or "flash" evaporation,
c.v.d., self-implantation etc... According to the wide scale of
elements which can be present it is worthwhile to remember that the
spectrum of analysis of nuclear reactions can match these requirements : this is particularly interesting for the analysis of glasses
(97)(102) which are often composed of medium-Z elements (As, Ge, Te,
Se, etc...). It is often the only way to make precise stoichiometry
determinations of such complexe compounds. Of course one must then
select a suitable substrate, which most often is not a serious
problem.

- In material implantation

A good depth resolution being necessary in implantation
profiles determination, one has advantage to use resonant reactions,
or sectioning techniques. Besides, as for backscattering, lattice
location is allowed. One again, the interest of an isotopic analysis
is pointed out. FELDMAN et al (126) have so investigated the position of carbon implanted into iron as ^{13}C in order to overcome the
problem of natural contamination by the residual vacuum. Of course,
resonant nuclear reaction allow important improvements in depth
resolution of implantation profiles. For oxygen implanted into
gallium phosphide, WHITTON et al. (68) have achieved a depth resolution of the order of 100 Å using (p,α) reaction and ^{18}O implantation. We have already reported the experiments of DUNNING
(15) for the most favorable case of aluminium into silicon carbide:
A 50 Å depth resolution is achieved at the surface. For silicon,
the doping elements boron and phosphorus can be easily determined
at very low concentrations. For the (n,α) reactions very specific
of boron a sensitivity as good as 9.10^{10} at / cm^2 (11) can be
reached, while (p,α) reactions associated with in-situ ionic etching
allow to determine as low as 10^{12} at/cm^2 for boron and 10^{14} at/cm^2
for phosphorus (43)(16). When depth resolution is more important
than sensitivity, other nuclear reactions than those listed in the
table can be used, as shown by SIMONS et al. (127) for implantation
profile determination of nitrogen using (p,γ) reaction.

Increasing development of the use of nuclear reactions must also be noticed in the field of biological materials studies. In situ isotopic tracing of oxygen in the brain and circulatory system of rats is one more example of the use of the ^{18}O (p,α) reaction (128), fluorine penetration into teeth, near the surface (76)(79) or more deeply (97) has been determined. Pollution by the same element of pine needles is another example (80). More recently (52) nitrogen depth distribution has been determined into cereals.

IV – NEW TRENDS : HYDROGEN AND HELIUM ANALYSIS BY NUCLEAR

REACTIONS

There is no particular boundary between light and medium-Z elements and we will only give some examples where the light mass elements play an important role.

If we consider the lighter of the elements, the hydrogen, it must be noticed that :

- The behaviour of a proton and the surrounding electronic structure in simple metals are of considerable interest since the proton is the simplest impurity, being a point ion with no complicating core-electron structure (130).

- In the field of the applications, numerous prospective reports suggest that hydrogen could be one of the future energetic source. Hydrogen production (electrochemical or thermochemical process), hydrogen storage, require better understanding of the embrittlement of the materials, hydrides formation etc...

Another energy concern where light elements are involved, is the building of the future fusion reactors. According to the fusion reaction the cladding of the plasma would be submitted to a low energy implantation of the following elements H, D, T, ^{3}He, ^{4}He. This could change the surface composition of the material. Moreover the fast neutrons emitted from the fusion reaction could lead to H or He implantation in the bulk of materials by (n,p) or (n,α) reactions. (This last point concerns likewise the fast breeder reactor problem). As far as fusion reaction are concerned, we can say that all the light elements of mass lower or equal to the lithium one are concerned. If we wish understand the problems of materials submitted to an implantation of light elements, we must study first the fundamental problem of the interaction between the

implantation defects and impurities. The implanted elements find
their final position in a surrounding rich in irradiation defects
(intersitials, monovacancies etc...).

A powerful method for the study of the light elements in solids
is the analysis by nuclear reactions which give both :

- The depth profile of the impurity in the near surface region
($\sim 1\mu$) without layer removal technique. Long range migration of the
impurity can be observed from the evolution of its distribution as
a function of the annealing temperature.

- The lattice location of the impurity using the channeling
technique. With the light elements implantation, the induced defects
rate is low and the single crystal quality is preserved.

In order to study the defect impurity interaction, comparisons
are made between the recovery stages temperature of the matrix and
the temperature for which variations appear in the depth profile
and the lattice location of the impurity.

Analysis method of the light elements

We have summarized in the table 2 some nuclear reactions which
have been used for hydrogen and helium isotopes analysis. From these
data several comments can be made.

Hydrogen being always present at the sample surface (contamination) an appropriate analysis method must permit to separate the
surface contribution from the bulk contribution. This can be typically obtained using a resonant nuclear reaction . The depth resolution, governed by the resonance width, depends strongly on the
straggling of the analysing beam. This term known with a little
precision for Li, B, N, F beams limits in fact the depth resolution
for a large depth analysis. The contamination problem does not exist
for d, T, ^{3}He, ^{4}He analysis and we can also use a non resonant
reaction. The depth profile is then obtained by the energy analysis
of the emitted particle spectrum (4). The depth resolution achievable
with the D(^{3}He,α)p reaction is about 800 Å (Table 2). This value
is a function of the detection solid angle (owing to the kinematics
broadening) and there is unfortunatly a balance between the depth
resolution and the sensitivity. This energy analysis method require
generally higher concentrations (1 or 2 orders) that for the resonance method. It is appropriate for hydrides study, blister forma -
tion...

The sensitivity given for each nuclear reaction depends on the
background of the detection system. If charged particles are detected the background comes chiefly from the pile-up due to the scattered particles induced pulses (using a low energy beam the level of
the activation and emission process (β or γ) remains generally weak).
The counting rate of the pile-up spectrum being a function of the
square of the scattered particles counting rate, to increas the sensitivity we must use a lower incident beam intensity, analyzing times are

TABLE II

Light elements analysis by nuclear reaction.

Elements Références	Nuclear Reactions	E_R (MeV)	Γ (keV)	σ mb	Incoming particle (MeV)	outgoing particle (MeV)	Résolution depth (Å) (1)	Probing depth (µ)	at/cm² (2)	p.p.m. (3)
H [131]	$H(^{7}Li,\gamma)^{8}Be$	3.07	70		^{7}Li 2.7-6	γ 14-17	1000 R	5	10^{12}	10
H [132] [133]	$H(^{19}F,\alpha\gamma)^{16}O$	16.4	88	500	^{19}F 16-18	γ 6.1-6.9	400 R	0.5	10^{13-14}	$10^{2}-10^{3}$
H [134] [135]	$H(^{11}B,\alpha)^{8}Be^{*}$	1.8	66	100	^{11}B 1.5-2.5	α 1-4	500 R	0.5	10^{13-14}	10^{3}
H [136]	$H(^{15}N,\alpha\gamma)^{12}C$	6.4	13.5	200	^{15}N 6-8	γ 4.4	100 R	3	10^{14}	10^{3}
D [137]	$D(d,n)^{3}He$		NR		d 2.5	n 5	2000	30	10^{15-16}	10^{3} (4)
D [138] [139]	$D(d,p)T$		NR	30	d ~0.2	t~1 p~3	1000 E.A.S	3		10^{4}
D [140]	$D(^{3}He,p)\alpha$	0.65	525	700	^{3}He 0.3-4	p~13	2.10^{4} R	~10	10^{12-13}	10 (5)
D [141]	$D(^{3}He,\alpha)p$	"	"	"	"	α~3.5	800 Å E.A.S	~1	10^{14-15}	$10^{3}-10^{5}$
T [137]	$T(p,n)^{3}He$		NR		p (pulsed) 2.5	n~1.5	2000 E.A.S	30		10^{3} (4)
T	β émission									
^{3}He [140]	$^{3}He(d,p)\alpha$	0.430	350	700	d >0.15	p 14	4 R	20-30	10^{12}	10
^{3}He [141] [142]	$^{3}He(d,\alpha)p$	"	"	"	"	α 5	500 F.A.S	1	10^{14-15}	$10^{3}-10^{5}$ (5)
^{4}He [143]	$^{4}He(^{10}B,n)^{13}N$	3.8	40	30	3.5-5	n	600 R	1		10^{4}

Ref	Reaction										Note
D [131]	D(d,n)³He	NR		30	d 2.5	n 5	2000	30		10³	(4)
D [132][133]	D(d,p)T	NR		700	d ~0.2	t~1 p~3	1000 E.A.S	3	10¹⁵⁻¹⁶	10⁴	
D [134][135]	D(³He,p)α	0.65	525	"	³He 0.3-4	p~13	2·10⁴ R	~10	10¹²⁻¹³	10	(5)
D [136]	D(³He,α)p	"	"	"	"	α~3.5	800 Å E.A.S	~1	10¹⁴⁻¹⁵	10³-10⁵	
T [137]	T(n,n)³He	NR			p (pulsed) 2.5	n~1.5	2000 E.A.S	30		10³	(4)
T	β émission										

Ref	Reaction										Note
³He [140]	³He(d,n)α	0.430	350	700	d >0.15	p 14	4 R	20-30	10¹²	10	
³He [141][142]	³He(d,α)p	"	"	"	"	α 3.5	500 F.A.S.	1	10¹⁴⁻¹⁵	10³-10⁵	(5)
⁴He [143]	⁴He(¹⁰B,n)¹³N	3.8	40	30	3.5-5	2-3.5 n	600 R	1		10⁴	

The final form of this bibliographic work has been modified by taking into account the work of
J. Bottiger, ST Picraux and N. Rud [143]

1) Value estimated for a material like Ni
   With the resonance method (R), the depth resolution is mainly defined by the resonance width. With the energy analysis spectrum (E.A.S.) the depth resolution depends strongly on the acceptance angle of the detector (Kinematics spreading)

2) Evaluation for the surface sample

3) The bulk sensitivity depends on the surface layer thickness

4) Time-of-flight neutron detection. Pulsed incident beam

5) This reaction is convenient for lattice location experiments

E_R = Resonance energy : Γ = resonance width : σ = Estimated cross section for resonance energy or incoming particle energy

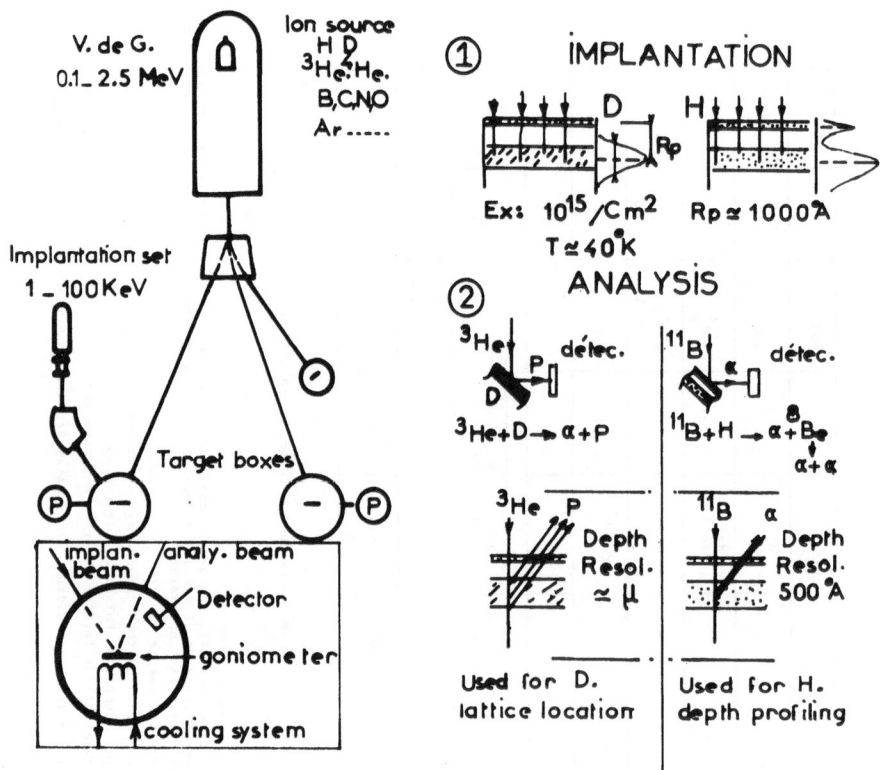

Fig. 11.   a) Experimental set-up. The implantation and
analysis beams are convergent on the same target
chamber.

b) Experimental process

1 - D and H implantation
2 - proper analysis method.

then unacceptably long. The background is drastically reduced when an absorber is placed in front of the detector, (ex : $H(^{11}B,\alpha)\alpha\alpha$ reaction), or when the ougoing particle energy is very high (for instance $D(^3He,p)\alpha$ reaction).

For γ rays or neutrons detection the background comes mainly from the natural background provided the analysing beam is correctly collimated. In this case the sensitivity is better when we use the strongest incident beam intensity ; a cooling target system is then sometimes useful.

The indications given in table 2 are reported references (131 to 143). The values of the depth resolution and the probing depth are relatively well defined by the different authors. Concerning the sensitivity, the values quoted in table 2 are sometimes questionable because they result from estimations made for a given experimental device. As a last comment it must be noticed that these analysis can be performed within a wide incident energy range. The D(d,p)T nuclear reaction requires only 200 KeV to take place. Then the same machine can be used for implantation and analysis of deuterium. For hydrogen profiling by resonant nuclear reaction the energy range may extend from about 2 MeV to 20 MeV.

Example of application : Study of hydrogen implanted in aluminium

For this purpose, not only hydrogen but also deuterium are implanted, the analysis being made in situ as indicated on the figure 11 a for the experimental set-up. Principles of this analysis are given on the figure 11 b summarizing the information obtained :

- The depth profile measurements for implanted hydrogen with the $^1H(^{11}B,\alpha)^8Be^*$ nuclear reaction.

- The lattice location (by channeling) of the implanted deuterium with the $D(^3He,p)\alpha$ nuclear reaction.

These measurements are performed for different implantation and annealing temperatures. From the excitation curve of the $^1H(^{11}B,\alpha)^8Be^*$ (Fig.12a) we have extracted the depth profiles (fig. 12 b) of the hydrogen implanted ; then we have plotted (fig. 12 c) the fraction of the remaining hydrogen in the implanted region as a function of the annealing temperature. These curves (fig. 12 c) point out a long range migration for hydrogen from about 300 K. Using the $D(^3He,p)\alpha$ nuclear reaction the angular scan yield is plotted through different axis or planes of aluminium (fig. 13 a). The comparison of these experimental results with what can be theoretically expected (fig. 13 b) shows unambiguously that the implanted deuterium (at T < 175 K) occupy a tetrahedral position in aluminium. This configuration is lost when the annealing temperature reaches about 300°K. More results and related interpretations being published elsewhere (144) our main conclusions will be

- In aluminium single crystals the implanted hydrogen occupies

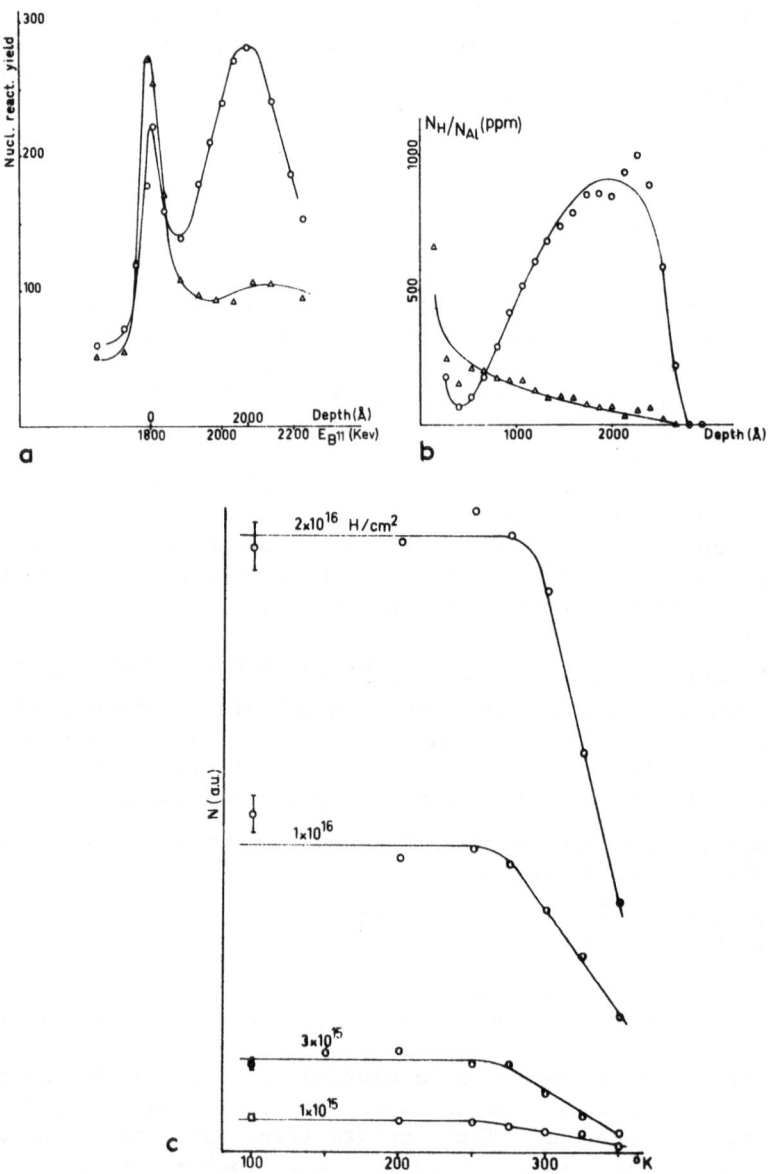

Fig. 12 : General conditions : Al, target implantations 10^{15} H/cm^2
10 keV- 33°K. a) Excitation curves of the ^{1}H(^{11}β,α)^{8}Be* nuclear
reaction. b) Depth profiles deduced from the excitation curves from
a mathematical deconvolution for the curve a and b. -o- annealing
cycles from 33 k. to 275 k. -Δ- annealing cycles beyond 300 K.
c) Isochronal annealing curves of hydrogen implanted in aluminium,
for several fluences.

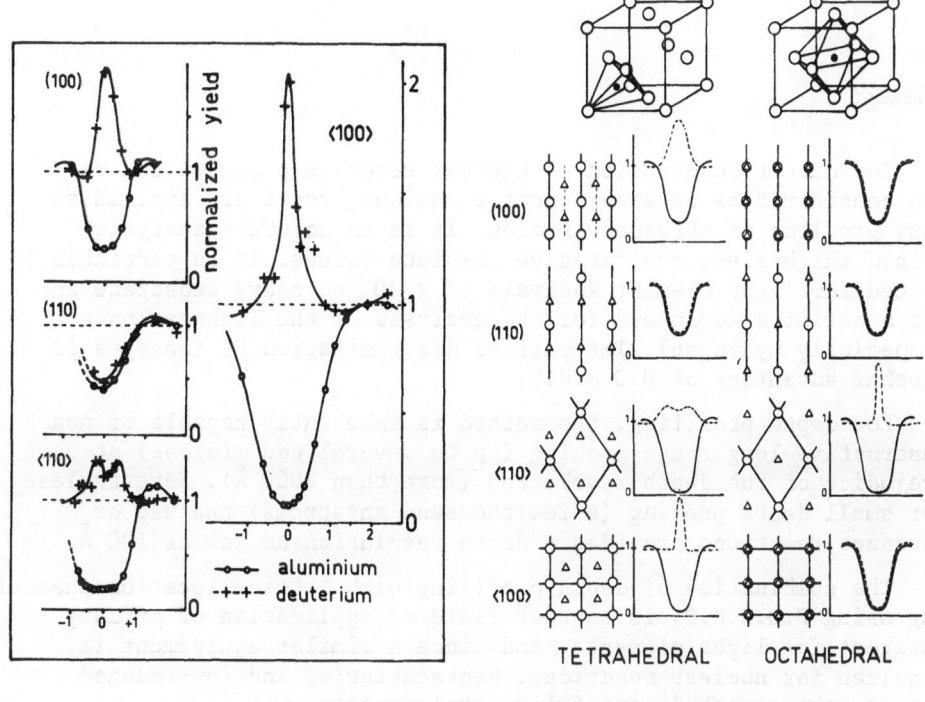

a   b

Fig. 13.   a) Experimental curves of angular scan yield through
               several axes and planes of Aluminium implanted
               with deuterium. 10^{15}D/cm^2, 15 keV, T = 33°K
               -o- ^{27}Al(^{3}He,^{3}He)^{27}Al backscattered yield
               -+- D (^{3}He,p)α nuclear reaction.

           b) Relative position of tetrahedral and octahedral
               sites for F.C.C. crystals and corresponding forms
               of the channeling curves.

the tetrahedral position when the implantation is lower than 175 K.

This site occupancy seems to be connected with the presence of monovacancies produced during the implantation.

Similar studies are presently in progress for other materials.

SUMMARY

The direct observation of nuclear reaction products can be now considered as an experienced technique, routinely applied to many problems of characterization. It is an accurate analysis method which gives quantitative absolute values. It is particularly suited for light element analysis (Z < 20) on heavy substrate and has a definite advantage for the analysis of the lightest ones (especially hydrogen). The perfect discrimination of isotopes is another advantage of D.O.N.R.P.

For depth profiling, the method is inherently capable of non destructive long range probing (up to several ten microns) at some prejudice of the depth resolution (more than 2000 Å). Nevertheless for small depth probing (a few thousand angströms) the use of resonant reactions provides a depth resolution as low as 100 Å.

The combination of depth profiling with lattice location channeling using D.O.N.R.P. is another field of application of primary interest for light elements. And since a similar equipement is required for nuclear reactions, backscattering and ion-induced X rays, these methods are fully complementary.

ACKNOWLEDGMENTS

Due to unexpected duties, Dr. S. RIGO from the "Group de Physique des Solides de l'ENS" of PARIS, who gave the lecture during the Summer School, has not been able to make-up his manuscript. The present authors, discussion leaders on the same subject at this time, acknowledge S. RIGO for having carefully read their manuscript, made pertinent suggestions and supplied a number of documents for illustration, which were used during the lecture .

REFERENCES

1. P. MARMIER, R. SZOSTAK, Schweiser Archiv.28, 2, (1962) 78
   R.F. SIPPEL, E.D. GLOVER, Nucl. Instrum. Methods 9, 1 (1960) 37
2. G. AMSEL, L.A.L. Report N° 1053, May 1963.
   G. AMSEL, D. SAMUEL, Anal. Chem. 39 (1967) 1689.
3. J. TARDY Thesis LYON (1975)
4. G. AMSEL, J.P. NADAÏ, E. D'ARTEMARE, D. DAVID, E. GIRARD,
   and J. MOULIN, Nuc. Inst. & Meth. 92, (1971) 481
5. M. BRUEL, M. BOISSIER, E. LIGEON, J. Radioanal. Chem. 17
   (1973) 79
6. A. CACHARD "Physico-Chemical analysis of thin films " In
   "Physics of non-metallic thin films" Ed. C.H.S. DUPUY and
   A. CACHARD   . New York Plenum Press (1976)
7. G. AMSEL, G. BERANGER, B. DE GELAS and P. LACOMBE, J. of Appl.
   Phys. 39, (1968) 2246
8. G. AMSEL, D. DAVID, G. BERANGER, P. BOISOT, Rev. Phys. Appl. 3
   (1968) 373
9. J.P. THOMAS, J. ENGERRAN, A. CACHARD, J. TARDY, Nucl. Inst.
   Meth. 119 (1974) 373
10. J.F. ZIEGLER, G.W. COLE, J.E.E. BAGLIN, J. Appl. Phys. 43
    (1972) 3809
11. K. MÜLLER, R. HENKELMANN, H. BOROFFKA Nucl. Inst. Meth. 129
    (1975) 557
12. K.L. DUNNING, NRL Report 7230, March 1971
13. J.P. NADAÏ Thesis PARIS (1971)
14. J. SIEJKA, J.P. NADAÏ, G. AMSEL, J. Electrochem. Soc. 118
    (1971) 727
15. K.L. DUNNING, G.K. HUBLER. J. COMAS, W.H. LUCKE and H.L. HUGHES
    Thin Solid Film 19, (1973)145
16. E. LIGEON, M. BRUEL, A. BONTEMPS, G. CHAMBERT, J. MONNIER,
    J. Radioanal. Chem. 16, (1973) 537
17. D. DAVID, G. AMSEL, P. BOISOT, G. BERANGER, J. Electrochem. Soc.
    122, (1975) 388
18. E.A. WOLICKI in "New Uses of ion accelerators" Ed. J.F. ZIEGLER
    Plenum Press. N.Y. (1975)
19. G. AMSEL, J. Radioanal. Chem. 17, (1973)15
20. J.W. MAYER, A. TUROS Thin Solid films 19   (1973) 1
21. W.K. CHU, J.W. MAYER, M.A. NICOLET, T.M. BUCK, G. AMSEL and
    F. EISEN. Thin Solid films 17 (1973) 1
22. T. LAURITSEN and F. AJZENBERG-SELOVE, Nuc. Phys. 78 (1968) 1

23. F. AJZENBERG-SELOVE and T. LAURITSEN, Nuc. Phys. 114 (1968) 1.
24. F. AJZENBERG-SELOVE Nuc. Phys. 152, 1 (1970) ; 166 (1971) 1.
     Al 90 1 (1972).
25. P.M. ENDT and C. VAN DER LEUN, Nuc. Phys. A 214 (1973) 1.
26. D.J. HOREN et al. Nuclear Data Sheets 11 ii (1974)
27. J.R. BIRD, B.L. CAMPBELL and P.B. PRICE, Atomic Energy Review
     12, (1974) 275.
28. "Handbook of Materials analysis" Edts J.W. MAYER and E. RIMINI
     Academic Press N.Y. 1977
29. J. LORENZEN, D. BRUNE, Aktiebolaget Atomenergi - AI 476
     Report (1973).
30. J.B. MARION, G. WEBER and F.S. MOZER, Phys. Rev. 104(1956) 1402.
31. R. PRETORIUS, Radiochem. Radioanal. Lett. 10(1972) 303.
32. R. PRETORIUS, P.P. COETZEE and M. PEISACH, J. Radioanal. Chem.
     16(1973) 551.
33. D. DIEUMEGARD Thesis PARIS (1971).
34. R. PRETORIUS, P. COETZEE J. Radioanal. Chem. 12(1972) 301.
35. W.E. SWEENY and J.B. MARION Phys. Rev. 182 (1968) 1007.
36. I. GOLICHEFF Thesis PARIS (1973).
37. J.W. BUTLER, "Table of (p,γ) Resonances" NRL Report 5282, Apr.
     1959.
38.   J.A. BIGGERSTAFF, R.F. HOOD, H. SCOTT, M.T. Mc ELLISTREM
     Nucl. Phys. 36 (1962) 631
39. R.V. POORE, P.E. SHEARIN, D.R. TILLEY, R.M. WILLIAMSON Nucl.
     Phys. A-92 (1967) 97
40. P.P. COETZE, R. PRETORIUS, and M. PEISACH, J. Radioanal. Chem.
     25 (1975) 283
41. J.P. THOMAS, J. ENGERRAN, J. TOUSSET, J. Radioanal. Chem. 25
     (1975) 163
42. J.P. LONGUEQUEUE, Thesis, GRENOBLE (1963)
43. E. LIGEON, A. BONTEMPS, J. Radioanal. Chem. 12 (1972) 335
44. I. GOLICHEFF, M. LOEUILLET and CH. ENGELMANN, J. Radioanal.
     Chem. 12 (1972) 233
45. E. KASHI, R.R. PERRY, J.R. RISSER, Phys. Rev. 117, 5 (1960)
     1289
46. M. HUEZ, L. QUAGLIA, G. WEBER Nucl. Inst. Meth. 105 (1972) 97
47. G. WEBER and L. QUAGLIA, J. Radioanal. Chem. 12 (1972) 323
48. T.B. PIERCE, J.W. McMILLAN, P.F. PECK and G. JONES, Nucl. Inst.
     Meth, 118 (1974) 115
49. D.A. CLOSE, J.J. MALANIFY and C.J. UMBARGER, Nucl.Inst. Meth.
     113, (1973) 561
50. M. BEAUMEVIELLE, M. LAMBERT, H. HAKEV and A. AMOKRANE Nuovo.
     Cim. XLII, 2 (1967) 139
51. G. AMSEL, D. DAVID, Revue de Phys. Appliquée 4 (1969) 383
52. B. SUNDQVIST, L. GÖNCZI, I. KOERSNER, R. BERGSMAN, U. LINDH
     "Ion beam surface layer analysis" 2 (1976) 945;edited by
     O. MEYER, G. LINKER, F. KAPPELER (Plenum Press. N.Y.)
53. C. OLIVIER, M. PEISACH, T.B. PIERCE, J. Radioanal. Chem. 32,
     (1976) 71
54. S. RIGO, G. AMSEL, M. CROSET, J. Appl. Phys. 47, 7 (1976) 2800.

55. M. CROSET, S. RIGO, G. AMSEL Appl. Phys. Lett. 19 (1971) 33
56. J. TARDY, A. CACHARD, J.P. THOMAS, J. ENGERRAN Le Vide 173 (1974) 359
57. S. GORODETZKY, J.C. ADLOFF, F. BROCHARD, P. CHEVALLIER, D. DISDIER, Ph. GORODETZKY, R. MODJTAHED-ZADEH, F. SCHEIBLING, Nucl. Phys. A 113 (1968) 221.
58. R.F. SEILER, C.H. JONES, W.J. ANZICK, D.F. HERRING and K.W. JONES Nucl. Phys. 45 (1963) 647
59. L. QUAGLIA, M. CUYPERS, G. ROBAYE, J.N. BARRANDON Nucl. Inst. Meth. 68 (1969) 315
60. M.Y. CHARTOIRE, J. ENGERRAN, J.P. THOMAS, J. TOUSSET, P. BUSSIERE, Thin Solid films 30 (1975) 311
61. A. CACHARD, J. PIVOT, A. ROGER, M. TALVAT, J.P. THOMAS, Rev. Phys. appl. 6, 3 (1971) 279
62. A. CACHARD, A. ROGER, J. PIVOT, C.H.S. DUPUY Phys. stat. sol. (a) 5 (1971) 637
63. L. QUAGLIA, G. WEBER, M. CUYPERS, Int. J. of Appl. Rad. and Is. 22 (1971) 449
64. G. AMSEL Anal. phys. 9 (1964) 297
65. M. CROSET, E. PETREANU, D. SAMUEL, G. AMSEL, J.P. NADAÏ, J. Electrochem. Soc. 118 5 (1971) 717
66. R. ROBIN, A.R. COOPER, A.H. HEUER J. Appl. phys. 44 8 (1973) 3770
67. E.C. LIGHTOWLERS, J.C. NORTH, A.S. JORDAN, L. DERICK and J.L. MERZ, J. Appl. Phys. 44, (1973) 4758
68. J.L. WHITTON, I.V. MITCHELL and K.B. WINTERBON, Can. J. Phys. 49 (1971) 1225
69. J.M. CALVERT, D.G. LEES, D.J. DERRY and D. BARNES, J. Radioanal. Chem. 12 (1972) 271
70. D.J. NEILD, P.J. WISE and D.G. BARNES, J. Phys. D : Appl. Phys. 5, (1972) 2292
71. W.A. RANKEN, T.W. BONNER, J.H. McCRARY Phys. Rev. 109 5 (1958) 1646
72. J.N. BARRANDON, R. SELTZ, C.R. Acad. Sc. PARIS 268 (1969) 1852
73. M. CROSET, D. DIEUMEGARD, J. Electrochem. Soc. 120 (1972) 526
74. B. MAUREL, D. DIEUMEGARD, G. AMSEL, J. Electrochem. Soc. 119 (1972) 1715
75. E. MÖLLER and N. STARFELT, Nucl. Inst. Meth. 50 (1967) 225
76. J.W. MANDLER, R.B. MOLER, E. RAISEN and K.S. RAJAN, Thin Solid films 19 (1973) 165
77. L. PORTE, J.P. SANDINO, M. TALVAT, J.P. THOMAS and J. TOUSSET J. Radioanal. Chem. 16 (1973) 493
78. I. GOLICHEFF and CH. ENGELMANN, J. Radioanal. Chem. 16 (1973) 503
79. J. STROOBANTS, F. BODART, G. DECONNINCK, G. DEMORTIER, G. NICOLAS, Ion beam surface layer analysis 2 (1976) 933 Edited by O. MEYER, G. LINKER, F. KÄPPELER (Plenum Press N.Y.)
80. J.P. GARREC, E. LIGEON, A. BONTEMPS, R. BLIGNI, A. FOURCY, J. Radioanal. Chem. 19 (1974) 339
81. Y. TAKEUCHI, Y. HIRATATE, K. MIURA, T. TONEI, S. MORITA, Nucl. Phys. A. 109 (1968) 105

82. G. DECONNINCK and G. DEMORTIER, J. Radioanal. Chem 12 (1972) 189
83. C. OLIVIER, M. PEISACH Radiochem. Radioanal. Lett. 19 3
     (1974) 217
84. P.M. ENDT, C. VAN DER LEUN, Nucl. Phys. A 105 (1969) 140
85. M.K. BERNETT, J.W. BUTTLER, E.A. WOLICKI and W.A. ZISMAN
     J. Appl. Phys. 42, (1971) 5826
86. R.O. BONDELID and C.A. KENNEDY Phys. Rev. 115 (1959) 1601
87. K.L. DUNNING and H.L. HUGHES, IEEE Trans. on Nucl. Sci. NS 19
     (1972) 243
88. G.J. McCALLUM Phys. Rev. 123 2 (1961) 568
89. E. GADOLI, G.M. MARCAZZAN, G. PAPPALARDO, Phys. Lett. 11
     2 (1964) 130
90. I.V. MITCHELL, W.N. LENNARD Ion beam surface layer analysis
     2 (1976) 925 Edited by O. MEYER, G. LINKER, F. KÄPPELER
     (Plenum Press N.Y.)
91. B.H. WILDENTHAL, R.W. KRONE, F.W. PROSSER Jr. Phys. Rev. 135
     3 B (1964) 680
92. J.F. CHEMIN, J. ROTURIER, B. SABOYA, G.Y. PETIT. J. Radioanal.
     Chem. 12 (1972) 221
93. G. GUERNET, E. LIGEON, N. LONGEQUEUE, TSAN UNG CHAN,
     P. LONGEQUEUE, J. de Physique 29 (1968) 9
94. H.R. SAAD, Z.A. SALEH, N.A. MANSOUR, E.M. SAYED, I.I. ZALUBOVSKY
     Nucl. Phys. 84 (1966) 629
95. E.A. WOLICKI and A.R. KNUDSON, Int. J. Appl. Rad. Isot. 18
     (1967) 429
96. J.W. OLNESS, W. HAEBERLI, M.W. LEWIS, Phys. Rev. 112 5 (1958)
     1702
97. M. COHEN, L. PORTE, J.P. THOMAS, J. TOUSSET, J. Radioanal.
     Chem. 17 (1973) 65
98. W.M. SANDERS, B.K. BARNES, D.M. HOLM, J. Radioanal. Chem. 16
     (1973) 525
99. B. COX, C. ROY J. Electrochem. Soc. 4 (1966) 121
100. R.W. OLLERHEAD, E. ALMQVIST, J.A. KUEHNER J. Appl. Phys. 37
     6 (1966) 2440

101. M. PEISACH, J. Radioanal. Chem. 12 (1972) 251
102. P. PEISACH and R. PRETORIUS, J. Radioanal. Chem. 16 (1973) 559
103. I.S. GILES, M. PEISACH, J. Radioanal. Chem. 32 (1976) 105
104. G.E. COOTE, N.E. WHITEHEAD and G.J. McCALLUM, J. Radioanal.
     Chem. 12 (1972) 491
105. G. DECONNINCK, J. Radioanal. Chem. 12 (1972) 157
106. C. OLIVIER, M. PEISACH, J. Radioanal. Chem. 5 (1970) 391
107. C. OLIVIER, M. PEISACH, J.S. Afric. Inst. 23 (1970) 77
108. J. LORENZEN, D. BRUNE and S. MALMSKOG, J. Radioanal. Chem. 16
     (1973) 483
109. E. MÖLLER, L. NILSON and N. STARFELT, Nucl. Inst. Meth. 50
     (1967) 270
110. R. HENKELMANN, H.J. BORN, J. Radioanal. Chem. 16 (1973) 473
111. J.A. MOORE, I.V. MITCHELL, M.J. HOLLIS, J.A. DAVIES, L.M. HOWE
     J. Appl. Phys. 46 1 (1975) 52

112. J. L'ECUYER, C. BRASSARD, C. CARDINAL, L. DESCHENES, Y. JUTRAS,
     J.P. LABRIE, Nucl. Inst. Meth. 140 (1977) 305
113. G. DEARNELEY in "Ion beam surface layer analysis" edited by
     O. MEYER, J. LINKER, F. KÄPPELER Vol. 2 885 Plenum Press N.Y.(1976)
114. J.A. COOKSON, A.T.G. FERGUSON, F.D. PILLING, J. Radioanal. Chem.
     12 (1972) 39
115. J.W. McMILLAN, T.B. PIERCE "Ion beam surface layer analysis"
     Edited by O. MEYER, G. LINKER, F. KÄPPELER, Vol 2, 913
     Plenum Press, N.Y. (1976)
116. A. TUROS, L. WIELONSKI, A. BAREZ, Nucl. Inst. Meth. 111 (1973)
     605
117. M. CROSET, S. RIGO, G. AMSEL, Proc. Intern. Conf. MIS Structures
     Grenoble (1969) 259
118. C. CHERKI, J. SIEJKA J. Electrochem. Soc. 120 (1973) 784
119. C. CHERKI, Electrochem. Acta 16, (1971) 1727
120. C. ORTEGA, J. SIEJKA, Proc. Electrochem. Soc. Meeting
     Cleveland (1971) 221
121. S. RIGO, J. SIEJKA, Solid. State Comm. 15 (1974) 259
122. G. AMSEL, D. DAVID, G. BERANGER, P. BOISOT , B. DE GELAS,
     P. LACOMBE, J. Nucl. mat. 29 (1969) 144

123. G. BERANGER, P. BOISOT, P. LACOMBE, G. AMSEL and D. DAVID
     Revue de Phys. Appl. 5 (1970) 383
124. L. FAURE-MAZAGOL, J. TOUSSET, M. BOISSIER, Analusis 2 4 (1973)
     287
125. C. DIAINE, J.A. ROGER, J. PIVOT, A. CACHARD, C.H.S. DUPUY
     A.V.I.S.E.M. Meeting PARIS (1971)
126. L.C. FELDMAN, E.N. KAUFMANN, J.M. POATE, W.M. AUGUSTYNIAK,
     in "Ion implantation in semiconductors and other materials"
     Edited by B.L. CROWDER (Plenum Press N.Y.) (1973).
127. D.G. SIMONS, D.J. LAND, J.G. BRENNAN, M.D. BROWN in Vol 2
     "Ion beam surface layer analysis" Edited by O. MEYER, G. LINKER
     and F. KÄPPELER (Plenum Press N.Y.) (1976) 863
128. A. MAYEVSKY, D. SAMUEL, C. ORTEGA, G. AMSEL, J. Appl. Physio-
     logy 39 (1975) 300.
129. Y. AKASAKA, K. HORIE, T. SAKURAI, H. NISHI, S. KAWABE and
     A. TOHI, J. Appl. Phys. 44 (1973) 145
130. Z.D. POPOVIC, M.Y. STOTT, Phys. Rev. Lett. 33 19 (1974) 1164
131. G.M. PADAWER, D.J. LAROON, J. RAND and P.N. ADLER. Met. Trans
     2 (1971) 2287
132. D.A. LEICH and T.A. TOMBRELLO. Nucl. Inst. Meth. 108 (1973) 67
133. J. BOTTIGER, J.R. LESLIE and N. RUD, J. Appl. Phys. 47 (1976)
     1672.
134. E. LIGEON, A. GUIVARCH, Rad. Effects 22, (1974) 101
135. E. LIGEON, A. GUIVARCH, Rad. Effects 27 (1976) 129
136. W.A. LANDFORD, H.P. TRAUTVETTER, J.F. ZIEGLER and J. KELLER.
     Appl. Phys. Lett. 28 9 (1976) 566
137. J.C. DAVIES and J.D. ANDERSON, J. Vac. Sci. Techno. 12 1
     (1975) 356

138. H.D. CARSTENJEN and R. SIZMANN, Phys. Lett. A 40 2(1972) 93

139. P.B. JOHNSON, Nucl. Inst. Meth. 114 (1974) 467

140. P.P. PRONKO, J.C. PRONKO, Phys. Rev. B 9 7 (1974) 2870

141. R.A. LANGLEY, S.T. PICRAUX and F.L. VOOK, J. of Nucl. Materials
     53 (1974) 257

142. W. ECKSTEIN, R. BEHRISCH and J. ROTH "Ion beam surface layer
     analysis" Edited by O. MEYER, G. LINKER and F. KÄPPELER
     Plenum Press. N.Y. and London 2 (1976) 821

143. J. BØTTIGER , S.T. PICRAUX and N. RUD, "Ion beam surface
     layer analysis" Edited by O. MEYER, G. LINKER and
     F. KÄPPELER. Plenum Press - N.Y. and London 2 (1976) 811

144. J.P. BUGEAT, A.C. CHAMI, E. LIGEON, Phys. Lett. A 58 2
     (1976) 127.

145. G. AMSEL, in Physics of Electrolytes, Vol.1, Edited by
     G. HLADIK, Academic Press, N.Y. (1972)

146. G. AMSEL, Proceeding of an International Colloquium on oxygen
     Isotopes, Cadarache (FRANCE), Sept. 1972, p. 163 and p. 138.

147. S. RIGO, Thesis, PARIS, (1977).

148. J. SIEJKA, C. CHERKI and J. YAHALOM, Electrochimica Acta,
     (1972) 161

Part IV

SOLID STATE STUDIES USING

CHANNELING EFFECTS

CHANNELING: GENERAL DESCRIPTION

J.A. Davies

Solid State Science Branch
Atomic Energy of Canada Limited
Chalk River, Ontario, Canada

1. INTRODUCTION

Whenever an ion beam penetrates a crystalline lattice, strong directional effects, commonly known as "channeling", may occur. Indeed, some of the more interesting and useful applications of ion beams in materials characterization are based on this channeling effect. The first lecture, therefore, will consist of a general description of the channeling phenomenon in order to provide a basis for the subsequent series of more specialized lectures by Mazzoldi, Rimini and myself on specific solid state applications. For those seeking a more detailed and comprehensive treatment of ion channeling and its applications, excellent reviews by Gemmell[1] and by Morgan[2] have recently been published.

What exactly do we mean by channeling? Basically, it is a gentle steering process arising from a correlated series of collisions whenever a beam of energetic charged particles moves through a crystal in a direction almost parallel to a major axis or plane. Provided the angle between the beam and the crystal row is sufficiently small, the increasing electrostatic repulsion between the ion and the screened field around each successive target nucleus (due to the gradually decreasing impact parameter at each successive collision) is sufficient to deflect the particle smoothly away from the row—as illustrated schematically in figure 1.

An obvious consequence of this steered motion is that it prevents the particle from having violent collisions with target atoms, and therefore the nuclear stopping is strongly attenuated. Consequently, the beam loses energy more slowly, it penetrates more

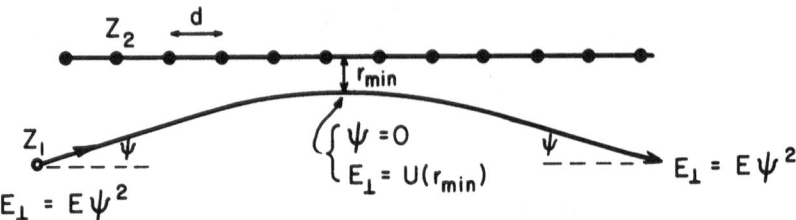

Figure 1 - Schematic diagram illustrating how a correlated sequence of collisions with an aligned row of atoms can gently steer (channel) a particle. Note that the transverse energy, $E_\perp$, is conserved throughout the collision sequence.

deeply, it creates much less damage along its track, and it has much less probability of wide-angle scattering, or of participating in nuclear reaction yields. In short, almost all physical proc-esses that can occur between moving particles and target atoms are drastically affected by such a steering process.

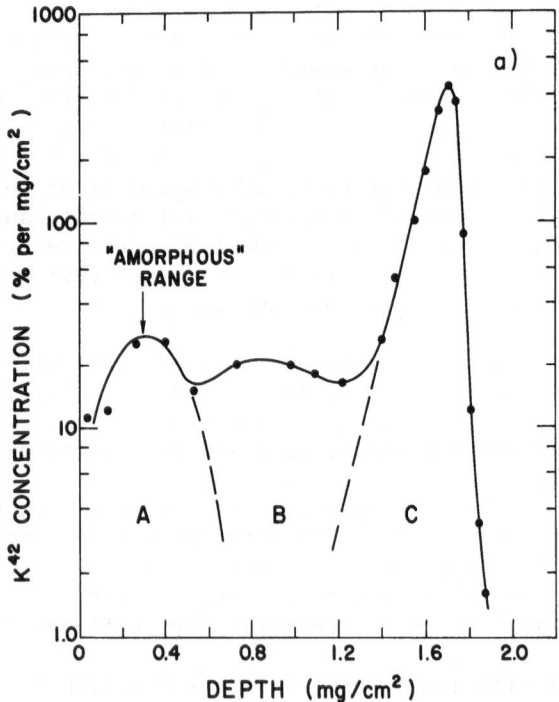

Figure 2 - Examples of channeling phenomena: a) Range profile of 500 kev ^{42}K ions injected at 25°C into tungsten along the <111> direction (from reference 3);

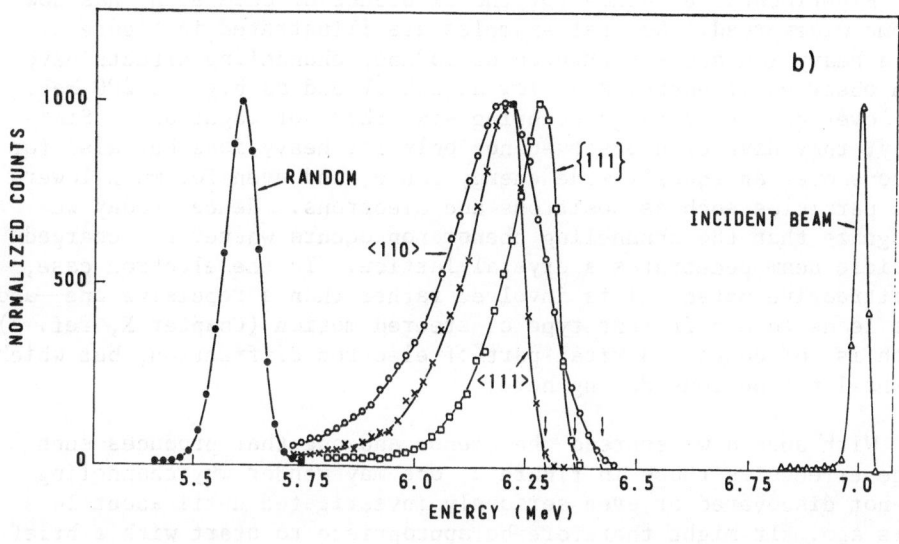

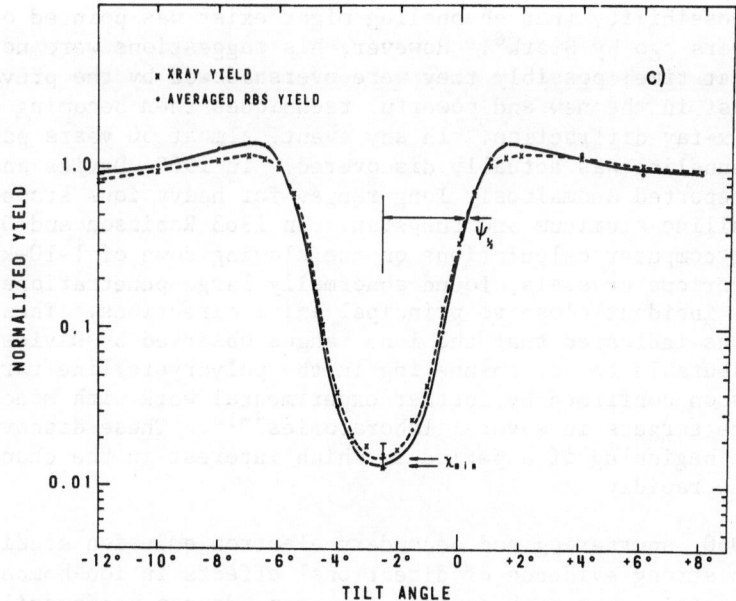

Figure 2 – Examples of channeling phenomena (cont'd); b) Energy spectra for 7.0 MeV ^{4}He ions transmitted through an 11.2 micron thick Si crystal along the directions indicated[4]; c) Comparison of Rutherford scattering and M X-ray yields in tungsten as a function of the angle between the incident 0.5 MeV He$^+$ beam and the <111> axis[5].

Experimental evidence for the existence of channeling has now
become widespread. Several examples are illustrated in figure 2.
For a heavy ion such as krypton or iodine, channeling effects have
been observed at energies as low as 200 eV and as high as 200 MeV:
i.e. over an energy range spanning six orders of magnitude. Similarly, they have been observed not only for heavy ions but also for
protons over an equally wide energy range, and even for much lower-
mass particles such as positrons and electrons. Hence, today we
recognize that the channeling phenomenon occurs whenever a charged-
particle beam penetrates a crystal lattice. In the electron case,
an attractive potential is involved rather than a repulsive one—and
this leads to a different type of steered motion (Chapter X, ref. 2)
which is, of course, a vital part of electron diffraction, but which
we shall not be considering here.

With such a widespread phenomenon and one that produces such
large effects as those in figure 2, one may wonder why channeling
was not discovered or even seriously investigated until about 14
years ago. It might therefore be appropriate to start with a brief
historical introduction.

The possibility that channeling might exist was pointed out
over 60 years ago by Stark[6]. However, his suggestions were not pur-
sued at that time; possibly they were overshadowed by the prevailing interest in the new and powerful techniques then becoming available with x-ray diffraction. In any event, almost 50 years passed
before channeling was actually discovered. In 1960, Davies and coworkers[7] reported anomalously long ranges for heavy ions stopping in
polycrystalline aluminum and tungsten. In 1963 Robinson and Oen[8],
performing computer calculations on the slowing down of 1-10-keV Cu
ions in various crystals, found abnormally large penetrations for
those ions incident close to principal axial directions. These
computations indicated that the long ranges observed by Davies et al,
were attributable to ion channeling in the polycrystalline targets.
This was soon confirmed by further experimental work with mono-
crystalline targets in several laboratories[9-12]. These discoveries
marked the beginning of a period in which interest in the channeling
effect grew rapidly.

By 1960, sputtering and secondary electron emission studies had
also shown strong evidence of directional effects in ion bombardment of crystals, but such surface phenomena do not necessarily
indicate that the ions are undergoing a governed or channeled
motion. Here, it is necessary to draw a rather important distinction between channeling and purely geometrical transparency. Channeling is defined as a more or less stable, steered trajectory per-
sisting for a significant depth into the crystal. In the absence
of channeling, purely geometrical effects could still exist. Consider, for example, a neutron beam entering a thin crystal almost
parallel to a low-index axis. Those neutrons entering near the

middle of an open channel would traverse the entire crystal without ever coming close to an atomic nucleus, even though the atomic nuclei exhibit no steering action on them. In other words, a purely geometric transparency effect can exist, especially at shallow depths. Even for channeled trajectories, such transparency effects often play a significant role at shallow depths ($\lesssim$ 10 nm), as we shall see in section 3.6.

For simplicity, most channeling studies of interest in the materials characterization field can be grouped into two widely separated energy regimes:

(i)   Relatively slow moving heavy ions ($M_1 \gtrsim$ 10, and $v \ll v_o$, the Bohr velocity).

(ii)   High velocity light ions, such as MeV helium ions and protons.

The first category covers the regime of interest in ion implantation experiments, i.e. in the use of ion bombardment to modify a solid by introducing foreign (dopant) atoms into the lattice. It is also the regime where radiation damage effects predominate and where quantitative prediction of channeling behaviour is almost impossible. Here, channeling is often an unwanted complication which must be controlled in order to achieve reproducible results.

The second category covers the regime where a sound theoretical framework exists for predicting channeling behaviour; such beams form the basis for most of the solid state applications of channeling which we shall be discussing in subsequent lectures.

2.   SLOW-MOVING HEAVY IONS

Such ions are widely used in ion implantation; hence our main interest centers on the extent to which channeling can modify the depth distribution (range) of an implanted ion and its associated damage distribution.

Figure 2(a) illustrates the type of range distribution that is observed when a heavy-ion beam is implanted into a well-aligned tungsten crystal. Tungsten is a rather "ideal" case in that its lattice vibrations at room temperature are small, and so dechanneling effects are minimized. We see that the distribution curve consists of three clearly divided regions, which may be identified roughly with the three types of trajectory discussed in section 3.2. The two peaks, A and C, correspond to the non-channeled and well-channeled ranges respectively; the latter is due to those particles that remain channeled throughout their entire path, and is characterized by a sharp cut-off or maximum range, R_{max}. This is true also in other crystals, provided the implantation temperature is low enough to minimize dechanneling.

In most crystals, however, dechanneling is the dominant effect
at room temperature even for particles starting out well channeled;
it causes most of them to end up somewhere in the intermediate region
B between the two peaks of the range distribution. The resulting
distribution then appears as a roughly exponential tail with an even-
tual cut-off at R_{max}; in some cases, the distribution falls to an
immeasurably low level long before reaching R_{max}, so that peak C is
not actually observed. Several such distributions are illustrated
in figure 3.

Even in an "ideal" crystal such as tungsten, there is often a
significant amount of dechanneling involved, and the observed distribution is sensitive to many factors, such as lattice temperature,
surface contamination, small misalignment, or angular divergence—
some of which are often difficult to control. The amount of dechanneling also depends very much on how far the ions must travel to
reach R_{max}—i.e., on the magnitude of the electronic stopping power.
Thus ^{42}K, for example, exhibits a much more pronounced channeled
peak in tungsten than does ^{64}Cu. The latter ion has a much smaller
$-dE/dx$ value; hence it penetrates to much larger depths then ^{42}K,
and so exhibits a greater amount of dechanneling.

2.1   The Maximum Range R_{max}

At present, the most reproducible experimental parameter for
characterizing range distributions in single crystals is the maximum
range R_{max}. It does not vary significantly with bombardment dose,
lattice temperature, surface contamination, small misalignment, or

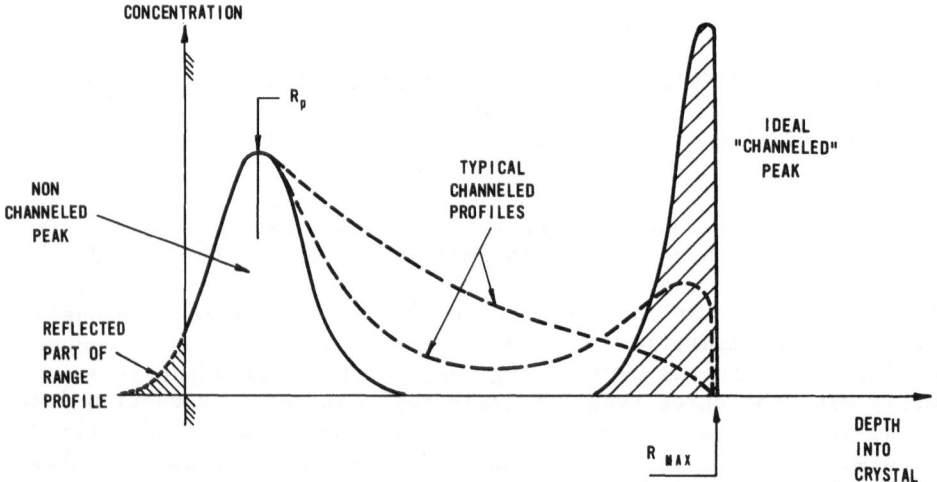

Figure 3 - Representative range profiles for an initially well-
channeled beam. The dashed curves represent cases where dechanneling effects are large.

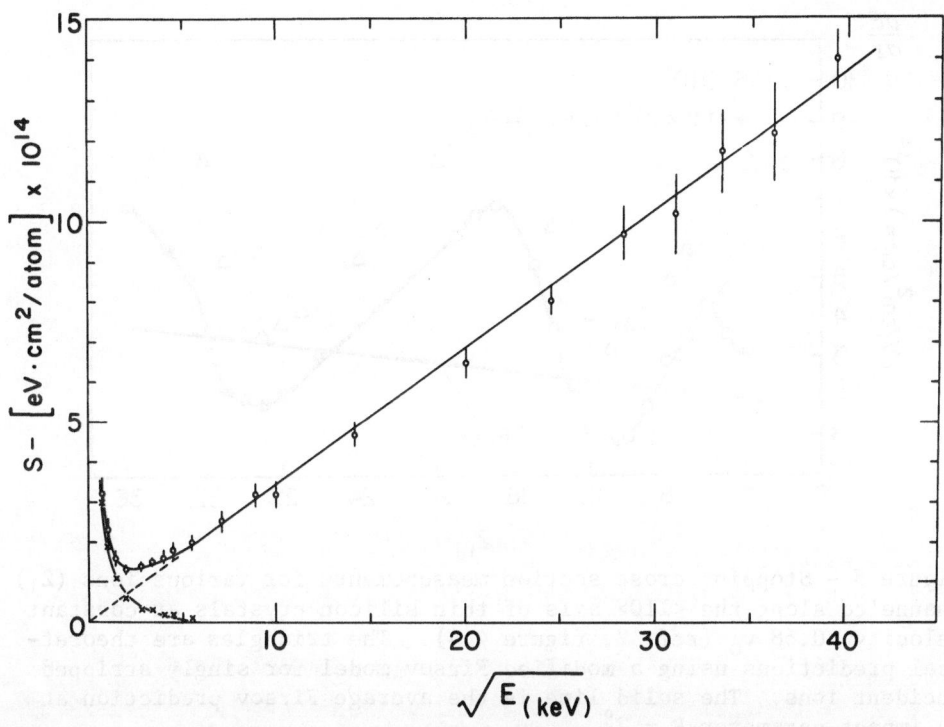

Figure 4 – Experimentally derived values of the stopping cross
section S versus $E^{1/2}$ for well-channeled xenon ions injected into
tungsten along the <100> direction[3]. The dotted line is an extrapolation of the electronic-stopping contribution S_e to lower
energies. Crosses indicate the nuclear stopping S_n obtained by sub-
tracting the extrapolated electronic stopping from the measured total
values.

beam divergence. On the other hand, the <u>number</u> of particles approaching R_{max}—i.e., the overall <u>shape</u> of the range distribution—
is quite sensitive to all these factors.

A detailed investigation[3] of the energy dependence of R_{max} for
various ions in tungsten has shown that R_{max} approximates closely
the $E^{0.5}$-dependence characteristic of electronic stopping down to
energies of a few keV, thus confirming that for a well-channeled ion,
electronic stopping is normally the dominant mechanism of energy
loss. This is illustrated in figure 4 in which the effective stopping cross section S of a well-channeled beam, obtained by differentiating the experimental R_{max} versus E curve, is plotted as a
function of $E^{0.5}$. For such a channeled beam, nuclear stopping is
evidently negligible, except at energies below ∿ 5 keV. For a <u>non</u>-
channeled xenon beam in tungsten, on the other hand, the LSS theory
predicts that the transition energy between nuclear and electronic

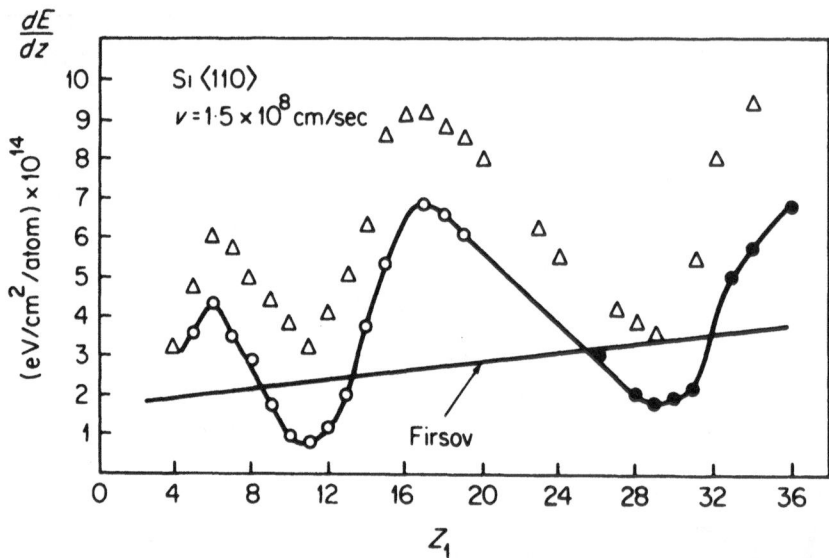

Figure 5 - Stopping cross section measurements for various ions (Z_1)
channeled along the <110> axis of thin silicon crystals at constant
velocity, 0.68 v_o (ref. 2, figure 4.4). The triangles are theoretical predictions using a modified Firsov model for singly stripped
incident ions. The solid line is the average Firsov prediction at
an impact parameter R = 2Å.

stopping occurs at 2.7 MeV—i.e., almost a factor of 10^3 higher in
energy. This leads to the observation that for well-channeled ions
of equal energy, the higher the mass the lower the velocity, and
hence the lower the stopping power. Therefore, contrary to the case
for non-channeled ions where nuclear stopping usually dominates,
channeled particles of high mass can penetrate deeper than lower-
mass particles of identical energy: for example, a well-channeled
uranium or bismuth beam at 100 keV actually has a longer R_{max} than
100-keV protons.

2.2 Z_1 Oscillations in R_{max}

There is one serious difficulty in attempting to predict R_{max},
as can be seen from figure 5, in which the electronic stopping cross
section S_e of a well-channeled ion is plotted at constant velocity
as a function of the atomic number Z_1 of the implanted atom. Not
only are the experimentally observed values of S_e significantly
smaller than those for an amorphous stopping medium, but also they
exhibit very large oscillations about a smooth Z_1 dependence. These
oscillations, which in amorphous targets rarely exceed 20-30%, are
evidently much enhanced for a channeled beam. Eisen's measurements

in silicon, for example, show peak-to-valley oscillations greater than a factor of 10.

Several theoretical groups have tackled the question of these Z_1 oscillations and have shown that the positions of the maxima and minima can be correlated roughly with ion size. Unfortunately, none of the theoretical treatments is yet able to predict accurately the magnitude of the oscillations, nor their dependence on energy or on channel size. It is hoped that further theoretical work will clarify the situation. In the meantime, the available evidence suggests that R_{max} is best obtained from experimental measurements.

2.3 Other Applications

Several significant consequences arise from the drastic reduction in the nuclear-stopping contribution:

(i) Electronic stopping can be investigated (for channeled ions) at much lower energies than would otherwise be possible.

(ii) A well-channeled ion creates no damage along its trajectory, and so may come to rest as an interstitial atom in a completely undamaged region of the crystal. This has interesting solid-state possibilities, including the observation of unusual types of interstitial diffusion, as shown by the very penetrating 'super-tail' in figure 6.

In summary, it is clear that channeling phenomena can have a dominant influence on the depth distribution of implanted ions in crystals. Unfortunately, the channeled part of a range profile is sensitive to many factors that may be difficult to control: for example, in most implantations a dose in excess of 10^{13} ions/cm^2 is desired; at low temperature this dose causes enough lattice disorder to markedly affect the channeled distribution, whereas at elevated temperature (where the disorder can be continuously annealed during the implantation) the enhanced vibrational amplitude of the lattice atoms makes dechanneling the dominant effect. Hence, the observed distribution is not easily reproducible, nor can it be accurately predicted. In many applications, it is therefore desirable to suppress the channeled component as much as possible. This may be achieved by carefully aligning the crystal so that the beam enters several degrees off-axis or, if the dose is large enough, it may become suppressed still further by the accumulation of lattice disorder during the implantation.

Fortunately, when one uses high energy, low-Z particles—such as the MeV protons and helium ions whose channeling behaviour we shall discuss in section 3—this sensitivity to surface contamination, bombardment dose, and lattice temperature is considerably reduced, and reproducible data are therefore not so difficult to obtain.

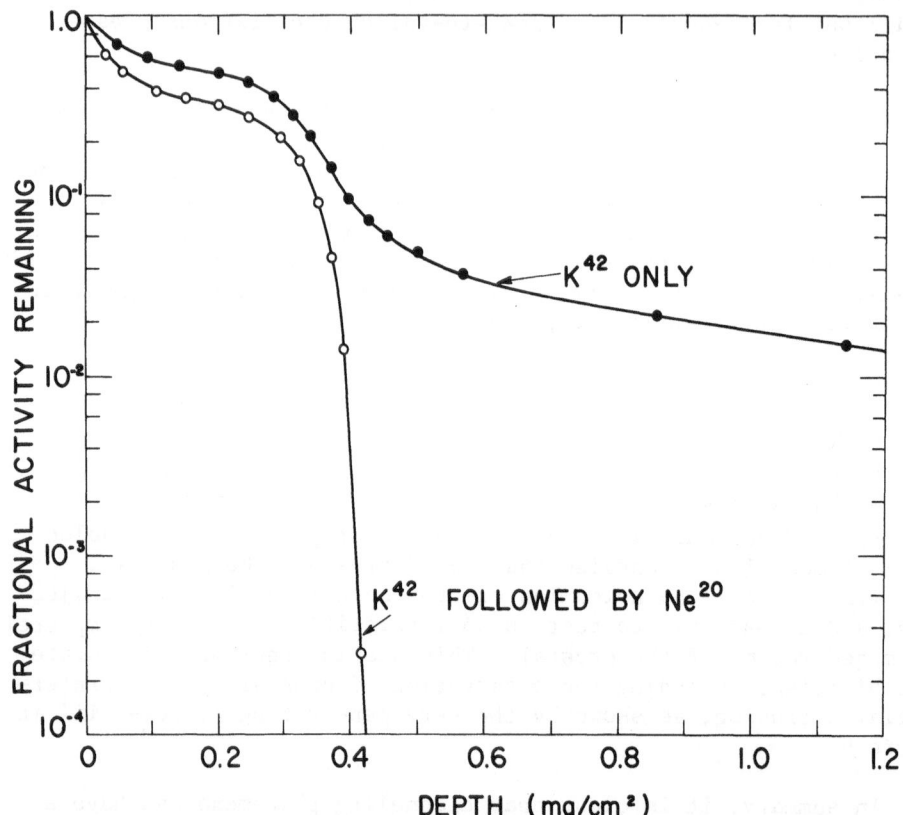

Figure 6 - Integral penetration profiles of 40-keV ^{42}K injected into tungsten along the <100> direction at 30 K, with (o) and without (•) post bombardment by a large dose (3×10^{15} ions/cm^2) of Ne ions to introduce trapping centers for interstitials. Note that in both cases the crystal was subsequently warmed to 300 K before measuring the range profile[13].

3.   CHANNELING OF MeV HELIUM IONS AND PROTONS

In this energy regime, there is a wide range of easily observable processes—for example, Rutherford backscattering (RBS), nuclear reactions, inner-shell X-ray production, and Coulomb excitation—all of which require fairly violent collisions between the incident ion and a target atom. For a well-channeled ion, the yield of such close-encounter processes falls almost to zero (as seen in figure 2(c)) and hence provides a rather sensitive probe for monitoring channeling behavior. Furthermore, as we shall see in section 3.1, a fairly quantitative theoretical framework exists for predicting channeled trajectories at these energies; also, the observed behavior is much more reproducible than at lower velocities, and is less

dependent on extraneous experimental factors such as surface contamination and dechanneling.

3.1 General Principles

One distinguishes three basically different types of particle
trajectories (figure 7).

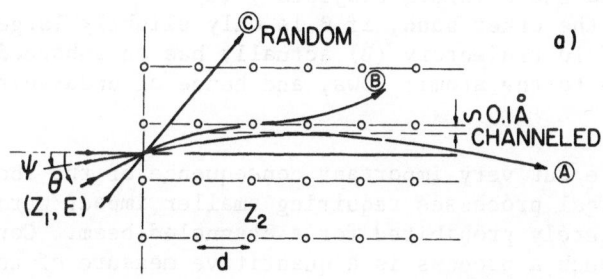

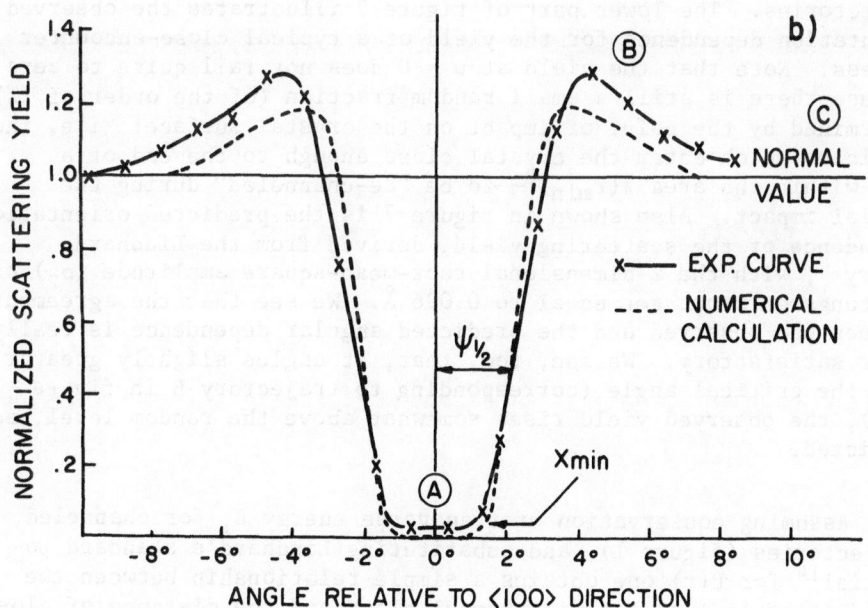

Figure 7 - a) Charged particle trajectories for three typical values
of the angle θ between the incident beam and a close packed row.
b) Experimental (x) and calculated(----) angular dependence of the
yield of a typical close encounter process (RBS of 480-keV protons
in <100> tungsten).

(i)   An energetic charged particle, entering a lattice with an angle θ less than the predictable critical angle $\psi_{1/2}$ of an atomic row (or plane), is steered by successive gentle collisions (trajectory A), and is thereby prevented from entering a forbidden region around each lattice row.   The radius r_{min} of this forbidden cylinder may be equated roughly to either the Thomas-Fermi screening distance, a, (typically 0.1 - 0.2 Å) or the vibrational amplitude ρ—whichever is larger.

(ii)   When the incident angle θ is much larger than $\psi_{1/2}$ the particle has no "feeling" for the existence of a regular atomic lattice, and so has a random trajectory (C).

(iii) On the other hand, if θ is only slightly larger than $\psi_{1/2}$, then the particle trajectory (B) actually has an enhanced probability of being close to the atomic rows, and hence of undergoing violent collisions.

One simple but very important consequence of the above model is that all physical processes requiring smaller impact parameters than r_{min} are completely prohibited for a channeled beam.   Consequently, the yield of such a process is a quantitive measure of the non-channeled fraction of the beam, and so provides a sensitive "detector" for studying the transition between channeled and non-channeled trajectories.   The lower part of figure 7 illustrates the observed orientation dependence for the yield of a typical close-encounter process.   Note that the yield at $\psi = 0$ does not fall quite to zero because there is still a small random fraction (of the order of 1%), determined by the point of impact on the crystal surface: i.e. those particles which enter the crystal close enough to the end of a row—within the area $\pi(r_{min})^2$—to be "de-channeled" during the initial impact.   Also shown in figure 7 is the predicted orientation dependence of the scattering yield, derived from the Lindhard theory[14], with the 2-dimensional root-mean-square amplitude (ρ_2) of the tungsten atoms set equal to 0.096 Å.   We see that the agreement between the observed and the predicted angular dependence is really quite satisfactory.   We see, too, that, at angles slightly greater than the critical angle (corresponding to trajectory B in figure 7(a)), the observed yield rises somewhat above the random level, as predicted.

Assuming conservation of transverse energy $E_\perp$ for channeled trajectories (figure 1), and substituting Lindhard's standard potential[14] for U(r) one obtains a simple relationship between the incident angle ψ at the mid channel plane and the distance of closest approach r_{min} to the lattice row:

viz. $E\psi^2 = U(r_{min})$

$$= \frac{Z_1 Z_2 e^2}{d} \cdot \log\left[\left(\frac{Ca}{r_{min}}\right)^2 + 1\right] \qquad (1)$$

where Z_1e and Z_2e are the nuclear charges of the moving ion and
target atom, d is the atomic spacing along the lattice row, 'a' is
the Thomas-Fermi scaling function and $C \sim \sqrt{3}$ is a numerical constant.
One of the necessary criteria for conserving $E_\perp$ and hence for maintaining a channeled trajectory is that r_{min} must not be smaller than
either 'a' or the vibrational amplitude ρ_2 of the lattice atoms.
Hence, a first-order estimate of the critical angle $\psi_{1/2}$ for axial
channeling is obtained by substituting ρ_2 for r_{min} in equation 1:

$$\psi_{1/2} = \left\{ \frac{Z_1 Z_2 e^2}{Ed} \quad \log\left[\left(\frac{Ca}{\rho_2}\right)^2 + 1\right] \right\}^{1/2} \tag{2}$$

More accurate estimates (from reference 15) are given below, with ψ
in degrees, E in MeV, lengths in Å:

Axial $\psi_{1/2} = 0.25 \ F_{ax}(\rho_2/a) \cdot (Z_1 Z_2/Ed)^{1/2}$ (3a)

Planar $\psi_{1/2} = 0.40 \ F_{pl}(\rho_1/a, \ 1/na) \left(\frac{Z_1 Z_2 na}{E}\right)^{1/2}$ (3b)

where n is the atomic density (atoms/Å^2) in the plane, and F_{ax} and
F_{pl} are weakly varying functions of the appropriate (ρ/a) ratio;
with typical F values for most lattices being in the 0.6 - 0.8 range.

Simple estimates of the minimum yield χ_{min} at $\psi = 0$ (figure
7(b)) can also be calculated directly from Lindhard's stability
criterion that $r_{min} \gtrsim$ the larger of 'a' and ρ:

Axial case: $\chi_{min} = Nd \ \pi \ (C_1 \ \rho_2{}^2 + C_2 a^2)$ (4a)

Planar case: $\chi_{min} = 2 \ (\rho_1{}^2 + a^2)^{0.5}/d_p$ (4b)

where d_p is the interplanar spacing, C_1 and C_2 are numerical
constants of the order of unity, and ρ_1 is the rms vibrational amplitude perpendicular to the plane. More accurate estimates[16]
for the axial χ_{min} are obtained by setting $C_1 \sim 3$ and $C_2 \sim 0.5$ in
equation 4a.

Values of $\psi_{1/2}$ vary from a few hundredths of a degree to a few
degrees. Values of χ_{min} are typically about 0.2 - 0.4 for planar
channeling and about 0.01 - 0.05 for axial channeling. Note that
both parameters depend significantly on ρ and hence channeling
effects are strongly enhanced at low temperature.

3.3 General Features of Channeling Experiments

A standard assembly for investigating the yield of backscattered particles is illustrated in figure 8. The incident beam (for

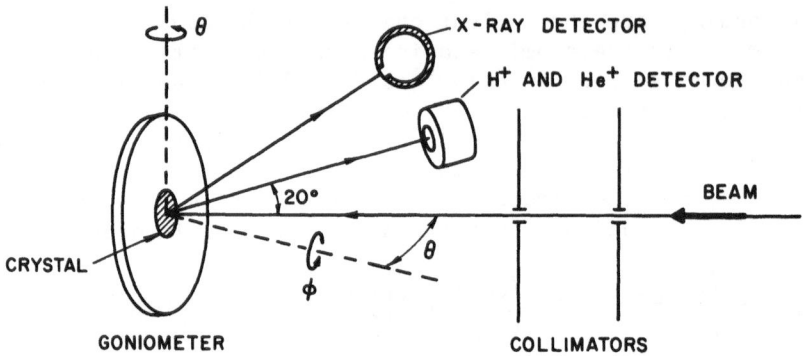

Figure 8 – Schematic diagram of the experimental assembly for a
typical channeling experiment.

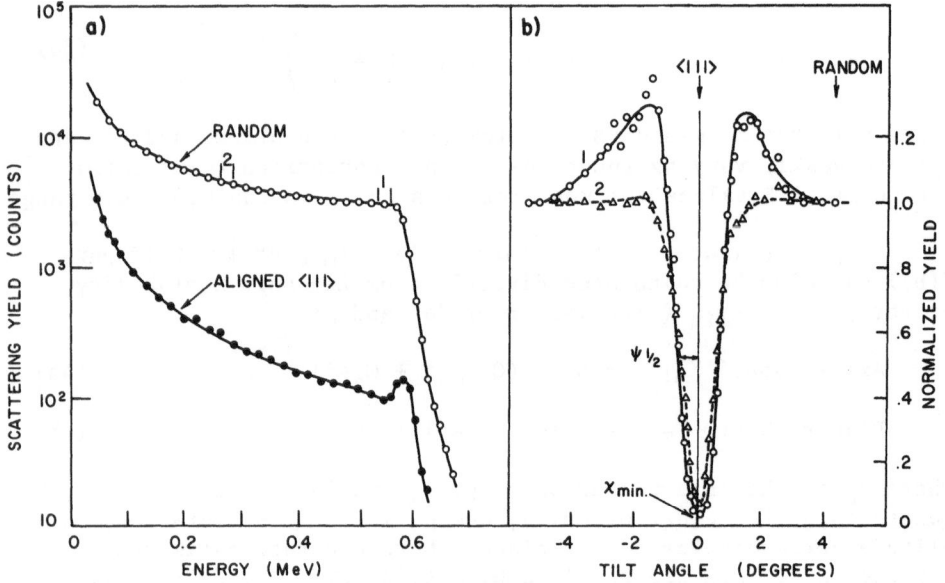

Figure 9 – a) Backscattered energy spectra for 1-MeV He ions incident
on a Si crystal for aligned (<111>) and non-channeled (random) direction of incidence.   b) Orientation dependence of the normalized yield
obtained from energy regions 1 and 2 in (a), i.e. from depths of 0.1
and 0.6 μm respectively.

example 1-MeV helium ions), collimated to 0.1° or better, is allowed
to strike a crystal mounted on a suitable double-axis goniometer,
which allows both a tilt motion (θ) and an azimuthal rotation (φ)

around an axis perpendicular to the base plate of the goniometer.
Particles that have undergone wide-angle ($\sim$ 150°) scattering are
detected and energy analyzed by means of a solid state detector.
For investigating other close-encounter processes (such as γ-rays
from a nuclear reaction, or X-ray emission) one merely includes an
appropriate γ- or X-ray detector. Rutherford scattering has been by
far the most widely used process to date.

Typical energy spectra for a 1-MeV helium beam backscattered
from a silicon crystal are shown in figure 9. The random spectrum is
obtained by orienting the crystal so that the incident beam is not
aligned with any crystal axis or plane. The aligned spectrum shows
the large reduction in backscattered yield when a crystal axis, <111>
in this case, is parallel to the beam direction.

The relation between such energy spectra and the corresponding
depth scale has already been discussed in an earlier lecture by
J.W. Mayer. For the aligned spectrum it is of course necessary to
know also the stopping power of the channeled beam which (as seen in
figure 2(b)) is always smaller than the non-channeled value. However, unlike the low velocity ions of section 2, for which nuclear
stopping is the dominant process, the difference in stopping power
between channeled and random trajectories at MeV energies (where
electronic stopping is dominant) is at most a factor of $\sim$ 2—as can

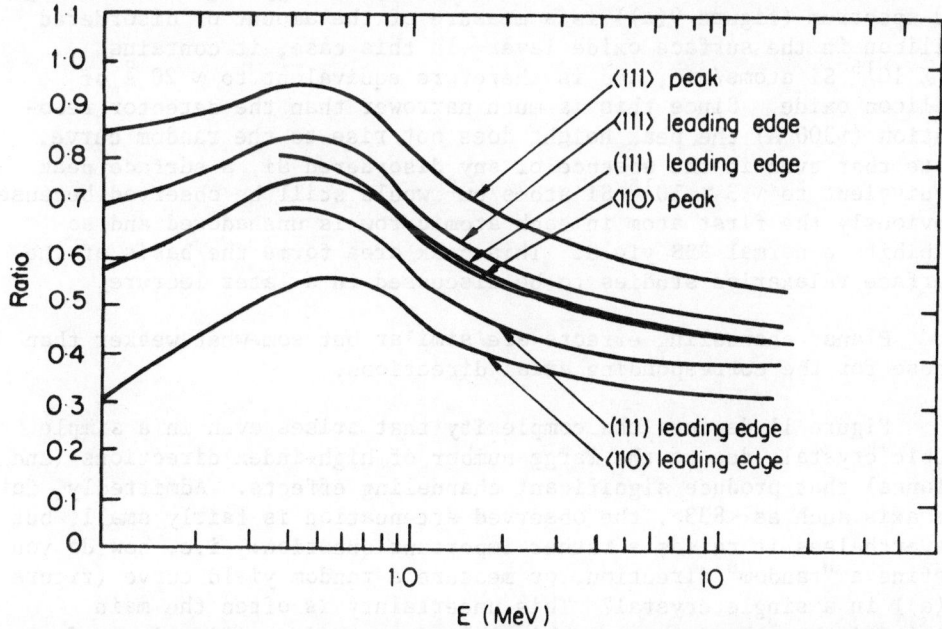

Figure 10 - Ratio of channeled to random stopping power for He ions
transmitted through Si as a function of ion energy (ref. 2, figure
5.8).

be seen in figure 10. Indeed, in the energy region of maximum stopping power (i.e. 0.3 - 1.0 MeV helium), the average <111> value is
only 5 - 10% less than the non-channeled value.

Detailed orientation scans from two different depths in silicon
are shown in figure 9(b). These are obtained by recording the yield
in the narrow energy regions 1 and 2 (figure 9(a)), while tilting the
<111> axis through the beam direction.

The observed critical angle ($\psi_{1/2}$) and minimum yield (X_{min}) both
depend on the depth beneath the surface at which the measurements are
made. The minimum yield X_{min} is defined as the ratio of the yield in
the perfectly aligned direction to that in a random direction; it is
therefore a direct measure of the un-channeled fraction of the beam.

From the results in figure 9, we see that even at a depth of
0.6 µm, more than 90% of the particles in the beam are still channeled, indicating that de-channeling effects are much less serious
than in the heavy-ion range distributions of section 2. On the other
hand, the region of enhanced yield at angles slightly greater than
$\psi_{1/2}$ (the so called "shoulder" region) decreases quite rapidly with
increasing depth—and indeed has completely disappeared in the
0.6 µm curve of figure 9(b).

The area of the small peak at the high-energy edge of the aligned spectrum (figure 9(a)) is a measure of the amount of disordered
silicon in the surface oxide layer—in this case, it contains
6 X 10^{15} Si atoms/cm^2, and is therefore equivalent to $\sim$ 20 Å of
silicon oxide. Since this is much narrower than the detector resolution ($\sim$300 Å) the peak height does not rise to the random curve.
Note that even in the absence of any disordered Si, a surface peak
equivalent to $\sim$ 3 x 10^{15} Si atoms/cm^2 would still be observed because
obviously the first atom in each atomic row is unshadowed and so
exhibits a normal RBS yield. This peak area forms the basis of the
surface relaxation studies to be discussed in a later lecture.

Planar channeling effects are similar but somewhat weaker than
those for the corresponding axial directions.

Figure 11 depicts the complexity that arises even in a simple
cubic crystal, due to the large number of high-index directions (and
planes) that produce significant channeling effects. Admittedly, for
an axis such as <833>, the observed attenuation is fairly small, but
nevertheless it raises a rather important question: i.e. how do you
define a "random" direction, or measure a random yield curve (figure
9(a)) in a single crystal? This uncertainty is often the main
limitation to the accuracy achievable in certain applications of
channeling. One fairly accurate solution to the problem involves
offsetting the target orientation so that the beam is incident at an
angle of about ten times $\psi_{1/2}$ relative to a major axis and then

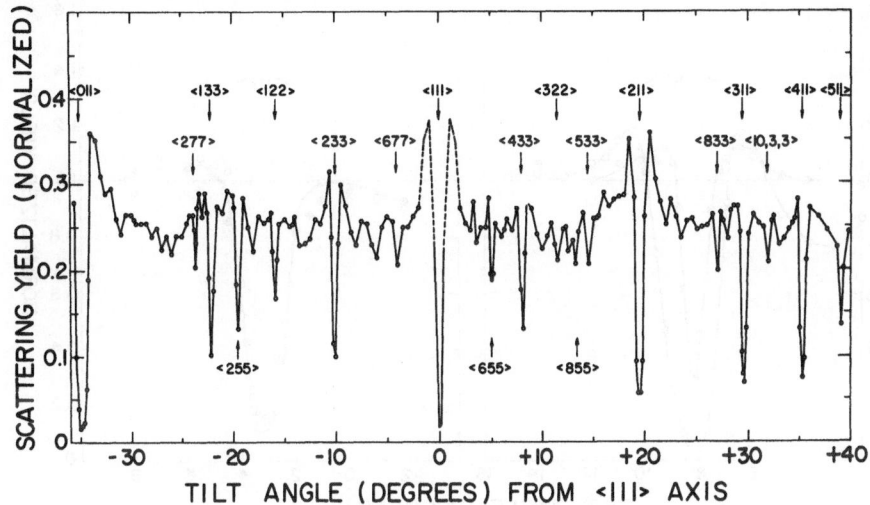

Figure 11 - Channeling of 1.0 MeV He ions in the (110) plane of Si,
illustrating the importance of high (Miller) index directions.

measuring the yield as the crystal is rotated about this axis.

3.4 Various Close-encounter Processes

According to theory, the yields of all processes requiring an
impact parameter less than $\sim$ 0.1 Å (figure 7(a)) should show exactly
the same orientation dependence. Experimental confirmation of this
point is provided in figure 2(c) in which simultaneous measurement of
Rutherford scattering and of M- X-ray yields in W were made. Note
that the impact parameters for these two processes (10^{-4} and 10^{-2} nm,
respectively) differ widely and yet they exhibit an almost identical
orientation dependence. Thus, we see that different close-encounter
processes can be used interchangeably in studying channeling
effects—provided the impact parameters are all less than $\sim$ 0.01 nm.

This flexibility is of particular advantage in investigating
crystals containing two or more different atomic species—since several different processes can be used to monitor simultaneously the
interaction of a channeled beam with each atomic species. For example, to study channeling in an UO_2 crystal, a beam of 0.975 MeV
deuterons is an ideal choice. At this energy, the $^{16}O(d,p)^{17}O$ reaction is selective for the oxygen sub-lattice, since the deuteron
energy of 0.975 MeV is much too low to induce a (d,p) reaction with
the heavy uranium atoms. Rutherford scattering, on the other hand,
is equally selective for the uranium sub-lattice, since scattering

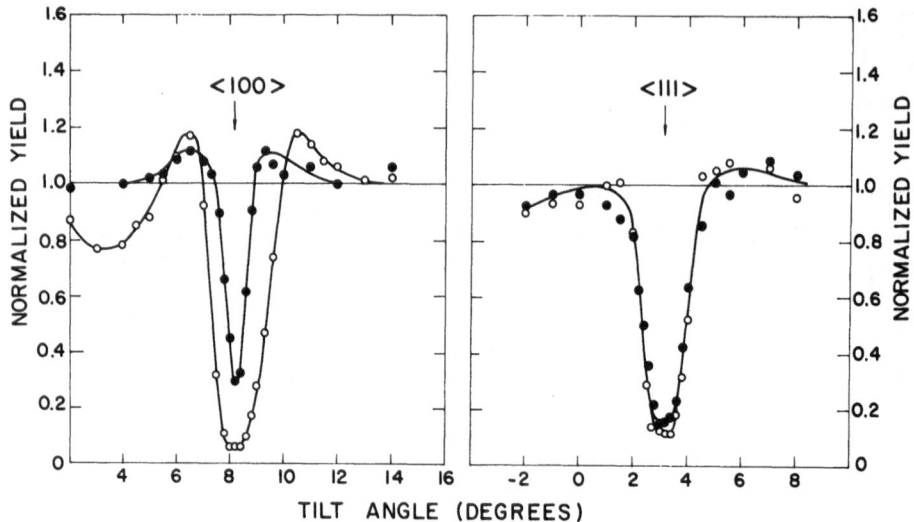

Figure 12 - Axial channeling of 0.975 MeV deuterons in UO_2 at a mean depth of 0.4 μm: o, RBS yield from the U atoms; •, proton yield from the $^{16}O(d,p)^{17}O$ reaction.

from the low-Z oxygen atoms is far too weak, both in energy and intensity, to contribute significantly to the observed yield.

The results of such a channeling study are illustrated in figure 12. Two completely different types of behaviour are seen, depending on the axis chosen. Along the <100> the backscattering yield off the U-atoms exhibits a much wider and deeper attenuation than that for the protons of the $^{16}O(d,p)$ reaction, whereas along the <111> both processes exhibit identical orientation dependence. The reason for this difference is that along the <100> there are in fact two different types of atomic rows (each containing only one atomic species), but along the <111> there is only one type of row (containing both oxygen and uranium atoms).

It is interesting to note that, along the <100> for example, one can tilt the crystal slightly so that the beam is no longer channeled with respect to the O-rows, and yet is still well within the critical steering angle of the U-rows. Under such conditions, the oxygen atoms behave essentially as interstitial scattering centres. Normal multiple scattering by these oxygen atoms causes a rather rapid dechanneling to occur with respect also to the uranium rows as the beam penetrates deeper into the crystal. Hence, at shallow depths the <100> direction exhibits considerably stronger channeling effects than the <111> as far as uranium atoms are concerned (as seen in figure 12); but at larger depths the <111> becomes the preferred

channeling direction, since it does not exhibit the rapid dechanneling effect described above.

3.5 "Blocking"

Up to now, we have considered the channeling behaviour of an
external beam of particles incident on a crystal surface (the exter-
nal-beam case). But, the same separation into allowed and forbidden
trajectories occurs if the charged particles originate from a position inside the lattice: e.g. as a consequence of radioactive decay
or of an elastic scattering event off lattice nuclei. If the trajectory originates from a position contained in a row of atoms (for
example, a lattice site), it cannot emerge from the crystal within
the critical angle characteristic of that row—i.e. it cannot become
channeled—since it started out from within one of the forbidden
regions (figure 7(a)) for channeled trajectories. Hence, a reduction in the number of emitted particles is observed in the direction
of the row, and this phenomenon is usually termed "blocking".

Reversibility arguments show that channeling and blocking experiments are completely equivalent, provided depth effects are negligible, and this equivalence is illustrated quantitatively in the
experimental work of Bøgh and Whitton (figure 13) in which the same
crystal was used to investigate both processes. Their blocking curve
was obtained by using a well-collimated movable detector to determine
the angular distribution of the emitted (backscattered) protons, with
the incident beam entering along a non-channeling direction.

In the case of RBS, it is possible to look at either the channeling or the blocking case by aligning the crystal either with the
incident-beam direction or with the detector, as seen in figure 13.
In fact, one can employ a double-alignment technique in order to
combine in a single measurement both the channeling of the incident
beam and the blocking of the emitted beam. The overall attenuation
in the Rutherford-scattering yield is then determined by a geometrical combination of two attenuation factors, i.e. the attenuation
factor due to channeling of the incoming beam, multiplied by the
comparable attenuation factor for the blocking effect. In this way,
attenuation factors of up to several thousand have actually been
produced. We shall see in later lectures how this enhanced attenuation due to double-alignment may be utilized in some of the solid-
state applications.

3.6 Depth Dependence of X_{min}

So far, throughout section 3 we have been assuming that channeled ions conserve their transverse energy, $E_\perp$: i.e. that X_{min}

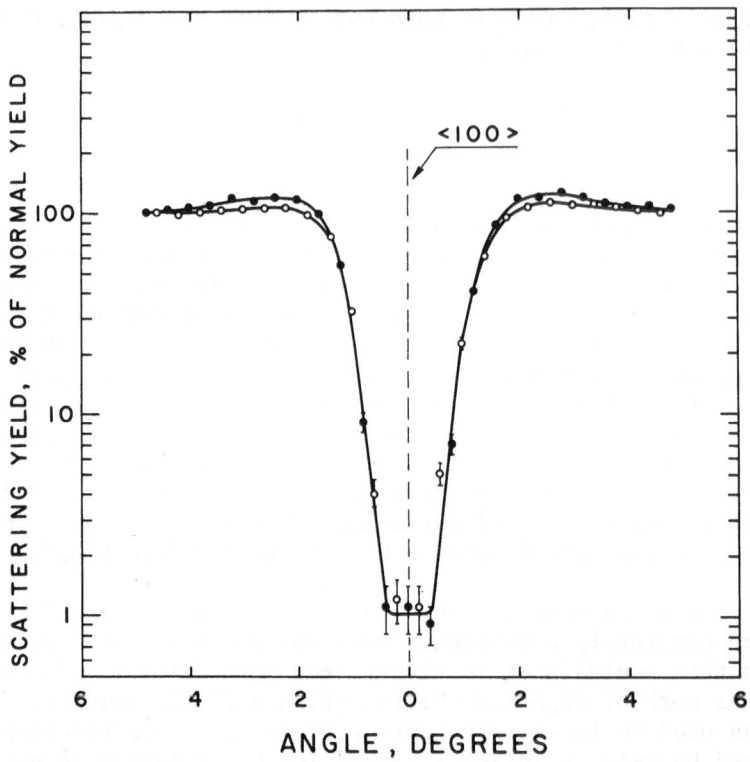

Figure 13 – Orientation dependence of the backscattered yield of 1.0 MeV protons in tungsten, measured at a depth of 0.3 – 0.4 μm beneath the surface: ●, channeling results; O, blocking results[17].

(and $\psi_{1/2}$) remain constant as the ion penetrates deeper into the crystal. In a well-aligned crystal this is a good first-order approximation for MeV He ions and protons, as can be seen in figure 9. However, at large depths, or in damaged crystals, χ_{min} is no longer constant and it becomes necessary to correct for the gradual dechanneling that occurs. RBS yield curves readily provide experimental information on χ_{min} as a function of depth and a large amount of dechanneling data have been obtained as a function of E, d, temperature, etc. Theoretical estimates of dechanneling are more difficult to generate, even in an undamaged crystal, since it is necessary to take account of multiple scattering by electrons as well as by lattice nuclei. Such estimates have been made at least for W and Si crystals and one finds rather good agreement with experimental data, as shown in figure 14. In ion-bombarded crystals dechanneling by lattice defects is often the dominant term, but this is even more complex and so is difficult to predict. It will be discussed in some detail by Rimini, since it represents an important correction

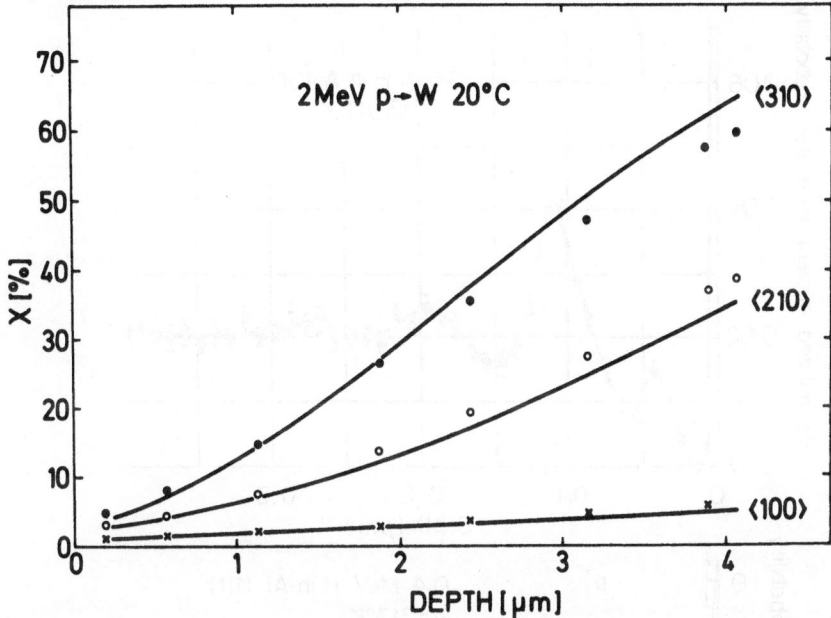

Figure 14 – Measured dechanneling along different axial directions in W (indicated by points) and the corresponding calculated curves[18].

in many of the solid state applications of channeling.

We conclude this discussion of X_{min} by considering briefly the depth region just beneath the crystal surface. Here, as shown in the computer simulations of figure 15, X_{min} exhibits several strong oscillations arising from the oscillatory nature of quasi-channeled ions (trajectory B, figure 7(a)).

At the surface the normalized yield obviously must approach unity, as the incident beam strikes the surface atoms with random impact parameter. This region constitutes the 'surface peak'. On penetrating the surface the ions sort themselves out into channeled trajectories, which do not undergo close collisions and so need not concern us for the moment, and the more relevant nonchanneled (quasi-channeled) trajectories which do undergo close collisions. The latter enter the surface in the vicinity of a row or plane from which they are immediately deflected, producing almost zero yeild just behind the surface.

In the planar case (figure 15(b)), these non-channeled ions cross the empty interplanar channel and produce a second maximum as they all reach the next plane in phase. If these ions may be char-

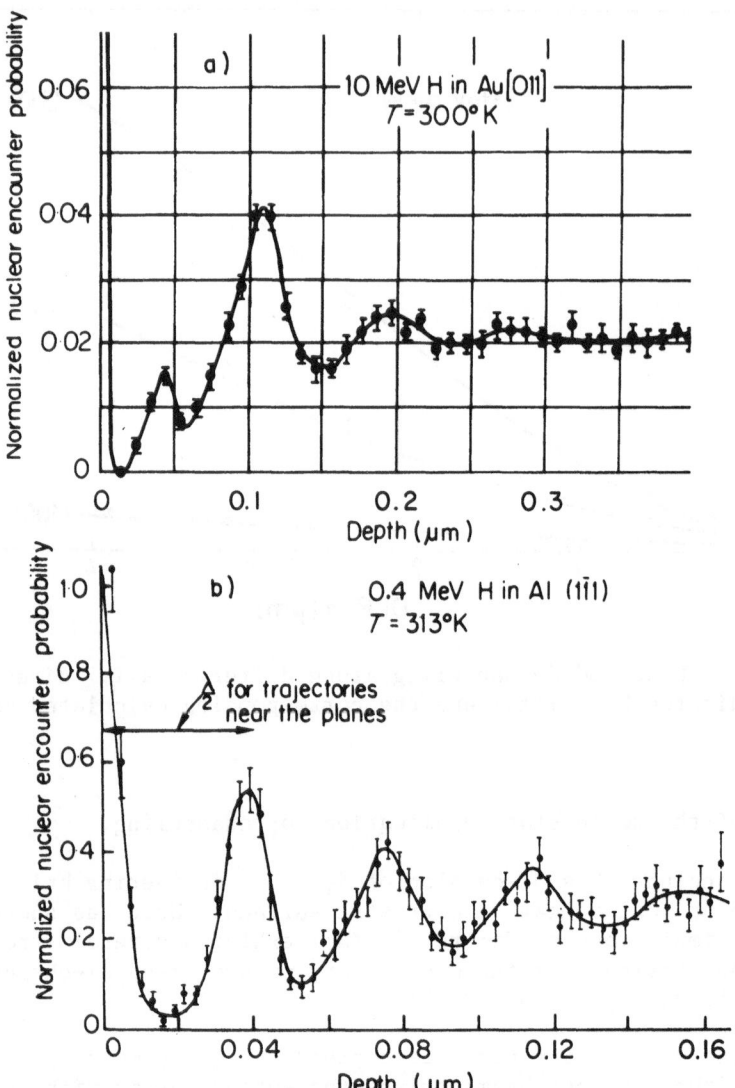

Figure 15 - Nuclear-encounter probability (χ_{min}) as a function of depth beneath the surface: a) for axial channeling: 10 MeV H^+ on <011> Au; b) for planar channeling: 0.4 MeV H^+ on (111) Al.[19]

acterized by a wavelength λ then successive maxima should appear at depths corresponding to $\lambda/2$, λ, $3\lambda/2$, etc. The oscillations tend to die away with increasing depth as the trajectories gradually grow out of phase due to the strong multiple scattering experienced by non-channeled ions and to variations in their initial wavelength.

The axial situation is more complex, since neither channeled nor quasi-channeled trajectories have a simple oscillatory motion. A computer simulation of the depth dependence of χ_{min} is illustrated in figure 15(a). The yield of unity at the surface and the first deep minimum are again present as in the planar case, but the subsequent peaks are irregular. The first maximum corresponds to quasi-channeled ions striking a nearest-neighbour row; the second, to second nearest-neighbour rows; and so forth. The damping is considerably faster in the axial case.

A simple estimate for the distance $(\lambda/2)$ to the first peak below the surface is given by $(d/\psi_{1/2})$ where d is the distance between neighbouring rows or planes. For the systems illustrated in figure 15, this would correspond to 325 Å and 400 Å for the axial and planar channeling cases, in fair agreement with the more detailed Monte Carlo simulation.

It is clear from this discussion that the equilibrium χ_{min} value must either be measured at a sufficiently deep depth (i.e. ≥ 0.2 μm in figure 15) or else be averaged over a depth interval large compared to the wavelength of these 'sub-surface' oscillations.

Closely related to these oscillations of χ_{min} near the surface is a similar non-uniformity of ion flux in the transverse plane: i.e. the so-called flux-peaking effect which occurs near the middle of an axial channel, and whose intensity again shows large depth oscillations at shallow depths. The origin of this flux peaking effect is illustrated in figure 16. It will be discussed in detail by Dr. Mazzoldi in the next series of lectures, since it is of major importance in interpreting the interaction yield of a channeled beam with interstitial (mid-channel) foreign atoms.

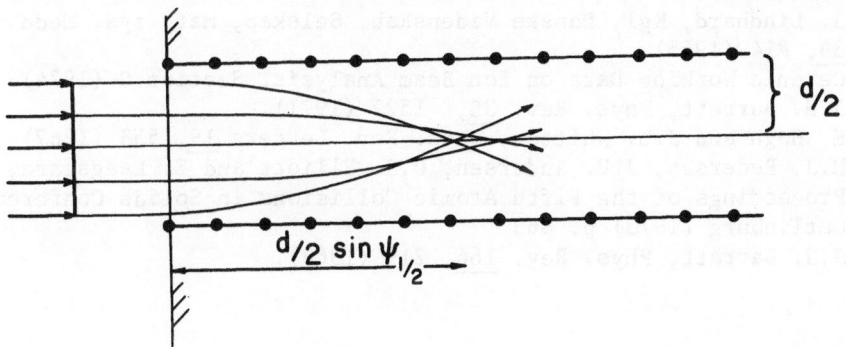

Figure 16 – Schematic diagram illustrating the origin of flux peaking and its oscillatory dependence on depth. Note that the flux peaking effect is zero at the crystal surface and reaches a maximum value at a depth of $\sim (d/2).\sin\psi_{1/2}$ (i.e. at ~ 0.02 μm, when $\psi_{1/2} = 0.5°$).

3.7 Summary

This concludes our general introduction to the channeling behavior of MeV ions. We will now consider some of its more important solid-state applications. These are all based on the unique property of a channeled beam that it cannot penetrate closer than ~ 0.1 Å to a lattice site—and hence has close-collision interactions only with those atoms which are displaced $>> 0.1$ Å from the normal lattice sites.

REFERENCES

1. D.S. Gemmell, Rev. Modern Phys. 46, 129 (1974).
2. Channeling—Theory, Observation and Applications, edited by D.V. Morgan, J. Wiley & Sons, London (1973).
3. L. Eriksson, J.A. Davies and P. Jespersgard, Phys. Rev. 161, 219 (1967).
4. F.H. Eisen, G.J. Clark, J. Bøttiger and J.M. Poate, Rad. Effects 13, 93 (1972).
5. J.U. Andersen and J.A. Davies, Nucl. Instr. & Meth. 132, 179 (1976).
6. J. Stark, Z. Phys. 13, 973 (1912).
7. J.A. Davies, J.D. McIntryre, R.L. Cushing and M. Lounsbury, Can. J. Chem. 38, 1535 (1960).
8. M.T. Robinson and O.S. Oen, Phys. Rev. 132, 2385 (1963).
9. G.R. Piercy, F. Brown, J.A. Davies and M. McCargo, Phys. Rev. Letters 10, 399 (1963).
10. H.O. Lutz and R. Sizmann, Phys. Lett. 5, 113 (1963).
11. R.S. Nelson and M.W. Thompson, Phil. Mag. 8, 1677 (1963).
12. G. Dearnaley, IEEE Trans. Nucl. Sci. NS 11, 249 (1964).
13. J.A. Davies, L. Eriksson and J.L. Whitton, Can. J. Phys. 46, 573 (1968).
14. J. Lindhard, Kgl. Danske Videnskab. Selskab, Mat. fys. Medd. 34, #14 (1965).
15. Catania Working Data on Ion Beam Analysis, Section C (1974).
16. J.H. Barrett, Phys. Rev. 3B, 1527 (1971).
17. E. Bøgh and J.L. Whitton, Phys. Rev. Letters 19, 553 (1967).
18. M.J. Pedersen, J.U. Andersen, D.J. Elliott and E. Laegsgard, Proceedings of the Fifth Atomic Collisions in Solids Conference, Gatlinburg (1973) p. 863.
19. J.H. Barrett, Phys. Rev. 166, 219 (1968).

FLUX PEAKING - LATTICE LOCATION

P. Mazzoldi

Istituto di Fisica dell'Università, Unità GNSM - CNR

Padova, Italy

A well known application of channeling is the location of impurities in crystal lattices. The lattice location has recently been comprehensively reviewed by J.A. Davies[1] and by S.T. Picraux[2].

In this paper the spatial distribution of particles during channeling will be briefly discussed. In particular, the presence of sufficiently high impurity concentration in discrete lattice sites will be illustrated. The basic principles involved in atom location studies will be described through examples of applications from current literature.

I. FLUX DISTRIBUTION

The determination of spatial distribution of channeled particles is essential for interpreting the atom location experiments. For a beam incident along a non-channeling direction (random direction) the particle distribution inside the crystal is uniform. When the beam is aligned along a low-index axis or plane the particle density shows an appreciable enhancement near the center of the channel relative to that for a beam incident along a random direction. The flux distribution of channeled particles can be calculated either by an analytical model[3,4,5] based on the continuum theory of channeling[6] or by a MonteCarlo computer simulation[5,7,8,9,10]. An illustration of two calculation methods will be presented for axial channeling.

I.1. Analytical Model

The total transverse energy of the channeled ion $E_\perp$, to first order in the deflection angle, is a conserved quantity $E_\perp = E\psi_i^2 + U(\underline{r})$ where E is the kinetic ion energy, ψ_i the angle of incidence relative to the channel symmetry axis and $U(\underline{r})$ the potential energy resulting from the screened Coulomb interaction with the lattice atoms, normalized to zero at the minimum in U. ($\underline{r}$ is the position vector relative to the channel axis). Using the continuum approximation[6], $U(\underline{r}) = \Sigma_i U_1(\rho_i)$, where ρ is the distance from a single axial row and $U_1(\rho)$ the average interatomic potential of string, experienced by an ion at ρ. The summation is taken over all neighbouring rows. Examples of an average potential contour plot for deuterons for the <110> direction in silicon and <0001> in titanium are shown in Fig. 1[11]. The standard Lindhard string potential[6] has been used in the calculations. In order to avoid divergence at the string an approximation for thermal vibration has been introduced[12]. Such correction is important only at distances of the order of the vibration amplitude, about 0.1 Å.

The trajectory of a particle which enters the crystal at the position $\underline{r}_i$, with an incident angle ψ_i, is confined to the area of the transverse plane $A(E_\perp)$, limited by a contour line of the potential given by

$$U(\underline{r}) = E_\perp(\underline{r}_i, \psi_i) = E\psi_i^2 + U(\underline{r}_i)$$

The minimum approach distance to the center of a string ρ_m, considering in the calculation only a single row, is defined by

$$E\psi_i^2 + U_1(\rho_i)_{min} = U_1(\rho_m)$$

where $(\rho_i)_{min}$ is the initial distance to the closest row. Lindhard[6] has shown that, once statistical equilibrium is reached, there is an equal probability of finding an ion anywhere within the accessible area. So that we may write for axial channeling the probability for a particule, with a minimum approach to a row ρ_m, to be at the position $\underline{r}$, relative to the channel axis[4,5]

$$P(E_\perp, \underline{r}) = \begin{cases} 1/A(E_\perp) & U(\underline{r}) \leqslant E_\perp \quad \text{and} \quad |\underline{r}_j - \underline{r}| \geqslant \rho_m, \rho_m > a_{TF} \\ 0 & U(\underline{r}) > E_\perp \quad \text{and} \quad |\underline{r}_j - \underline{r}| < \rho_m, \rho_m > a_{TF} \\ 1/A_o & \rho_m \leqslant a_{TF} \end{cases}$$

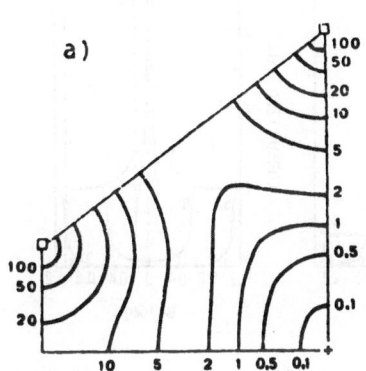
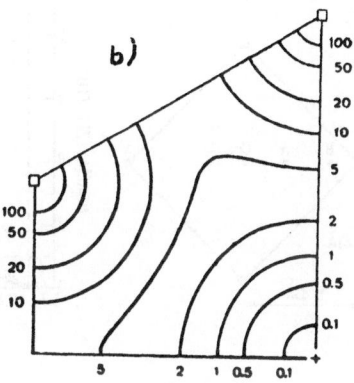

Fig. 1. Equipotential contour plot for standard Lindhard string
potentiel for 1-MeV deuterons incident a) along <110> axis in silicon and b) <0001> axis in titanium. The values at the lattice site
are respectively ∿130 eV and ∿115 eV. Only a quarter of channel is
shown. (From Ref.11).

where A_0 is the total channel area, $\underline{r_j}$ is the position vector relative to the channel axis of the closest row, a_{TF} the Thomas-Fermi
screening parameter. It is assumed that when $\rho_m \leqslant a_{TF}$ the particule
contributes to the random fraction of the beam. The total flux at
the position $\underline{r}$, averaged over all incident ions and taken relative to
the random flux, $1/A_0$ is given by

$$F(\underline{r}) = \int_{E_\perp(\underline{r_i},\psi_i) \geqslant U(\underline{r})} dA/A(E_\perp)$$

This expression determines that the mid. channel flux should have
logarithmic divergence, when $E_\perp$ goes to zero, that is for those ions
entering the channel near the potential minimum with small value of
ψ_i. However, at large distances from the atomic strings, the potential $U(\underline{r})$ becomes small and flat, so that fluctuations in transverse
energy can no longer be neglected. Various experimental factors,
such beam divergence, electron scattering, and thin surface oxide,
introduce a finite spread in $E_\perp$ and determine a limiting value for
the flux near the center of a channel.

R.B. Alexander et al.[5] calculated, using the analytical treatment, the particle distribution for the case of 3.5 MeV ^{14}N ions
incident along a <100> axial direction of the bcc Fe lattice for 4
different lattice sites. The results are reported in Fig. 2.

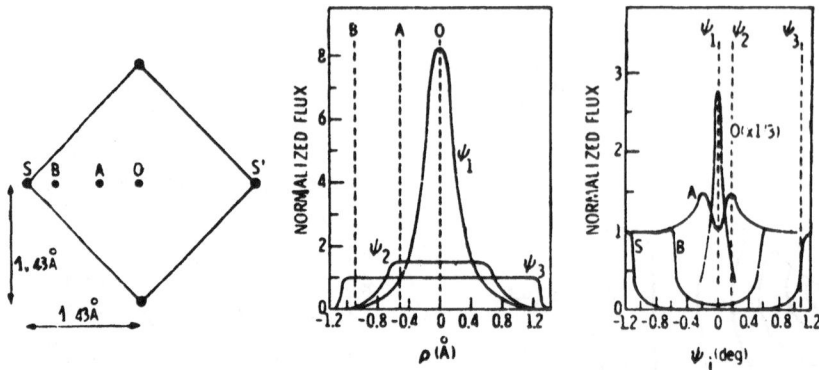

Fig. 2. Calculated spatial and angular distributions by analytical method for 3.5 MeV ^{14}N incident along the <100> axial direction in Fe. Left panel shows the cross-section of the <100> axial channel in the bcc Fe lattice, showing four distinct sites O,A,B and S. The filled circles represent rows of Fe atoms. Center panel shows the spatial flux distributions for ^{14}N ions along the line SOS', at three different incident angles relative to the channeling axis; $\psi_1 = 0$ and ψ_3 is the critical angle for channeling ($\psi_1 < \psi_2 < \psi_3$). Right panel shows the angular distribution calculated for the four positions O,A,B and S as a function of the angle of incidence ψ_i. (From Ref. 5).

Multiple scattering by electrons and by the thermally vibrating atoms increases the root mean square transverse energy and smoothes out the spatial distribution of particles with increasing depths. We have to replace the term $E\psi_i^2$ by $E\psi_i^2 + \delta E_\perp$, where $\delta E_\perp$ is the increase in $E_\perp$ associated with the interaction process. Depth oscillations of the particle density occur at small depth, since statistical equilibrium is reached only at appreciable depths into the crystal. Calculations of such effects need more detailed treatments.

Recently Kumakhov[13], developing a quantitative theory of channeling based on the Lindhard statistical approach[14],[15] analyzed the distribution of particle flux across the channel and determined the characteristic depths, at which statistical equilibrium is established. Fig. 3 shows the calculated depth dependence of the relative flux in the center of the channel for 700-KeV protons in silicon (Curve 1). Curve 2 represents a calculation in which the statistical equilibrium was assumed. The calculation gives partially identical results at a depth about 1000 atomic layers, suggesting that at such depth the statistical equilibrium distribution may be

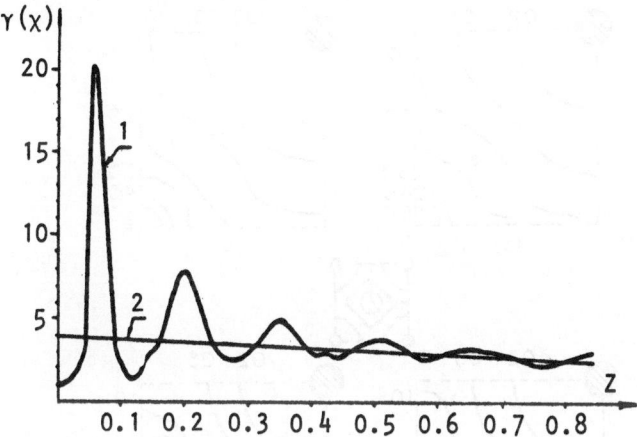

Fig. 3. The depth dependence of the relative flux of 700-KeV
protons in Si in the center of the channel. Curve 1: no statistical
equilibrium is assumed. Curve 2: statistical equilibrium is assumed.
(From Ref. 13).

assumed. The depth oscillation of the flux observed at the center
of the channel is quite distinct from the surface peak and subsequent
minima generally found in the close collision yield from lattice
atoms[16,17]. Indeed the latter is connected with those ions of high
transverse energy, which enter the crystal near the atomic rows and
are immediately deflected towards the empty channel center[8].

I.2. Computer Simulation

Computer simulation presents some advantages with respect to
the analytical calculation, because many experimental factors can be
taken into account, such as electronic multiple scattering, beam
divergence, surface oxide films, lattice damage.

Computer simulation models can be divided into two categories:
the Vineyard model and the binary collision model. The former is
generally used for studies of radiation damage when the relaxation
of the surrounding lattice influences the scattering process. In
the channeling experiments, the study is particularly focused on
the trajectory of the incident ion, moving with an energy of ~ 1
MeV. The relaxation around a struck atom is negligible and therefore does not influence the interaction process, so that the binary
collision model can be used. In this model the interactions between
the energetic ions and the lattice atoms are treated as a series of

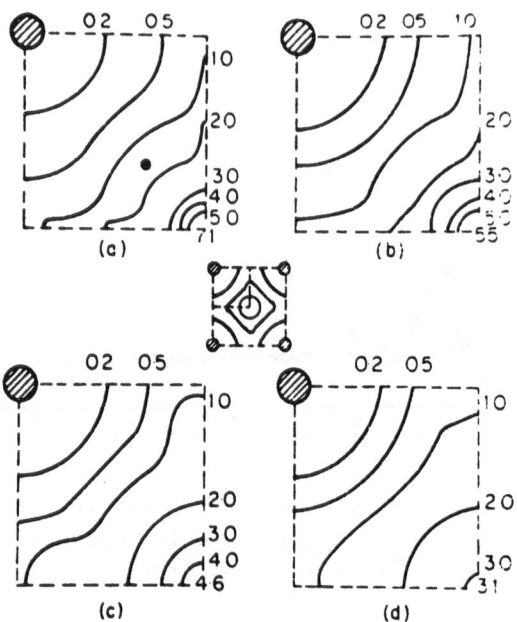

Fig. 4. Computed contour plots of the flux of 1-MeV α-particles in
the <001> Cu transverse plane for perfect alignment and averaged
between 0 and 1080 Å.   (a) With no multiple scattering; (b) lattice
vibrations correspondig to 273°K; (c) 273°K plus inelastic multiple
scattering; (d) 273°K plus a beam collimation of ±0.23°. (From
Ref. 7).

independent two-body collisions described by classical mechanics.
A general review of the argument is given in Ref. 7 and 18.

     The computed contour plots[7] of the flux of 1-MeV α-particles
in the <001> Cu channel for a rigid lattice, introducing thermal
vibrations, inelastic scattering and finite beam collimation respectively is shown in Fig. 4 (a), (b), (c), (d).   Due to the statistical
nature of the results, the contours are rather qualitative but the
large variation of the flux across the channel is evident.   The
computed flux profile shows the existence of strong depth oscillations
in the mid-channel flux along the axial channels.   This effect is
illustrated in Fig. 5 for 1-MeV α-particles along <001> axis in
Cu[7].   With no multiple scattering, a very pronunced depth oscillation is present, the normal flux at the surface is followed by a
peak of more than 20 times normal and then a dip of just over twice
normal.   The spacing between the minima and maxima is ∼250 Å.

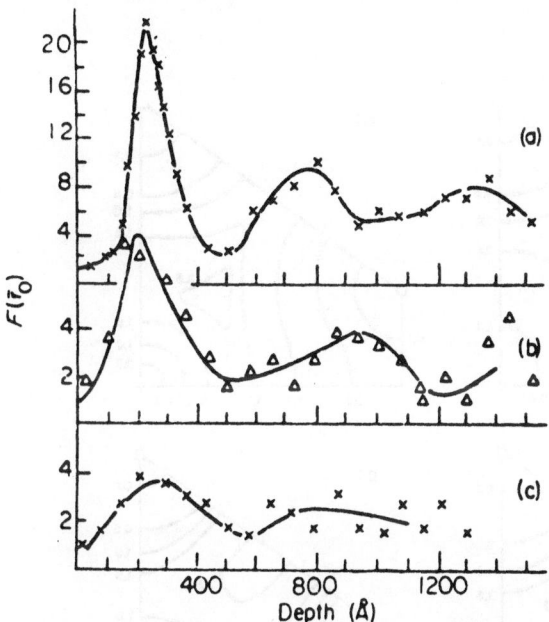

Fig. 5. Computed flux profile as function of depth, for 1-MeV α-particles in <001> Cu. (a) No multiple scattering; (b) lattice vibrations corresponding 273°K, inelastic multiple scattering and beam collimation of ±0.06°; (c) as (b) but with a beam collimation of ±0.23°. (From Ref. 7).

The multiple scattering reduces the magnitude of the initial oscillation, without removing it entirely. Also in the cases (b) and (c) where the effects of introducing inelastic scattering, thermal vibrations and two different values of beam collimation are studied, the first maxima and minima are still quite distinct, and even in the latter case they differ by a factor two. This calculation illustrates the advantages of a computer simulation with respect to analytical methods, although not as elegant, particularly in situations where the continuum model is inapplicable. Furthermore, the physical phenomena which cause the variations of particle flux can be isolated and studied.

For sufficiently high concentration of impurities located on discrete lattice sites, the flux distribution of channeled particles should be modified with respect to the behaviour for the host lattice. A study of this "impurity effect" has been developed using a computer simulation by G. Della Mea et al.[11]. In the center of the channel along the direction <110> in Si lattice interstitial B atoms were

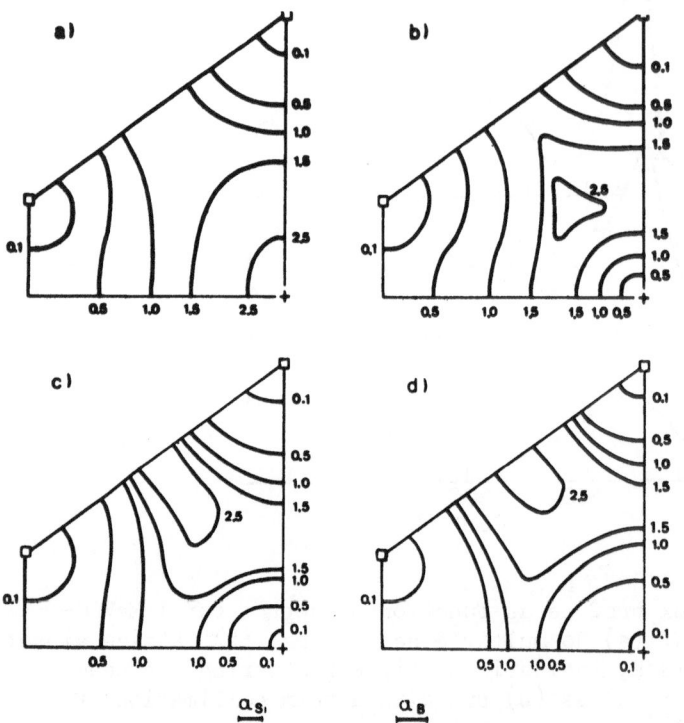

Fig. 6. Flux contour plot for 1-MeV deuterons incident parallel to
the <110> axis in Si containing B atoms at the mid channel axis, for
different compositions SiB_x. $\square$, Si site; + B site. (a) x = 0;
(b) x = 0.1; (c) x = 0.25; (d) x = 0.5. a_{Si} and a_B, the Thomas-
Fermi radii relative to the two atoms, are shown for comparison on
the same scale. (From Ref. 11).

located. Calculations were performed for 1-MeV deuterons in order
to compare with experiments on TiO_x system[19]. Fig. 6 shows the flux
contour plot for different concentration of interstitial B atoms.
The sequence shows the progressive displacement of the maximum flux
towards the center of the new channel. We observe that at the composition x = 0.1, the B interstitial row is able to repel the ion flux
to some extent, also if the nuclear encounter probability within a
distance of about 1.5 a_{TF} is still of the order of the random value.
At x = 0.5 the symmetry of the new axial channel is built on, even
though the strength of Si and B strings is not the same. Fig. 7
shows the corresponding Lindhard[6] continuous string potential. The
potential contour of Si lattice without impurities is reported in
Fig. 1. The close resemblance between Fig.s 6 and 7 can easily be
noticed.

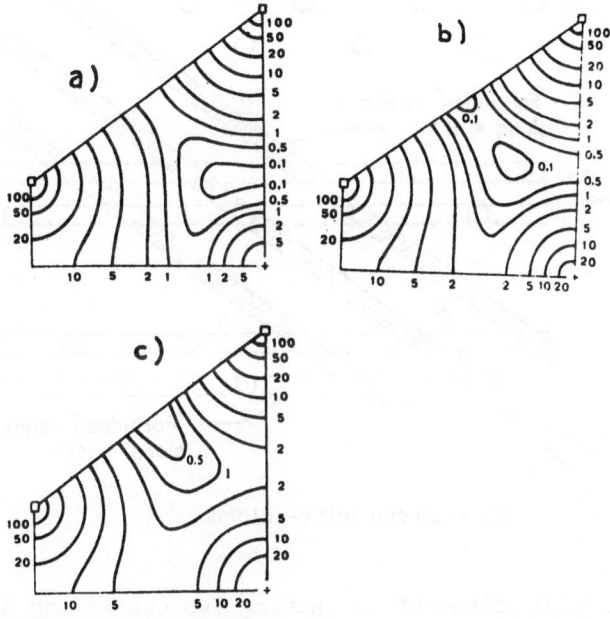

Fig. 7. Equipotential contour plot for standard Lindhard string potential for the same conditions as Fig. 6. (a) x = 0.1; (b) x = = 0.25; (c) x = 0.5. (From Ref. 11).

II. ATOM LOCATION

Lattice location analysis is carried out by the comparison between the interaction yield from the impurity and lattice atoms, for angular scans along axial or planar directions. Strong yield attenuation of the channeled particles with the foreign atoms prove that they lie within the shadow of the aligned set of atomic rows (impact parameter smaller than the mean vibrational amplitude are completely prohibited in channeling conditions). Successive measurements along two or more different channeling directions can be used to locate the exact position of the impurity in the lattice.

Fig. 8 schematically illustrates[20], for a simplified two dimensional lattice, how the channeling effect can be used to locate an impurity in Si lattice for three different impurity positions (● substitutional site; X tetrahedral interstitial site; ▲ another non-substitutional position). The substitutional impurity (●) would give a channeling behaviour similar to the lattice atoms for channeling along both the <110> and <111> directions. The tetrahedral interstitial impurity (X) would exhibit the same channeling behaviour as the substitutional impurity for the <111> direction, but due to

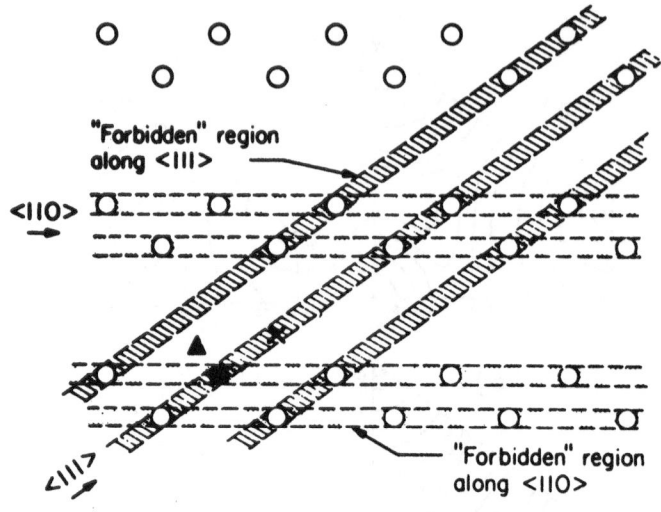

O - silicon lattice atoms

Fig. 8. The {110} plane of Si showing two channeling directions:
<110> and <111>. The filled circle corresponds to a substitutional
site, the x to the tetrahedral interstitial site and the filled
triangle to another nonsubstitutional position. (From Ref. 20).

the flux distribution in the channel, would give an enhanced yield
along the <110> direction. The other nonsubstitutional position
(▲) would present an impurity yield for the channeling along the
<110> direction rather similar to that observed for tetrahedral
interstitial impurity (X). However, along the <111> direction we
expect an increase in the yield relative to "random" level, when
the beam becomes well channeled, due to flux enhancement.

We observe that measurements along different low-index directions are, in general, required to assign the lattice location.
Complete angular scans are needed for reliable interpretation of
foreign atoms location, expecially when the impurities occupy several
different sites.

The detection of foreign atoms, usually in small concentration
in the presence of a large excess of lattice atoms, is obtained by
selecting the suitable close-impact parameter process. The most
used process in atom-location studies is the wide angle scattering
(normally called backscattering) particularly applicable to cases
in which the impurities are significantly heavier than the lattice
atoms. This measurement method gives simultaneously the amount of

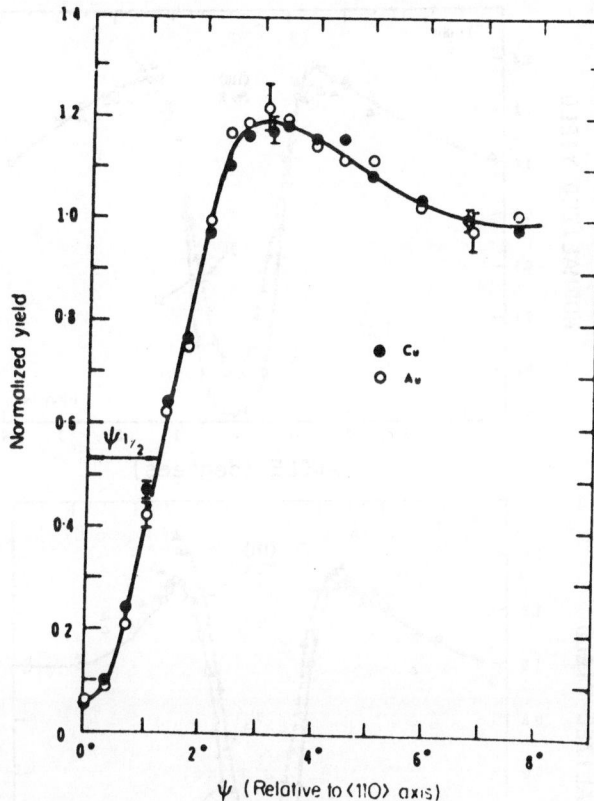

Fig. 9.   Channeling normalized yield for 1.2-MeV ^{4}He ions incident along the <110> axis in Cu, containing approximately 2 at. % Au. Open circles denote ^{4}He backscattering yield from Au atoms, full circles from Cu atoms.   (From Ref. 21).

lattice disorder and the distribution of the foreign atoms, as fuction of depth in the crystal. For low-Z impurity atoms, nuclear reactions are often suitable. Alternatively, the yields of inner-shell-X-rays may be detected. Techniques of impurity detection have been reviewed in some papers of this book.

Now we want to illustrate some examples relative to particular site location of impurities: II.1. substitutional; II.2. "almost" substitutional; II.3. interstitial positions.

II.1. Substitutional Case

Fig. 9[21] shows the angular dependence of the normalized back-

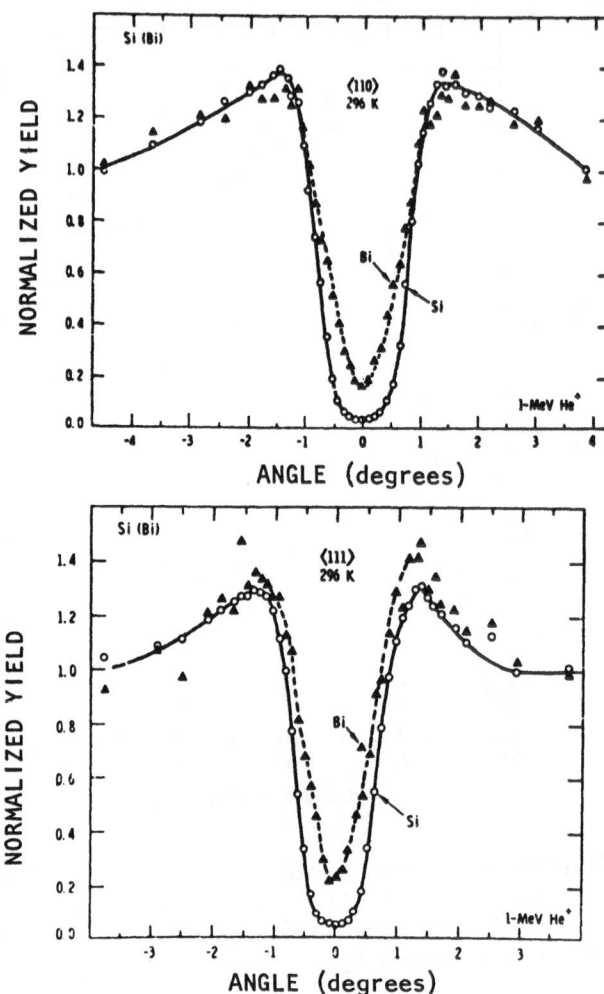

Fig. 10.   Channeling angular distributions for 1-MeV ^{4}He ions scattering from Bi (triangle) and Si (circle) for a Bi-implanted Si sample.
Left panel: <110> axis, right panel <111> axis. (From Ref. 22).

scattering yields of 1.2 MeV ^{4}He ions from Au and Cu atoms in a Cu
single crystal with an impurity concentration 2 at % Au, which is a
typical example of substitutional solid solution. The incident beam
was channeled along the <110> axis. Both types of atoms are shadowed along the<110> direction and therefore the Au and Cu angular
curves are indistinguishable. This is true as long as the mean
vibrational amplitude of two species are of the same order (0.1 Å in
this case). Virtually 100% of the Au atoms must be located within

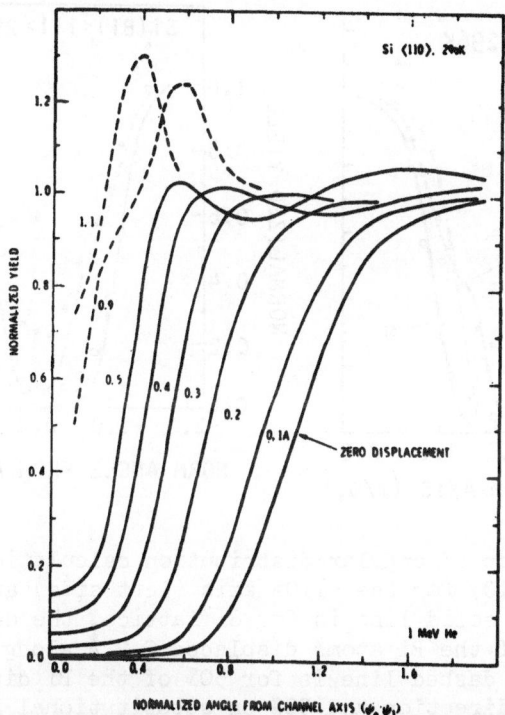

Fig. 11.   Calculated single-alignment angular distributions as a
function of equilibrium displacement distance from the row in Å,
for 1-MeV ⁴He ions along the <110> axis in Si at 296°K. (From Ref.22).

approximately 0.1 Å, of the substitutional sites. This illustrated
experimental situation is not always observed.   Often, the normalized minimum yield for the impurity atoms is higher than that for
the substrate atoms. Some of the foreign atoms are not exactly on
substitutional sites.

II.2. "Almost" Substitutional Case

The <110> and <111> axis angular distributions for Bi in Si are
shown in Fig. 10 for 1 MeV ⁴He ions channeling[22]. The target
temperature was 300°K. The Si sample has been implanted with 150
KeV Bi at room temperature and then annealed for half an hour at
900°K. The implantation dose was $1.8 \times 10^{14}/cm^2$. The measurements
for both angular distributions were made on the same sample. We
observe that the Si and Bi yields are similar in shoulder region,
but at smaller angles the Bi dip is narrower than that for the Si
lattice. The corresponding minimum yield is greater. In order to

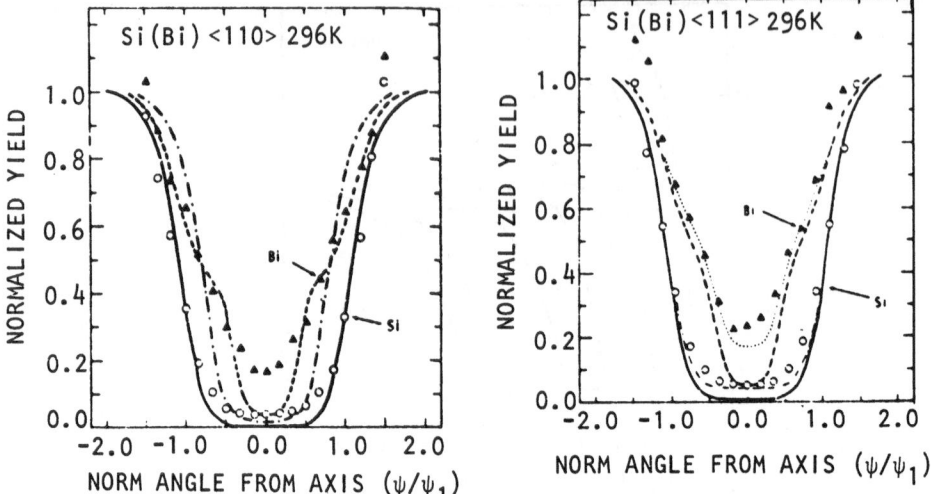

Fig. 12. Comparison of angular-distribution calculations with measurements (see Fig. 10) for the <110> axis (left side) and <111> axis
(right side). The solid line is for Si lattice, the dashed curve for
Bi, assuming all of the Bi atoms displaced 0.2 Å along the <110>
direction, and the dashed line is for 50% of the Bi displaced 0.45
Å along the <110> direction and 50% on substitutional lattice sites.
(from Ref. 22).

interpret the experimental results, the authors calculated the angular
distribution for single-alignement axial channeling as a function of
the "equilibrium displacement" of an atom from a lattice row. The
calculation, based on the continuum potential model developed by
Lindhard[6], was similar to that carried out by J.U. Andersen[23] for
substitutional atoms. The calculation of angular distribution for
1-MeV [4]He ions along the <110> axis in Si is reported in Fig. 11, as
a function of equilibrium displacement from the row. The angular
width of the dip decreases rapidly with increasing displacement
distance. Moreover, the minimum yield increases. The transition to
flux peaking as the "displaced" atoms approach the center of the
channel (the <110> channel radius is 2 Å) is qualitatively evident.
Fig. 12 shows the comparison between the experimental angular
distributions as in Fig. 10 and the calculated dips. The meaning
of the symbols are reported in the caption. The agreement between
calculations and experiments for the width and slope of the dip sides
is good, while in the region of the minimum yield a better quantitative approach should be needed.

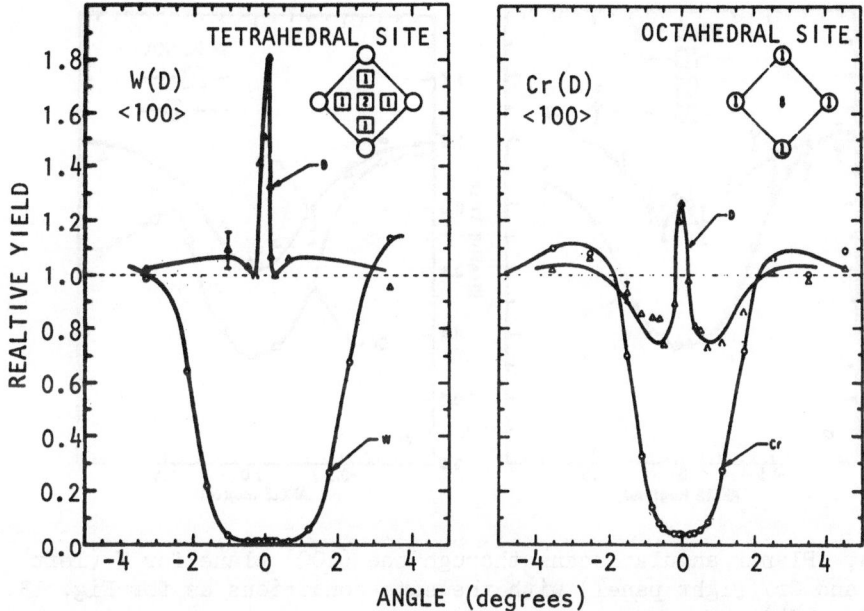

Fig. 13.   Angular scans through the <100> axis for W (left panel) and Cr (right panel) for $3.10^{15}/cm^2$ 30 and 15 KeV D implants, respectively.   A 750-KeV ^{3}He analysis beam was used; circles correspond to backscattering yields for W and Cr, and triangles correspond to proton nuclear-reaction yields from D. (From Ref. 24).

II.3. Interstitial Case

The accurate location of interstitial impurities is usually limited by the need for correct information of the flux distribution in the mid-region of channel, as discussed in Sect. I.   Extensive experiments with angular scans across the major crystal axis and planes, are needed to determine the location in rather complicated distribution of "nonsubstitutional" impurities[7].   The interpretation of experimental results need the use of sophisticated computer simulations to obtain the flux distribution.   The yield of a reaction Y between channeled particles and a given bulk distribution of impurities can be calculated as $Y(E_o, \psi_i, z) = \int_E \int_r F(E,r,z)\sigma(E,z)dEdz$.   Here $\sigma(E,z)$ denotes the reaction cross section; z the penetration depth, E_o the incident energy.   The integrations extend   over the available energy range and the area of the unit cell.

However, interstitial foreign atoms are often located on well-defined interstitial sites along certain axial or planar directions, as tetrahedral or octahedral sites respectively in diamond-type or

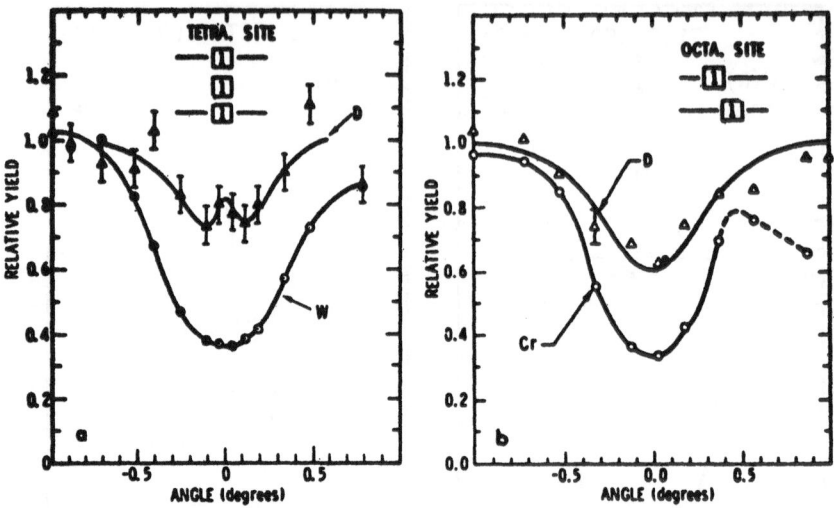

Fig. 14. Planar angular scans thorugh the {100} plane for W (left panel) and Cr (right panel) with the same conditions as for Fig. 13. (From Ref. 24).

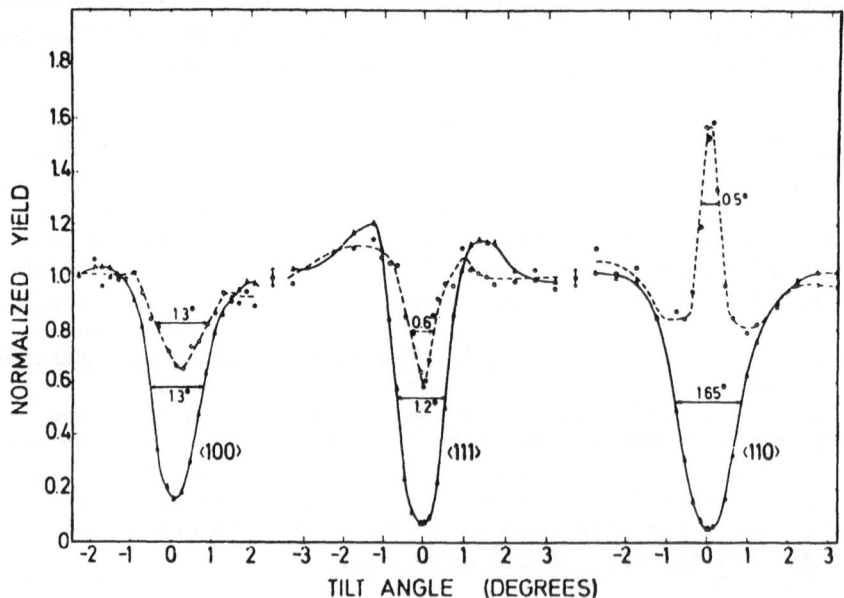

Fig. 15. Channeling normalized yields by backscattering from 5.10^{14} Yb/cm^2 implanted at 60 KeV, 750°K into Si for 1 MeV-^{4}He ions incident along the <100>, <111> and <110> axis. (From Ref. 4).

body-centered cubic lattices.

Fig. 13 shows the channeling angular distributions along the <100> axis for deuterium (D) implanted into bcc crystals W and Cr. The implantation energies were 15 and 30-KeV, respectively, for the implants in Cr and W samples[24]. The dose was 3.10^{15} D atoms/cm^2. The deuterium was detected by the protons emitted from the nuclear reaction $D(^3He,p)^4He$ at 750 KeV, and the lattice atoms by ^3He ions backscattering. For W (left side) the strong enhancement in the D yield suggests that the impurity is located in the tetrahedral interstitial sites, which are not shielded by metal atoms along the <100> lattice rows. The data for Cr (right side) show a sharp flux peak superimposed on a broad dip with an angular width similar to that observed for the Cr lattice. The minimum yield of the "extrapolated" dip is about 70%. 1/3 of the octahedral sites are shielded by metal atoms along the studied <100> direction. The minimum yield value and the dip width are consistent with 1/3 of the D along the <100> lattice row. The central flux corresponds to the remaining D atoms in the center of <100> channel. This interpretation has been checked by angular scan along the {100} plane, as shown in Fig. 14. In this configuration, the octahedral sites lie within {100} planes, while the tetrahedral sites are: 2/3 in {100} plane, 1/3 between the planes. We observe only a dip for D in Cr and a small flux peak superimposed on a dip for D in W.

A second example of the location of an interstitial impurity is shown in Fig. 15 for Yb implanted in Si[4]. The channeling angular scans were performed along the <100>, <111> and <110> axial directions. We can observe: a) a dip along the <100> with an angular width similar to that for the Si lattice, which suggests that about 30% (the value of the Yb minimum yield is 70%) of the impurities are located along this row, while the remaining fraction is in some intermediate sites; b) a narrower dip for the <111> direction in comparison with the Si dip, which suggests that a fraction of impurities are "almost" substitutional along the considered channel; c) a flux peak along the direction <110>, which indicates that a fraction of the Yb is located near the center of this channel. The authors concluded, using further information by scans along planar directions, that the Yb atoms are located along the <100> row in the mid-point of the <110> channel.

We want now to discuss an experiment of atom location in the system TiO_x, varying the concentration of O-atoms from 0.11 to 0.39[11,19]. The angular distribution appears concentration dependent in agreement with the consideration made on the influence of high concentration impurities on flux distribution. For the <0001> direction in Ti lattice, the O atoms lie in the center of the channel formed by Ti rows. The authors studied the channelling angular distribution along the <0001> axis for scans within the {1210} plane, which is a mixed plane containing both O and Ti atoms.

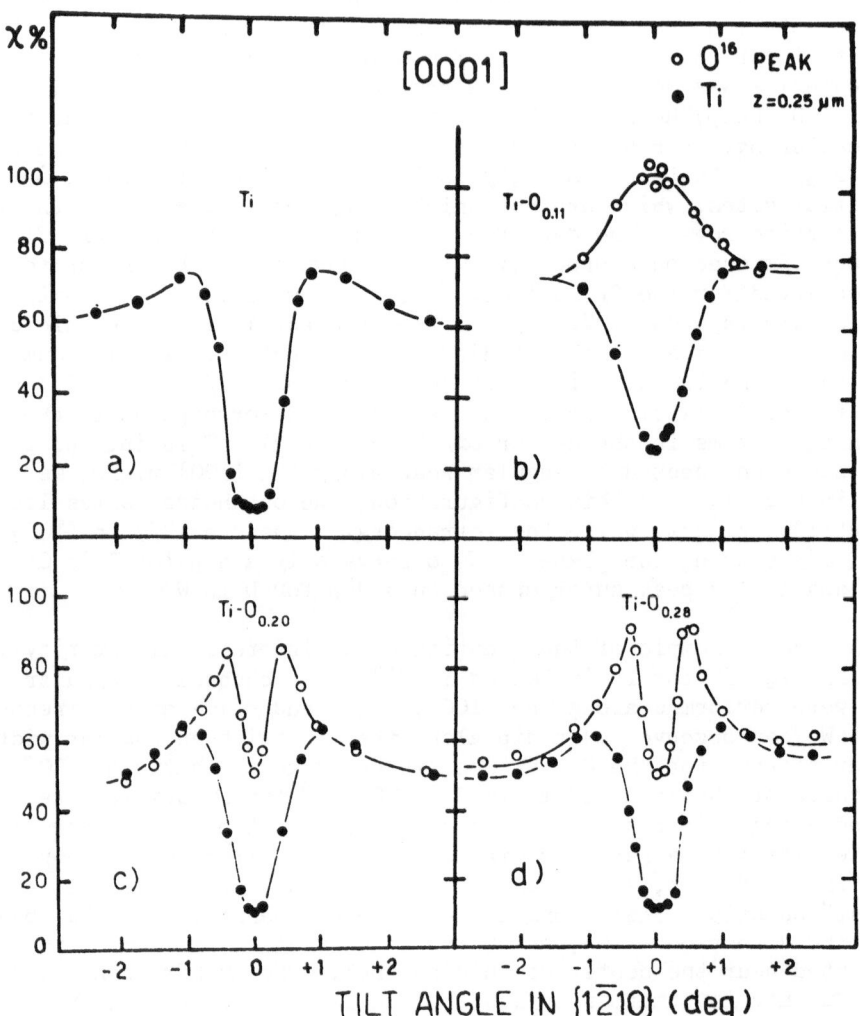

Fig. 16. Channeling normalized yields for <0001> axis at various
TiO$_x$ compositions. Tilt angle is within a {1$\overline{2}$10} plane, which
contains both Ti and O atoms. Open circles denote proton yield
from nuclear reaction and full circles deuterons backscattering
yield from Ti atoms. (a) x = 0; (b) x = 0.11; (c) x = 0.20; (d)
x = 0.28. (From Ref. 11).

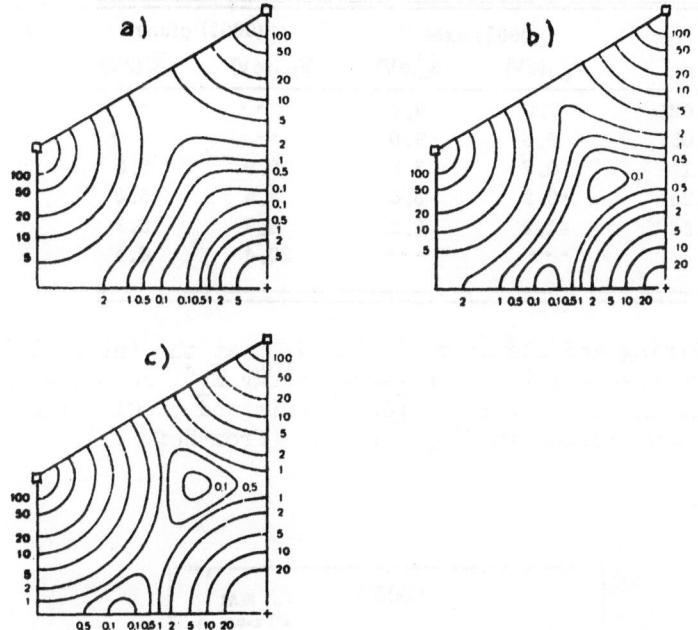

Fig. 17. Equipotential contour plot for standard Lindhard[6] string potential for deuterons along the <0001> axis in Ti ($\square$ site) containing O atoms at the center of the channel (+ site) for different TiO$_x$ compositions. (a) x = 0.1; (b) x = 0.25; (c) x = 0.5. The case x = 0 is reported in Fig. 1. The potential value ($\sim$115 eV) at the lattice string depends slightly on impurity concentration. (From Ref. 11).

Fig. 16 shows the angular distribution for different compositions, reported in the caption. A beam of 1-MeV deuterons was used to permit simultaneous observations of the channeling behaviour both with respect to Ti atoms, by means of wide-angle Rutherford scattering, and to O atoms, by means of the $^{16}O(dp)^{17}O^*$ reaction. At low concentration, the O sublattice does not yet give rise to channeling effect and we observe a weak flux peaking, due to the balanced repulsive effect of Ti and O strings on incident ions. At higher concentration the O-string is strong enough to display a yield dip in the nuclear reaction. The angular width of the dips depends on the different strength of Ti and O-strings. Fig. 17 shows the continuum potential for the <0001> axial channel as a function of O concentration. The equipotential contour for Ti lattice without O impurities, has been also reported in Fig. 1. By the comparison between the O string potential value, V_{int} with the deuterons mean initial transverse energy, $\bar{E}_\perp^o$ evaluated from the continuum potential making a computer average over many starting points (Table I),

	[0001] axis		(0001) plane	
x	V_{int}(eV)	$\overline{E}_{\perp}^{0}$(eV)	V_{int}(eV)	$\overline{E}_{\perp}^{0}$(eV)
0.025	1.0	9.4	...	...
0.05	2.8	9.0	...	...
0.1	6.7	8.5	0.9	5.5
0.25	19.2	8.0	3.5	5.0
0.5	41.2	8.2	8.7	5.1
1	...	...	20.4	6.4

Table I. String and planar potential V_{int} at the interstitial
position and mean initial transverse energy $\overline{E}_{\perp}^{0}$, computed as
indicated in the text, for the [0001] axis and (0001) plane for
different compositions in TiO_x system. (From Ref. 11).

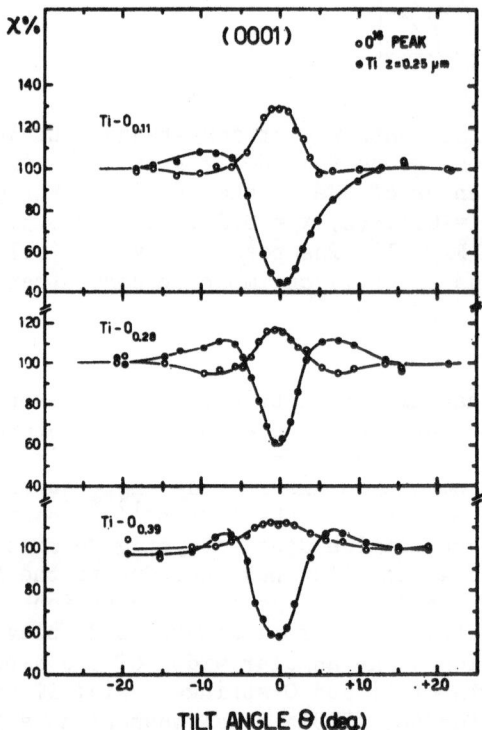

Fig. 18. Channeling normalized yield for the (0001) plane at various
TiO_x compositions. The open and full circles have the same meaning
as in Fig. 16. (a) x = 0.11; (b) x = 0.28; (c) x = 0.39. (From
Ref. 11).

we observe that the potential at interstitial string becomes higher than $E_\perp^\circ$ value only for x> 0.1. Then we should find the O dip only at concentrations higher than 0.1, in agreement with the experimental results. Fig. 18 shows the normalized yield for the (0001) plane at various compositions. The planes (0001) consists of a single atomic species. The curves relative to scattering by Ti sublattice always show the expected channeling effect; while the nuclear-reaction yield from the O-sublattice gives a peak higher than random value for planar alignement at all O concentrations. The O impurities can be considered as single interstitial atoms. However, the experiments also show that planar flux peaking is progressively destroyed at the interstitial site with increasing interstitial concentration. From Table I we observe that the planar V_{int} value is lower than the average transverse energy up to the highest O concentration studied (x = 0.39).

We conclude this review with an example of the determination of impurity atom location, observing the channeling dependence of the yield of inner-shell X rays, induced by bombardment with 1.5 MeV ^{4}He.

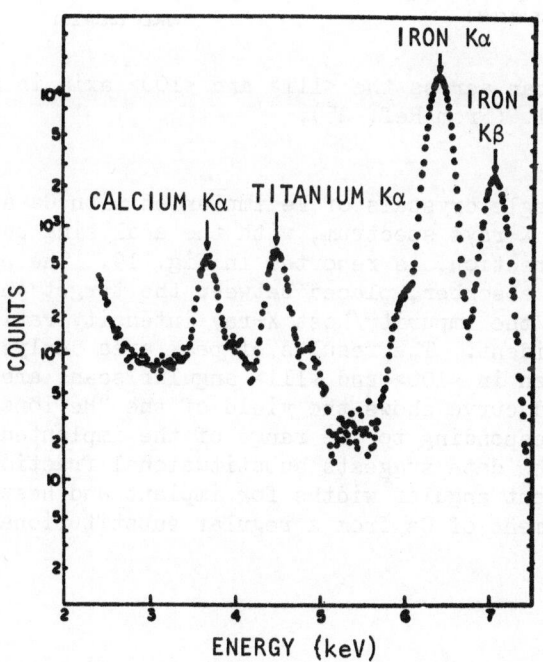

Fig. 19. X-rays spectrum obtained by bombardment of a Ca-implanted Fe single crystal with 1.5-MeV ^{4}He ions incident in a <111> axial channeling direction. The Ca dose was $3.10^{15}/cm^2$. (From Ref. 25).

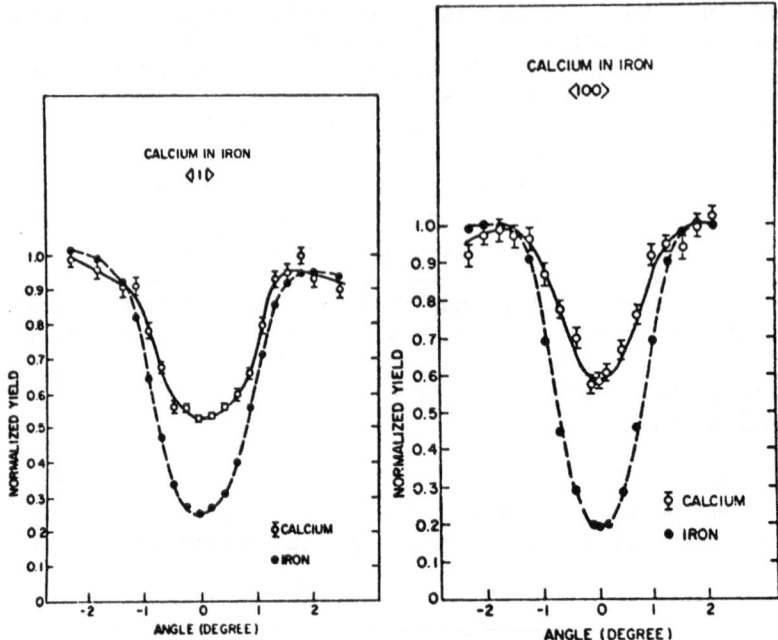

Fig. 20.   Angular scan across the <111> and <100> axis in a Ca-implanted Fe single crystal. (From Ref. 25).

The samples were single crystals of Fe implanted with Ca at room
temperature[25].   The X-rays spectrum, with the analysing beam incident
in a <111> axial direction, is reported in Fig. 19.   The peaks due to
the Fe host, the Ti absorber, placed between the target and X-ray
detector to improve the impurity/host X-ray intensity ratio, and the
implanted Ca are evident.   The results of peak area analysis for
similar spectra taken in <100> and <111> angular scans are shown in
Fig. 20.   The dashed curve shows the yield of the ^{4}He ions backscattered from depths corresponding to the range of the implanted ions.   An
interpretation of the data suggests substitutional fraction of 55%
for Ca.   The different angular widths for implant and host scans may
indicate a displacement of Ca from a regular substitutional location.

III. CONCLUSIONS

        The main limitations in the use of ion channeling to lattice
location studies can be summarized: a) crystal perfection; b) detection sensitivity; c) radiation induced damage by the analysis beam;
d) difficulty of a detailed analytic estimate of the channeling
effect; c) ambiguous site determination for low-symmetry crystals.
The detection sensitivity is about 10^{-4} atom fraction, quite adequate

for many solid state investigations. An improvement in this limit
might be obtained with the use of inner-shel X-ray production resulting from light ion bombardment. With regard to analytical estimates,
additional MonteCarlo computer simulation studies appear to be very
useful to establish the field of validity of the theoretical models,
such as assumption of statistical equilibrium. In fact, many
experiments refer to location of impurity, at depths less than that
at which equilibrium is achieved. Experiments on compounds are needed to clarify the interaction between sublattice of different strengths.

REFERENCES

1. J.A. Davies, Channeling: Theory, Observation and Applications,
    ed. by D.V. Morgan (John Wiley, 1973), p. 391.

2. S.T. Picraux, New Uses of Ion Accelerators, ed. by J.F. Ziegler
    (Plenum Press, N.Y., 1975), p. 229.

3. M.A. Kumakhov, Phys. Lett. A32, 558 (1970); Radiat. Eff. 15,
    85 (1972).

4. J.U. Andersen, O. Andreasen, J.A. Davies, and E. Uggerhoj,
    Radiat. Eff. 7, 25 (1971).

5. R.B. Alexander, P.T. Callaghan, and J.M. Poate, Phys. Rev. B9,
    3022 (1974).

6. J. Lindhard, K. Dan. Vidensk. Selsk. Mat.-Fys. Medd. 34, No. 14
    (1965).

7. R.B. Alexander, G. Dearnaley, D.V. Morgan, J.M. Poate, and D.
    Van Vliet, Proc. Europ. Conf. on Ion Implantation, (Peregrinus
    Stevenage, England, 1970), p. 181; D. Van Vliet, Radiat. Eff.
    10, 137 (1971).

8. J.H. Barrett, Phys. Rev. B3, 1527 (1971); Atomic Collision in
    Solids, Vol. II, ed. by S. Datz, B.R. Appleton and C.D. Moak,
    (Plenum Press, N.Y., 1975), p. 841.

9. H.D. Carstanjen, and R. Sizmann, Radiat. Eff. 12, 225 (1972).

10. A. De Salvo, R. Rosa, and F. Zignani, J. Appl. Phys. 43, 3755
    (1972); G. Della Mea, A.V. Drigo, S. Lo Russo, P. Mazzoldi,
    G.G. Bentini, A. De Salvo, and R. Rosa, Phys. Rev. B7, 4029
    (1973).

11. G. Della Mea, A.V. Drigo, S. Lo Russo, P. Mazzoldi, S. Yamaguchi,
    G.G. Bentini, A. De Salvo, and R. Rosa, Phys. Rev. B10, 1836
    (1974).

12. G. Foti, F. Grasso, R. Quattrocchi and E. Rimini, Phys. Rev. B3,
    2169 (1971).

13. M.A. Kumakhov, Radiat.Eff. 26, 43 (1975).

14. V.V. Beloshitsky, M.A. Kumakhov, and V.A. Muralev, Radiat. Eff.
    13, 9 (1972); 20, 95 (1973).

15. E. Bonderup, H. Esbensen, J.U. Andersen, and H.E. Shiøtt, Radiat. Eff. 12, 261 (1972).

16. F. Abel, G. Amsel, M. Bruneaux, C. Cohen, and A. L'Hoir, Phys. Rev. B12, 4617 (1975).

17. F. Abel, M. Bruneaux, C. Cohen, G. Della Mea, A.V. Drigo, S. Lo Russo, and P. Mazzoldi, Nucl. Instr. and Meth. 132, 197 (1976).

18. D.P. Jackson, Atomic Collisions in Solids, Vol. I, ed. by S. Datz, B.R. Appleton, and C.D. Moak, (Plenum Press, N.Y., 1975), p. 185.

19. G. Della Mea, A.V. Drigo, S. Lo Russo, P. Mazzoldi, S. Yamaguchi, G.G. Bentini, A. De Salvo,and R. Rosa, Atomic Collisions in Solids, Vol. II, ed. by. S. Datz, B.R. Appleton, and C.D. Moak, (Plenum Press, N.Y., 1975), p. 791.

20. J.W. Mayer, L. Eriksson, and J.A. Davies, Ion Implantation in Semiconductors, (Ac. Press, N.Y.,1970).

21. R.B. Alexander, and J.M. Poate, Radiat. Eff. 12, 211 (1972).

22. S.T. Picraux, W.L. Brown, and W.M. Gibson, Phys. Rev. B6, 1382 (1972).

23. J.U. Andersen, K. Dan. Vidensk. Selsk. Mat.-Fys. Medd. 36, 7 (1967).

24. S.T. Picraux,and F.L. Vook, Phys. Rev. Lett. 33, 1216 (1974).

25. J.R. MacDonald, R.A. Boie, W. Darcey, and R. Hensler, Phys. Rev. B12, 1633 (1975).

ANALYSIS OF DEFECTS BY CHANNELING

E.Rimini

Istituto di Struttura della Materia

Corso Italia 57 - 95129 Catania,Italy

1. INTRODUCTION

Energetic ions impinging along rows or planes of a single crystal are steered by means of correlated series of small angle collisions. The decrease of yield occurs because the energetic particles that are steered along the "channels" in the crystal do not approach the lattice atoms closely enough to undergo the wide-angle elastic scattering processes[1]. The channeled component of the beam can act as a probe to detect atoms, either host or impurity, which are displaced from substitutional lattice position by distances exceeding about 0.1 - 0.2 Å.

Channeled particles during their motion inside a perfect crystal suffer multiple scattering by electrons and thermally vibrating lattice atoms[2]. This small angle scattering causes dechanneling when the transverse energy of the channeled particles overcomes the barrier potential of the atomic row or plane. The dechanneled particles can then interact with the non-displaced host atoms and give rise to a backscattering yield which increases with the traversed depth. The dechanneled component of the beam is also called random component.

Displaced atoms can interact with the channeled component in both wide-angle collisions and forward elastic scattering events as illustrated in Fig.1. The large-angle deflections are those leading to the direct detection of displaced atoms through backscatte-

ring analysis. The small-angle scattering can cause the channeled
particles to be scattered at angles larger than the critical angle
for channeling. This dechanneling mechanism in addition to that

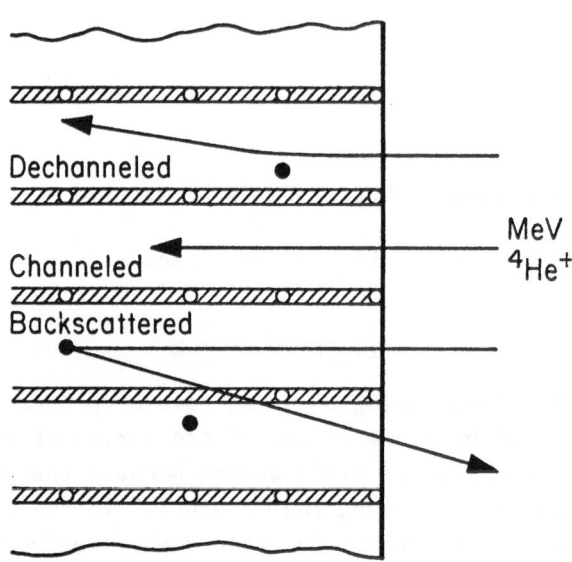

Fig.1 - Schematic representation of direct backscattering and dechanneling event of channeled particles on defects (from reference 3 - Chapter VIII).

experienced in a perfect crystal is responsible for the high backscattering yield. A proper treatment of dechanneling processes is
necessary in order to extract the number of displaced atoms   and
their depth distribution from the aligned spectra.

        The relative contribution of direct backscattering and dechan-
neling depends on the nature of defects. If an "amorphous" structure is imbedded in a perfect crystal and the displaced atoms are
randomly distributed, the channeling effect technique provides
with accuracy the number and the depth profile of disordered atoms.
If the defects are not randomly distributed or if they agglomerate
as dislocation loops, clusters, stacking fault etc., the extraction procedure is ambiguous. The interaction of channeled particles with all these types of defects and their influence on de-

channeling is not completely understood at the present. A channeling experiment can provide only an indication of the lattice disorder. The number and the nature of defects are not directly related to the backscattering spectra. The analysis beam energy dependence of the dechanneling on different type of defects can be used to obtain information on the predominant defect structure. The channeling technique is not a sensitive tool in damage analysis. This is associated with the fact that a finite yield is observed already in the aligned spectrum of the perfect crystal. The magnitude of this yield determines the smallest amount of damage which can be observed. This amounts to the displacement of at least 1% of the atoms of the crystal using single alignment.

As a first example (Sect.2) of the application of channeling effect to disorder we will consider the influence of a thin amorphous layer on a single crustal[4]. The disorder created by ion implantation in semiconductors will be then analysed in Sect.3 assuming a random distribution of displaced atoms. The single and double alignment measurements[5], and the off-axis analysis[6] will be described in Sect.4 to obtain information on the distribution of defects across the channel. Dislocations due to the associated strain field interact with channeled particles causing mainly dechanneling. A simple analytical treatment of dechanneling[7] together with a comparison with available experiments[8] will be presented. A brief discussion of the influence of other types of defect (mosaic spread and stacking faults) on dechanneling will be also considered in Sect.6.

2. CRYSTALS OVERLAID WITH AN AMORPHOUS FILM

Consider a well collimated beam of particles impinging along a low index axis of a single crystal covered with a thin amorphous layer. The beam particles suffer single scattering with the atoms of the amorphous layer and if they are deflected at angles larger than the critical angle, $\psi_{\frac{1}{2}}$, for channeling the crystal is seen as a random medium. Assuming Rutherford scattering law, the cross section for dechanneling is given by

$$\sigma_d(\psi_{\frac{1}{2}}) = \int_{\psi_{\frac{1}{2}}}^{\infty} \frac{d\sigma}{d\Omega} \, d\Omega = \frac{\pi \, Z_1^2 Z_2^2 e^4}{E^2 \, \psi_{\frac{1}{2}}^2} \tag{1}$$

where Z_1e and Z_2e are the atomic charge of the projectile and target atoms respectively, E the beam energy and $\psi_{\frac{1}{2}}$ the critical angle.
For MeV He and for lattice spacings d, of 4 Å, eq.(1) becomes

$$\sigma_d(\psi_{\frac{1}{2}}) = 3.5 \times 10^{-20} \, Z_2/E \text{ cm}^2 \text{ (E in MeV)} \qquad (2)$$

The probability that a particle will be scattered in a single collision when traversing a film of Nt atoms/cm^2 is given by

$$P(\psi_{\frac{1}{2}}) = \sigma_d(\psi_{\frac{1}{2}}) \text{ Nt} \qquad (3)$$

For an amorphous Ge layer of 10^{17} atoms/cm^2 on a Ge single crystal,
$\sigma_d \approx 10^{-18}$ cm^2 for 1 MeV He and $P \approx 0.1$ i.e. 10% of the collimated beam
particles are deflected beyond the critical angle. As a result of
these deflections the minimum yield of the underlying single crystal
increases and as first approximation is given by $P(\psi_{\frac{1}{2}}) + \chi_v$, where
χ_v is the minimum yield in the uncovered crystal.

If the thickness of the amorphous layer increases one must account for the occurence of double, triple, etc., scattering. The
beam at the exit of the layer is characterized by an angular distribution which has been described correctly by the plural scattering
regime[9]. To relate the minimum yield of the underlying crystal with
the amorphous thickness it is sufficient to determine the particles
scattered outside the axial half-angle $\psi_{\frac{1}{2}}$. This procedure assumes
a step function or square well approximation to the angular yield
profiles as illustrated in Fig.2.

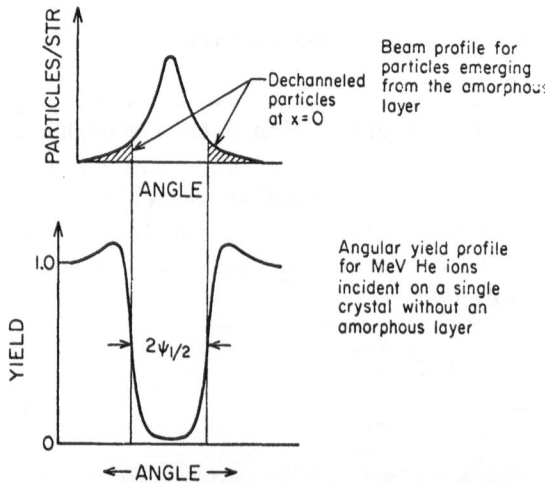

Beam profile for
particles emerging
from the amorphous
layer

Angular yield profile
for MeV He ions
incident on a single
crystal without an
amorphous layer

Fig.2 – The upper part
represents schematically
the angular distribution
of particles after tra-
versing a reduced tick-
ness m. An angular yield
profile is shown in the
lower part with the step-
function probability
(dashed-line) approxima-
tion. This yield profile
is used as the probabi-
lity that a particle in-
cident at an angle θ is
dechanneled (from refe-
rence 3 – Chapter VIII).

These plural scattering distributions are expressed in terms of a reduced thickness, m, and of reduced angles $\tilde{\theta}$, as shown in formula 4

$$m = \pi \, a^2 \, Nt \quad \text{and} \quad \tilde{\theta} = \frac{aE}{2 \, Z_1 Z_2 e^2} \, \theta \qquad (4)$$

where a is the Thomas-Fermi screening radius. Physically, m is the mean value of the number of collisions of the particles with the atoms in the thin film for a cross section of $\pi \, a^2$. For He in Ge m=0.1 corresponds to Nt=1.45x10^{16} atoms/cm^2=32.9 Å anf for He→Si it corresponds to 0.85x10^{16} atoms/cm^2=17 Å.

The probability of scattering beyond the critical angle $\psi_{\frac{1}{2}}$ is given by

$$P(\tilde{\theta}_c, m) = \int_{\tilde{\theta}}^{\infty} 2\pi \, f_m(\tilde{\theta}) \, \tilde{\theta} \, d\tilde{\theta}$$

where

$$\tilde{\theta}_c = \frac{aE}{2 \, Z_1 Z_2 e^2} \, \psi_{\frac{1}{2}},$$

and $f_m(\tilde{\theta})$ is the angular distribution of the particles. These distributions are available from m=0.001 to m=2.000 as tables[9]. For large $\tilde{\theta}$ values or for small reduced thickness m values, the plural scattering approachs the Rutherford single scattering distribution.

The minimum yield observed in experiments with Al and Au thin film on Si has been found to compare well with the $P(\tilde{\theta}_c, m)$ values obtained in the plural scattering regime as illustrated in Fig.3a and 3b. Another feature of interest is the analysis beam energy dependence of the minimum yield. As shown in Fig.4 the measured minimum yield decreases with an increase in the beam energy. This energy dependence arises from two factors: the critical angle decreases as $E^{-\frac{1}{2}}$ while the beam angular distribution narrows more than the critical angle decreases.

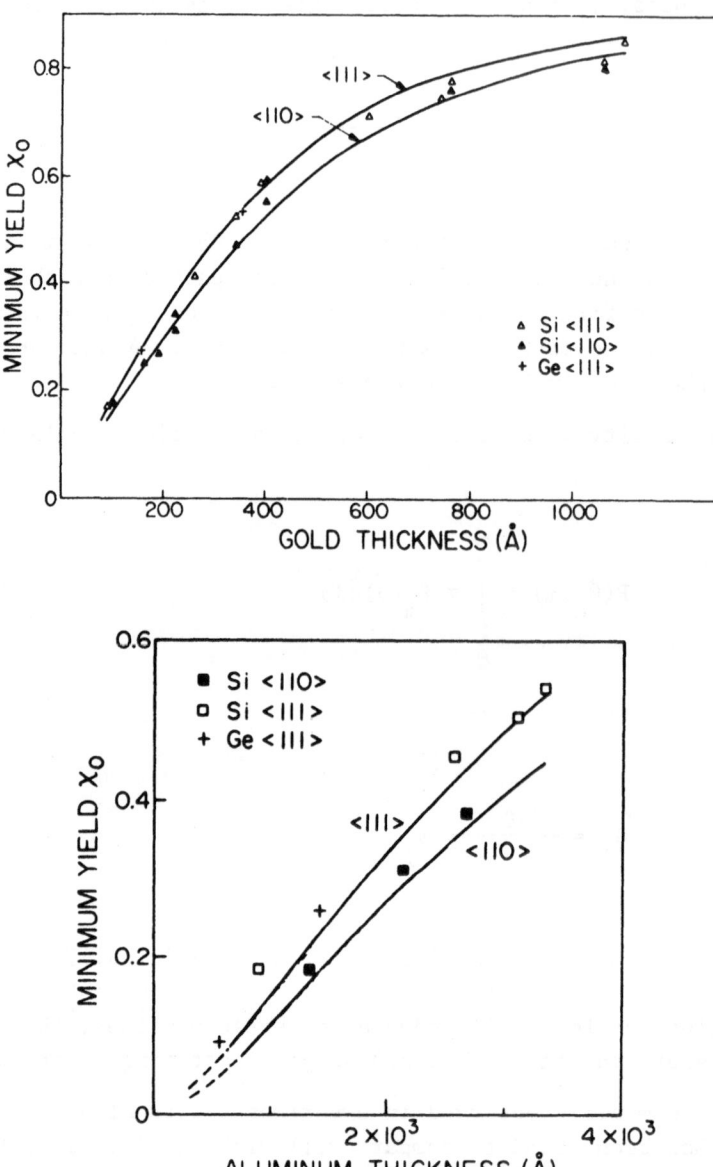

Fig.3 - Minimum yield for 1.8 MeV He+ impinging along the <110> and
<111> axes of Si covered with different thicknesses of Au (a) and Al
(b). The lines show the calculated values using the step-function
approximation and plural scattering distribution (from reference 4).

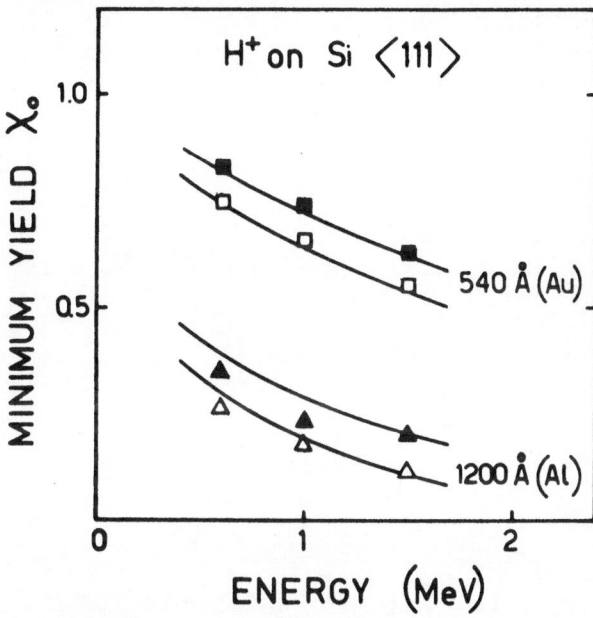

Fig.4 - Proton beam energy dependence of the minimum yield, χ_{min}, along the <111> axis of Si covered with 540 Å of Au and 1200 Å of Al and for target temperature of 80K(Δ and □) and 300K(Δ and □). The solid lines represent the calculated χ by the convolution method (from reference 10).

3. RANDOM DISTRIBUTION OF DISPLACED ATOMS

In this section we treat the analysis of an implanted semiconductor and we assume that the displaced atoms are randomly distributed and located within a perfect crystal which has a well defined value of $\psi_{\frac{1}{2}}$.

The aligned axial and random backscattering spectra for MeV He ions incident on a crystal containing disorder near the surface are schematically illustrated in Fig.5. For comparison the aligned yield for the same crystal without disorder is also shown. The higher yield at depths larger than the implanted region is due to the fact that some of the channeled ions have been dechanneled by small angle deflection (>$\psi_{\frac{1}{2}}$) from defects. These dechanneled ions, can be backscattered by all of the atoms of the crystal and so they contribute to the aligned yield. This is the same problem considered previously for the dechanneling caused by an amorphous layer on

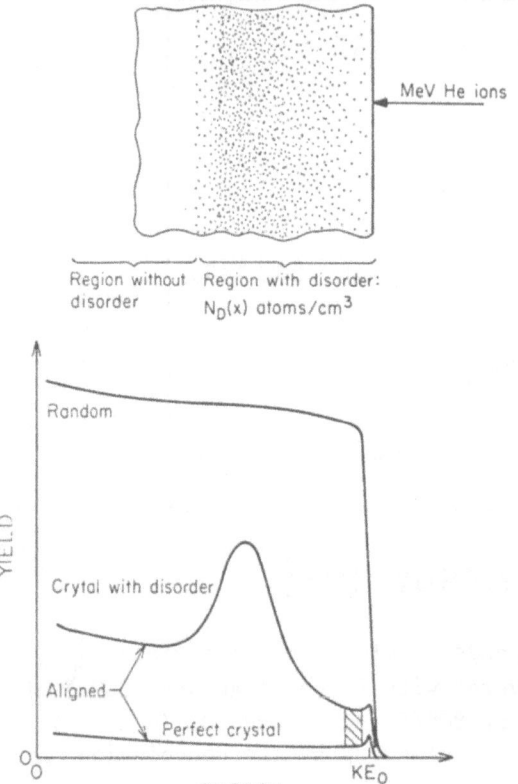

Fig.5 - Typical backscattering energy spectra of a crystal containing displaced atoms randomly distributed (from reference 3, Chapter VIII).

a single crystal. In the present case the disorder is contained in the region of the crystal in which dechanneling occurs.

The problem consists now in the separation of the two contributions: backscattering of the channeled component from defects and backscattering from the dechanneled component from all the crystal atoms.

Just behind the surface peak the contribution of dechanneling is minimal and so the disorder concentration is given by

$$N_D(o) = N \frac{\chi(o) - \chi_v(o)}{1 - \chi_v(o)} \qquad (6)$$

where N is the bulk density in atoms/cm^3 and χ is the ratio of aligned to random yields at the energy where the disorder is evaluated.

χ_v is the yield ratio for the virgin crystal and $\chi_v(o)$ is equal to χ_{min}. In the above equation $(1-\chi_v)$ represents the channeled component and the term $\chi(o)-\chi_v(o) = (1-\chi_v(o))-(1-\chi(o))$ represents the difference between the channeled fractions in the virgin and in the damaged crystal.

The analysis for the depth profile inside the crystal can be accomplished by an iterative procedure shown in Fig.6 for a sample

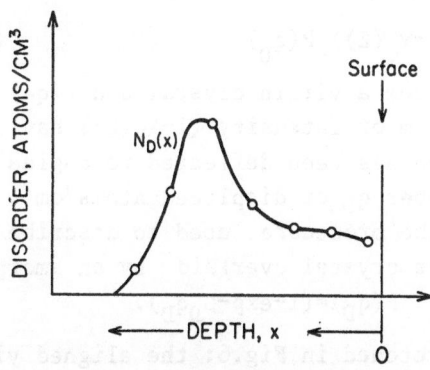

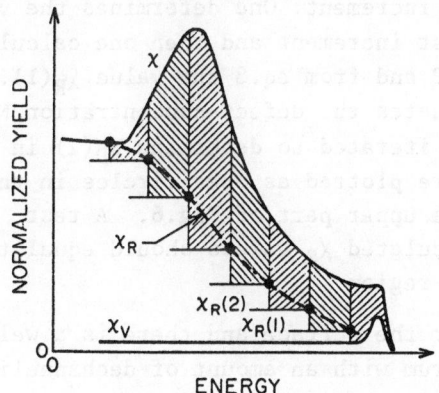

Fig.6 - Illustration of the iterative procedure used to extract from the normalized aligned yield in the implanted and unimplanted crystal the defect profile shown in the upper part (from reference 3, Chapter VIII).

containing a disorder distribution $N_D(Z)$. Assuming that the beam particles can be divided into two parts: a channeled component which can interact with displaced atom and a random component which is scattered by all the atoms of the crystal we have[11]

$$\chi(Z) = \chi_R(Z) + (1-\chi_R(Z)) \frac{N_D(Z)}{N} \qquad (7)$$

where $\chi(Z)$ is the normalized yield measured at a depth Z from the surface, $\chi_R(Z)$ is the random component, $(1-\chi_R(Z))$ is the channeled component which interacts with the displaced atoms. The basis problem is the determination of the random component $\chi_R(Z)$ as function of depth. Since dechanneling is produced by the same disordered atoms that cause the backscattering of the aligned component, it is possible to relate the random fraction to the total number of scattering centers encountered by the beam up to a depth Z. It is assumed moreover that a scattering center contributes proportionally to the backscattering as it does to dechanneling. The random component is then usually approximated by

$$\chi_R(Z) = \chi_v(Z) + (1-\chi_v(Z)) \, P(q_D) \qquad\qquad (8)$$

where $\chi_v(Z)$ is the aligned yield from a virgin crystal and $P(q_D)$ represents the probability that a beam of intensity $(1-\chi_v(Z))$ having initially a δ function distribution has been deflected to angles larger than $\psi_{\frac{1}{2}}$ after traversing a number q_D of displaced atoms/cm^2. The value of $P(q_D)$ can be found from the procedure used to describe the amount of dechanneling in a perfect crystal overlaid by an amorphous layer, e.g. for single scattering[12] $P(q_D)=(1-\exp-\sigma_D q_D)$.

The iterative procedure is sketched in Fig.6: the aligned yield is divided into equivalent thickness increments Δt and the disorder has been assumed constant in each increment. One determines the value $N(o) \, \Delta t$ (see eq.7) in the first increment and then one calculates the dechanneling probability P and from eq.8 the value $\chi_R(1)$. From eq.7 by measuring χ one evaluates the defect concentration $N_D(1)$ and hence $\chi_R(2)$. The procedure is iterated to determine $N_D(2)$ in the next layer. The values of N (i) are plotted as open circles in the disorder distribution curve in the upper part of Fig.6. A test of consistency requires that the calculated χ_R values should equal to χ at depths below the disordered region.

If the damage is located near the surface and there is a well defined peak in the aligned spectrum with an amount of dechanneling not too large, a much simple procedure may be used to determine the number of displaced atoms. The procedure (see Fig.7) consists in approximating the dechanneling with a straight line and from the shaded area the number of disordered atoms can be obtained by comparison with the random yield[13].

In the extraction procedure there are several assumptions:
1) Displaced atoms are randomly distributed. 2) The critical angle $\psi_{\frac{1}{2}}$ is unchanged by the introduction of disorder. 3) Dechanneled particles are not scattered back into channels. 4) The dechanneling due to defects is additive to that caused by electrons and thermally vi-

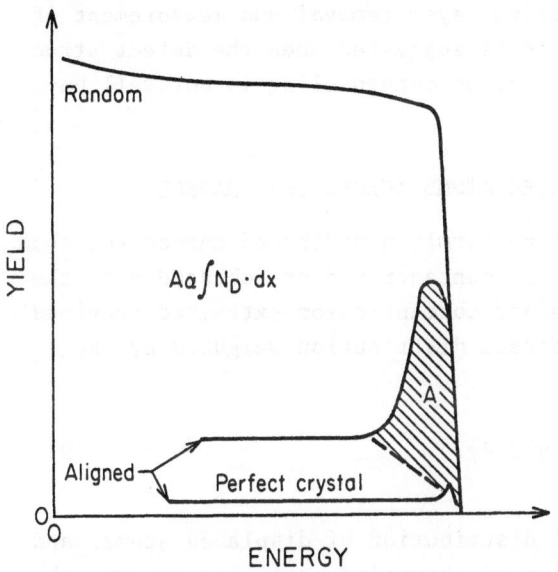

Fig.7 - Analysis of a
crystal with disorder
near the surface. The
dashed line represents
the dechanneling con-
tribution while the
area A under the peak
is proportional to the
number of displaced
atoms per unit area
(from reference 3, Chap-
ter VIII).

brating nuclei. 5) Stopping power of channeled particles does not
depend on the disorder.

Assumption 4 neglects the simultaneous occurence of small angle
scattering by defects and by thermal vibrations and electrons. This
interference term if properly taken into account, should decrease
the extracted defect density. Two different approaches have been u-
sed to include this term: one is based on a description of dechanneling as a diffusion process in transverse momentum space[14] and
the second on an analytical approximation of dechanneling with the
model of the steady increase of channeled ion transverse energy[15]
The stopping power of channeled particles near the center of the
channel is lower than that of the random beam. Different stopping
powers imply different depth scales for the random and aligned yield,
moreover the change in the Rutherford scattering law for the particle energy should be also taken into account[16]. To overcome all
these difficulties the approach in the disorder analysis is often
to use the same stopping power for the aligned and for the random
yield. This is justified by experiments which show that the stopping
cross section for dechanneled particles is close to that of the random beam[17].

A truly experimental approach to obtain depth profile of disorder regions consists in successive layer removal and measurement of surface disorder. This procedure is suggested when the defect structure is not known or its influence on dechanneling is not well understood.

4. DISTRIBUTION OF DISPLACED ATOMS ACROSS THE CHANNEL

The channeled beam is not uniformly distributed across the channel between the axial rows. It is concentrated or focussed near the center of the channel[18]. The defect concentration extracted previously represents an average of the defect distribution weighted by the channeled ion flux.

$$N_D(z) = \int_A N_D(x,y,z)F(x,y)dxdy \tag{9}$$

where $N_D(x,y,z)$ is the spatial distribution of displaced atoms and $F(x,y)$ is the flux distribution of channeled particles across the channel. The integration is performed over the area of the channel.

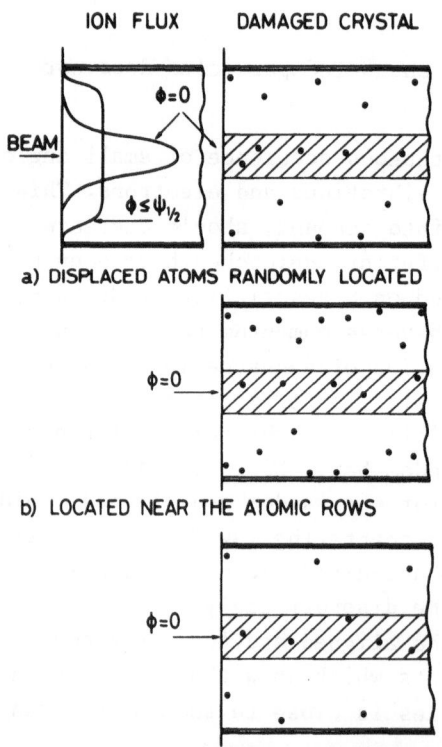

a) DISPLACED ATOMS RANDOMLY LOCATED

b) LOCATED NEAR THE ATOMIC ROWS

c) LOCATED IN THE CENTER OF CHANNEL

Fig.8 – Different defect distributions with the same number of displaced atoms in the shaded region at the channel center. The channeled ion flux is sketched for two different tilting angles.

In Fig.8 three damage distributions are shown all having the same defect density (displaced atoms) near the center of the channel, but different density near the atomic rows. The channeled beam probes mainly the low potential region near the center. The yield due to scattering on defects is then the same for all the three distributions although the total number of disordered atoms is different.

Flux distributions can be changed by tilting the crystal axis of an angle Φ (within $\psi_{\frac{1}{2}}$) with respect to the beam direction. The yields (see eq.9) are now different for the three distributions. The variations in the measured yields as a function of the tilting angles are then related to the distribution across the channel of the displaced atoms. The method represents an extension of the flux-peaking effect used for impurity atom location to host displaced atom location[18,19].

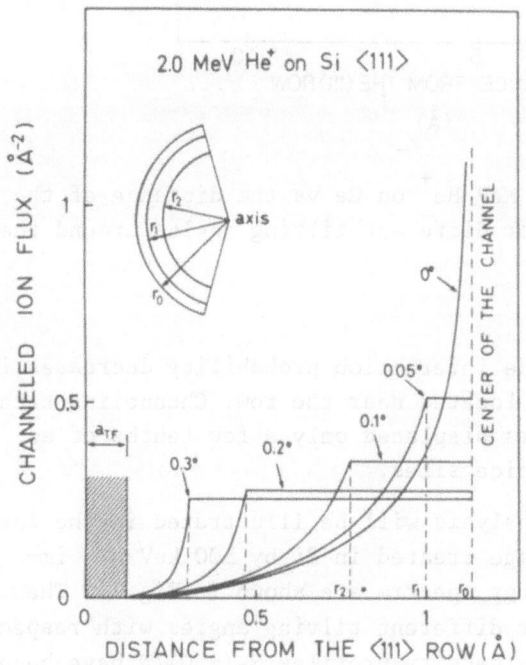

Fig.9 – Ion flux for 2.0 MeV He[+] on Si for different tilting angles around the <111> axis (from reference 20).

As an example in Fig.9 the ion fluxes computed on the basis of Lindhard continuum potential approximation are shown for 2.0 MeV He[+] on Si around the <111> axis. At distances larger than r_i given by $U(r_i)-U(r_o) = E\Phi_1^2$ the fluxes are constant and they decrease parabolically near the row. The interaction yield of channeled particles

with displaced atoms depends then on the flux distribution and on
the location of defects. The yields in a.u. are shown in Fig.10 for
a 2.0 MeV He$^+$ analyzing beam in Ge vs distance from the <111> row

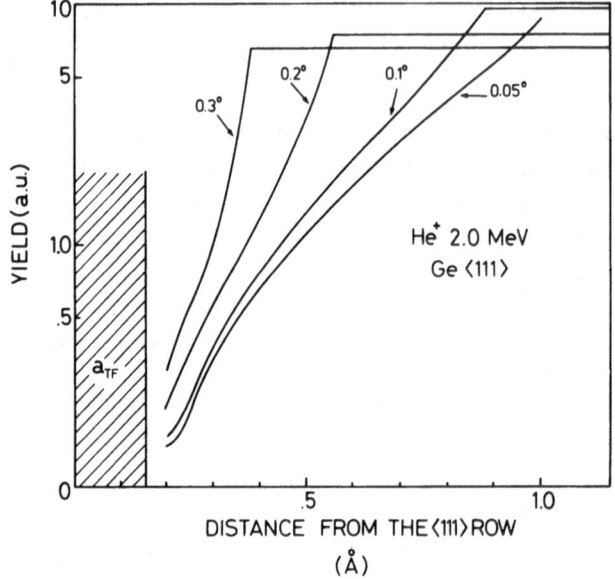

Fig.10 - Yield in a.u. of 2.0 MeV He$^+$ on Ge vs the distance of the
displaced atom from the row for different tilting angles around the
<111> axis.

for several tilting angles. The interaction probability decreases in
all cases for displaced atoms located near the row. Channeling technique is not sensitive to atoms displaced only a few tenths of an
angstrom off their normal lattice sites.

      The method of off-axis analysis will be illustrated in the fol-
lowing by considering the damage created in Si by 300 keV N   implants[20]. Backscattering energy spectra are shown in Fig.11. The
spectra have been recorded for different tilting angles with respect
to the <111> axis. The defect density profiles $N_D(z,\Phi)/N$ have been
extracted according to the procedure outlined in Sect. 3   and are
reported in Fig.12. The profiles are angle dependent and indicate
that the scattering centers are not uniformly distributed across
the channel. A random location of defects would give rise to a profile independent of the tilting angle. Assuming that the channeled
particles are distributed uniformly over the accessible areas A_i
(which depends on the tilting angle), the distribution of defects

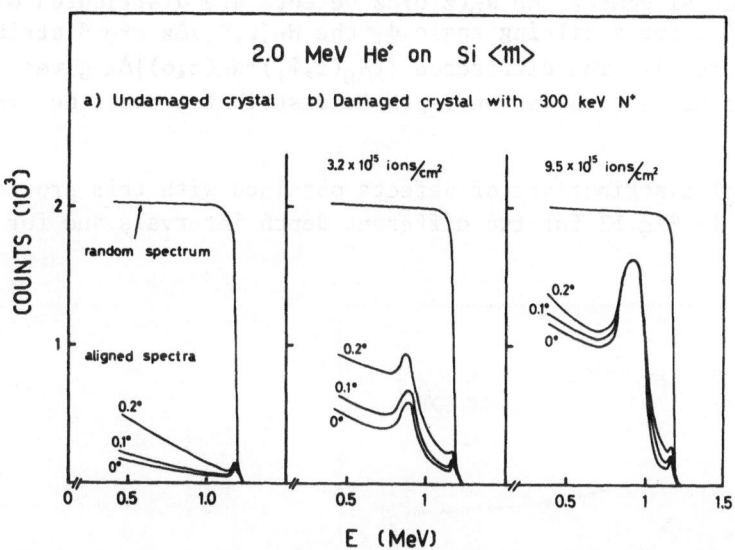

Fig.11 - Energy spectra of 2.0 MeV He particles backscattered from
a) undamaged silicon crystal, b) damaged with 3.2×10^{15} and 9.5×10^{15}
ions/cm² of 300 keV N⁺ fluences respectively. The aligned spectra
have been recorded for 0°,0.1° and 0.2° beam angles with the <111>
axis (from reference 20).

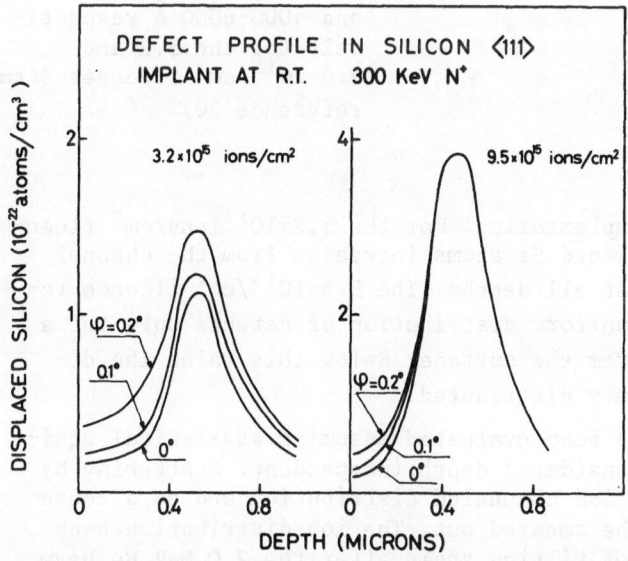

Fig.12 - Defect pro-
files in Si implanted
with 3.2×10^{15} and
9.5×10^{15} ions/cm² fluences of 300 keV N⁺
(from reference 20).

across the channel can be obtained by a subtraction procedure. For perfect alignment the $N_D(z_1o)\Delta z$ defects are distributed over the area A_o, for a tilting angle Φ_1 the $N_D(z,\Phi_1)\Delta z$ are distributed over the area A_1. The difference $|(N_D(z,\Phi_1)-N_D(z,o)|\Delta z$ gives the the number of defects in the depth Δz distributed over the area (A_1-A_o).

Radial distributions of defects obtained with this procedure are shown in Fig.13 for two different depth intervals and for two

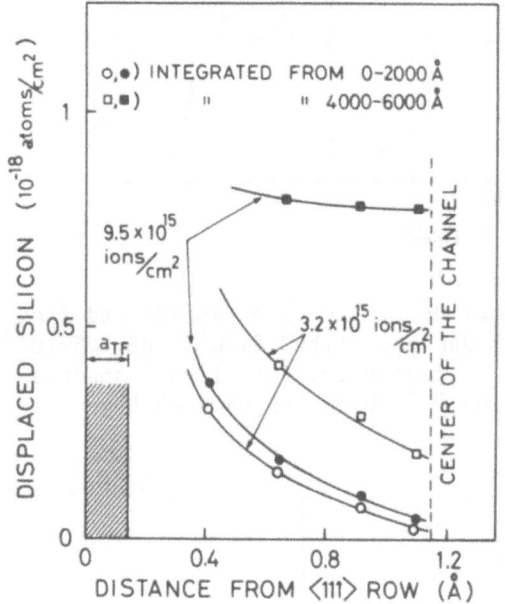

Fig.13 – Radial distributions of displaced atoms integrated over two different depth ranges: 0-2000Å and 4000-6000 Å respectively for the 3.2 and $9.5\times10^{15}/cm^2$ fluences (from reference 20).

fluences of 300 keV N implantation. For the 3.2×10^{15} ions/cm² fluence, the density of displaced Si atoms increases from the channel center towards the row at all depths. The $9.6\times10^{15}/cm^2$ fluence results instead in a non-uniform distribution of defects only to a depth of about 4000 Å from the surface. Below this value the displaced atoms are randomly distributed.

The ion fluxes have been evaluated assuming statistical equilibrium and have been considered depth independent. Scattering by defects will spread the ion channeled distribution and as a consequence the fluxes will be smeared out. The ion distribution becomes nearly independent of tilting angle after the 2.0 MeV He beam has traversed a disordered region of 5×10^{17} displaced atoms/cm². Hence, the random distribution of displaced atoms obtained in the

$9.5 \times 10^{15}/cm^2$ implant of fig.13 below a depth of about 4000 Å can be either due to a true random distribution or to flux profiles smeared out by small angle scattering on defects.

Comparison of the scattering data obtained in single and double alignment provides also information on the defect distribution across the channel. In single alignment the beam impinges along a low index axis and the backscattered particles are detected along a non channeled direction. In double alignment the trajectories of both the incident and backscattered particles are parallel to low index axes.

In double alignment the extracted defect density represents the actual one weighted by the square of the ion-flux distribution in formula

$$\overline{N}_{double} = \int N_D(x,y,z)F^2(x,y)dxdy \qquad (10)$$

If the ratio between the scattering center values in double and single alignment is higher than one, the displaced atoms are located mainly in the central region of the channel. If the ratio is less than one the displaced atoms are situated near the atomic row. A ratio of one indicates a random distribution of scattering centers.

This method provides the mean square distance from the row of displaced atoms[21]. The comparison cannot be performed at high defect level due to the spreding of the channeled beam. Double alignment enhances the sensitivity of the channeling technique to disorder detection. The increase in sensitivity will be proportional to the decrease in the minimum yield in the undamaged crystal and should be at least an order of magnitude. However, the measurements require a longer exposure to the analyzing beam and possible beam annealing and/or damage effects should be considered.

5. DECHANNELING BY DISLOCATIONS

Damage in metals usually consists of a complex structure of defects which give rise mainly to dechanneling with a negligible contribution to direct backscattering. Among the several types of defects, we consider the influence of dislocations on the channeled particles.

An edge dislocation is shown in Fig.14. Particles impinging normal to the dislocation line see a distorted planar or axial chan-

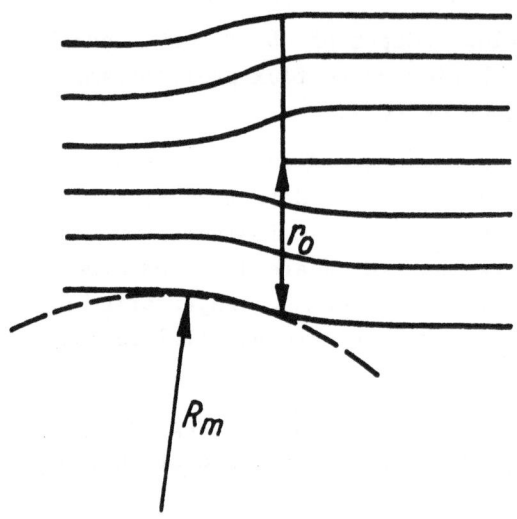

Fig.14 - Edge dislocation

nel due to the bending of the atomic planes. The distortion gives
rise to a centrifugal force acting in the transverse motion of the
channeled particle. If R_m is the minimum value of the radius of curvature the maximum transverse force is given by

$$F_\perp = \frac{2E}{R_m} \, .$$

The equilibrium between $F_\perp$ and the force due to Lindhard's string
potential is reached at a distance, r, from the row given by

$$\frac{2E}{R_m} = \frac{2 \, Z_1 Z_2 e^2}{dr} \tag{11}$$

For r values of the order of a or of the thermal vibrational amplitude, u, dechanneling will occur. This r_c value defines a critical
radius of curvature $R_{m,c}$. The distance r_o from the dislocation at
which the radius of curvature reaches the value $R_{m,c}$ gives the dechanneling radius. On the basis of the elastic theory of the strain
field surrounding a dislocation line the calculation[7] gives the following quantities for the dechanneling cylinder diameter $\lambda = 2r_o$

$$\overline{\lambda}(E) = \left(\frac{b \, d \, a \, E}{\alpha \, Z_1 Z_2 e^2} \right)^{\frac{1}{2}} \tag{12}$$

b, Burgers vector and

α = 12.5 for a straight screw dislocation
α = 4.5 for a straight edge dislocation
α = 7.2 for an equi-number of screw and edge dislocation

This value has been obtained averaging on all the possible orientation of the dislocation line with the axial channel direction.

The same type of evaluation may be derived for planar channeling and in this case we obtain for a screw dislocation[22]

$$\bar{\lambda}_p(E) = \left(\frac{Eb}{8.6\ Z_1 Z_2 e^2 N_p}\right)^{\frac{1}{2}}$$

with N_p atomic density of the planes.

For MeV He the $\bar{\lambda}$ values are about 10 Å and 100 Å for axial and planar channeling respectively. The above formulae should be considered with some caution because the calculations have been done for non-oscillating channeled particles. In addition increase of dislocation density or deviation from straight line decreases the stress field and then the dechanneling diameter. Despite these limitations, the functional dependence of λ on the analysis beam energy and on the atomic number of the projectiles is probably correct.

If dislocations are the predominant defect, the dechanneling increases with the analysis beam energy; the opposite behaviour is shown by displaced atoms. The experimental effort on dechanneling by other types of defects then amorphous zones has been scarce[23].

The structure of defects produced by implantation, stress, or quenching is more complex than a simple straight dislocation. In metals of low stacking fault energy (Au, Cu, Ag e.g.) the dislocations are dissociated and a ribbon of stacking fault is present. Annealing of quenched metals produces stacking fault tetraedra, dislocation loops etc. In Al, due to the high stacking fault energy, dislocations are not dissociated. Attenuation of α-particles in a polycristalline stressed Al films has been measured (canaligraphy technique)[22]. The dechanneling diameter for a planar case has been obtained by comparison with transmission electron microscopy(TEM) and its value 140 Å compares well with the calculated 175 Å value[24].

Recently, the backscattering technique has been applied to investigation of damage in Al single crystals implanted with Zn ions[8]. The analysis has been performed at several analysis beam energies to check the predicted functional dependence of the dechanneling cross section. In Fig.15 are shown the backscattering energy

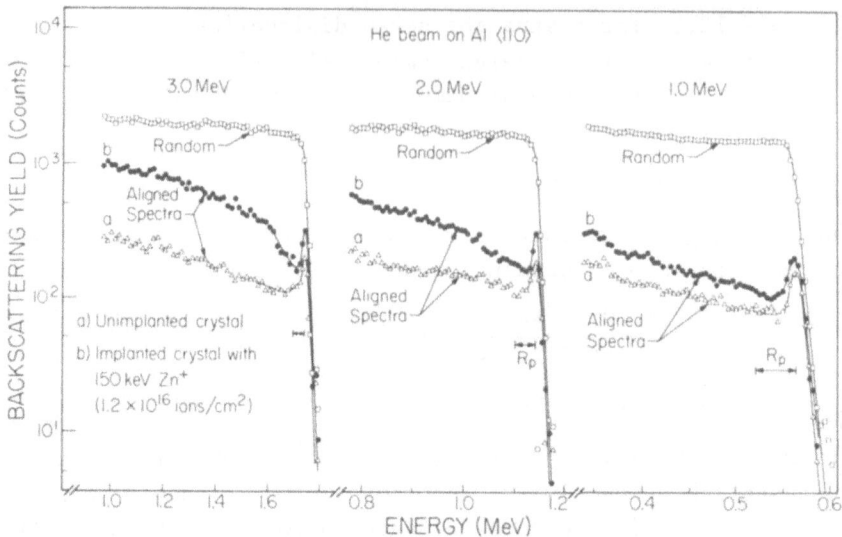

Fig.15 - Backscattering energy spectra of 3.0, 2.0 and 1.0 MeV He beam incident along the <110> direction of Al. Aligned spectra are taken both for the unimplanted and the 150 keV, 1.2×10^{16}/cm^2 Zn-implanted crystal (from reference 8).

spectra for unimplanted and 150 keV Zn-implanted Al to a fluence of 1.2×10^{16}/cm^2. In the implanted crystal the aligned yield increases rapidly up to a depth of about twice the projected range indicating the presence of high disorder. Beyond this depth the rate of dechanneling decreases, as is seen from the "knee" in the 3.0 MeV <110> spectrum.

The most important aspect shown in Fig.15 is the relative increase in the aligned yield with increasing energy of the He probing beam. This is opposite to the trend observed in the unimplanted crystal, in implanted semiconductors and in crystals overlaid with an amorphous layer. The energy dependence for the implanted crystals indicates that the disorder causing dechanneling can not be described in terms of randomly - distributed atoms. This is consistent with the absence of a disorder peak in the aligned yield. Defects characterized by small displacements or distortions of the lattice rows result in appreciable dechanneling without sufficient direct scattering to cause a disorder peak.

For quantitative comparison with dechanneling data the structure of defects and their amount has been determined by transmission electron microscopy (TEM) measurements. A dense network of
dislocations is observed in Zn-implanted Al and a typical dark
field micrograph in shown in Fig.16. The entire depth of the disor-

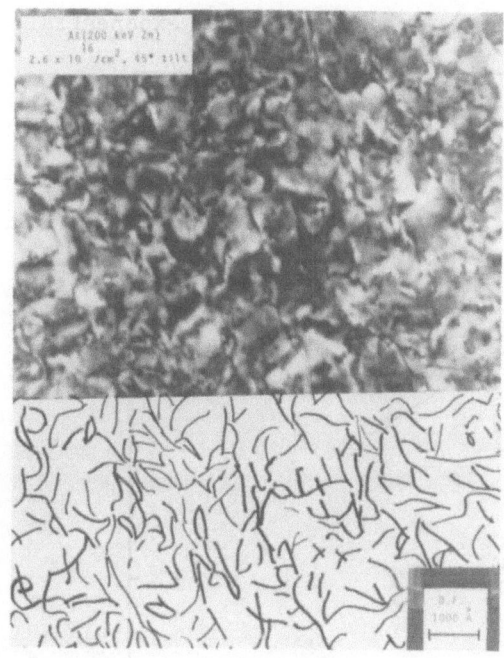

Fig.16 - Dark field micrograph
using g=111 for 200 keV,
$2.6 \times 10^{16}/cm^2$ Zn-implanted Al.
Example of the interpreted dislocation network is shown in
the lower part of the figure
(from reference 8).

dered region is included in the micrograph. The total projected
length of dislocation lines in the plane normal to the <110> direction has been estimated on the basis of the map shown in the lower
part of the figure. In the present case the total projected length
amounts to 9.6×10^5 cm/cm^2. Due to the dark regions resulting from
strain around dislocations, the measured value is accurate within
a factor 2.

In this case as shown by the absence of a peak in the aligned
yield the direct contribution is negligible. The dechanneled fraction coincides with the random component of the beam i.e. $\chi_D \simeq \chi_R$ and
Eq.8 can be used for analysis. The dechanneling cross section σ is
not related to N_D by coulomb interaction as would be the case for
randomly distributed displaced atoms. From eq.8 we obtain

$$\frac{1 - \chi_D}{1 - \chi} = \exp(-\sigma \; q_D) \tag{14}$$

In the dislocation case $\sigma \ N_D \approx \lambda \ell$, where ℓ is the total projected
length of dislocation lines and the dechanneling width λ is the
cross section per unit length.

The ratio between the channeled fractions in the Zn-implanted
and unimplanted Al crystals, left hand side of Eq.14, taken at a
depth $Z=R_p+\Delta R_p (\approx 1450 \ \text{Å}$ for 200 keV Zn in Al), is plotted vs the
square root of the analysis beam energy in Fig.17. The exponential

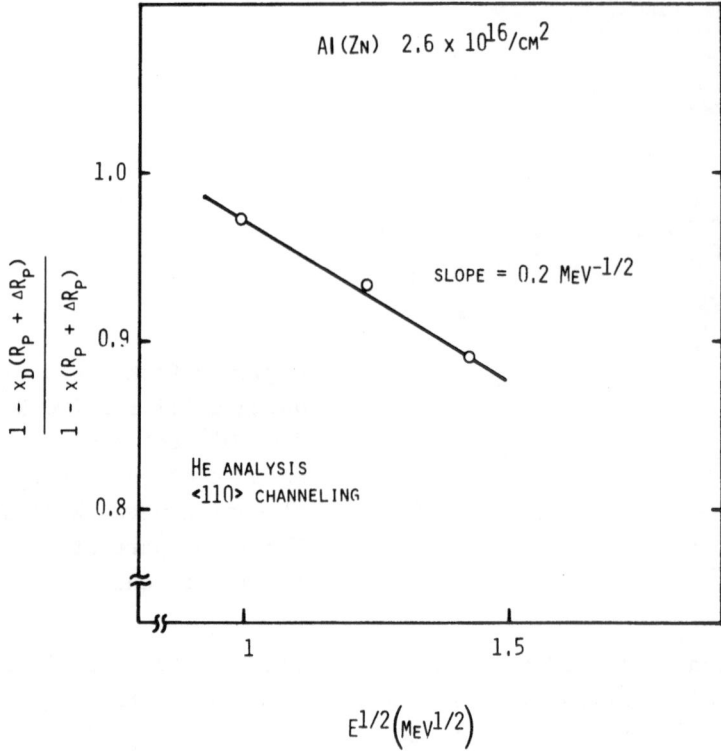

$$E^{1/2} \left(\text{MeV}^{1/2} \right)$$

Fig.17
The ratio between the channeled fraction in the implanted and unimplanted crystal at a depth $R_p+\Delta R_p=1450$ Å is plotted vs the square
root of the analysis beam energy for 200 keV, $2.6 \times 10^{16}/\text{cm}^2$ Zn-implanted (from reference 8).

exp-$\lambda \ell$ has been approximated by 1-$\lambda \ell$. The linear decrease with $E^{\frac{1}{2}}$ is
in agreement with the predicted functional dependence (see Eq.12).
From the slope of the straight line together with the measured total

projected length of dislocation lines, one obtains $\lambda=20$ $E^{\frac{1}{2}}$ where λ in Å and E in MeV. The value predicted by theory above reported is 21 Å $E^{\frac{1}{2}}$ which compares extremely favorably with the experimental value, considering the accuracy of the measured dislocation length by TEM.

As a final remark it must be pointed out that a high density of dislocation is required for detection by dechanneling measurements. In the case illustrated in Fig.16, the dislocation density in the implanted region is $\simeq 7 \times 10^{10}$ cm length of line/cm^3 (= lines/cm^2). This density is comparable to that found in cold worked metals. A minimum density of 10^9 to 10^{10} lines/cm^2 is required for detection by channeling effect measurements.

6. DECHANNELING BY MOSAIC-SPREAD AND STACKING-FAULT

Other types of structure like mosaic or textured polycrystalline films are characterized by increase of the minimum yield with increasing the analysis beam energy.

The aligned yield falls between that for a single crystal and that for a layer composed of random polycrystallites. In this case the high minimum yield is due to the fact that only a fraction of the crystalline grains is aligned with the beam while the rest is misoriented. The value of the minimum yield depends on the orientation distribution of the grains and on the angular yield profile of the analysis beam particles in the corresponding perfect crystal. If the angular width of the distribution is much larger than the $\psi_{\frac{1}{2}}$ values, the minimum yield is characteristic of a random distribution of crystallites. The minimum yield approaches the value of the perfect crystal for distribution narrower than $\psi_{\frac{1}{2}}$. Information on the width of the distribution can be obtained if the two angular widths are comparable[25].

As an example in Fig.18 are shown the backscattering energy spectra of a thin film of Au deposited on a <111> Ge substrate[25]. The analysis beam energies range between 0.675 and 2.5 MeV He. The minimum yield decreases with decreasing beam energy, while the dechanneling rate increases as in a perfect crystal. For this case an angular spread of about 1° has been deduced from a detailed analysis of the energy dependence of both the minimum yield and critical angle, which compares very well with the value inferred by X-ray diffraction.

As a last case of simple defect, we consider the influence of

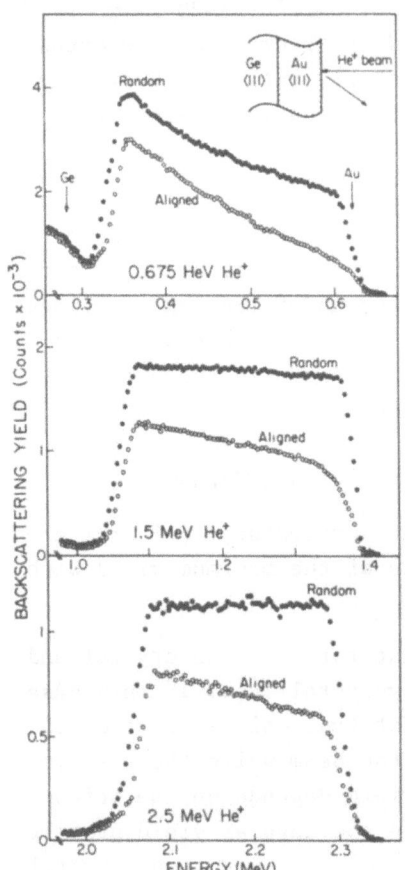

Fig.18 - Backscattering energy
spectra of 0.675,1.5 and 2.5 MeV
He beam incident along the <111>
direction of a Au layer grown
on Ge <111> substrate. Note the
increase in the minimum yield
with increasing analysis beam
energy (from reference 26).

stacking fault on dechanneling. Randomly distributed displaced atoms
obstruct the channel and enhance dechanneling by increasing the kinetic transverse energy component of the channeled beam particles.
Stacking-faults obstruct also the channel but they change the potential transverse energy component of channeled ions. The stacking
a b c a b c a b c   in a perfect f.c.c. crystal is shown in Fig.19 a.
The introduction of a fault changes the sequence, and we may have
an intrisic fault with the a b c b c a b c   sequence characterized
by the missing plane a or an extrinsic fault a b c b a b c a b c
by the extra plane b. Particles impinging along a direction parallel or normal to the fault plane for both axial and planar channeling are not disturbed by the fault. Along other directions (like
(1) in Fig.19) the channeled particles see the same lattice structure but shifted.

    As a first approximation the dechanneling due to a stacking-

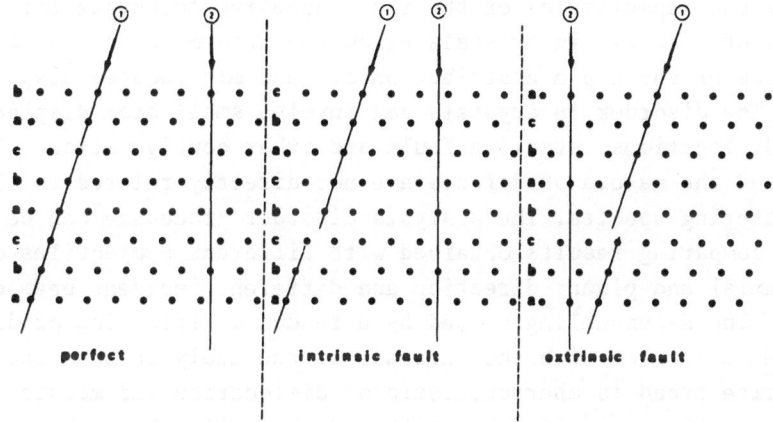

Fig.19 – Stacking sequence in a f.c.c. perfect crystal (a), in a crystal with an intrinsic fault (b) and with an extrinsic fault (c).

fault can be related to the minimum yield for a beam entering a crystal parallel to a low index axis or plane. The dechanneling probability per row will be of the order of $\pi u_1^2/A$ or according to Barret[27] $18.8 u_1^2$ and $2u_1/d_p$ for the axial and planar case respectively u_1 is the unidimensional vibrational amplitude, A the cross sectional area associated to a row and d_p the interplanar spacing. For the extrinsic fault one must add to this contribution tha due to the amorphous layer (the extra plane) one atomic layer thick.

The dechanneling cross section per row is in almost all cases a factor 10 higher than that due to Rutherford single scattering, i.e. one must compare $18.8 u_1^2$, with

$$\frac{\pi Z_1^2 Z_2^2 e^4}{E^2 \psi_{\frac{1}{2}}^2}$$

(see Eq.1). The direct backscattering may be also neglected if compared with the relevant dechanneling contribution. Another important characteristic of dechanneling due to stacking-fault is its energy independence. At high beam energy the minimum yield is in fact nearly energy independent. From the analysis point of wiew, in this case one may also neglect the direct contribution to the aligned yield i.e. $\chi_D \simeq \chi_R$ and can assume eq.8 using for σ the value $18.8 u_1^2$.

7. CONCLUSION

Current capabilities of the ion channeling technique for the analysis of disorder in crystals allow the determinaton in the best cases of the depth distribution of randomly located displaced atoms. The disorder in crystals can involve small atom displacements, dislocations, stacking-fault and other configurations. The number and the nature of defects are not directly related to the backscattering spectra. The analysis disorder procedure can be tested by comparing results obtained with different projectiles,different axial and planar direction and different incident beam energies. The dechanneling caused by a random distribution of displaced atoms decreases with the increase of the analysis beam energyes, an opposite trend is characteristic of dislocation and mosaic spread. Stacking-faults are associated with a dechanneling beam energy independent. In cases of ambiguity one must use another complementary technique or layer removal to provide quantitative information about the disorder. The advantage of the channeling technique in backscattering measurements is that it gives a fast and simple evaluation of the crystalline quality of the sample and an indication of relative changes due to heat treatment or other process steps.

The author is greatly indebted to G.Foti, S.U.Campisano, J.W. Mayer, S.T.Picraux, J.A.Davies and W.Brown for many clarifying discussions.

REFERENCES

1. J.A.Davies, "Channeling: general description", These proceedings. F.H.Eisen, "Channeling", Ed.by D.V.Morgan (John Wiley and Sons, London 1973) chap.XIV, p.415.
2. F.Grasso in "Channeling", Ed.by D.V.Morgan (John Wiley and Sons, London 1973) chap.VII, p.181.
3. W.K.Chu, M.A.Nicolet and J.W.Mayer,"Backscattering Spectrometry" (Academic Press, N.Y.1977) chap.VIII.
4. E.Rimini, E.Lugujjo and J.W.Mayer, Phys.Rev.B $\underline{6}$(1972)718.
5. W.L.Brown in Radiation Damage and Defects in Semiconductors Conf.Ser. n.16 (The Institute of Physics, London, 1973) p.416.
6. P.Baeri, S.U.Campisano, G.Ciavola, G.Foti and E.Rimini, Appl. Phys.Lett. $\underline{28}$(1976)9.
7. Y.Quéré, Phys.Stat.Sol. $\underline{30}$(1968)713.

8.   G.Foti, S.T.Picraux, S.U.Campisano, E.Rimini and R.A.Kant:
     "Dechanneling by dislocation in Zn - implanted Al", V Int.
     Conf. Ion Implantation in Semiconductors and other Materials
     (Plenum Press, N.Y. 1977).

9.   P.Sigmund and W.K.Winterbon, Nucl.Instr.Meth. 119(1974)541.

10. S.U.Campisano, G.Foti, F.Grasso, E.Rimini, Phys.Rev.B 8(1973),
     1811.

11. L.C.Feldman and J.W.Rogers, J.Appl.Phys. 41(1970)3776.

12. E.Bogh, Can.J.Phys. 46(1968)653.

13. J.W.Mayer "Backscattering of MeV energy ions" These proceedings.

14. N.Matsunami and N.Itoh in "Atomic Collisions in Solids" Edited
     by.S.Datz, B.R.Appleton and C.D.Moak (Plenum Press, N.Y. 1975),
     Vol.1 pag.175.
     N.Matsunami, J.Phys.Soc.Jap. 38(1975)848.

15. S.U.Campisano, G.Foti, F.Grasso, and E.Rimini in "Atomic Colli-
     sions in Solids (Plenum Press, N.Y.1975) edited by S.Datz, B.R.
     Appleton and C.D.Moak, Vol.2. p.905.
     P.P.Pronko, Nucl.Instr.Meth. 132(1976)249.

16. J.F.Ziegler, J.Appl.Phys. 43(1972)2973.

17. J.Bøttiger and F.H.Eisen, Thin Solid Films 19(1973)239

18. P.Mazzoldi "Flux peaking-Lattice location" These proceedings.

19. S.T.Picraux in "New Uses of Ion Accelerators" Ed.by J.Ziegler
     (Plenum Press, N.Y.1975) chap.IV, p.229.

20. P.Baeri, S.U.Campisano, G.Ciavola, and E.Rimini, Nucl.Instr.
     Meth. 132(1976)237.

21. J.K.Hirvonen, W.L.Brown and P.M.Glotin in"Ion Implantation in
     Semiconductors" ed.by I.Ruge and J.Graul (Springer Verlag-Ber-
     lin 1971) p.8.

22. Y.Quéré, Ann.Phys. 5(1970)1005.

23. P.P.Pronko and K.L.Merkle, in "Ion Beams in Metals" (ed. S.T.
     Picraux, E.P.EerNisse and F.L.Vook; Plenum Press, N.Y.1974),
     p.481.
     K.L.Merkle, P.P.Pronko, D.S.Gemmel, R.C.Mikkelson and J.R.
     Wrobel, Phys.Rev.B 8(1973)1002.
     L.Howe, Nucl.Instr.Meth. 132(1976)247.

24. J.Mory and Y.Quéré, Rad.Eff. 13(1972)57.

25. D.Sigurd, R.W.Bower, W.F.Van der Weg and J.W.Mayer, This Solid
     Films, 19(1973)319.

26. G.Foti, S.U.Campisano and E.Rimini (unpublished work).

27. J.H.Barrett, Phys.Rev.B 3(1971)1527.

APPLICATION OF MeV ION CHANNELING TO SURFACE STUDIES

J.A. Davies

Atomic Energy of Canada Limited
Chalk River Nuclear Laboratories
Chalk River, Ontario, Canada

1. INTRODUCTION

Previous lectures by Mazzoldi and Rimini have discussed the use
of channeling to pinpoint the location of foreign atoms and displaced
lattice atoms within the bulk of a single crystal. In this final
lecture, we will consider a simple extension of these atom-locating
techniques that permits similar studies to be performed on surface
atoms[1,2]. Of particular interest is the possibility of using single-
and double-alignment techniques for studying changes in structure and
lattice spacing of surface atoms relative to the underlying lattice.
Such information is of basic interest in understanding the nature of
surface adsorption on various crystal surfaces and hence has obvious
application in the field of catalysis and in testing theoretical
models for estimating interatomic potentials.

Although no new principles are involved, this application of MeV
ion channeling has not yet been extensively tested, mainly because
of the technical difficulties of performing channeling experiments
under sufficiently high quality vacuum conditions (i.e. < 10^{-10}
torr.) for meaningful comparison with other surface studies. However, during the past year or so, at least four groups[3-7] have resolved the uhv problems and already some exciting and controversial
results have been obtained.

2. BASIC PRINCIPLES

Figures 1 and 2 illustrate the basic concept involved in using
channeling to study structural changes (lateral re-ordering, surface

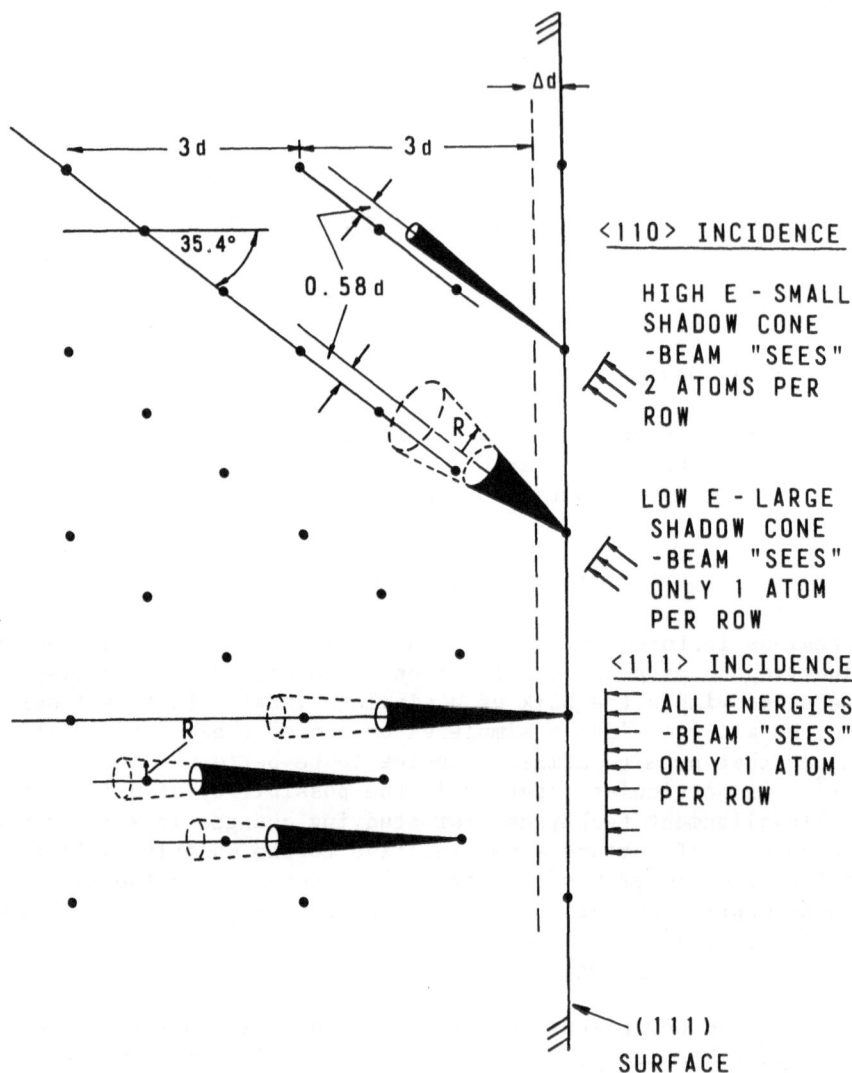

Figure 1 – Atomic configuration near the surface of a (111) platinum crystal illustrating how the channeling effect may be used to measure the surface relaxation Δd, where d is the bulk (111) planar spacing.

relaxation, etc.) at a crystal surface. Whenever a beam of MeV He ions enters a well-aligned single crystal parallel to a low-index direction, a clearly resolvable surface peak is obtained in the RBS energy spectrum, as seen in figure 2. This peak area provides a quantitative measure—either by comparing it to the corresponding non-channeled yield, or by using a calibrated detector geometry—of the number of unshadowed lattice atoms per cm^2 in the surface region.

For perpendicular incidence (<111> in figure 1), the second atom in each row is perfectly shadowed and hence the surface peak area should be equivalent to 1 atom per row (i.e. 4.5 X 10^{15} Pt atoms/cm^2 at all energies, provided the 2-dimensional vibrational amplitude ρ_2 is much smaller than the collisional shadow-cone radius R. Note that a surface relaxation Δd (in figure 1) has no affect on this <111> shadowing.

On the other hand, for non-perpendicular axes (<110> in figure 1), a surface displacement Δd shifts the shadow cone relative to the underlying row of atoms, thus causing the surface peak to increase towards a value of 2 atoms/row. Since the cone radius R varies inversely with incident beam energy E, the surface peak should increase from a value of 1 atom/row at low E to $\sim$ 2 atoms/row at high E. Hence, the magnitude of Δd can be obtained by measuring the surface peak area N as a function of E for various low-index directions.

To determine the sign of Δd—i.e. whether the relaxation is outward (+) or inward (-)—additional measurements of the surface peak

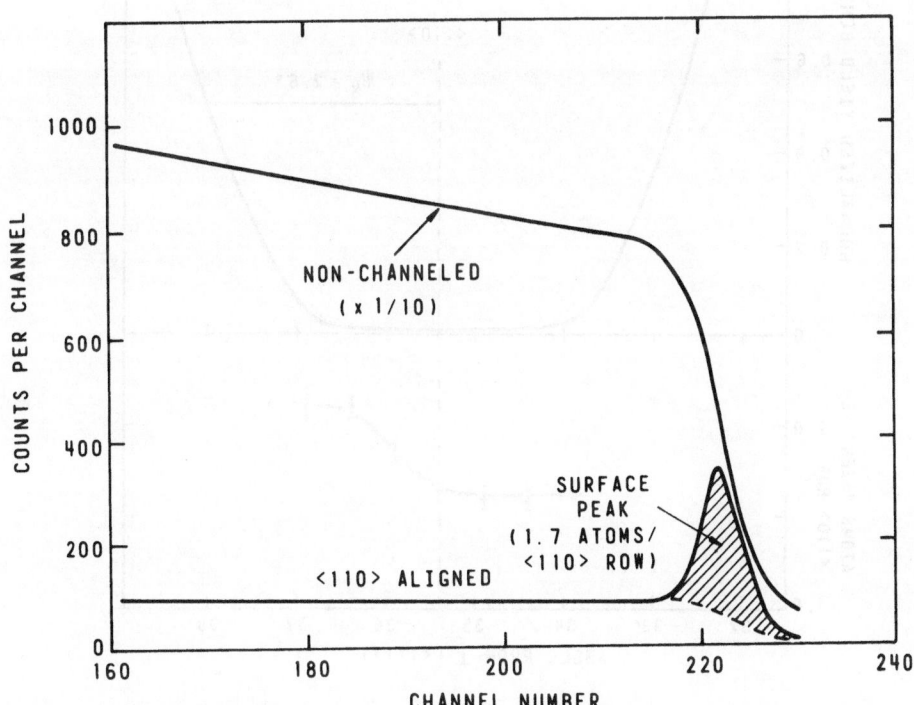

Figure 2 - Backscattered energy spectrum for 1.0 MeV He$^+$ on a (111) Pt crystal, showing the well-resolved surface peak for <110> (channeled) incidence.

are required, in which the tilt angle θ is varied in a narrow range within $\psi_{1/2}$ (the critical channeling angle) of the <110>. If Δd is positive, then as θ increases, the second atom will tend to move out of the shadow cone (figure 1) and N will increase towards a value of 2.0; as θ decreases, the second atom will become better shadowed and N will fall towards the limiting value of 1.0. On the other hand, if Δd if negative, then N should increase as θ decreases and vice versa. An example of such an angular scan is given in figure 3. It does indeed exhibit a marked asymmetry, with the value of N approaching 2.0 at slightly larger tilt angles than 35.4°, thus indicating

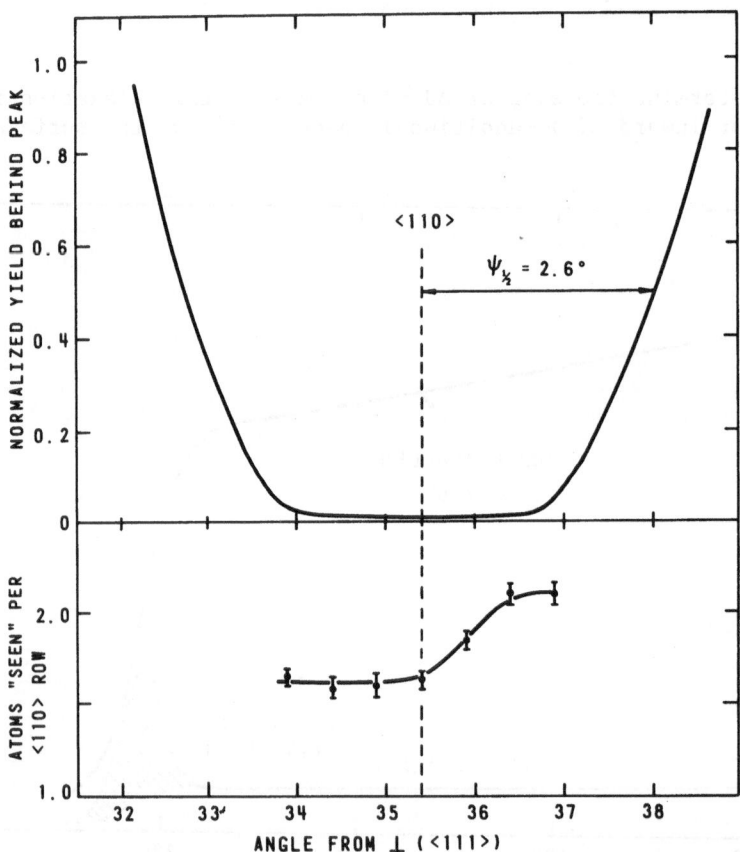

Figure 3 – Angular scan with 1.0 MeV He[+] through the <110> direction in a (111) Pt crystal. Upper part: normalized yield (X_{min}) behind the surface peak (channels 210–215 in figure 2). Lower part: the number (N) of atoms in the surface peak per <110> row[4].

an outward relaxation of the surface plane. It is a bit surprising however that N does not fall completely to its limiting value of 1.0 as θ decreases.

A simpler case to investigate is one in which one or more surface planes of atoms have been laterally displaced with respect to the underlying crystal—as for example in the reconstructed surface region of certain crystals. In such a case, even for perpendicular incidence, the surface peak would correspond to much more than one atomic plane; again, this increase in peak area provides a direct measure of the number of extra atoms contributing to the reconstructed layer. A good example was reported recently for (100) Au by Appleton et al[3]. They observed a large increase (∿ 1 _extra_ atom/row) in the Au surface peak after an identical surface-cleaning treatment as that required previously to produce the LEED pattern characteristic of the reconstructed surface.

3. QUANTITATIVE INTERPRETATION

So far we have ignored lattice vibrations. If these are comparable in magnitude to the shadow cone radius R, then even in the absence of any surface relaxation or lateral re-ordering, the channeled beam will be able to "see" more than just the first atom per row: i.e. N can be significantly greater than 1.0. In such cases, the interpretation becomes considerably more complex and requires computer simulation of ion trajectories over the first few atomic planes. Such Monte Carlo simulations are now becoming available, due to the work of Barrett[8] and Jackson[9]; however their value is still somewhat limited in that it requires accurate knowledge of the vibrational amplitude of the surface atoms. Bulk Debye temperatures are usually available; however, atoms in the surface plane probably have a significantly enhanced vibrational mode perpendicular to the surface. There is an additional complication in that correlated vibrations may also occur between adjacent atoms in a rew, especially at low temperature.

Fortunately, for high Z lattices such as Pt, the shadow cones are often much larger than ρ_2 throughout the optimum energy region for RBS measurements (i.e. 0.5 - 2.5 MeV), so that the interpretation does not depend too heavily on the exact value of ρ_2 chosen. This can be clearly seen in figure 4, where the computer simulation results (for Δd = 0) do not increase significantly above N = 1.0, except at very high beam energies. Included in figure 4 are our experimental N values for various axes as a function of energy. For perpendicular (<111>) incidence, both the measured and the simulated data agree well with the limiting value of N = 1.0, indicating that no lateral re-ordering was present. Note that a large Δd (i.e. 0.25Å) has a negligible effect on the simulated result, as one

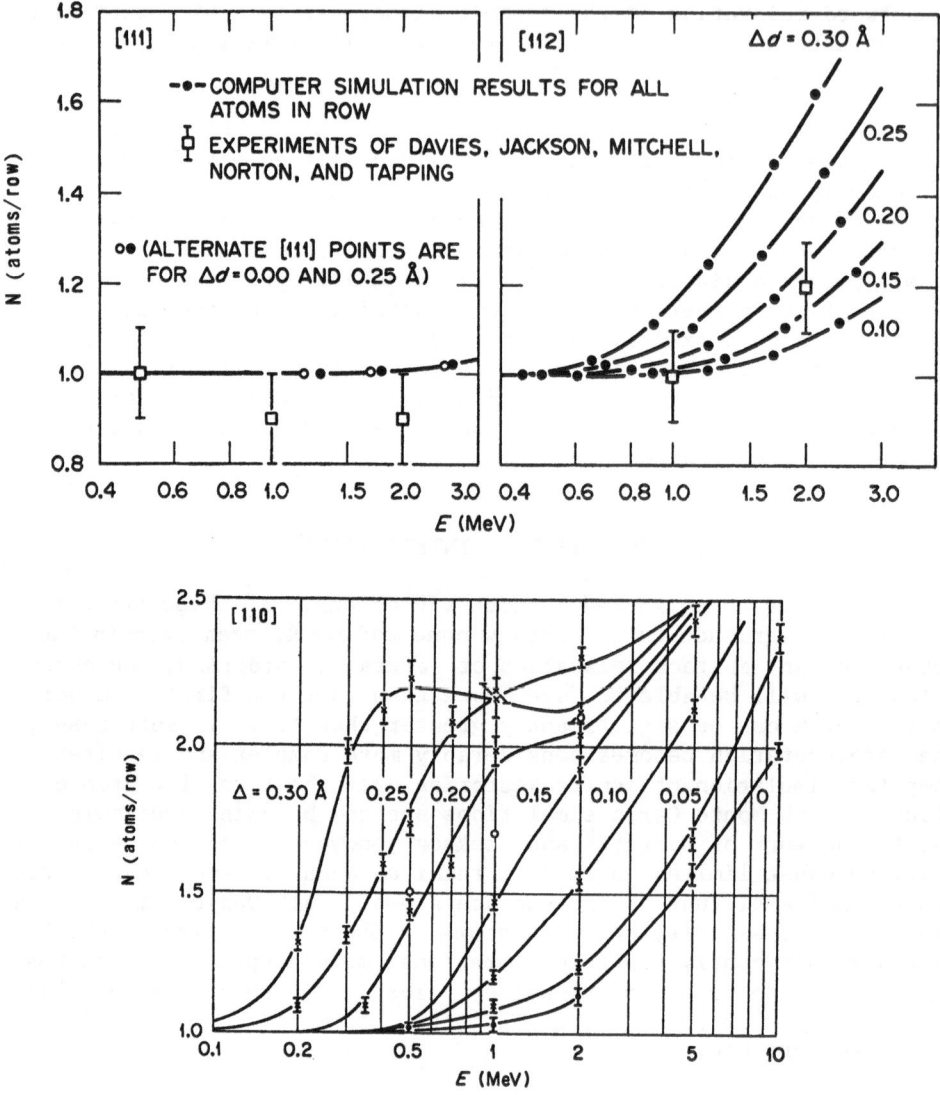

Figure 4 – Pt (111) surface relaxation study showing the observed and simulated[8] number of atoms/row as a function of energy and incidence direction of the He ion beam. Best fit to the <110> data indicates that Δd = 0.20Å.

would expect. For non-perpendicular (<110> and <112>) incidence, the simulation curves are seen to depend strongly on the value of Δd; the experimental data agree well with the Δd = 0.20Å simulation, indicating rather clearly that an outward (figure 3) displacement of

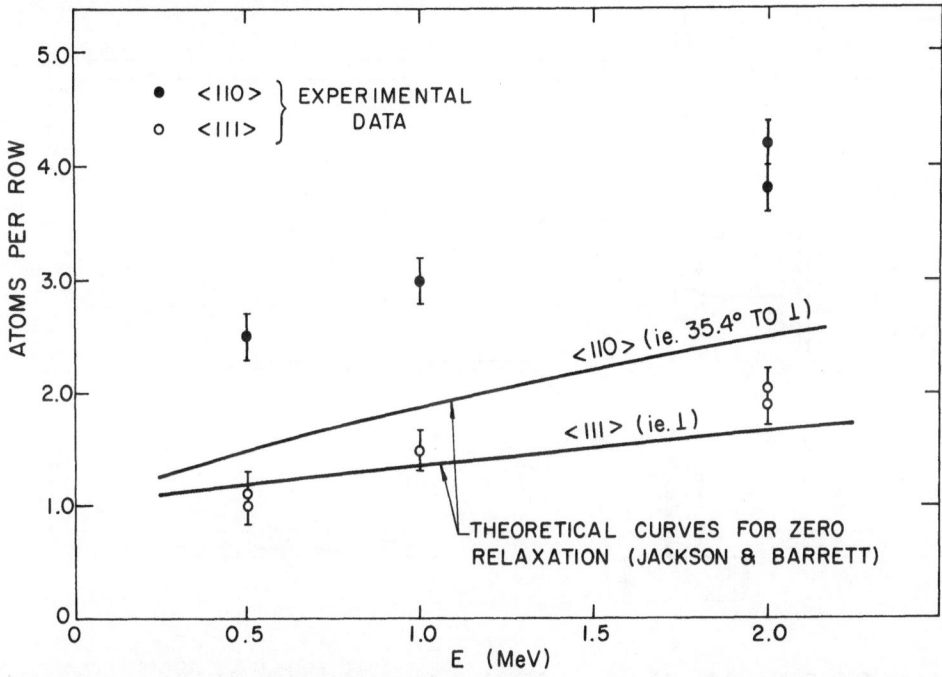

Figure 5 - Ni (111) surface relaxation study showing the observed and
simulated number of atoms/row as a function of energy and incidence
direction of the He ion beam. Best fit to the <110> data indicates
that $\Delta d = 0.16$Å.

0.20 $\pm$ 0.03Å has taken place with respect to the underlying lattice.

A somewhat less favourable situation exists for Ni crystals, as
shown in figure 5. Here, even the perpendicular incidence case shows
considerably more than 1 atom/row, since the shadow cone radius
behind an Ni surface atom is no longer large enough (relative to ρ_2)
to shadow completely the vibrating second, third, etc. atoms. Note
however that the Jackson and Barrett theoretical curve still fits the
observed <111> energy dependence rather nicely, indicating that there
is no lateral re-ordering of the surface atoms. For non-perpendicular (<110>) incidence, the discrepancy between the data and the
theoretical curve suggests that significant surface relaxation has
again taken place; Dave Jackson[9] has recently generated a set of simulation curves in Ni for various values of Δd (similar to those in
figure 4), from which he obtains a best estimate of $\Delta d \sim 0.16 \pm$
0.03Å for (111) Ni.

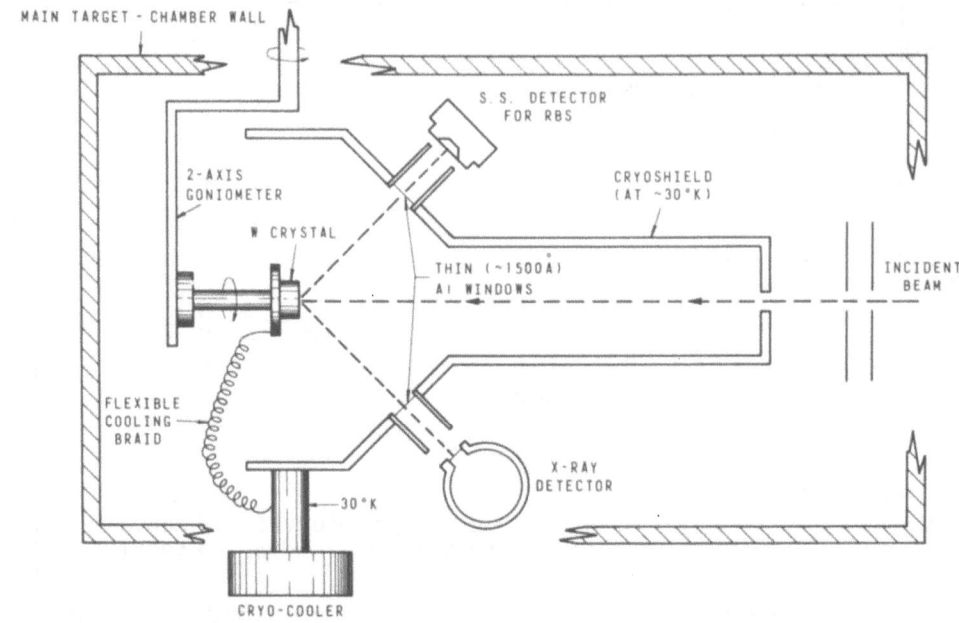

Figure 6 - Schematic diagram of a low-temperature goniometer and cryoshield assembly for clean surface studies.

It is worth noting that the only previous technique for studying surface relaxation has been low-energy electron diffraction (LEED), which unfortunately has a rather large interpretive content associated with it. In the above two cases—Pt(111) and Ni(111)—recent LEED studies[10,11] had concluded that the surface relaxation was negligibly small, i.e. $\lesssim$ 0.05 Å. Obviously, this is an interesting field for further study!

4. EXPERIMENTAL COMPLICATIONS

The most serious limitation to widespread use of MeV ion-channeling techniques for surface studies has been the problem of putting a high-precision goniometer into an ultra-high vacuum (uhv) chamber ($< 10^{-10}$ torr) which in turn must be connected via a suitable differential pumping stage to the ordinary vacuum system ($\sim 5 \times 10^{-6}$ torr.) of the Van de Graaf beam line. One fairly simple solution to this problem is the low-temperature goniometer/cryoshield[12] illustrated in figure 6. By completely surrounding the target with a 20 K cryoshield, except for a small ($\sim 10^{-3}$ steradians) entrance aperture for the incident beam, one maintains an effective pressure of

$\sim 5 \times 10^{-11}$ torr at the target surface with respect to all gases except H, He and perhaps Ne. All other gases have vapour pressures $\ll 10^{-11}$ torr at 20 K and cannot reach the crystal surface without first encountering the cryoshield wall. Nuclear microanalysis for C, O and N—viz. $^{12}C(d,p)$, $^{16}O(d,p)$, and $^{14}N(d,\alpha)$—confirm that the rate of condensation of O_2, N_2 or CO on a clean, cold Pt crystal inside the cryoshield is < 0.1 monolayers/hour; this sets an upper limit of $\sim 5 \times 10^{-11}$ torr to the effective pressure within the shield. Note that in this system the target chamber and the main part of the goniometer are in an ordinary vacuum system and so do not have to satisfy the stringent requirements of uhv.

By providing an extra detector port (figure 6), other close-encounter processes such as inner-shell X-rays or nuclear reactions also can be observed in addition to the standard RBS yield. This permits in situ analysis of various low-Z surface impurity atoms (C, N, O, etc.) and so extends considerably the versatility of MeV ion beams for surface studies.

Another experimental problem is the question of the proper background subtraction (dotted line in figure 2) in order to obtain the surface peak area. In my previous lecture (figure 15), we saw that χ_{min} is not constant at shallow depths beneath a crystal surface; instead, it falls almost to zero immediately behind the surface peak and then rises strongly and exhibits several oscillations before levelling out at the so-called equilibrium value. For 1 MeV He on <111> Pt, the wave length of these oscillations (~ 8 nm) is somewhat smaller than the normal detector resolution and so they are not clearly resolved. Hence, it is dangerous to obtain the background correction by a simple extrapolation.

One solution is to improve the depth resolution sufficiently—for example, by using a magnetic spectrometer[7]—that the depth oscillations can be resolved. Another solution is to use uni-axial double alignment (channeling + blocking)[13] to suppress the background correction to an almost negligible level, as shown in figure 7. Note that the double-alignment value of χ_{min} (2.4×10^{-4}) is almost exactly twice the square of the single-alignment value (1.0×10^{-2}) in figure 2, as one would predict from theoretical considerations (reference 13, and also section 3.5 in my previous lecture).

5. SURFACE BLOCKING TECHNIQUE

Recently, Turkenburg[5] has developed a rather elegant extension of this RBS/ion-channeling technique of surface analysis. He still uses channeling of the incident beam in order to obtain a well-resolved surface peak (figure 2), but combines this with blocking of the emitted beam along some other direction, as shown in figure 8. Note that, along the incident beam direction, only 50% of the atomic

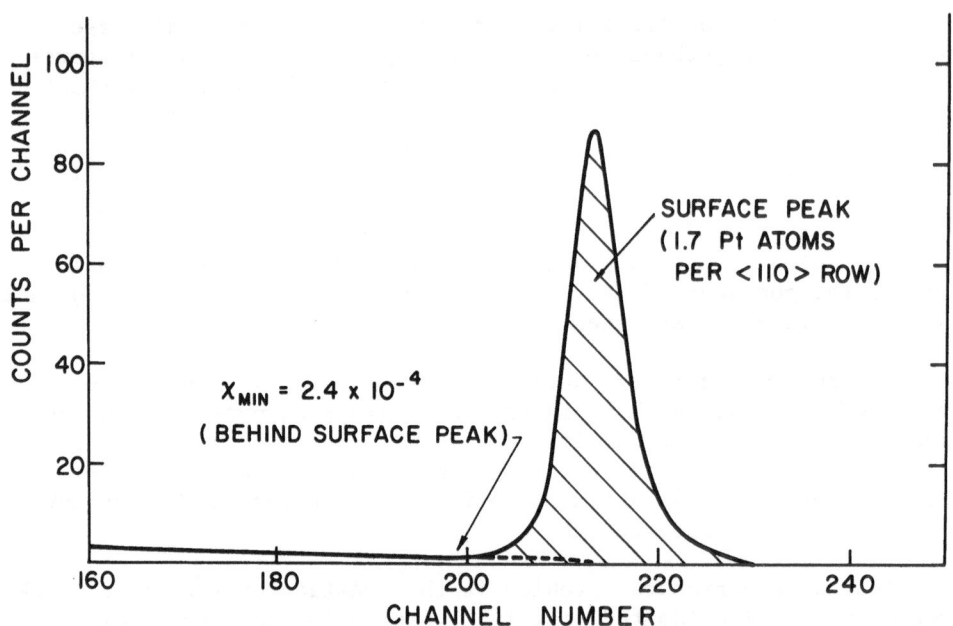

Figure 7 - Backscattered energy spectrum for 1.0 MeV He$^+$ on a (111) Pt crystal along the <110> direction, using uniaxial double alignment to suppress the background correction.

rows originate in the surface plane; the other 50% originate in the second plane. Consequently, in this case, half the backscattered ions contributing to the surface peak can be blocked along certain emission directions by the surface plane of atoms, thus producing a 2-fold blocking minimum. Any relaxation Δd of the surface plane of atoms will rotate the position of this blocking dip through an angle $\Delta\theta$ relative to the underlying bulk blocking direction. Geometrical considerations show that in figure 8 $\Delta d = (d_{<110>}.\sin \Delta\theta)/\sin \theta$. Hence the magnitude and sign of Δd can be obtained without requiring a detailed knowledge of the scattering potential. The estimated sensitivity of the method is extremely high.

Already, this method has been applied to an extensive study[6] of Cu(110) and Ni(110) surfaces, with and without various adsorbed contaminant atoms such as carbon and sulphur. A typical result is shown in figure 9; the small (0.4°) shift in blocking minima between the surface and bulk curves indicates that the surface plane has relaxed

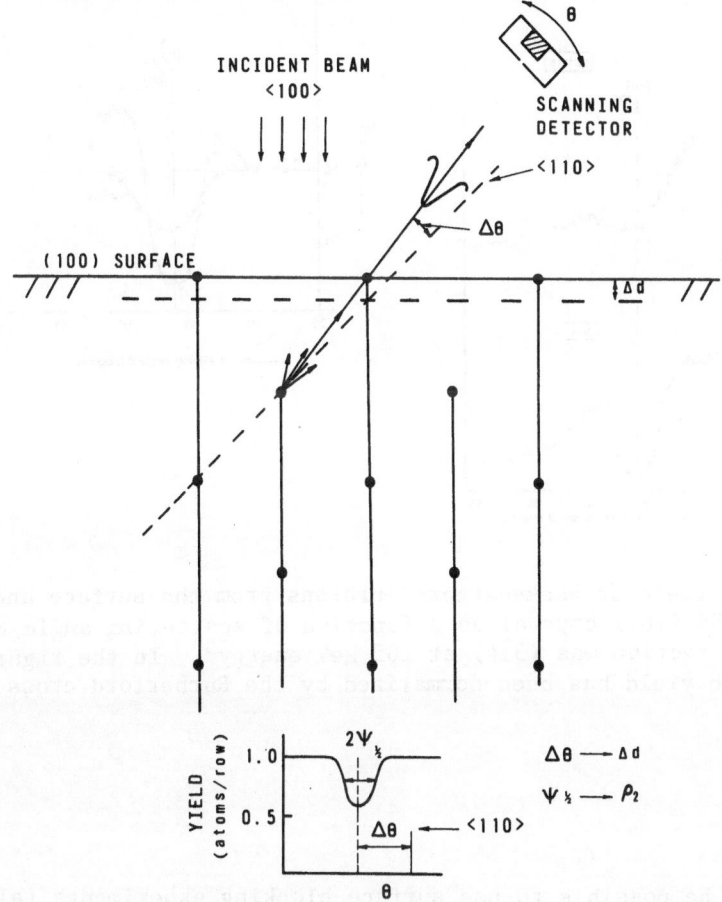

Figure 8 - An illustration of the combined use of channeling and blocking to study surface relaxation.

outwards by ∿ 1.6%, i.e. by ∿ 0.002 nm.

However, this blocking method suffers from the same experimental complications as the single alignment method: i.e. difficulty of background subtraction, need for uhv, etc. Also, if the shadow-cone radii are small enough that more than 1 atom/row can contribute to the surface peak (as in figure 5), blocking minima such as in figure 9 become much more complex to interpret.

An additional advantage of the surface-blocking method is that the angular width of such blocking dips is a function not only of energy, atomic number and lattice spacing, but also of the vibrational amplitude ρ_2 in the transverse plane (see, for example, the critical angle formula 3a in lecture 1). Consequently, it should

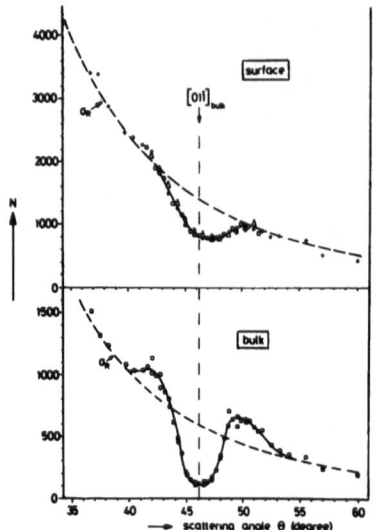
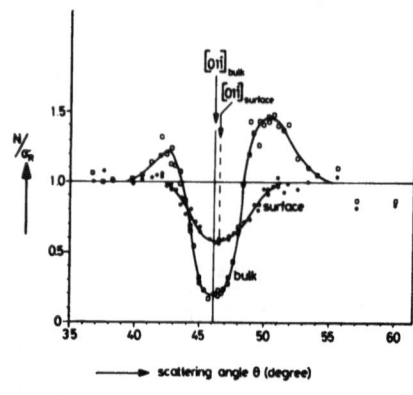

Figure 9 - Yield of backscattered protons from the surface and the bulk of a Ni (101) crystal as a function of scattering angle θ. Incident direction was (3$\bar{1}$4) at 165 keV energy[5]. In the right hand figure, the yield has been normalized by the Rutherford cross section.

eventually be possible to use surface blocking experiments (along several different emission directions) to determine the magnitude and anisotropy of the vibrational modes of the surface plane of lattice atoms relative to the underlying lattice: i.e. to measure the surface Debye temperature.

6.   SUMMARY

In the last few lectures we have discussed some of the more important applications of MeV ion channeling to solid state studies— in investigating radiation damage and bulk lattice defects, in locating foreign impurity atoms within the lattice, and finally in determining the structure of the crystal surface itself. Hopefully, as our knowledge of channeled ion trajectories improves (especially in the near-surface region), these applications will become established on a more quantitative basis and perhaps other equally useful new types of application will emerge.

REFERENCES

1. E. Bøgh in Channeling, ed. D.V. Morgan, (Wiley & Sons, London, 1973) p. 435.
2. J.A. Davies, J. Vac. Sci. Technol. 9, 487 (1971).
3. B.R. Appleton, D.M. Zehner, J.H. Barrett, T.S. Noggle, J.W. Miller, C.H. Jenkins and O.E. Schow, Proceedings of the second Ion Beam Surface Layer Analysis conference (Plenum Press, New York, 1976) p. 607.
4. J.A. Davies, D.P. Jackson, J.B. Mitchell, P.R. Norton and R.L. Tapping, Nucl. Instr. Meth. 132, 609 (1976).
5. W.C. Turkenburg, W. Soszka, F.W. Saris, H.H. Kersten and B.G. Colenbrander, Nucl. Instr. Meth. 132, 587 (1976).
6. W. Turkenburg, FOM Institute, Amsterdam. Private Communication and doctoral thesis (1976).
7. E. Bøgh, University of Aarhus. Private Communication (1976).
8. J. Barrett, ORNL. Private Communication (1976).
9. D.P. Jackson, Nucl. Instr. Meth. 132, 603 (1976).
10. L.L. Kesmodel and G.A. Somorjai, Phys. Rev. B11, 630 (1975).
11. J.E. Demuth and T.N. Rhodin, Surf. Sci. 42, 261 (1974).
12. J. Bøttiger, J.A. Davies, J. Lori and J.L. Whitton, Nucl. Instr. Meth. 109, 579 (1973).
13. L.C. Feldman and B.R. Appleton, Appl. Phys. Lett. 15, 305 (1969).

GENERAL CONCLUSIONS

GENERAL CONCLUSIONS

This summary article has been redacted by A. Cachard and J.P.
Thomas upon the comments expressed during the panel discussions and
the review session by the lecturers and the following discussion
leaders : S.U. Campisano, G.P. Ceasar, C.E. Christodoulides, G.
Dearnaley, D. Dieumegard, C.A. Evans, J. Gyulai, S. Kalbitzer, R.
Laubert, E. Ligeon, G. Linker, M.A. Nicolet, J.C. Poizat, Y. Quere,
J. Remilleux, H. Von Seefeld, A. Turos, J.F. Ziegler. It is a
pleasure to acknowledge here their unvaluable contribution.

The characterization of materials involves the determination
of their elemental composition, their chemical nature and their
structure. Among the numerous techniques available, those dealing
with ion beams were considered in the summer school. As a first
characteristic, due to the limited path of ions in solids the analysis will take place from the surface to a few microns.

We can roughly divide the methods presented in this book in
two categories : the first one involves five techniques (ion scattering spectroscopy, secondary ion mass spectroscopy, beam induced
light emission, Auger electron spectroscopy and photoelectron
spectroscopy) which are typical surface analysis methods in the
sense that only the outermost layers of the sample contribute to
the informations. The second group (Rutherford backscattering,
nuclear reactions and ion-induced X-rays) concerns typical in
depth methods for which detected species are initiated in a few
thousand Å (or a few microns) under the surface. They are inherently
capable of giving depth profiles.

In fact such a division is somewhat too restrictive as it appears
from the lectures. Surface analysis methods when combined with
proper erosion techniques are able to give depth profiles. On the
other hand surface analysis is feasible with the methods of the
second group when the surface contribution can be separated from
the bulk. Nevertheless for convenience we will always refer to
surface analysis and in depth analysis for the two groups. The aim
of this overview is only to summarize the merits and limitations of
these methods.

I - Typical surface analysis methods

As far as the common feature of ISS, SIMS, BILE, AES, and ESCA is to provide analysis of the outer most atomic layers, a stringent requirement is the use of high vacuum techniques for meaningfull measurements. In fact this problem will be the same with other methods when applied to surface studies.

The other common characteristics of these methods is that depth profile is only possible when erosion techniques are used such as sputter etching. They are thus destructive and the question in depth profiling is how sputtering departs from an ideal (atomic layer by atomic layer) microsectionning technique ? This question has been recently reviewed by J.W. Coburn (1). The following processes can introduce interpretational problems in data analysis. For multicomponent systems, differences in the sputtering yields cause the elemental composition of a sputter etched   surface to be in general different from that of the bulk material. In single crystals, polycrystalline targets or multiphased systems, the sputtering process often tends to developp extensive microtopographical changes which will lead to significant degradation in sensitivity and depth resolution. Also ion bombardment can induce motion of atoms in the sample by direct momentum transfer, cascade mixing, enhanced diffusion or, in the case of ionized species, enhanced drift motion. In addition, for SIMS and BILE, chemical effects must be considered in the emission efficiency of ionized or excited species giving rise to strong changes in the relationship between the yield and the concentration.

In spite of these problems the sputter-etching process has the advantage of being applicable to essentially all solid materials and is compatible with the ultrahigh vaccuum required by surface-sensitive methods. Especially for amorphous target significant depth profiles are obtained with a depth resolution of a few percent of depth removed.

I.1 - I.S.S. : Due to the low energy of the beam ions the elastically detected ions come from the top atomic layer only. This fact makes I.S.S. to be the surface analysis method "par excellence". The composition of the outer most layer is obtained through its mass spectrum but a calibration is necessary as far as no satisfactory theory exists. The sensitivity of 10^{-3} monolayer is of the same order of magnitude for all elements except for hydrogen where it is lower. In addition masking effects between layers make surface structure determinations feasible, which is an advantage compared to LEED.

I.2 - SIMS : The observation of the mass spectrum of the ion sputtered from a solid target under   keV ion beam bombardment gives qualitatively the local chemical and isotopic composition of the surface. The main interests of this method is a good lateral resolution ($\simeq$ 1 μm) and a great sensitivity (10^{-6} range). As

pointed out previously quantitative analysis is possible but attention must be given to the topography (grain boundaries, multiphased
systems,...) and to the variations of ionization probabilities.
Quantitative analysis can be applied with reasonable confidence to
trace elements profiling and to dilute alloys analysis.

I.3 - BILE : The detection of radiations emitted by excited
sputtered and reflected particles allows the qualitative analysis
of a solid. This technique is closely related to SIMS, but is
extremeley cheap and simple. Quantitative analysis is complicated
by very strong chemical effects. The basic mechanisms are still
under investigations and no systematic study of sensitivity has
been done so far.

I.4 - AES : The energy of Auger electrons is characteristic
of the atoms from where they originate. These electrons can be
produced by several methods though the conventional one is electron bombardment at energies of a few keV. Due to their limited
escape depth (< 50 Å) Auger electrons allow chemical analysis of
the near surface of solids.

It is possible to analyse all elements with the exception of
hydrogen and helium. The sensitivity, better for low Z elements,
is in the range of 0.1 at %. Quantitative analysis is possible by
means of calibration with standards. The lateral resolution is
better than 0.5 µm. In principle chemical informations are provided
by the measurements of Auger transition shifts.

I.5 - ESCA : Monochromatic photons (soft X rays or ultra-violet
radiations) can eject electrons from solids. Immediate atomic identification is obtained from the spectroscopy of the photoelectrons
which gives their binding energy. The escape depth of detected electrons is approximatively the same as for Auger electrons. Chemical
shifts are more straightforward for ESCA than for AES and ESCA
provides a better chemical information. The sensitivity is of the
same order of magnitude but the spatial resolution is very poor.
The quantitative analysis suffers of the same limitations as AES
especially when the sample composition is not uniform within the
escape depth region.

II - In-depth analysis methods

When increasing the energy of an ion beam the penetration
depth increases up to the order of µm or more in the MeV range,
and the path becomes a straight line. These two features give the
in-depth analysis capabilities of MeV ion beams.

When detecting charged particles coming from the interaction
between the ion beam and the target (backscattering or nuclear
reactions), the in-depth analysis is based on the knowledge of the
stopping power dE/dx of the target relative to the beam ions which
provides a biunivoque relation between the energy of the detected
species and the location of the interaction event. Clearly the

in-depth separation possibility of two events (so called in-depth resolution R) is dependent on the energy resolution of the detection system ΔE and can be roughly written $R = \Delta E/[dE/dx]$.

When detecting electromagnetic radiations (ion-induced X-rays or nuclear reaction induced γ rays) their energy is independent on the event location. Only the cross section σ of the interaction is energy dependent. Nevertheless in-depth informations can be provided by observing the variations of the reaction yield when varying energy. But a significative depth resolution R is achievable only near sharp variations of $\sigma(E)$ (threshold energy or resonance). R is then mainly dependent on the sharpness of $\sigma(E)$ variations and on the beam energy spread.

In both cases the in-depth analysis is non destructive. Several technological improvements permit very good resolution with these methods and can make them competitive for surface studies.

II.1 - <u>Backscattering</u> : As for ISS the energy analysis of backscattered particles in the MeV range, gives the mass spectrum of the solid. But due to the excellent agreement with the Rutherford law such analysis is quantitative and the substrate (when known) provides a built-in calibration for foreign atoms analysis. This method is simple and direct and gives with a good precision absolute concentrations and depth profiles. The best sensitivity is found for heavy foreign atoms in a lower mass substrate, and can be of the order of 10^{18} at/cm^3. For MeV ^{4}He ions the depth resolution is about 200 Å. It can be improved by use of higher mass ions. On another hand for 200 keV ^{4}He ions and with electrostatic analyser a resolution between 3 and 10 Å is obtained which is sufficient for surface analysis. Other technical improvements such as grazing incidence or electromagnetic detection can provide good depth resolution near the surface.

II.2 - <u>Nuclear reaction</u> : The nuclear reactions induced by MeV ions (mainly protons or deuterons) give background free detection of light elements in heavier matrix. Nuclear reactions and backscattering appear as complementary techniques. As for backscattering nuclear reaction yields are insensitive to the matrix and absolute quantities may be obtained by comparison to reference standards. An advantage compared to backscattering is that isotopic tracer experiments are easy and precise. Depth profiles can be determined, but the depth resolution is comparable to that of backscattering only when using reactions with sharp resonance in cross section. All nuclei from hydrogen up to chlorine are detected with typical sensitivities of the order of 10^{14} at/cm^2. The use of heavy ions such as ^{11}B or ^{15}N and even of neutrons presents interesting improvements for the method as for hydrogen analysis which is an up to date problem in energy production.

II.3 - <u>Ion induced X rays</u> : The emission of characteristic
X rays from a target under ion bombardment sets a simple method
for a multi element analysis. It is fast, non destructive and has
a good concentration sensitivity (10^{-6} to 10^{-7} g/g). Its main
advantages on electron microprobe are due to the reduction of continuous background radiation and the selective X-ray excitation by
use of heavy ions - Quantitative analysis (< 5% accuracy) is possible especially for thin samples but in depth resolution is rather
poor even when combined with sputtering.

II.4 - <u>Channeling</u> : When a beam of energetic charged particles
moves through a crystal in a direction almost parallel to a major
axis or plane the steering effect of the crystal rows prevents
the particles from having violent collisions with target atoms.
Consequently almost all physical processes which involves low
impact parameters such as wide angle scattering, nuclear reactions
or X ray emission are strongly reduced. Hence a channeled beam has
close collision interactions only with those atoms which are displaced more than 0.1 Å from the normal lattice sites. Thus channeling provides atom location with a precision as good as 0.02 Å. It
has been used to pinpoint the location of foreign atoms and displaced lattice atoms within the bulk of a single crystal. Also surface
structure may be quantitatively studied using channeling. Channeling associated to backscattering, nuclear reactions or ion induced
X rays appears as a promising tool. But although the physical processes has been rigorously investigated, analytical applications
require great care and computer model calculations for relevant
experiments.

III - <u>General comments</u>

It would be interesting from such a review summary to be able
to indicate the more appropriate method for given characterization
problem. It does exist, of course, as often shown through this book,
typical areas of investigation where a best choice seems easy :
Catalysist has to study the absorption of a monolayer on the surface of a solid ; metallurgist wants to know the early stage of
formation of an oxyde on a metal or alloy ; electrochemist has to
determine the nature of an oxide-layer and its kinetics of growth ;
semiconductors people must choose the adequate implantation parameters for a device fabrication and thus has to know the implantation profile, etc... Nevertheless even suited to a specific need,
every method may not fit specific requirements and the problem may
be the exception where the method does not work.

Keeping so in mind that exceptions are numerous, it remains
worthwhile, instead of general appreciations, to give some indications on the possible choices :

- <u>Problems of surface reactivity</u> : in catalysis where the
outermost layer is of prime importance, the ISS technique has the
most potential interest but SIMS and AES have not to be omitted.

When chemical informations are wanted ESCA is without concurrence in this field. Nevertheless the high sensitivity and the quantitativity of nuclear methods can also make them very attractive in some case, especially when connected with channeling experiments.

- <u>Problems of interfaces</u> : Despite some restriction in the use of erosion techniques when the interface is sharp SIMS, AES or ESCA can be considered as very valuable. When resonant nuclear reaction exist for light elements they can be recommended.

- <u>Thin film kinetics</u> : this is a typical field of application of MeV ion beam technique exemplary illustrated by backscattering analysis of metal silicide layer formation or by nuclear reactions and isotope tracer experiments for the study of oxidation kinetics.

- <u>Diffusion or implantation profiles</u> : if MeV ion beam techniques are widely applied to the subject, SIMS has the definite advantage to allow profile determination down to the ppm range for almost all the elements.

- <u>Atom location</u> : the channeling technique is unique in this field where it can also provides the study of defects and disorder induced by implantation.

Such a list cannot be exhaustive as already mentioned, and the reader may find trough this book stimulating examples and ideas helping to solve his problem. As a matter of fact there is a consensus between all who participate to these lectures and discussions on the necessity of a combination of techniques to insure the validity of the results obtained.

REFERENCE

(1)   J.W. COBURN, J. Vav. Sc. Technol. 13 (1976) 1037.

PARTICIPANTS

DIRECTOR

J.P. THOMAS
Institut de Physique Nucléaire
Université LYON I
43 Bd du 11 Novembre 1918
69 621 Villeurbanne FRANCE

ORGANIZING COMMITTEE

A. CACHARD
Département de Physique des
Matériaux
Université LYON I
43 Bd du 11 Novembre 1918
69621 Villeurbanne FRANCE

G. CHASSAGNE
Département de Physique des
Matériaux
Université LYON I
43 Bd du 11 Novembre 1918
69621 Villeurbanne FRANCE

C.H.S. DUPUY
Département de Physique des
Matériaux
Université LYON I
43 Bd du 11 Novembre 1918
69621 Villeurbanne FRANCE

M. FALLAVIER
Institut de Physique Nucléaire
Université LYON I
43 Bd du 11 Novembre 1918
69621 Villeurbanne FRANCE

J. REMILLEUX
Institut de Physique Nucléaire
Université LYON I
43 Bd du 11 Novembre 1918
69621 Villeurbanne FRANCE

ADVISORY COMMITEE

G. DEARNALEY
Nuclear Physics Division
AERE Harwell, Oxfordshire
OX II ORA ENGLAND

J. GYULAI
Central Research Institute for
Physics.
Hung. Acad. Sciences
1525 BUDAPEST, P.O. Box 49

J.W. MAYER
California Institute of Technology
Department 116-81
Pasedena , CA 91125 USA

O. MEYER
Kernforschungszentrum Karlsruhe
Institut für Angewandte Kern-
physik
75, Karlsruhe, Posfach 3640
GERMANY

I.V. MITCHELL
Solid State Science Branch
Atomic Energy of Canada
Chalk River, Ontario KOJ IJO
CANADA

Y. QUERE
C.E.N., B.P. N°6
92260 Fontenay aux Roses FRANCE

E. RIMINI
Istituto di Struttura della
Materia
University of Catania
Corso Italia 57 I 95129 Catania
ITALY

LECTURERS

G. BLAISE
Laboratoire de Physique du Solide
Bât 510 Université Paris Sud
91405 Orsay   FRANCE

H.H. BRONGERSMA
Philips Research Labs,
Eindhoven
NETHERLANDS

J.A. CAIRNS
Metallurgy Division
Building 393
Aere Harwell Oxfordshire OX 11 ORA
UNITED KINGDOM

W.K. CHU
IBM SPD 300-095
East Fishkill
Hopewell Junction
N.Y. 12533   USA

J.A. DAVIES
Solid State Science Branch
Atomic Energy of Canada
Chalk River, Ontario KOJ IJO
CANADA

J.H. FREEMAN
Marine Technology Support Unit
AERE Harwell, Oxfordshire
OX 11 ORA U.K.

F. FOLKMANN
Gesellschaft für Schwerionen-
forschung.
Planckstrasse
Postfach 541, D-6100 Darmstadt
GERMANY

J.W. MAYER
California Institute of
Technology
Department 116-81
Pasadena, CA 91125   USA

P. MAZZOLDI
Istituto di Fisica "G. Galilei"
Via Marsolo 8 - 35100 Padova
ITALY

S. RIGO
Groupe de Physique des Solides
Ecole Normale Supérieure
Université Paris VII, Tour 23
2 Place Jussieu
75221 Paris   FRANCE

E. RIMINI
Istituto di Struttura della
Materia
University of Catania
Corso Italia 57 I 95129 Catania
ITALY

J. TOUSSET
Insitut de Physique Nucléaire
Université Claude Bernard
43 Bd du 11 Novembre 1918
69621 Villeurbanne FRANCE

W.F. VAN DER WEG
Philips Research Laboratories
Department Amsterdam
Oosterringdijk 18, Amsterdam
NETHERLANDS

H. VERBEEK
Max-Planck Institut für Plasma-
physik.
D-8046 Garching/München
GERMANY

A. AMOKRANE
Institut d'Etudes Nucléaires
Bd Frantz Fanon   Alger
ALGERIA

P. BAERI
Istituto di Struttura della
Materia
University of Catania
Corso Italia 57 I-95129 Catania
ITALY

M. BENMALEK
Université Claude Bernard
43 Bd du 11 Novembre 1918
69621 Villeurbanne   FRANCE

L. BUENE
Institute of Physics
University of Oslo, P.O. Box 1048
Blindern Oslo 3   NORWAY

S.U. CAMPISANO
Istituto di Struttura della
Materia
University of Catania
Corso Italia 57 I-95129 Catania
ITALY

A. CARNERA
Istituto di Fisica G. Galilei
University di Padova
35100 Padova   ITALY

G.P. CEASAR
Dept. of Chemistry, Univ. of
Rochester
Rochester, N.Y. 14627   and
Xerox Joseph C. Wilson Center
for Technology
Rochester, N.Y. 14580   USA

G. CENBALDI
C.N.R. Lab. Lamel
    Via de Castagnoli 1
40126 Bologna
ITALY

C.E. CHRISTODOULIDES
Department of Electrical Enginee-
ring
University Of Salford
Salford M5 4WT CANCS U.K.

C.A.N. CONDE
Laboratorio de Fisica
Universidade de Coimbra
Coimbra PORTUGAL

N. CUE
Physics Department
State University of New York at
Albany
Albany, New York 12222 USA

A. DE CHATEAU THIERRY
Institut National Des Sciences
et Techniques Nucléaires
CEN Saclay
91190 Gif/Yvette FRANCE

F. DEGREVE
Aluminium Pechiney
Centre de Recherche de Voreppe
B.P. n°24
38340 Voreppe FRANCE

P. DEMONCY
I.B.M. France
224 Bd Kennedy
91 Corbeil Essones FRANCE

S. DENAGBE
C.E.N.B.G. Le Haut Vigneau
33170 Gradignan FRANCE

D. DIEUMEGARD
Lab. Central de Recherches
Thomson -CSF, Domaine de
Corbeville
91401 Orsay FRANCE

C.A. EVANS Jr.
Materials Research Laboratory
Univ. of Illinois, Urbana
Illinois, 61801 USA

G.R. FENSKE
Argonne National Laboratory
9700 South Cass Avenue
Argonne, Illinois 60439   USA

L.F.R. FERREIRA
Laboratorio de Fisica
Universidade de Coimbra
Coimbra PORTUGAL

A. GENOUX LUBIN
I.E.N. Bd Frantz Fanon
Alger ALGERIA

A. GUIVARC'H
Centre National d'Etude des
Télécommunications
Route de Trégastel
22300 Lannion FRANCE

F.L. HAASE
Institut für Angewandte Kern-
physic
Postfach 3640 Kernforschungszen-
trum
D-7500 Karlsruhe GERMANY

G. HEINE
Ruhr Universität Bochum
Institut für Experimentalphysic III
Universitätstrasse 150, Gebäude NB
463 Bochum GERMANY

D. HEITMANN
Institut für Angewandte Physik
2 Hamburg 36
Jungiusstrasse 11   GERMANY

G. IMME
Univ. of Catania
Corso Italia 95129 Catania
ITALY

R.A. JARJIS
University of Manchester
Manchester M13 9PL ENGLAND

S. KALBITZER
Max Planck Institut für Kernphysik
69 Heidelberg 1, Postfach 103980
GERMANY

J. KEMPF
IBM Deutsland GmbH
Dept. 0135, Bldg. 7032-47
German Manufacturing Technology
Center
D - 7032 Sindelfingen   GERMANY

A.R. KNUDSON
Naval Research Laboratory
Washington D.C. 20375   USA

M. KWADOW
Centre de Recherches Nucléaires
de Strasbourg
Laboratoire de Chimie Nucléaire
23 Rue de Loess
67 Strasbourg Cronenbourg FRANCE

R. LAUBERT
Physics Dept. New York Univ.
4, Washington Place N.Y. 10003
USA

T. LAURSEN
Institute of Physics
University of Aarhus
DK-8000 Aarhus C DENNMARK

J. L'ECUYER
Laboratoire de Physique Nuclaire
Université de Montréal
C.P. 6128 Montréal   CANADA

E. LIGEON
Dept. de Recherche Fondamentale
C.E.N.G. 85 X
38041 Grenoble Cedex   FRANCE

G. LINKER
Kernforschungszentrum Karlsruhe
Institut für Angewandte Kernphysik
75 Karlsruhe, Posfach 3640
GERMANY

M.E. LUOMAJARVI
Accelerator Laboratory
Siltavuorenpenger 20 e,
HELSINKI 17   FINLAND

G. LUZZI
Dipartimento di Fisica
Università della Calabria
Arcavacata di Rende (Cosenza)
ITALY

J.R. McGINLEY
Texas A&M University
Center for Trace Characteriza-
tion.
Dpt. of Chemistry
College Station Texas 77843 USA

G.M. MARIN
Max-Planck Institut für Plasmaphysik. Abt. QP
8046 Garching bei Munchen
West GERMANY

G. MAJNI
Istituto di Fisica dell'università
Via Vivaldi, 70
41100 Modena   ITALY

M. MEYER
Laboratoire de Physique des
Matériaux
C.N.R.S. 1 Pl. A. Briand
92190 Meudon Bellevue FRANCE

T. MIHAC
Institut de Physique et de
Chimie
Faculté des Sciences
Vojvode Putnika 43
Sarajevo   YUGOSLAVIA

I.V. MITCHELL
Central Bureau for Nuclear
Measurements
Euratom B. 2440 Geel   BELGIUM

H. MOMMSEN
Institut für Strahlen und
Kernphysik der Universität Bonn,
Nussallee 14-16
D-53 Bonn   GERMANY

P. MÜLLER
Physikalisches Institut II
D. 8520 Erlangen
Erwinrommel Strasse 1, GERMANY

M.A. NICOLET
California Institute of Technolo-
gy Dpt. 116-81
Pasedana, C.A. 91125   USA

J.C. OBERLIN
Insitut d'Etudes Nucléaires
Bd Frantz Fanon Alger
ALGERIA

R.C. PILLER
A.E.R.E. Harwell Oxfordshire
Manchester M13 9PL
UNITED KINGDOM

J.C. PIVIN
Laboratoire de Métallurgie
Physique
Bâtiment 413 - Université Paris-
Sud. Centre d'Orsay
91405 Orsay   FRANCE

J.C. POIZAT
Institut de Physique Nucléaire
Université Claude Bernard
43 Bd du 11 Novembre 1918
69621 Villeurbanne FRANCE

G. POSPIECH
Ruhr Universitat Bochum
Institut für experimental
Physik IV
Postfach 2148   463 Bochum
GERMANY

A.J. RATKOWSKI
Institut de Physique Nucléaire
43 Bd du 11 Novembre 1918
69621 Villeurbanne   FRANCE

I.M. REID
Newcastle upon Tyne Polytechnic
Dept. of Physics & Physical
Electronics Ellison Building
Ellison Place, Newcastle upon
Tyne
NE 1, 8ST   UNITED KINGDOM

M.R. RISCH
Lehrstuhl Fuer Festkoerperphysik
Prof. Sizmann Amalienstrasse 54
D-8000 Munchen 40   GERMANY

J. ROTURIER
Centre D'Etudes Nucléaires de
Bordeaux-Gradignan
33170 Gradignan   FRANCE

F. RUDOLF
Insitut de Physique
Rue A.L. Breguet 1
CH-2000 Neuchatel   SWITZERLAND

H. Von SEEFELD
Max-Planck Institut für Plasmaphysik
8046- Garching   GERMANY

A. SHOAIB
School of M.A.P.S.
University of Sussex, Falmer
Brighton BNI 9QH   UNITED KINGDOM

J.F. SINGLETON
A.E.R.E. Harwell
Abingdon Oxon OX11 ORA
UNITED KINGDOM

A. STEPANESCU
Istituto Elletro Tecnico Nazionale
10125 Torino   ITALY

A. TUROS
Institute of Nuclear Research UL
Hoza 69, 00-681 Warsaw
POLAND

J.A. VAN DEN BERG
University of Salford
The Crescent
Salford M5 4 WT
UNITED KINGDOM

P. WILLIAMS
Material Research Laboratory
University of Illinois
Urbana, Illinois 61801   USA

A. ZARTNER
Max-Planck Insitut für Plasma-
physik
Garching/München
GERMANY

J.F. ZIEGLER
I.B.M. Research Box 218
Yorktown Heights
N.Y. 10598
U.S.A.

INDEX